Stem Cells: From Basic to Advanced Principles

Stem Cells: From Basic to Advanced Principles

Edited by Jack Collins

hayle
medical

New York

Hayle Medical,
750 Third Avenue, 9ᵗʰ Floor,
New York, NY 10017, USA

Visit us on the World Wide Web at:
www.haylemedical.com

ISBN: 978-1-63241-464-9

Cataloging-in-Publication Data

Stem cells : from basic to advanced principles / edited by Jack Collins.
    p. cm.
Includes bibliographical references and index.
ISBN 978-1-63241-464-9
1. Stem cells. 2. Cells. 3. Cytology. I. Collins, Jack.
QH588.S83 S74 2017
616.027 74--dc23

Table of Contents

Preface

This book aims to highlight the current researches and provides a platform to further the scope of innovations in this area. This book is a product of the combined efforts of many researchers and scientists from different parts of the world. The objective of this book is to provide the readers with the latest information in the field.

Stem cells are cells that are undifferentiated but can become specialized cells through cell division. This book on stem cells discusses the fundamental concepts in the harvest and implementation of stem cells for medical research. Clinical and experimental research of stem cells has focused on their restorative capability which could potentially regenerate damaged and diseased organs. Current stem cell therapy includes bone marrow transplantation as well as cure for neurodegenerative diseases. This book will help new researchers by foregrounding their knowledge in this branch of medicine. It covers in detail some existent theories and innovative concepts revolving around stem cells. Coherent flow of topics, student-friendly language and extensive use of examples make this book an invaluable source of knowledge.

I would like to express my sincere thanks to the authors for their dedicated efforts in the completion of this book. I acknowledge the efforts of the publisher for providing constant support. Lastly, I would like to thank my family for their support in all academic endeavors.

Editor

Treatment of periodontal intrabony defects using autologous periodontal ligament stem cells: a randomized clinical trial

Fa-Ming Chen[1*†], Li-Na Gao[1,2†], Bei-Min Tian[1†], Xi-Yu Zhang[1], Yong-Jie Zhang[2], Guang-Ying Dong[1], Hong Lu[1], Qing Chu[1], Jie Xu[1], Yang Yu[1,2], Rui-Xin Wu[1], Yuan Yin[1], Songtao Shi[3*] and Yan Jin[2*]

Abstract

Background: Periodontitis, which progressively destroys tooth-supporting structures, is one of the most widespread infectious diseases and the leading cause of tooth loss in adults. Evidence from preclinical trials and small-scale pilot clinical studies indicates that stem cells derived from periodontal ligament tissues are a promising therapy for the regeneration of lost/damaged periodontal tissue. This study assessed the safety and feasibility of using autologous periodontal ligament stem cells (PDLSCs) as an adjuvant to grafting materials in guided tissue regeneration (GTR) to treat periodontal intrabony defects. Our data provide primary clinical evidence for the efficacy of cell transplantation in regenerative dentistry.

Methods: We conducted a single-center, randomized trial that used autologous PDLSCs in combination with bovine-derived bone mineral materials to treat periodontal intrabony defects. Enrolled patients were randomly assigned to either the Cell group (treatment with GTR and PDLSC sheets in combination with Bio-oss®) or the Control group (treatment with GTR and Bio-oss® without stem cells). During a 12-month follow-up study, we evaluated the frequency and extent of adverse events. For the assessment of treatment efficacy, the primary outcome was based on the magnitude of alveolar bone regeneration following the surgical procedure.

Results: A total of 30 periodontitis patients aged 18 to 65 years (48 testing teeth with periodontal intrabony defects) who satisfied our inclusion and exclusion criteria were enrolled in the study and randomly assigned to the Cell group or the Control group. A total of 21 teeth were treated in the Control group and 20 teeth were treated in the Cell group. All patients received surgery and a clinical evaluation. No clinical safety problems that could be attributed to the investigational PDLSCs were identified. Each group showed a significant increase in the alveolar bone height (decrease in the bone-defect depth) over time ($p < 0.001$). However, no statistically significant differences were detected between the Cell group and the Control group ($p > 0.05$).

Conclusions: This study demonstrates that using autologous PDLSCs to treat periodontal intrabony defects is safe and does not produce significant adverse effects. The efficacy of cell-based periodontal therapy requires further validation by multicenter, randomized controlled studies with an increased sample size.

(Continued on next page)

* Correspondence: cfmsunhh@fmmu.edu.cn; songtaos@dental.upenn.edu; yanjin@fmmu.edu.cn
†Equal contributors
[1]State Key Laboratory of Military Stomatology, Department of Periodontology, School of Stomatology, Fourth Military Medical University, Xi'an, Shannxi, P. R. China
[3]Department of Anatomy and Cell Biology, School of Dental Medicine, University of Pennsylvania, 240 South 40th Street, Philadelphia, PA 19104, USA
[2]State Key Laboratory of Military Stomatology, Research and Development Center for Tissue Engineering, School of Stomatology, Fourth Military Medical University, Xi'an, Shannxi, P. R. China

(Continued from previous page)

Trial Registration: NCT01357785 Date registered: 18 May 2011.

Keywords: Stem cell-therapy, Periodontitis, Periodontal regeneration, Cell sheet, Tissue engineering, Translational medicine

Background

Periodontitis is an inflammatory disease that causes pathological alterations in tooth-supporting tissues, which can lead to tooth loss if left untreated. National surveys have shown that the majority of adults suffer from moderate periodontitis, and up to 15 % of the population is affected by severe generalized periodontitis at some stage of their lives [1, 2]. The significant burden of periodontal disease and its impact on general health and patient quality of life suggest a clinical need for the effective management of this condition [3–5]. The ultimate goal of periodontal therapy is the predictable regeneration of the functional attachment apparatus that is destroyed by periodontitis, which involves at least three unique tissues, including the cementum, periodontal ligament (PDL), and alveolar bone. To date, several regenerative procedures have been developed in an attempt to treat periodontitis, including guided tissue regeneration (GTR), bone graft placement, and the use of bioactive agents, such as growth factors (reviewed in [5–8]). However, the current therapeutic techniques used either alone or in combination have limitations in producing complete and predicable regeneration, especially in advanced periodontal defects. In these cases, remaining deep intraosseous defects following periodontal therapy are high-risk sites for the further progression of periodontitis (reviewed in [9–11]).

According to histological evidence, the GTR technique combined with grafting materials, such as Bio-oss˙ (Geistlich Pharm. AG, Volhusen, Switzerland) and autologous bone, is partially effective at treating periodontal defects; however, the currently available GTR-based therapies remain rudimentary and show poor clinical predictability (reviewed in [7, 8, 11]). Recent advances in stem cell biology and regenerative medicine have enabled the use of cell-based therapy in periodontal diseases (reviewed in [5, 12]). To date, a large number of studies have indicated that ex vivo-manipulated stem cells derived from either bone marrow or the PDL can be used in conjunction with different physical matrices (autografts, xenografts, allografts, and alloplastic materials) to regenerate periodontal tissues in vivo (reviewed in [13–15]).

Although controversy remains regarding which tissues provide the most appropriate donor source for cell isolation, there is evidence that the cells of PDL tissues have the capacity to form a complete periodontal attachment apparatus (reviewed in [13–15]). The regenerative capacity of the PDL is attributed to a few progenitor cells within

the PDL that maintain their proliferation and differentiation potential; thus, regeneration of the periodontium depends on the participation of these mesenchymal stem/stromal cells (MSCs) (reviewed in [5, 12, 13]). PDL-derived progenitors are committed to several developmental lineages, i.e., osteoblastic, fibroblastic and cementoblastic [16], which suggests that these cells are capable of regenerating multiple periodontal tissues. Indeed, positive preclinical results have been achieved in a wide range of in vitro and in vivo models [17–28]. The next phase of study requires the clinical application of these advanced therapies.

Worldwide, periodontitis remains highly prevalent and leads to a loss of the affected teeth. This disease threatens the quality of life of the middle-aged population as far as oral functioning is concerned. Unfortunately, no current clinical periodontal treatments can heal the defects in the affected region or regenerate lost periodontal tissue to a normal structure and functionality. It is clear that there is a clinical need for such treatments and a vast patient demand. Importantly, several groups have commenced small-scale pilot/feasibility studies in humans [21, 29–31]; thus, there is now sufficient information to support endeavors to move cell-based periodontal therapy into the clinical arena. We established a clinical protocol to further test the safety, feasibility, and potential efficacy of stem cells for the treatment of periodontal deep intraosseous defects.

Methods

Study design

This study had a randomized design involving one dental facility (Translational Research Center, School of Stomatology, Fourth Military Medical University) and was conducted in compliance with Good Clinical Practice (GCP) guidelines according to the schedule shown in Fig. 1. This clinical trial, including the recruitment of subjects, was performed from 1 June 2011 to 30 December 2013, and the study was completed at the end of 2014 with a 1-year follow-up of the patients. The study protocol for this trial is provided in Additional file 1.

Ethics

This study was approved by the ethical committees of the School of Stomatology, Fourth Military Medical University (2011-02) and is registered with the ClinicalTrials.gov database (reference no. NCT01357785). This study was conducted according to the Declaration of Helsinki, and

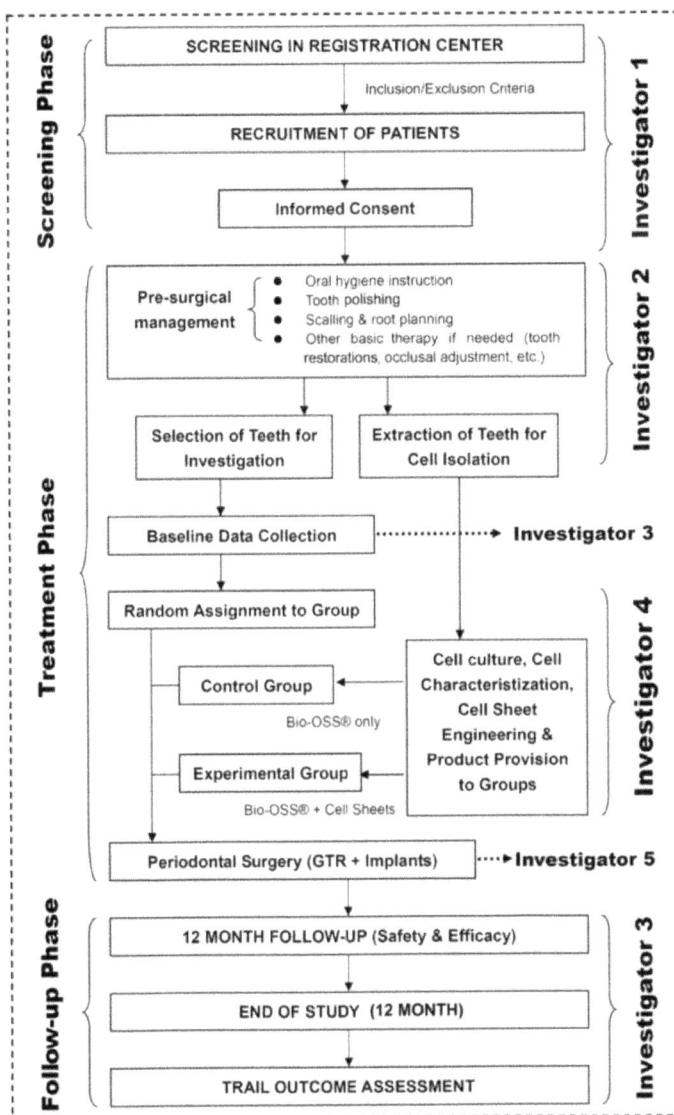

Fig. 1 The flow chart of the trial describes the selection, randomization, treatment and follow-up process. A physician (Investigator 3) performed the follow-up examination of the patients and remained blinded to the treatment conditions until the clinical trial was completed. *GTR* Guided tissue regeneration

all recruited patients consented to participate in this trial and contribute their trial data for noncommercial purposes. The protocol of this trial was externally reviewed and approved by an anonymous independent ethical review committee to ensure no serious ethical concerns.

Patients, enrollment and randomization

Patients with periodontitis visiting our dental institution were requested to participate in the study. In compliance with GCP guidelines, prospective patients who provided written informed consent underwent clinical inspection

and an oral cavity diagnosis. We selected subjects who satisfied the inclusion and exclusion criteria (recorded as the date of recruitment). The majority of these criteria were used in previous similar periodontal clinical trials [32, 33]. The inclusion and exclusion criteria and methods for randomization are provided in Additional file 2 (Appendices 1 and 2).

Study products and interventions

The third molars of the patients in the Cell group were extracted and subjected to cell isolation and transplant production according to the Good Laboratory Practice and Good Manufacturing Practice (GMP) guidelines. The cells were assessed for cell colony-forming ability and osteogenic/adipogenic differentiation (Fig. 2A). Prior to the extraction surgery, at least two independent assessors concluded that a tooth (or teeth) extraction was required due to impacted or nonfunctional reasons. The methods for cell isolation and characterization are presented in Additional file 2 (Appendix 3). The PDL cell sheets obtained from the patient's own tooth/teeth (see inclusion criteria) were produced using the Good Laboratory Practice and GMP guidelines using a standardized procedure in the Research and Development Center for Tissue Engineering (Fourth Military Medical University, 145th West Chang-le Road, Xi'an 710032, Shaanxi, People's Republic of China). The detailed method is described in Additional file 2 (Appendix 3). Bio-Oss* and Bio-Guide* were purchased from Geistlich Pharma AG (Volhusen, Switzerland). Both transplants (Bio-oss* only or Bio-oss*/cell sheets) were freshly prepared by laboratory researchers (Fig. 2B a–c). Investigator 3, who performed the follow-up study, was kept blinded to the treatment conditions until the study was completed. For the surgical treatment, Bio-oss* only (Control group) or Bio-oss*/cell sheets (Cell group) were administered only to the bony defect region (Fig. 2B d–f). Each subject received a standard initial preparation, including oral hygiene instruction, full-mouth scaling, and root planning before surgical treatment, in order to minimize the bacterial insult and reduce variability between lesions at baseline. The operations were performed using GCP procedures. A 12-month postoperative follow-up was performed for each patient.

Safety assessment

A cell safety assessment was performed to detect chromosomal karyotype changes between freshly isolated periodontal ligament stem cells (PDLSCs) and those obtained from the cell sheets (the latter underwent approximately 30-day ex-vivo cultures). The detailed methods are described in Additional file 2 (Appendix 3). Complications and adverse events during postoperative healing were recorded, and we examined the extent of adverse event occurrence. In addition, blood was obtained from patients preoperatively and at 2 weeks, 3 months and 12 months postoperatively. Blood examinations included: (i) a decrease in the white blood cell count; (ii) an increase in the red blood cell count; (iii) a decrease/increase in the percentage of neutrophils; (iv) a decrease/increase in the percentage of lymphocytes; (v) an increase in blood bilirubin; (vi) a decrease in blood lactate dehydrogenase; (vii) an increase in C-reactive protein; and (viii) an increase in creatinine phosphokinase. Moreover, the levels of IgA, IgG, IgM, C3 and C4

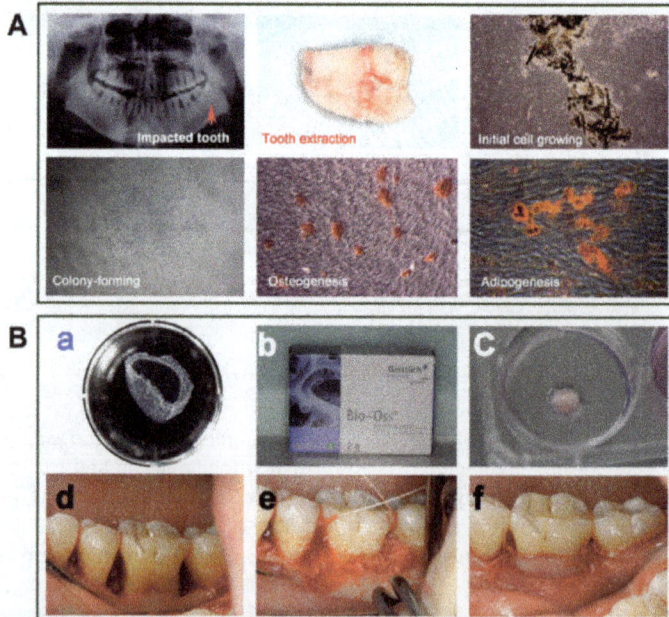

Fig. 2 Cell isolation, characterization and surgery. **A** The impacted third molar of patients was extracted and subjected to cell isolation and cell characterization for cell colony-forming ability and osteogenic/adipogenic differentiation. **B** The production of cell sheet/scaffold transplants and in vivo transplantation, including: (*a*) cell sheet formation; (*b*) Bio-Oss* particulates; (*c*) cell sheet/scaffold transplants; (*d*) exposure of bone defects; (*e*) placement of transplants; and (*f*) closure of the flap

were measured in the serum using enzyme-linked immunosorbent assay (ELISA) at the Department of Clinical Laboratory, Fourth Military Medical University School of Stomatology. At the time of blood collection, the urine of each patient was collected and assessed for: (i) a positive test for glucose/albumin; (ii) an increase in β-N-acetyl-D-glucosaminidase; and (iii) an increase in β2 microglobulin.

Efficacy assessment
The main outcome measure in the study protocol was the rate of increase in alveolar bone height at 3, 6 and 12 months postoperation (primary outcome). The bone-defect depth (the distance in millimeters from the deepest part of the defect to the cementoenamel junction of the tooth) was measured as described in Additional file 2 (Appendix 4) [34]. The clinical attachment level (CAL), probing depth (PD) and gingival recession (GR) measured in millimeters are generally used to assess pathology in periodontal disease. However, these parameters do not directly assess the efficacy of cells in periodontal tissue regeneration and were selected as secondary outcome measures to ascertain if the cells caused abnormal periodontal healing following periodontal surgery. The methods for the determination of these parameters at baseline and 3 months postoperation are described in Additional file 2 (Appendix 4) [32, 33].

Statistics
This study was performed using a per-protocol analysis. In this analysis, all of the randomized teeth received at least one therapy, but the teeth that did not receive treatment were excluded (modified per-protocol analysis). The last-observation-carried-forward method was used for the per-protocol analysis. The missing data points were input into the postbaseline follow-up visits from the last observation available for each patient. For analysis, we employed SAS version 8.2 software (SAS Institute Inc., Carey, North Carolina, USA). The per-protocol set analysis was performed for the primary outcome. The baseline between-group comparisons in age and clinical examination indices were performed using independent group t tests. The between-group comparison of sex was performed using the Fisher's exact probability test. The changes in clinical examination indices were tested using a repeated-measures analysis of variance. The level of statistical significance was set at $p < 0.05$ prior to analysis.

Results
Enrollment and teeth
The flow diagram for the study is shown in Fig. 3. A total of 48 screened teeth were randomly assigned to either the Control group or the Cell group. However, only 41 teeth

received surgery (21 teeth in the Control group and 20 teeth in the Cell group). The baseline measurements of the teeth are shown in Table 1. A Fisher's exact probability test found no significant between-group differences in the donors who provided teeth for randomization and testing.

Cell culture and surgery
In this trial, patients who had at least one tooth (e.g., wisdom tooth) that needed to be extracted due to impaction or nonfunctional reasons and agreed to the tooth extraction were enrolled. Prior to extraction surgery, at least two independent assessors concluded that a tooth or teeth required extraction. The extracted teeth were used for cell isolation. Only two teeth failed during the cell isolation step, and the corresponding two patients were excluded from further study. All of the cells exhibited colony-forming ability. In addition, these cells were positive for the MSC markers STRO-1, CD146, CD105, CD29, and CD90 and negative for the hematopoietic markers CD31 and CD45. The cells were successfully differentiated in osteogenic and adipogenic microenvironments and subsequently used for cell sheet production and periodontal surgery (refer to Additional file 2 (Appendix 3) for more information).

Safety evaluation
Postoperative healing occurred without significant problems, and none of the patients reported any complications/adverse events other than medium-sized swelling and pain. None of the pain experienced by patients required therapy. All of the patients underwent blood and urine tests preoperatively and at 2 weeks, 3 months and 12 months postoperatively. Changes in the white/red blood cell count, percentage of neutrophils/lymphocytes, and blood bilirubin/lactate dehydrogenase/C-reactive protein/creatinine phosphokinase levels were within the clinically accepted range (no measurement exceeded its clinical reference value). Importantly, no significant changes in IgA, IgG, IgM, C3 or C4 concentrations were found in the serum of any of the patients. A urine test showed that one patient (with one tooth that received GTR and Bio-oss* therapy without stem cells) was positive for glucose (this patient was ultimately not diagnosed with diabetes mellitus) and two patients were positive for albumin (each patient had two teeth involved in this trial, and one tooth per patient received cell therapy). No significant changes in urinary β-N-acetyl-D-glucosaminidase or β2 microglobulin were found for any of the patients.

Evaluation of efficacy
Patient demographic data and the baseline measurements of the affected teeth are shown in Table 1. All of the treated teeth in both groups adequately recovered

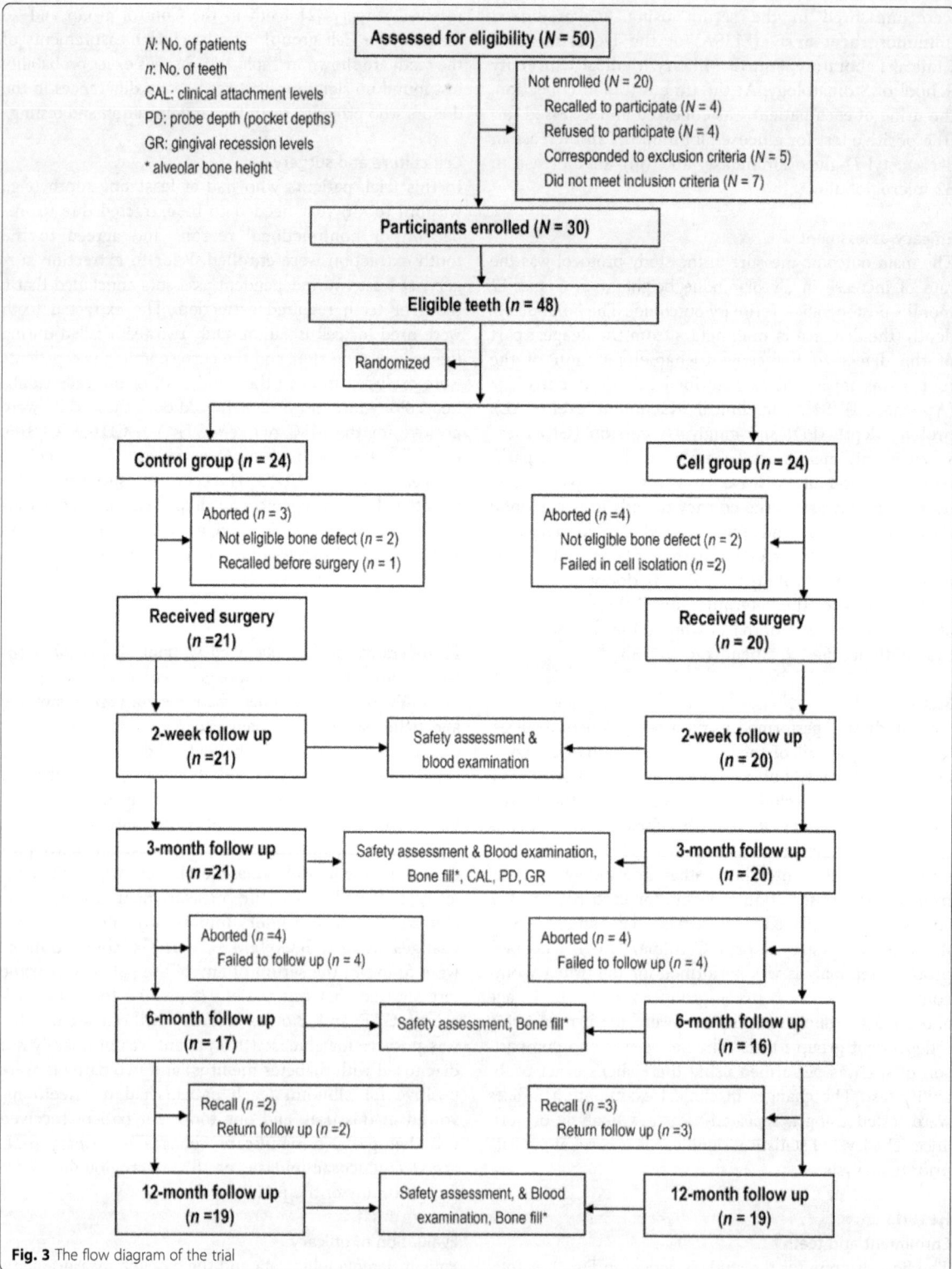

Fig. 3 The flow diagram of the trial

Table 1 Teeth and baseline

	Control group	Cell group	p value
Sex (n)	21	20	0.134[a]
Male (n)	6	2	
Female (n)	15	18	
Age (years, mean ± SE)	30.04 ± 7.90	26.05 ± 4.44	0.053[b]
CAL (mm, mean ± SE)	5.28 ± 1.60	5.15 ± 1.52	0.795[b]
BDD (mm, mean ± SE)	7.19 ± 1.87	7.20 ± 2.65	0.990[b]
PD (mm, mean ± SE)			
Facial	5.68 ± 1.59	6.43 ± 1.92	0.185[b]
Lingua (palatal)	5. 86 ± 1.43	6.25 ± 1.36	0.373[b]
GR (mm, median (interquartile range))			
Facial	0.33 (1.0)		0.692[c]
Lingua (palatal)	0.33 (0.83)		0.320[c]

[a]Fisher's exact probability test; [b]independent group *t* test; [c]Mann-Whitney test.
BDD bone-defect depth, *CAL* clinical attachment levels, *GR* gingival recession, *PD* probe depth, *SE* standard error

following surgery. There was no loss of treated teeth during this trial. X-ray examinations showed significant bone fill in both groups (Fig. 4). The magnitude of increase in alveolar bone height at 3, 6 and 12 months (bone fill over time) was determined as the decrease in the bone-defect depth. Each group showed a significant increase in the alveolar bone height over time ($p < 0.001$). However, no statistically significant differences were found between the Cell group and the Control group ($p > 0.05$) (Table 2). Regarding the clinical periodontal parameters, no statistically significant differences were found for the increased CAL, PD or GR between the Cell and Control groups at 3 months postsurgery ($p > 0.05$) (Table 3).

Discussion

Although there are a number of clinical techniques available for the management of periodontal intrabony defects, clinicians continue to seek more predictable regenerative therapies that are less technique-sensitive, lead to rapid tissue regeneration, and applicable to the broad array of periodontal conditions that are encountered daily in the clinic. Recent evidence from animal models [17–28] and several small-scale pilot/feasibility studies [21, 29–31] indicates that ex vivo-cultured PDL cells may serve as a powerful tool for periodontal therapy. A number of animal studies have provided an overwhelming body of evidence that MSCs can be safely and effectively used for periodontal regeneration (reviewed in [12]). As a consequence of these successful animal studies, the clinical application of

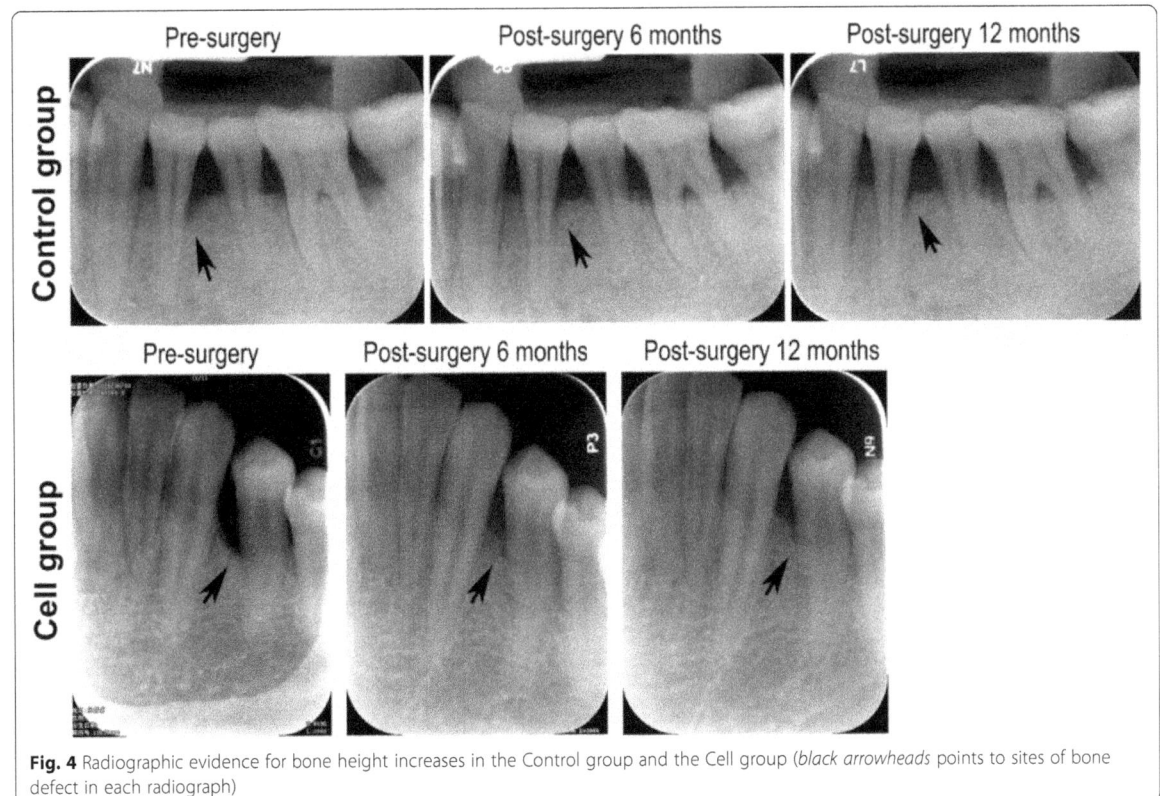

Fig. 4 Radiographic evidence for bone height increases in the Control group and the Cell group (*black arrowheads* points to sites of bone defect in each radiograph)

Table 2 Bone-defect depth with time (the distance from the deepest part of the defect to the cementoenamel junction of the tooth in mm, mean ± standard error)

	No. of teeth	Baseline	3 months	6 months	12 months	F value	p value
Control group	21	7.19 ± 1.87	4.81 ± 1.93	5.11 ± 1.53	4.80 ± 1.41	0.11	0.742[a]
Cell group	20	7.20 ± 2.65	4.89 ± 1.73	4.61 ± 1.87	4.49 ± 2.03		

[a]Repeated-measures analysis of variance

stem cells for the regeneration of periodontal tissue has begun [30, 31].

Substantial evidence suggests that it is time is to move cell-based periodontal therapy from animal studies to human clinical trials. However, there are critical steps in moving this field towards human clinical utility. In addition to clinical efficacy, the safety of cell-based therapies has not been fully evaluated, and the risks of stem cell therapies have been underscored by several clinicians and researchers. Moreover, issues, such as cell delivery, cell immunogenicity, use of autologous cells or allogeneic cells, control of cell fates in vitro and in vivo, and cost-effectiveness, are all important considerations that should be addressed before this therapy can move forward [35–39]. The next critical phase requires the identification of tissues that provide the most appropriate donor source(s) and the systematic validation of these specific MSCs as reliable for periodontal cytotherapeutic use. Furthermore, the establishment of large-scale preparation facilities incorporating the stringent protocols of GMP will be an absolute necessity. Regulatory agencies need to define new criteria to evaluate the risk associated with specific stem cells and their differentiated progeny (reviewed in [35, 36]). The purpose of this trial was to provide evidence for the use of ex vivo-cultured cells to treat periodontitis and determine the best approach to treat this disease.

For incurable and life-threatening diseases, such as diabetes, Parkinson's, muscular dystrophy, Alzheimer's, neural and cardiac diseases and refractory systemic lupus erythematosus, cell-based therapy is more likely to be warranted and accepted by the government and patients [37–39]. However, periodontal tissue regeneration using cell therapy is not yet economically viable or competitive with current root canal therapies and dental implants. Due to the non-life threatening nature of periodontitis, periodontal tissues have not been considered to be a major target for stem cell-based regenerative medical research. Nevertheless, affected teeth are ideal for the evaluation of new therapies because the patients are not usually ill. Thus, if anything goes wrong with the treatment, the situation is far less likely to be life threatening. Furthermore, the accessibility of teeth facilitates treatments that do not require major surgery [40].

In a previous study by Feng et al., periodontitis patients were treated with the local administration of autologous PDL or gingival stem cells. Neither adverse reactions nor an increase in any autoantibodies was observed; however, only three cases were observed long-term [30]. To move cell-based periodontal therapy from preclinical study and case report into clinical trials, we conducted this study to evaluate the safety and primary efficacy of this treatment. The results of this study have implications for oral health care resourcing and may facilitate improvements in the

Table 3 Changes in clinical examination indices over time (mm, mean ± standard error)

	No. of teeth		Baseline	3 months	F value	p value
CAL						
Control group	21		5.28 ± 1.60	5.07 ± 1.48	0.817	0.371[a]
Cell group	20		5.15 ± 1.52	4.42 ± 1.19		
PD						
Control group	21	Buccal	5.68 ± 1.59	3.88 ± 0.77	0.962	0.333[a]
Cell group	20	Buccal	6.43 ± 1.92	3.80 ± 1.03		
Control group	21	Lingual or palatal	5. 86 ± 1.43	3.79 ± 0.55	2.191	0.147[a]
Cell group	20	Lingual or palatal	6.25 ± 1.36	4.20 ± 0.86		
GR						
Control group	21	Buccal	0.62 ± 0.89	1.54 ± 0.96	0.133	0.728[a]
Cell group	20	Buccal	0.70 ± 1.09	1.28 ± 0.82		
Control group	21	Lingual or palatal	0.52 ± 0.85	1.38 ± 1.37	0.012	0.915[a]
Cell group	20	Lingual or palatal	0.73 ± 0.87	1.23 ± 0.92		

[a]Repeated-measures analysis of variance. *CAL* clinical attachment levels, *PD* probe depth, *GR* gingival recession

treatment of people with periodontitis. No clinical safety problems attributable to the investigational PDLSCs were identified in our study. Each group showed a significant increase in alveolar bone height over time ($p < 0.001$; Table 2). Although no statistically significant differences were found for the outcomes in the bone and clinical parameters between the Cell and Control groups ($p > 0.05$; Tables 2 and 3), the results of this study were the first to show that cell-based interventions are safe for human dental use in trials. Geistlich Bio-Oss˙ is a natural bone substitute with osteoconductive properties; however, there is no substantial evidence that this product will lead to effective or predictable bone regeneration in periodontal intrabony defects. Because Bio-Oss˙ particles may become an integral part of the newly formed bone framework and preserve volume over time [41], the use of this material as a scaffold for PDLSCs requires further investigation. Previous studies showed that allogeneic bone matrix may regenerate new bone, new cementum, and a new PDL around teeth that were previously contaminated by bacterial plaque [42, 43]. Human-derived biomaterials would be the first choice as a cell carrier in similar future trials [44]. Nevertheless, the results that were obtained in this study may have important implications for the design of more appropriate cell-delivery materials to test cell therapies in patients with periodontitis.

The study presented here used central randomization, which is a strict and complete randomization method that ensures adequate concealment. The surgeon and the investigator who collected the baseline and follow-up data worked independently in this trial. Throughout the entire trial, the patients were not aware of which group they were assigned to. The patients were only informed that they would receive a periodontal surgical treatment that potentially included cell products. This trial was a single-center, randomized controlled study of 30 patients with 12 months of follow-up. In a phase I clinical trial, safety assessments should be the primary outcome. The methodology in this study was designed to assess both safety and efficacy because the use of PDL-derived cells for periodontal therapy was previously reported several years ago [30]. Indeed, several groups worldwide have completed small-scale pilot/feasibility studies that indicated no adverse reactions after local cell administration [29, 31]. Although these studies did not have a randomized designed, the reported safety of the treatment offers substantial evidence to justify the use of cell-based periodontal therapy in the clinic. To date, this study was one of the largest randomized controlled trials addressing the safety and effectiveness of cell therapy in combination with GTR and bone replacement for periodontitis. However, this study does not deliver substantial evidence to confirm the safety of cell-based periodontal therapy. Similar to the clinical use of stem cells in the oral cavity (excluding the periodontium) for bone regeneration [45, 46], cell-based techniques in periodontal regenerative medicine should be further investigated in more challenging clinical scenarios with well-designed and standardized randomized controlled trials. In addition, these therapies should be tested in combination with bioactive molecules and new materials in an attempt to improve the final outcome. We plan to design a phase II clinical trial for cell-based periodontal therapy aimed at selecting more suitable scaffolding materials, determining appropriate cell doses, and providing evidence for multicenter, randomized controlled trials.

Conclusions

Stem cell therapy is a promising new therapeutic avenue that may enable the regeneration of lost periodontal tissue, and regenerative dentistry is at the forefront of the transition from basic science research to the clinical reconstructive arena. Although there are many issues that need to be resolved before stem cell therapies become commonplace, clinicians should continue to monitor the progression of these technologies. The data obtained in this study showed that autologous PDLSC-based treatment for periodontal intrabony defects was safe; however, more rigorous clinical trials are recommended to evaluate the efficacy of this therapy. Future clinical endeavors in cell-based periodontal therapy should identify more suitable scaffolding materials and define safe and effective cell dosing procedures based on well-designed, multicenter, randomized controlled trials.

Consent to publish

The authors confirm that they have obtained consent from the participants to publish the trial data.

Additional files

> **Additional file 1:** The study protocol (2011-06) for the trial. (DOC 6959 kb)
>
> **Additional file 2:** Appendices (inclusion and exclusion criteria of the trial; detailed methods for randomization; detailed methods for cell isolation, characterization and cell transplant preparation; and detailed methods for determining bone fill). (DOCX 2833 kb)

Abbreviations
CAL: Clinical attachment level; GCP: Good Clinical Practice; GMP: Good Manufacturing Practice; GR: Gingival recession; GTR: Guided tissue regeneration; MSC: Mesenchymal stem/stromal cell; PD: Probing depth; PDL: Periodontal ligament; PDLSC: Periodontal ligament stem cell.

Competing interests
The authors declare that they have no competing interests.

Authors' contributions
F-MC contributed to the study conception and design, data acquisition, analysis, and interpretation, and drafted and critically revised the manuscript. L-NG and B-MT contributed to the study design, data acquisition, analysis, and interpretation, and drafted and critically revised the manuscript. Y-JZ contributed to the study conception and design,

data acquisition, analysis, and interpretation, and drafted the manuscript. X-YZ, G-YD, HL, QC, and JX contributed to the study conception, data acquisition and interpretation, and critically revised the manuscript. YYu, R-XW, and YYi contributed to data acquisition, analysis, and interpretation, and drafted the manuscript. SS and YJ contributed to the study conception and design, data acquisition, analysis, and interpretation, and critically revised the manuscript. All of the authors read and approved the manuscript.

Authors' information
F-MC, L-NG, and B-MT are co-first authors. F-MC, SS, and YJ are co-corresponding authors.

Acknowledgements
This project was supported by a translational research grant from the Fourth Military Medical University School of Stomatology. The funding bodies played no role in the study design or the decision to submit the manuscript for publication. The authors also acknowledge our previous basic and animal research in this field that was supported by grants from the National Natural Science Foundation of China (81471791, 81500853 and 81530050).

References
1. Burt B. Position paper: epidemiology of periodontal diseases. J Periodontol. 2005;76:1406–19.
2. Pihlstrom BL, Michalowicz BS, Johnson NW. Periodontal diseases. Lancet. 2005;366:1809–20.
3. Williams RC, Barnett AH, Claffey N, Davis M, Gadsby R, Kellett M, et al. The potential impact of periodontal disease on general health: a consensus view. Curr Med Res Opin. 2008;24:1635–43.
4. Chen FM, Zhang J, Zhang M, An Y, Chen F, Wu ZF. A review on endogenous regenerative technology in periodontal regenerative medicine. Biomaterials. 2010;31:7892–927.
5. Lu H, Xie C, Zhao YM, Chen FM. Translational research and therapeutic applications of stem cell transplantation in periodontal regenerative medicine. Cell Transplant. 2013;22:205–29.
6. Chen FM, Jin Y. Periodontal tissue engineering and regeneration: current approaches and expanding opportunities. Tissue Eng Part B Rev. 2010;16:219–55.
7. Reynolds MA, Aichelmann-Reidy ME, Branch-Mays GL. Regeneration of periodontal tissue: bone replacement grafts. Dent Clin North Am. 2010;54:55–71.
8. Villar CC, Cochran DL. Regeneration of periodontal tissues: guided tissue regeneration. Dent Clin North Am. 2010;54:73–92.
9. Cortellini P, Labriola A, Tonetti MS. Regenerative periodontal therapy in intrabony defects: state of the art. Minerva Stomatol. 2007;56:519–39.
10. Bosshardt DD, Sculean A. Does periodontal tissue regeneration really work? Periodontol 2000. 2009;51:208–19.
11. Trombelli L. Which reconstructive procedures are effective for treating the periodontal intraosseous defect? Periodontol 2000. 2005;37:88–105.
12. Chen FM, Sun HH, Lu H, Yu Q. Stem cell-delivery therapeutics for periodontal tissue regeneration. Biomaterials. 2012;33:6320–44.
13. Lin NH, Gronthos S, Mark BP. Stem cells and future periodontal regeneration. Periodontol 2000. 2009;51:239–51.
14. Bartold PM, McCulloch CA, Narayanan AS, Pitaru S. Tissue engineering: a new paradigm for periodontal regeneration based on molecular and cell biology. Periodontol 2000. 2000;24:253–69.
15. Catón J, Bostanci N, Remboutsika E, De Bari C, Mitsiadis TA. Future dentistry: cell therapy meets tooth and periodontal repair and regeneration. J Cell Mol Med. 2011;15:1054–65.
16. Seo BM, Miura M, Gronthos S, Bartold PM, Batouli S, Brahim J, et al. Investigation of multipotent postnatal stem cells from human periodontal ligament. Lancet. 2004;364:149–55.
17. Doğan A, Ozdemir A, Kubar A, Oygür T. Assessment of periodontal healing by seeding of fibroblast-like cells derived from regenerated periodontal ligament in artificial furcation defects in a dog: a pilot study. Tissue Eng. 2002;8:273–82.
18. Doğan A, Ozdemir A, Kubar A, Oygür T. Healing of artificial fenestration defects by seeding of fibroblast-like cells derived from regenerated periodontal ligament in a dog: a preliminary study. Tissue Eng. 2003;9:1189–96.
19. Nakahara T, Nakamura T, Kobayashi E, Kuremoto K, Matsuno T, Tabata Y, et al. In situ tissue engineering of periodontal tissues by seeding with periodontal ligament-derived cells. Tissue Eng. 2004;10:537–44.
20. Bruckmann C, Walboomers XF, Matsuzaka K, Jansen JA. Periodontal ligament and gingival fibroblast adhesion to dentin-like textured surfaces. Biomaterials. 2005;26:339–46.
21. Yamada Y, Ueda M, Hibi H, Baba S. A novel approach to periodontal tissue regeneration with mesenchymal stem cells and platelet-rich plasma using tissue engineering technology: a clinical case report. Int J Periodontics Restorative Dent. 2006;26:363–9.
22. Yang Y, Rossi FM, Putnins EE. Periodontal regeneration using engineered bone marrow mesenchymal stromal cells. Biomaterials. 2010;31:8574–82.
23. Yang ZH, Zhang XJ, Dang NN, Ma ZF, Xu L, Wu JJ, et al. Apical tooth germ cell-conditioned medium enhances the differentiation of periodontal ligament stem cells into cementum/periodontal ligament-like tissues. J Periodontal Res. 2009;44:199–210.
24. Ding G, Liu Y, Wang W, Wei F, Liu D, Fan Z, et al. Allogeneic periodontal ligament stem cell therapy for periodontitis in swine. Stem Cells. 2010;28: 1829–38.
25. Park CH, Rios HF, Jin Q, Bland ME, Flanagan CL, Hollister SJ, et al. Biomimetic hybrid scaffolds for engineering human tooth-ligament interfaces. Biomaterials. 2010;31:5945–52.
26. Washio K, Iwata T, Mizutani M, Ando T, Yamato M, Okano T, et al. Assessment of cell sheets derived from human periodontal ligament cells: a pre-clinical study. Cell Tissue Res. 2010;341:397–404.
27. Tsumanuma Y, Iwata T, Washio K, Yoshida T, Yamada A, Takagi R, et al. Comparison of different tissue-derived stem cell sheets for periodontal regeneration in a canine 1-wall defect model. Biomaterials. 2011;32:5819–25.
28. Bright R, Hynes K, Gronthos S, Bartold PM. Periodontal ligament-derived cells for periodontal regeneration in animal models: a systematic review. J Periodontal Res. 2015;50:160–72.
29. d'Aquino R, De Rosa A, Lanza V, Tirino V, Laino L, Graziano A, et al. Human mandible bone defect repair by the grafting of dental pulp stem/progenitor cells and collagen sponge biocomplexes. Eur Cell Mater. 2009;18:75–83.
30. Feng F, Akiyama K, Liu Y, Yamaza T, Wang TM, Chen JH, et al. Utility of PDL progenitors for in vivo tissue regeneration: a report of 3 cases. Oral Dis. 2010;16:20–8.
31. McAllister BS. Stem cell-containing allograft matrix enhances periodontal regeneration: case presentations. Int J Periodontics Restorative Dent. 2011; 31:149–55.
32. Kitamura M, Nakashima K, Kowashi Y, Fujii T, Shimauchi H, Sasano T, et al. Periodontal tissue regeneration using fibroblast growth factor-2: randomized controlled phase II clinical trial. PLoS One. 2008;3:e2611.
33. Kitamura M, Akamatsu M, Machigashira M, Hara Y, Sakagami R, Hirofuji T, et al. FGF-2 stimulates periodontal regeneration: results of a multi-center randomized clinical trial. J Dent Res. 2011;90:35–40.
34. de Molon RS, Morais-Camillo JA, Sakakura CE, Ferreira MG, Loffredo LC, Scaf G. Measurements of simulated periodontal bone defects in inverted digital image and film-based radiograph: an in vitro study. Imaging Sci Dent. 2012;42:243–7.
35. Hynes K, Menicanin D, Gronthos S, Bartold PM. Clinical utility of stem cells for periodontal regeneration. Periodontol 2000. 2012;59:203–27.
36. Yoshida T, Washio K, Iwata T, Okano T, Ishikawa I. Current status and future development of cell transplantation therapy for periodontal tissue regeneration. Int J Dent. 2012;2012:307024.
37. Daley GQ, Scadden DT. Prospects for stem cell-based therapy. Cell. 2008; 132:544–8.
38. Chen FM, Zhao YM, Jin Y, Shi S. Prospects for translational regenerative medicine. Biotechnol Adv. 2012;30:658–72.
39. Lalu MM, McIntyre L, Pugliese C, Fergusson D, Winston BW, Marshall JC, et al. Safety of Cell Therapy with Mesenchymal Stromal Cells (SafeCell): a systematic review and meta-analysis of clinical trials. PLoS One. 2012;7: e47559.
40. Volponi AA, Pang Y, Sharpe PT. Stem cell-based biological tooth repair and regeneration. Trends Cell Biol. 2010;20:715–22.
41. Traini T, Valentini P, Iezzi G, Piattelli A. A histologic and histomorphometric evaluation of anorganic bovine bone retrieved 9 years after a sinus augmentation procedure. J Periodontol. 2007;78:955–61.
42. Gantes B, Martin M, Garrett S, Egelberg J. Treatment of periodontal furcation defects. (II). Bone regeneration in mandibular class II defects. J Clin Periodontol. 1988;15:232–9.

43. Mellonig JT. Histologic and clinical evaluation of an allogeneic bone matrix for the treatment of periodontal osseous defects. Int J Periodontics Restorative Dent. 2006;26:561–9.

44. Chen FM, Liu X. Advancing biomaterials of human origin for tissue engineering. Prog Polym Sci. 2016;53:86–168.

45. Kaigler D, Pagni G, Park CH, Braun TM, Holman LA, Yi E, et al. Stem cell therapy for craniofacial bone regeneration: a randomized, controlled feasibility trial. Cell Transplant. 2013;22:767–77.

46. Padial-Molina M, O'Valle F, Lanis A, Mesa F, Dohan Ehrenfest DM, Wang HL, et al. Clinical application of mesenchymal stem cells and novel supportive therapies for oral bone regeneration. Biomed Res Int. 2015;2015:341327.

2

The role of long-term label-retaining cells in the regeneration of adult mouse kidney after ischemia/reperfusion injury

Xiangchun Liu, Haiying Liu*, Lina Sun, Zhixin Chen, Huibin Nie, Aili Sun, Gang Liu and Guangju Guan*

Abstract

Background: Label-retaining cells (LRCs) have been recognized as rare stem and progenitor-like cells, but their complex biological features in renal repair at the cellular level have never been reported. This study was conducted to evaluate whether LRCs in kidney are indeed renal stem/progenitor cells and to delineate their potential role in kidney regeneration.

Methods: We utilized a long-term pulse chase of 5-bromo-2'-deoxyuridine (BrdU)-labeled cells in C57BL/6J mice to identify renal LRCs. We tracked the precise morphological characteristics and locations of BrdU$^+$LRCs by both immunohistochemistry and immunofluorescence. To examine whether these BrdU$^+$LRCs contribute to the repair of acute kidney injury, we analyzed biological characteristics of BrdU$^+$LRCs in mice after ischemia/reperfusion (I/R) injury.

Results: The findings revealed that the nuclei of BrdU$^+$ LRCs exhibited different morphological characteristics in normal adult kidneys, including nuclei in pairs or scattered, fragmented or intact, strongly or weakly positive. Only 24.3 ± 1.5 % of BrdU$^+$ LRCs co-expressed with Ki67 and 9.1 ± 1.4 % of BrdU$^+$ LRCs were positive for TUNEL following renal I/R injury. Interestingly, we found that newly regenerated cells formed a niche-like structure and LRCs in pairs tended to locate in this structure, but the number of those LRCs was very low. We found a few scattered LRCs co-expressed Lotus tetragonolobus agglutinin (LTA) in the early phase of injury, suggesting differentiation of those LRCs in mouse kidney.

Conclusions: Our findings suggest that LRCs are not a simple type of slow-cycling cells in adult kidneys, indicating a limited role of these cells in the regeneration of I/R injured kidney. Thus, LRCs cannot reliably be considered stem/progenitor cells in the regeneration of adult mouse kidney. When researchers use this technique to study the cellular basis of renal repair, these complex features of renal LRCs and the purity of real stem cells among renal LRCs should be considered.

Keywords: Label-retaining cells, Renal stem cells, Renal progenitor cells, Kidney regeneration, Ischemia/reperfusion injury, AKI

Background

Ischemia/reperfusion (I/R) is one of the leading causes of acute kidney injury (AKI) [1]. Approximately two million people die of AKI worldwide every year, despite the widely available renal replacement therapies [2, 3]. Interestingly, after acute injury, the kidney undergoes a quick regeneration to recover from renal functional failure by producing new cells. Thus, to develop targeted treatments for AKI, a clear understanding of the cell type contribution to the repair of injured kidney is a prerequisite. Whether the cells responsible for kidney regeneration are terminally differentiated tubular cells, pluripotent progenitor cells, or adult renal stem cells is still controversial [4–8].

Adult stem cells contribute to tissue repair in many organs such as skin, brain, heart, and gastrointestinal mucosa [9–11]. Adult nephron progenitors have been identified in zebrafish, but the role of adult stem/progenitor-like cells in mammalian renal repair remains unclear [12, 13]. The evidence supporting the role of adult stem/progenitor cells in the restoration of renal tubules after ischemic injury is derived from the presence of renal

* Correspondence: haiyingl@sdu.edu.cn; guanguangju@163.com
Department of Nephrology, The Second Hospital of Shandong University, Shandong University, Jinan, PR. China

label-retaining cells (LRCs) in adult rodents. In 2003, Mae-shima et al. first demonstrated the existence of renal progenitor-like tubular cells and their role in tubular re-generation by a 2-week chase using the label-retaining cell (LRC) technique [14]. Since then, the LRC technique has been improved by a long chase period to search for stem/progenitor cells in the adult mouse kidney. Using this tech-nique, studies have been reported to demonstrate the role of renal stem/progenitor cells in the regeneration of ische-mic renal injury at specific locations, such as renal papilla [15–17], S3 segment of proximal tubules [18], and the junction of cortex and outer medulla [19]. However, based on a lineage-tracing technique, recent studies showed that the surviving native tubular epithelial cells repaired the injured kidney without involving specialized progenitors [7, 20]. Thus, further studies are needed to resolve this controversy and to judge the real existence of renal stem/progenitor cells in the regeneration of adult kidney.

LRCs were recognized as rare stem and progenitor-like cells based on their slow-cycling, quiescent nature [21]. However, label retention by itself does not distin-guish potential quiescent stem/progenitor cells from other less proliferative cell types that are labeled during the pulse period. Using the LRC technique in kidney, different studies demonstrated varying localization and function of LRCs during the repair of AKI [14–16, 18, 19, 22]. Rangarajan et al. revealed inconsistent localization of LRCs in adult kidney following deoxyuridine treatment at different dosages [23]. However, the complex biological features of LRCs in renal repair have never been eluci-dated. Considering the limitations of the LRC technique [24], we sought to evaluate whether these LRCs in kidney were indeed renal stem/progenitor cells and to delineate their potential role in kidney regeneration.

Methods
BrdU labeling
Male C57BL/6J mice were purchased from Vital River Laboratories (China). All animal procedures were given approval by the animal ethics committee of Shandong University (Jinan, China) and followed the Guide for the Care and Use of Laboratory Animals published by the US National Institutes of Health (NIH Publication No. 85–23, revised 1996). To investigate the long-term dis-tribution of LRCs in mouse kidneys, postnatal mice were injected intraperitoneally with 5-bromo-2'-deoxyuridine (BrdU) (Sigma-Aldrich) at a dose of 50 μg/g twice daily, starting from 12 hours after birth to 3 days [25, 26]. The labeled pups were allowed to grow without a medical operation until eight-weeks old.

Renal ischemia/reperfusion injury model
Eight-week-old BrdU-labeled C57BL/6J mice were sub-jected to renal bilateral I/R injury. Each group contained

at least four mice. Briefly, mice were anesthetized with chloral hydrate (125 mg/kg, intraperitoneally) [27], and kidneys were exposed through a midline incision. The renal artery and vein were isolated from surrounding tis-sues and kidneys were subjected to ischemia by clamping bilateral renal pedicles with non-traumatic mirovascular clamps (Roboz Surg Instruments, USA) for 22 min [28, 29]. In the sham group, the renal pedicles were isolated without using clamps. After removing the clamps, reper-fusion was verified and the incision was sutured and 1 ml of 37 °C saline was injected into the abdomen to supple-ment fluid loss. Mouse body temperature was maintained at 37 °C using a warming pad. Mice were sacrificed and blood and tissue samples were collected at 1, 2, 3, 4, 5, 6, 9, 14, and 28 days after I/R injury.

Assessment of renal function
To assess renal function, serum creatinine (Scr) and blood urea nitrogen (BUN) levels were monitored. Serum was isolated from blood samples by centrifuga-tion at 3000 rpm for 5 minutes. Scr and BUN were mea-sured at the clinical laboratory of the Second Hospital of Shandong University using an automatic biochemistry analyzer (Beckman Instruments Inc., USA).

Histology and immunohistochemistry
Renal tissues were fixed in 4 % paraformaldehyde, dehy-drated and embedded in paraffin, and 5-μm-thick sec-tions were cut using a microtome. For general histology, sections were stained with hematoxylin and eosin (H&E). BrdU immunohistochemistry was performed to determine the staining of LRCs [14, 25]. Primary anti-bodies against BrdU (Merck Millipore, USA, 1:50) were incubated with the sections overnight at 4 °C. Sections were washed with PBS, incubated with biotinylated goat anti-mouse IgG for 30 minutes, washed again and cov-ered with horseradish peroxidase (HRP)-conjugated streptavidin (Zymed laboratories, USA) for 30 minutes. Next, a diaminobenzidine (DAB) solution was used to obtain a visible brown staining. Staining of Ki67-positive cells was performed in a similar manner. For the negative control, the primary antibody was omitted to estimate the specificity. Morphological images and examination of sec-tions were captured by a light microscope (Nikon Instru-ment Inc.).

Immunofluorescence
Renal tissues were embedded in OTC freezing medium (Sakura Finetek, Torrance, CA, USA) and sliced into 5-μm sections. The tissues were washed and blocked with 10 % goat serum for 30 minutes at room temperature, incubated with both the mouse monoclonal anti-BrdU (1:100) antibody and the rabbit polyclonal to Ki-67 antibody (Abcam, Cambridge, UK;1:100) or Lotus tetragonolobus

agglutinin (LTA) antibody (Vector Labs, Berlingame, CA, USA;1:200) overnight at 4 °C. After washing with PBS, fluorescent secondary antibodies (rhodamine-conjugated goat anti-mouse IgG and FITC-conjugated goat anti rabbit IgG) were added to the sections in the dark for 1 hour at room temperature, and 4′,6-diamidino-2-phenylindole (DAPI) was used to stain nuclei. To observe cellular death in kidney, terminal deoxynucleotidyl transferase dUTP nick end labeling (TUNEL) (FragEL™ DNA Fragmentation Detection Kit-Fluorescent, Merck, Germany) was used by fluorescent staining the free 3′ -OH termini of DNA [30]. For negative controls, the primary antibody was replaced with normal goat serum, which did not show positive staining, thus confirming specificity. Pictures of each section were obtained at × 200 or × 400 magnification with a fluorescence microscope (ECLIPSE-TS100, Nikon, Japan).

Quantification of BrdU or Ki 67 or TUNEL-positive cells

The quantification of BrdU-retaining cells was performed by counting the number of positive nuclei in five selected fields of sections in a blinded manner under a light microscope at × 200. The average number was calculated as the number of BrdU-retaining nuclei per field. The fields of sections were generated from 3 ~ 4 separated BrdU-labeled mice per group, and the quantitative analysis was expressed by the average number of BrdU-labeled cells in the five random fields [14, 25]. Quantification of Ki 67 or TUNEL-positive cells was performed in a similar manner at × 400 magnification.

Statistical analysis

Data are presented as means ± SEM. Differences between groups were analyzed by one-way ANOVA using SPSS 16.0. $P < 0.05$ was considered statistically significant.

Results
Restoration of mouse renal function and structure after I/R injury

AKI activates pathways of cell death and cell proliferation. To investigate the natural healing process of injured kidneys, we observed the renal functional and structural changes in a mouse model of ischemic AKI. Scr and BUN were measured as markers of renal function

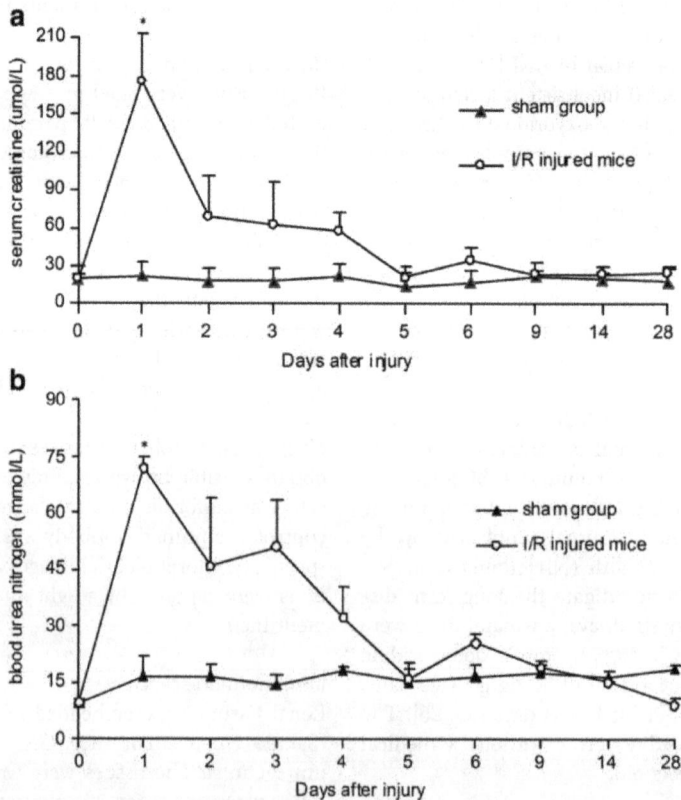

Fig. 1 Restoration of renal function in C57BL/6J mice after I/R injury ($n = 4$ in each group). **a** Serum creatinine (Scr) level significantly increased one day after injury and was restored to normal gradually over time. **b** The peak value of blood urea nitrogen (BUN) was found at the end of the first day after injury, followed by return to normal levels ($P < 0.05$, I/R injured mice vs sham group). I/R ischemia/reperfusion

at 0 (baseline), 1, 2, 3, 4, 5, 6, 9, 14, and 28 days after ischemia. We found significantly increased Scr levels on day 1. The peak value of Scr was 9-fold greater than the mean baseline value (Scr in µmol/L, 1 day after injury vs baseline, 175.93 ± 36.61 vs 19.53 ± 5.03, $P < 0.05$); it returned to normal within 9 days (Fig. 1a). The trend of BUN after I/R injury was consistent with Scr (Fig. 1b).

To verify and localize the renal injury and regeneration directly, we examined the histological changes of kidney tissues in ischemic AKI by H&E staining. Twenty-four hours after I/R injury, we observed the typical renal tubular damage, such as severe tubular dilatation, loss of brush border, sloughed debris in tubular lumen space, and denuded basement membrane. The typical renal tubular damage lasted 3 days in renal cortex, medulla, and papilla (Fig. 2). On day 4, the newly generated tubular cells increased rapidly and formed a special niche-like structure of new cells (arrows in Fig. 2 and Fig. 3). The renal structure tended to restore rapidly within the next 3 ~ 5 days. Nine days after I/R injury, kidney tissues clearly showed normal histological architecture of renal cortex, medulla, and papilla (Fig. 2). These results indicated that the adult mouse kidney manifested natural regenerative capacity to repair I/R injured tubular cells.

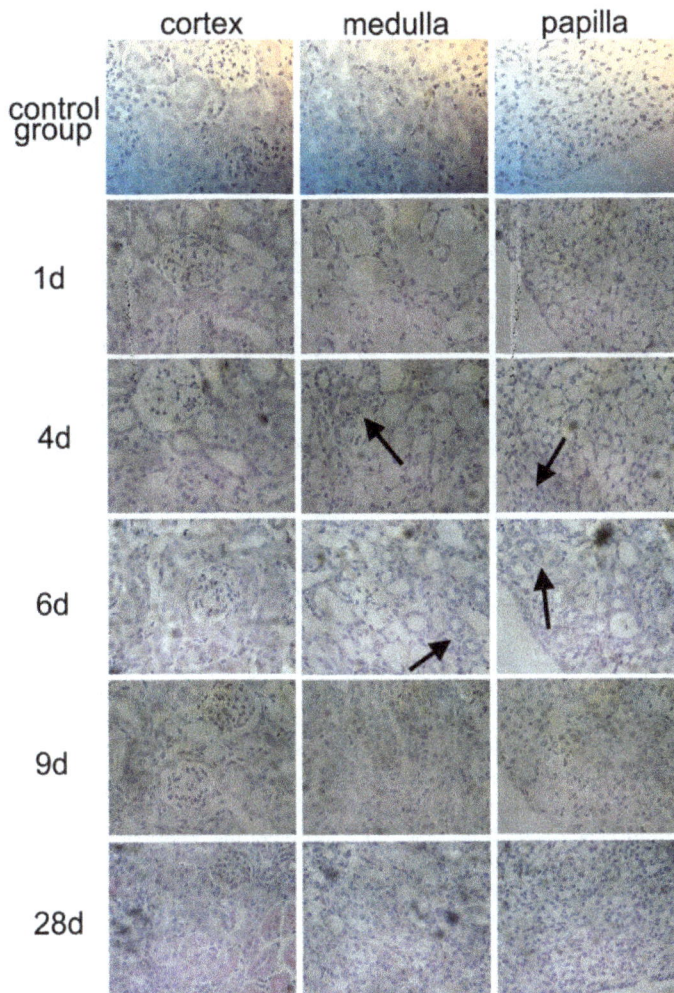

Fig. 2 Renal histology in I/R injury. Kidney tissues from C57BL/6J mice with 22 min of bilateral renal ischemia or sham operation were collected on 1, 2, 3, 4, 5, 6, 9, 14, and 28 days and stained with H&E for histological examination. During the first 3 days of injury, kidney tissues showed severe tubular dilatation, loss of brush border, and sloughed debris in tubular lumen space. On day 4, the newly generated tubular cells increased rapidly. Nine days after I/R injury, the structure of damaged tissues showed normal histological architecture in cortex, medulla, and papilla (*arrows*, on day 4 and 6, indicate the special niche-like structure of newly restored tubular cells; magnification × 400). *I/R* ischemia/reperfusion, *H&E* hematoxylin and eosin

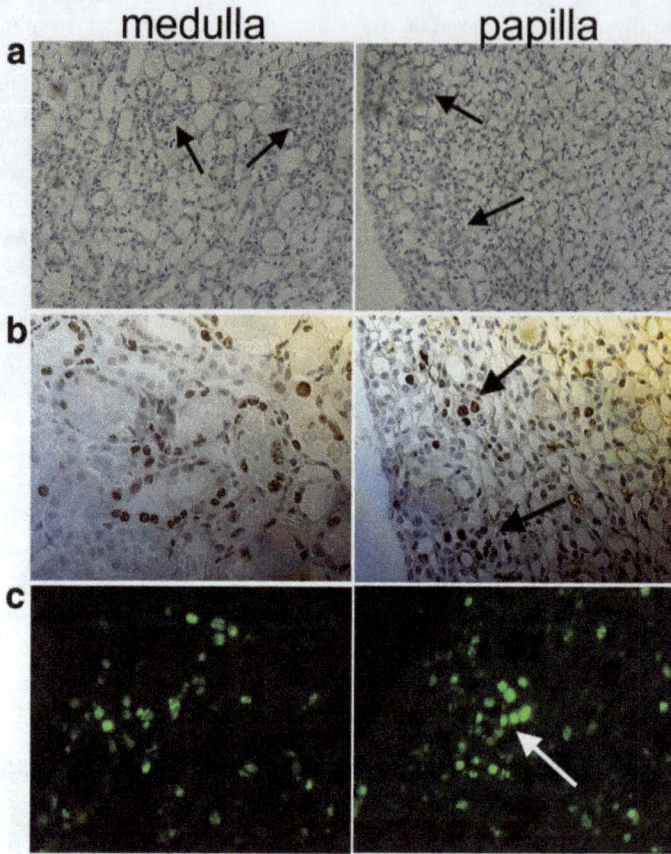

Fig. 3 The newly regenerated cells formed a niche-like structure on day 4 during the repair process of I/R injury. **a** Kidney tissues stained with H&E for histological examination by magnification × 200; **b** Ki67-positive cells in medulla and papilla by immunohistochemistry (magnification × 400); **c** Ki67-positive cells in medulla and papilla by immunofluorescence (magnification × 400). *I/R* ischemia/reperfusion, *H&E* hematoxylin and eosin

Distribution and morphological characteristics of long-term LRCs in normal mouse kidneys

To investigate the nature and potential role of long-term LRCs further, we first identified the precise location of these cells in normal adult mouse kidneys. After an 8-week chase period with BrdU, we found that BrdU$^+$ LRCs scattered among adult kidneys (Fig. 4 and Fig. 5). The difference in the number of these cells in cortex, medulla, and papilla was statistically significant (Fig. 3b $P < 0.05$). Most BrdU$^+$ LRCs were located on the tubular cells, with a few scattered on renal interstitium (Fig. 5). These results indicated that the adult mouse kidney retained original cells that were quiescent in DNA synthesis within the first 3 days after birth.

In addition, to study the morphological characteristics of these cells directly, we observed BrdU$^+$ LRCs under a microscope immunohistochemically. A few of the BrdU$^+$ LRCs were located in pairs, and the number of these cells in total BrdU$^+$ LRCs was 21.7 ± 1.5 % (Fig. 5 solid arrows, d). We observed that the nuclei of long-term

BrdU$^+$ LRCs exhibited different morphological characteristics in normal adult kidneys. Some of their nuclei were fragmented and disassembled, in approximately 10.1 ± 0.9 % of the BrdU$^+$ LRCs (Fig. 5 dotted arrows, d). These fragmented and disassembled nuclei showed the variation of long-term LRCs in adult mouse kidneys, suggesting DNA damage or defect in these LRCs. In addition to slow-cycling cells, the damaged and non-proliferating cells may also be labeled by long-term BrdU$^+$ LRCs in adult mouse kidney.

Limited role of long-term LRCs in the regeneration of I/R injured kidney

Eight weeks after BrdU-labeled mice underwent the renal I/R injury, we observed that the distribution of BrdU$^+$ LRCs diminished over time with renal functional recovery. In the first 3 days after injury, the number of BrdU$^+$ LRCs in medulla and papilla dramatically decreased while the majority of these cells in cortex decreased 4 days after injury (Fig. 6). For the repair process

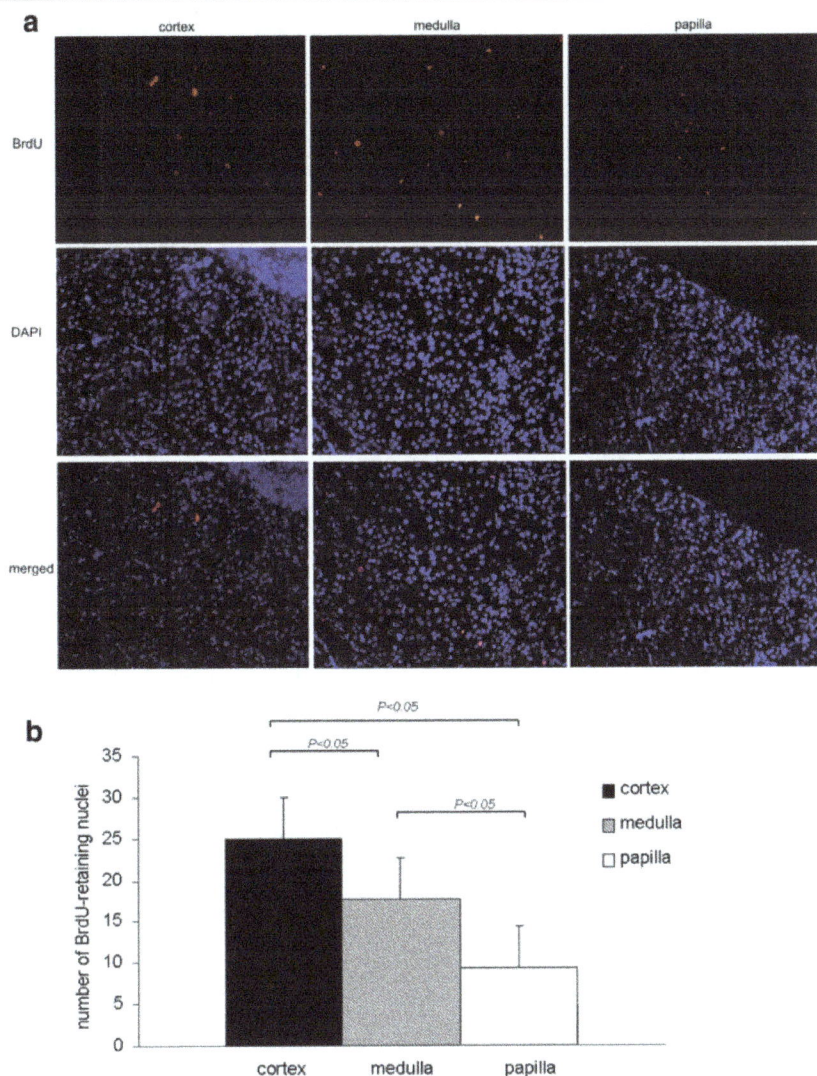

Fig. 4 Localization of BrdU-retaining cells in 8-week-old mouse kidneys using immunofluorescence. **a** BrdU$^+$ LRCs (red) scattered among normal adult kidneys; **b** BrdU-retaining cells in cortex, medulla, and papilla. BrdU-retaining cells were quantified by counting the number of positive nuclei in five randomly selected fields of sections under a fluorescence microscope (magnification × 200). *BrdU* 5-bromo-2'-deoxyuridine, *LRCs* label-retaining cells

of acute renal I/R injury, both cell death and proliferation are essential components in the remodeling of tissue structures [31]. Thus, to investigate the contribution of LRCs in renal tubule regeneration after I/R injury, we first stained newly proliferating cells with Ki67, which specifically recognizes the proliferating cells [32], and stained apoptotic cells by TUNEL kit in our murine model. TUNEL$^+$ cells peaked at 2 days after ischemic injury and a number of Ki67$^+$cells were observed among tubular cells in ischemic kidneys 24 h after reperfusion (Fig. 7). At 4 days after the initial injury, Ki67$^+$cells were significantly increased. Quantitative analysis

of the number of TUNEL$^+$ cells or Ki-67$^+$ cells is shown in Fig. 7. These results indicated that apoptosis of tubular cells peaked at 2 days after ischemic injury, which is followed by a well-established repair phase of tubule regeneration via proliferation of tubular cells 3 ~ 5 days after injury. Therefore, we speculated that the quick disappearance of LRCs in the first 4 days after I/R injury may be due to both ablation of superfluous apoptotic cells and newly proliferating cells in the process of remodeling injured renal structures.

To verify this speculation, we further stained the co-localization of long-term LRCs and Ki-67$^+$cells or

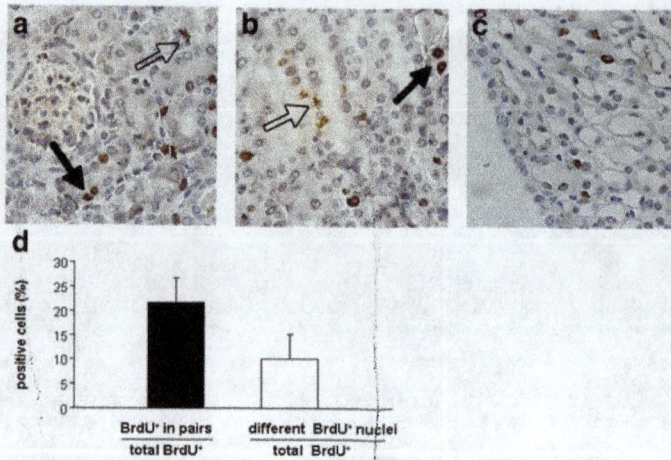

Fig. 5 Localization of BrdU-retaining cells by staining with the DAB lectin in 8-week-old mouse kidneys. **a** In the cortex, most of the BrdU-retaining cells were located among the tubular cells, and a few of them were scattered in renal interstitium; **b** Localization of BrdU-retaining cells in the medulla; **c** BrdU-retaining cells on the renal papilla; **d** Quantitative analysis of BrdU+ LRCs located in pairs (*solid arrows*) and BrdU+ LRCs revealed nuclear fragmentation and disassembly (*dotted arrows*, magnification × 400). *BrdU* 5-bromo-2'-deoxyuridine, *DAB* diaminobenzidine, *LRCs* label-retaining cells

TUNEL$^+$cells in murine kidney after renal ischemia. On day 4, a period of renal recovery with more tubule proliferation and fewer cell deaths, we observed that not all BrdU$^+$ LRCs were positive for Ki67 (Fig. 8a). BrdU$^+$Ki67$^+$ cells constituted only 24.3 ± 1.5 % of the total BrdU$^+$ LRCs, and the majority of these BrdU$^+$ LRCs occurred in pairs (arrows in Fig. 8a). BrdU$^+$ LRCs located in pairs tended to distribute in the niche-like structure formed by newly proliferating cells. These results indicated that only a few of the total BrdU$^+$ LRCs re-entered the cell cycle to contribute to the regeneration of I/R injured kidneys. In addition, we found that 9.1 ± 1.4 % of BrdU$^+$LRCs co-expressed with TUNEL at day 4 after injury, and TUNEL$^+$BrdU$^+$cells were observed in isolated scattered BrdU$^+$LRCs (arrows in Fig. 8b). This result demonstrated that long-term LRCs included cells with DNA fragmentation in the mouse kidney after injury and further indicated that not all the LRCs played a positive role in the regeneration process of I/R injured kidney.

Phenotype of long-term LRCs in mouse kidney after I/R injury

To further characterize the differentiation of LRCs, we examined the expression of Lotus tetragonolobus agglutinin (LTA), a marker of proximal tubular cell differentiation, in controls and I/R injured renal tissues. We found that some isolated scattered BrdU$^+$ LRCs co-expressed LTA both in normal and I/R injured kidney, demonstrating that a few LRCs were in a differentiated state in the early phase of renal injury (arrowheads in Fig. 9). These results

indicated that the scattered LRCs might not be the real stem-cell pool, and the regenerated immature cells might not be derived from long-term renal LRCs completely. Interestingly, BrdU$^+$LRCs located in pairs did not co-express LTA in mouse kidney after I/R injury (arrows in Fig. 9). With the serious damage of I/R injured kidney on day 2, BrdU$^+$LRCs seldom co-expressed LTA, indicating dedifferentiation of these LRCs at time points after I/R injury or migration of LRCs in kidney.

Discussion

In this study, we demonstrated that the adult mouse kidney was capable of spontaneously restoring its functions from I/R injury, with only a small part of the total long-term BrdU$^+$LRCs contributing to this regeneration. During the process, we found a few scattered LRCs exhibiting the characteristics of apoptotic cell death and cell differentiation in the early phase of tubular damage, which is inconsistent with the properties of stem/progenitor-like cells. Additionally, we first observed that the nuclei of long-term BrdU$^+$ LRCs exhibited different morphological characteristics in normal adult kidneys, including fragmentation and disassembly of nuclei. Although we cannot rule out that LRCs in pairs are the real stem/progenitor cells in mouse kidney, we demonstrated that not all long-term LRCs represent the real niche of stem/progenitor cells in adult kidneys. Thus, our novel findings can serve as a reminder that putative renal stem/progenitor cell niches identified by the LRC technique in previous studies, such as renal papillae, need to be re-evaluated.

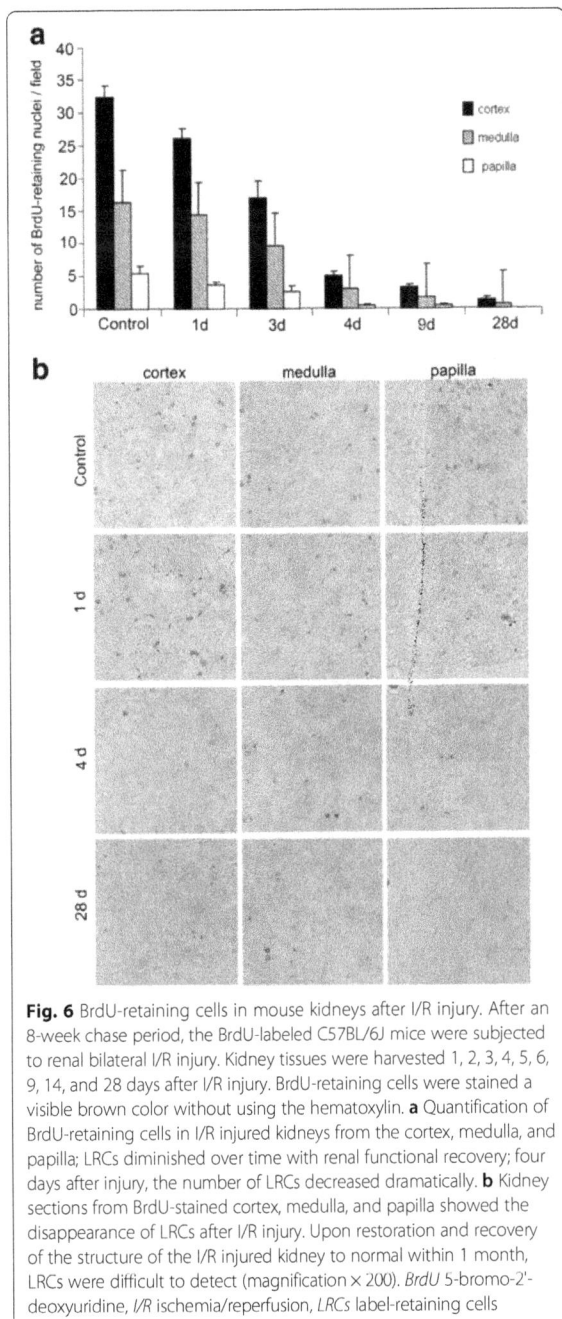

Fig. 6 BrdU-retaining cells in mouse kidneys after I/R injury. After an 8-week chase period, the BrdU-labeled C57BL/6J mice were subjected to renal bilateral I/R injury. Kidney tissues were harvested 1, 2, 3, 4, 5, 6, 9, 14, and 28 days after I/R injury. BrdU-retaining cells were stained a visible brown color without using the hematoxylin. **a** Quantification of BrdU-retaining cells in I/R injured kidneys from the cortex, medulla, and papilla; LRCs diminished over time with renal functional recovery; four days after injury, the number of LRCs decreased dramatically. **b** Kidney sections from BrdU-stained cortex, medulla, and papilla showed the disappearance of LRCs after I/R injury. Upon restoration and recovery of the structure of the I/R injured kidney to normal within 1 month, LRCs were difficult to detect (magnification × 200). *BrdU* 5-bromo-2'-deoxyuridine, *I/R* ischemia/reperfusion, *LRCs* label-retaining cells

Although label-retaining cells have been observed in adult rodent kidney since 2003, the exact localization and the number of those cells are still controversial [23]. In the present study, we found that the number of long-term BrdU$^+$ LRCs in cortex or medulla is much higher than that in papilla of 8-week old male C57BL/6J mice (Figs. 4 and 6). However, Rangarajan et al. showed a lower number of neonatal LRCs located in cortex or

medulla and a higher number in the papilla [23]. One of the possible reasons for this discrepancy is mainly due to the sex differences in mouse between the studies. Rangarajan et al. used female C57BL/6J mice, while we used male mice for the present study. As we know, sex difference is found in responses to ischemia/reperfusion-induced acute kidney injury and females may attenuate the kidney injury [33–36]. In our study, the distribution of long-term LRCs also showed sex differences in the differing compartments of murine kidney compared with Rangarajan et al.'s study. Given that long-term LRCs were identified as somatic stem/progenitor-like cells, diverse distributions of those cells between different genders may be one potential explanation for the greater susceptibility to I/R renal injury in the male gender. Additionally, differences in methods of labeling administration, time interval of injection, species, and labeling markers may also result in inconsistent localization of LRCs between each of these studies [14, 17, 37]. For example, Oliver et al. labeled long-term LRCs by giving BrdU subcutaneously to 3-day-old Sprague-Dawley rats, while we labeled LRCs intraperitoneally, as in most studies on neonatal mice, twice daily from 12 hours after birth to 3 days. Renal tissue in neonatal rodent undergoes active development every day, especially within 7 days after birth [38, 39]. So the timing and methods of administration of deoxyuridine in rodent have an important effect on the localization of LRCs. In short, the above differences that exist in different studies should be considered when researchers estimate the localization of LRCs in adult kidney.

Among our results was the novel finding of the variation of long-term LRCs in adult mouse kidneys which was directly observed by a microscope immunohistochemically. Our results showed that the nuclei of long-term BrdU$^+$LRCs were in pairs or scattered, fragmented or intact, strongly or weakly positive (Fig. 5), indicating different classes of long-term LRCs in adult mouse kidney. Thus, we observed biological characteristics of these LRCs in responses to I/R injury to clarify their roles in adult kidney. Although previous studies had observed that LRCs contributed to renal repair, the exact percentage of those LRCs counted by statistical analysis or the complex characteristics of long-term LRCs in the regeneration period of adult mouse kidney after I/R injury has never been reported [14, 15, 17, 19, 22]. In our study, we found that only 24.3 ± 1.5 % of total BrdU$^+$LRCs co-expressed with Ki67 and 9.1 ± 1.4 % of BrdU$^+$LRCs were positive for TUNEL at day 4 after injury (Fig. 8). Results showed newly regenerated cells did not completely derive from renal long-term LRCs in adult mouse kidneys. These results also showed only a small part of LRCs were responsible for repair, and most of the re-cycle LRCs were in pairs (Fig. 8a). Long-term LRCs exhibited complex characteristics,

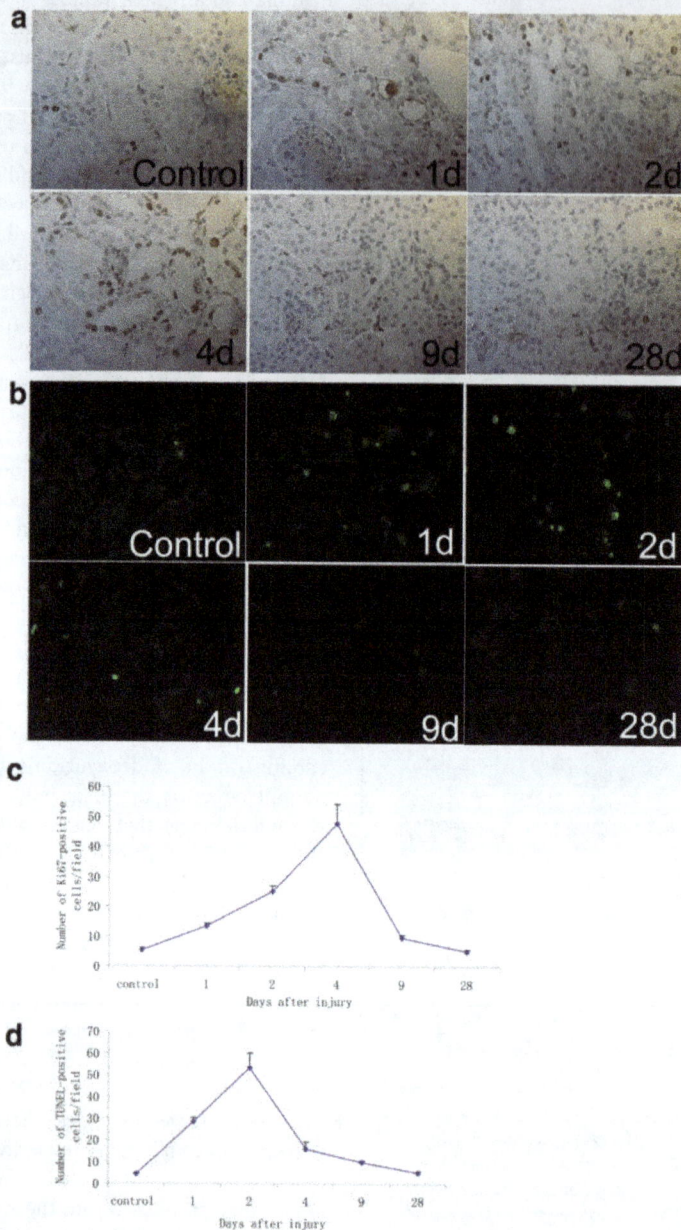

Fig. 7 Analysis of the expression of Ki67-positive cells (**a**, **c**) and TUNEL-positive cells (**b**, **d**) in mouse kidneys after IRI (magnification × 400)

including cell proliferation, apoptosis and others that are unknown. Indeed, most of the BrdU+LRCs did not reenter the cell cycle after injury. Therefore, the proportion of long-term LRCs which can represent true stem/progenitor cells in the I/R injured kidney is also very low, consistent with the finding observed in the hematopoietic system [24]. Further studies will be needed to test the differentiation potency of these LRCs by isolated culture or biological markers to redefine these cells.

The majority of published studies about LRCs in renal repair did not report the severity of renal injury by monitoring renal function or structure in their animal models of ischemic AKI. Given that different severity of renal injury affects the spontaneous remodeling process in adult kidney [18, 29], we estimated both renal function and histological changes and analyzed both proliferating cells and apoptotic cells in our murine model. Results showed apoptosis of tubular cells peaked at 2

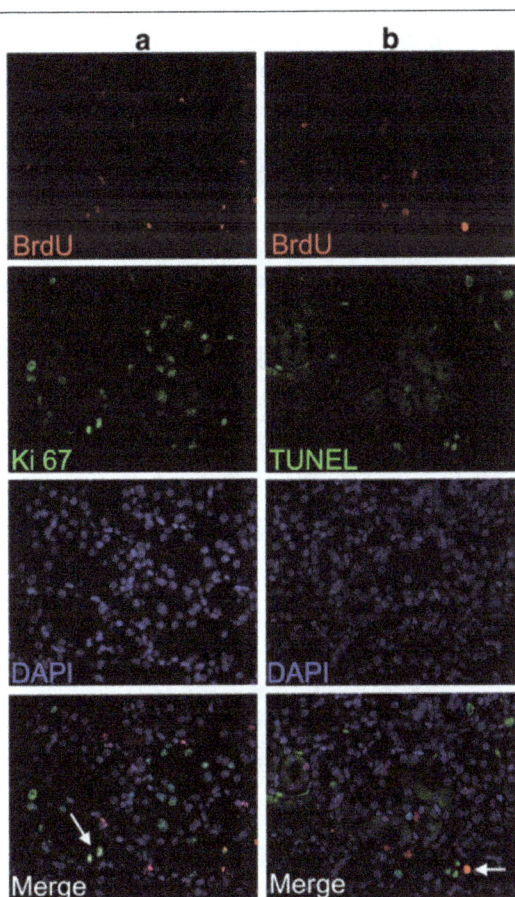

Fig. 8 Contribution of BrdU$^+$ LRCs in the regeneration of I/R injured mouse kidneys. **a** Double-immunofluorescence staining for BrdU (red) and Ki67 (green) to visualize the role of BrdU$^+$ LRCs in cell proliferation 4 days after IRI. Newly generated cells formed a special niche-like structure (green), and only some of the BrdU$^+$ LRCs co-expressed Ki67 in this area (arrows). **b** Double immunofluorescence was used to determine BrdU$^+$ LRCs undergoing cellular apoptosis on day 4. Scattered BrdU$^+$ LRCs co-expressed TUNEL (arrows) (arrows indicate the double-stained location; magnification × 400). BrdU 5-bromo-2'-deoxyuridine, LRCs label-retaining cells, I/R ischemia/reperfusion

decreased with the repair process of I/R injured kidney, especially during the first 4 days after injury (Fig. 6). The causes of the loss of long-term LRCs after injury are still controversial. Oliver et al. attributed this phenomenon to reentry of LRCs into the cell cycle during the repair phase [15], whereas Humphreys et al. recently showed that cell division could not explain the loss of LRCs [7, 15]. We speculated that ablation of superfluous apoptotic cells may be one of the reasons for the quick disappearance of LRCs in the first 4 days after I/R injury.

Interestingly, we found that newly regenerated cells formed a niche-like structure (Fig. 3), and LRCs in pairs tended to locate in this structure (Fig. 8a). Most isolated scattered BrdU$^+$ LRCs did not co-express Ki67 in the renal repair process. In this study, we also observed that a terminal differentiated proximal tubule marker, LAT, co-expressed in scattered BrdU$^+$ LRCs in the early phase of renal injury (Fig. 9). This result further revealed the characteristic differentiation of a few long-term LRCs. Previous studies explained these differentiated LRCs as transit–amplifying (TA) cells, which are the early descendants of stem cells [14]. In our study, these differentiated LRCs were observed before the proliferating peak. Thus, these LRCs cannot be considered as TA cells.

The LRC technique is based on the assumption that somatic stem cells are quiescent or retain their original template DNA strands as they undergo slow cell cycle, compared with the majority of cells in the tissue. Although this technique has been used to identify stem cells in the mammary gland [40], neural stem cells [41] and the skin [42], it has its own limitations. On the one hand, stem cells in tissues cannot be completely detected by LRCs, challenging the recognition and quantification of stem cells using this method [43, 44]. On the other hand, not all LRCs are recognized as stem/progenitor cells. For example, Kiel et al. reported that fewer than 0.5 % of LRCs in the hematopoietic system are stem cells [24]. Another study showed that other long-lived, rarely cycling cells, such as leukocytes and endothelial cells, may be identified as LRCs [21]. Thus, label retention is neither sensitive nor specific for identifying stem/progenitor cells. Further, in our study, immunofluorescence showed that 9.1 ± 1.4 % of BrdU$^+$ LRCs co-expressed cellular apoptosis marker in adult kidneys after the I/R injury (Fig. 8b). Immunohistochemical analysis revealed fragmentation and disassembly of 10.1 ± 0.9 % of the nuclei (Fig. 5). Therefore, we speculated that those LRCs may be arrested in the early phase of the cell cycle as a consequence of DNA damage or defect. The damaged and non-proliferating cells were also labeled, showing prolonged BrdU retention. Thus, our results further demonstrate that LRCs are not simple slow-cycling cells in adult kidneys, including cells with DNA damage or defect in the early phase.

days after ischemic injury, which is followed by a repair phase of tubule proliferation that peeked on the 4th day (Fig. 7). During the process, the number of LRCs

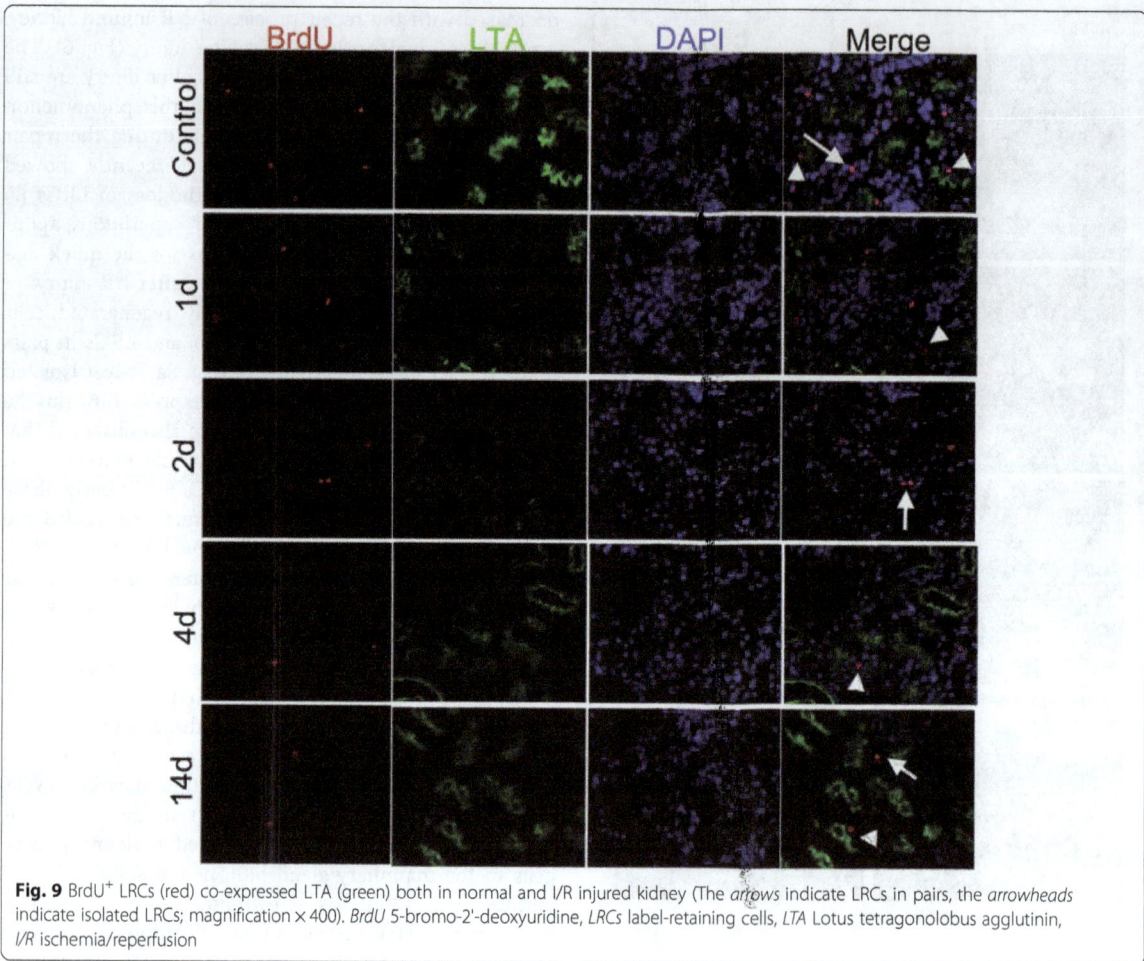

Fig. 9 BrdU+ LRCs (red) co-expressed LTA (green) both in normal and I/R injured kidney (The *arrows* indicate LRCs in pairs, the *arrowheads* indicate isolated LRCs; magnification × 400). *BrdU* 5-bromo-2'-deoxyuridine, *LRCs* label-retaining cells, *LTA* Lotus tetragonolobus agglutinin, *I/R* ischemia/reperfusion

Renal papilla is the structure composed of collecting ducts, loops of Henle, vasculature, and interstitial stromal cells [26]. Previous studies reported that LRCs of the renal papilla play a role in repair after ischemic injury and papillary LRCs were regarded as a stem cell niche in adult kidney [15–17, 45]. Recently, these findings have been questioned by using the genetic lineage analysis in kidneys [46]. Humphreys et al. have demonstrated that papillary LRCs do not contribute to the repair of I/R injured tubular cells. We also observed the papillary LRCs in our study, and our findings about complex biological characteristics of LRCs may provide a rational explanation for the inconsistent results.

The reason for the manifestation of different biological features by renal LRCs in the regeneration of I/R injured mice is also an unresolved issue. We speculate that since BrdU was injected over a 3.5-day period, characteristics of BrdU+LRCs might be varied during the active renal development in the first 7 days after birth. Renal LRCs may also include cells that are arrested in the early phase

of the cell cycle as a consequence of DNA damage or defect. A recent study presented similar findings, suggesting that renal LRCs were not the same population with respect to different loading and chase periods [23].

In summary, we report novel findings that long-term renal LRCs exhibited multiple biological characteristics during the recovery from I/R injury in mouse kidneys. Renal LRCs do not totally represent the adult stem-cell population based on our observations of apoptotic LRCs and terminally differentiated LRCs in mouse kidneys. Although we cannot rule out the possibility that a small number of LRCs located in a special structure formed by newly proliferating cells in pairs are genuine stem progenitor cells, our findings demonstrate that long-term LRCs are not uniform in adult kidneys. This diversity in the LRC population should be considered in studies searching for real stem/progenitor cells in adult kidneys by the LRC technique. The purity of real stem cells among LRCs should be assessed in studies exploring the cellular basis of renal repair using this technique. Furthermore, our results also show that adult kidneys exhibit spontaneous recovery from acute renal injuries. Therefore, further studies are needed to clarify the types of renal cells mediating the repair and the remodeling of acutely injured kidneys to develop targeted interventions.

Conclusions

Overall, this study revealed that long-term LRCs were not simple slow-cycling cells in adult kidneys, including cells with DNA damage or terminal differentiation in both normal and early I/R injury phase. Newly regenerated cells formed a niche-like structure to spontaneously repair injured renal cells, while only a small part of the LRCs in pairs tended to locate in this structure. Most isolated scattered LRCs did not reenter the cell cycle after injury, suggesting a limited role of long-term LRCs in the regeneration of I/R injured kidney. Thus, when researchers use the LRC technique to locate true stem/progenitor cells and to study the cellular basis of renal repair, these complex features of renal LRCs and the purity of real stem cells among renal LRCs should be considered.

Abbreviations
AKI: acute kidney injury; BrdU: 5-bromo-2'-deoxyuridine; BUN: blood urea nitrogen; DAB: diaminobenzidine; I/R: ischemia/reperfusion; LRCs: Label-retaining cells; LTA: Lotus tetragonolobus agglutinin; Scr: serum creatinine; TUNEL: terminal deoxynucleotidyl transferase dUTP nick end labeling.

Competing interests
The authors declare that they have no competing interests.

Authors' contributions
XL carried out the mouse renal ischemia/reperfusion (I/R) injury model, participated in assessment of renal function in mice, and was involved in drafting the manuscript. HL participated in the design of the study, performed the statistical analysis, and revised the manuscript critically for important intellectual content. LS carried out the immunohistochemistry, analyzed data, participated in the design of the study, and drafted the manuscript. ZC and HN participated in the immunofluorescence study, analyzed data, and revised the manuscript. AS and GL performed counting the quantification of BrdU or Ki 67 or TUNEL-positive cells. GG conceived of the study, participated in its design and coordination. AS, GL and GG also revised the manuscript critically for important intellectual content. All authors agreed to be accountable for all aspects of the work in ensuring that questions related to the accuracy or integrity of any part of the work are appropriately investigated and resolved. All authors read and approved the final manuscript.

Acknowledgments
We are grateful to Central Research Laboratory of the Second Hospital of Shandong University for technical assistance and generous support. This study was supported by an innovation fund of Shandong University (2012DX004), and a National Natural Science Foundation of China (81100509).

References
1. Bonventre JV, Yang L. Cellular pathophysiology of ischemic acute kidney injury. J Clin Invest. 2011;121(11):4210–21. doi:10.1172/jci45161.
2. Murugan R, Kellum JA. Acute kidney injury: what's the prognosis? Nat Rev Nephrol. 2011;7(4):209–17. doi:10.1038/nrneph.2011.13.
3. Li PK, Burdmann EA, Mehta RL. Acute kidney injury: global health alert. Kidney Int. 2013;83(3):372–6. doi:10.1038/ki.2012.427.
4. Benigni A, Morigi M, Remuzzi G. Kidney regeneration. Lancet. 2010; 375(9722):1310–7. doi:10.1016/S0140-6736(10)60237-1.
5. Pleniceanu O, Harari-Steinberg O, Dekel B. Concise review: kidney stem/progenitor cells: differentiate, sort out, or reprogram? Stem Cells. 2010;28(9):1649–60. doi:10.1002/stem.486.
6. Kramann R, Kusaba T, Humphreys BD. Who regenerates the kidney tubule? Nephrol Dial Transplant. 2015;30(6):903–10. doi:10.1093/ndt/gfu281.
7. Humphreys BD, Czerniak S, DiRocco DP, Hasnain W, Cheema R, Bonventre JV. Repair of injured proximal tubule does not involve specialized progenitors. Proc Natl Acad Sci U S A. 2011;108(22):9226–31. doi:10.1073/pnas.1100629108.
8. Rinkevich Y, Montoro DT, Contreras-Trujillo H, Harari-Steinberg O, Newman AM, Tsai JM, et al. In vivo clonal analysis reveals lineage-restricted progenitor characteristics in mammalian kidney development, maintenance, and regeneration. Cell Rep. 2014;7(4):1270–83. doi:10.1016/j.celrep.2014.04.018.
9. Lavker RM, Sun TT. Epidermal stem cells: properties, markers, and location. Proc Natl Acad Sci U S A. 2000;97(25):13473–5. doi:10.1073/pnas.250380097.
10. Korbling M, Estrov Z. Adult stem cells for tissue repair - a new therapeutic concept? N Engl J Med. 2003;349(6):570–82. doi:10.1056/NEJMra022361.
11. Beltrami AP, Barlucchi L, Torella D, Baker M, Limana F, Chimenti S, et al. Adult cardiac stem cells are multipotent and support myocardial regeneration. Cell. 2003;114(6):763–76.
12. Diep CQ, Ma D, Deo RC, Holm TM, Naylor RW, Arora N, et al. Identification of adult nephron progenitors capable of kidney regeneration in zebrafish. Nature. 2011;470(7332):95–100. doi:10.1038/nature09669.
13. Yeagy BA, Cherqui S. Kidney repair and stem cells: a complex and controversial process. Pediatr Nephrol. 2011;26(9):1427–34. doi:10.1007/s00467-011-1789-x.
14. Maeshima A, Yamashita S, Nojima Y. Identification of renal progenitor-like tubular cells that participate in the regeneration processes of the kidney. J Am Soc Nephrol. 2003;14(12):3138–46.
15. Oliver JA, Maarouf O, Cheema FH, Martens TP, Al-Awqati Q. The renal papilla is a niche for adult kidney stem cells. J Clin Invest. 2004;114(6):795–804. doi:10.1172/jci20921.
16. Al-Awqati Q, Oliver JA. The kidney papilla is a stem cells niche. Stem Cell Rev. 2006;2(3):181–4. doi:10.1007/s12015-006-0046-3.
17. Oliver JA, Klinakis A, Cheema FH, Friedlander J, Sampogna RV, Martens TP, et al. Proliferation and migration of label-retaining cells of the kidney papilla. J Am Soc Nephrol. 2009;20(11):2315–27. doi:10.1681/ASN.2008111203.
18. Fujigaki Y, Goto T, Sakakima M, Fukasawa H, Miyaji T, Yamamoto T, et al. Kinetics and characterization of initially regenerating proximal tubules in S3 segment in response to various degrees of acute tubular injury. Nephrol Dial Transplant. 2006;21(1):41–50. doi:10.1093/ndt/gfi035.
19. Kim J, Kim JI, Na YK, Park KM. Intra-renal slow cell-cycle cells contribute to the restoration of kidney tubules injured by ischemia/reperfusion. Anat Cell Biol. 2011;44(3):186–93. doi:10.5115/acb.2011.44.3.186.
20. Humphreys BD, Valerius MT, Kobayashi A, Mugford JW, Soeung S, Duffield JS, et al. Intrinsic epithelial cells repair the kidney after injury. Cell Stem Cell. 2008;2(3):284–91. doi:10.1016/j.stem.2008.01.014.
21. Morris RJ, Potten CS. Slowly cycling (label-retaining) epidermal cells behave like clonogenic stem cells in vitro. Cell Prolif. 1994;27(5):279–89.
22. Miya M, Maeshima A, Mishima K, Sakurai N, Ikeuchi H, Kuroiwa T, et al. Age-related decline in label-retaining tubular cells: implication for reduced regenerative capacity after injury in the aging kidney. Am J Physiol Renal Physiol. 2012;302(6):F694–702. doi:10.1152/ajprenal.00249.2011.
23. Rangarajan S, Sunil B, Fan C, Wang PX, Cutter G, Sanders PW, et al. Distinct populations of label-retaining cells in the adult kidney are defined temporally and exhibit divergent regional distributions. Am J Physiol Renal Physiol. 2014;307(11):F1274–82. doi:10.1152/ajprenal.00213.2014.
24. Kiel MJ, He S, Ashkenazi R, Gentry SN, Teta M, Kushner JA, et al. Haematopoietic stem cells do not asymmetrically segregate chromosomes or retain BrdU. Nature. 2007;449(7159):238–42. doi:10.1038/nature06115.
25. Chan RW, Gargett CE. Identification of label-retaining cells in mouse endometrium. Stem Cells. 2006;24(6):1529–38. doi:10.1634/stemcells.2005-0411.
26. Adams DC, Oxburgh L. The long-term label retaining population of the renal papilla arises through divergent regional growth of the kidney. Am J Physiol Renal Physiol. 2009;297(3):F809–15. doi:10.1152/ajprenal.90650.2008.

27. Patel NS, Sharples EJ, Cuzzocrea S, Chatterjee PK, Britti D, Yaqoob MM, et al. Pretreatment with EPO reduces the injury and dysfunction caused by ischemia/reperfusion in the mouse kidney in vivo. Kidney Int. 2004;66(3):983–9. doi:10.1111/j.1523-1755.2004.00847.x.

28. Melnikov VY, Faubel S, Siegmund B, Lucia MS, Ljubanovic D, Edelstein CL. Neutrophil-independent mechanisms of caspase-1- and IL-18-mediated ischemic acute tubular necrosis in mice. J Clin Invest. 2002;110(8):1083–91. doi:10.1172/JCI15623.

29. Wei Q, Dong Z. Mouse model of ischemic acute kidney injury: technical notes and tricks. Am J Physiol Renal Physiol. 2012;303(11):F1487–94. doi:10.1152/ajprenal.00352.2012.

30. Gavrieli Y, Sherman Y, Ben-Sasson SA. Identification of programmed cell death in situ via specific labeling of nuclear DNA fragmentation. J Cell Biol. 1992;119(3):493–501.

31. Price PM, Safirstein RL, Megyesi J. The cell cycle and acute kidney injury. Kidney Int. 2009;76(6):604–13. doi:10.1038/ki.2009.224.

32. Starborg M, Gell K, Brundell E, Hoog C. The murine Ki-67 cell proliferation antigen accumulates in the nucleolar and heterochromatic regions of interphase cells and at the periphery of the mitotic chromosomes in a process essential for cell cycle progression. J Cell Sci. 1996;109(Pt 1):143–53.

33. Wei Q, Wang MH, Dong Z. Differential gender differences in ischemic and nephrotoxic acute renal failure. Am J Nephrol. 2005;25(5):491–9. doi:10.1159/000088171.

34. Park KM, Kim JI, Ahn Y, Bonventre AJ, Bonventre JV. Testosterone is responsible for enhanced susceptibility of males to ischemic renal injury. J Biol Chem. 2004;279(50):52282–92. doi:10.1074/jbc.M407629200.

35. Elliot SJ, Berho M, Korach K, Doublier S, Lupia E, Striker GE, et al. Gender-specific effects of endogenous testosterone: female alpha-estrogen receptor-deficient C57Bl/6J mice develop glomerulosclerosis. Kidney Int. 2007;72(4):464–72. doi:10.1038/sj.ki.5002328.

36. Kang KP, Lee JE, Lee AS, Jung YJ, Kim D, Lee S, et al. Effect of gender differences on the regulation of renal ischemia-reperfusion-induced inflammation in mice. Mol Med Rep. 2014;9(6):2061–8. doi:10.3892/mmr.2014.2089.

37. Wang J, Lin G, Alwaal A, Zhang X, Wang G, Jia X, et al. Kinetics of label retaining cells in the developing rat kidneys. PLoS One. 2015;10(12):e0144734. doi:10.1371/journal.pone.0144734.

38. Rumballe BA, Georgas KM, Combes AN, Ju AL, Gilbert T, Little MH. Nephron formation adopts a novel spatial topology at cessation of nephrogenesis. Dev Biol. 2011;360(1):110–22. doi:10.1016/j.ydbio.2011.09.011.

39. Hartman HA, Lai HL, Patterson LT. Cessation of renal morphogenesis in mice. Dev Biol. 2007;310(2):379–87. doi:10.1016/j.ydbio.2007.08.021.

40. Smith GH. Label-retaining epithelial cells in mouse mammary gland divide asymmetrically and retain their template DNA strands. Development. 2005;132(4):681–7. doi:10.1242/dev.01609.

41. Karpowicz P, Morshead C, Kam A, Jervis E, Ramunas J, Cheng V, et al. Support for the immortal strand hypothesis: neural stem cells partition DNA asymmetrically in vitro. J Cell Biol. 2005;170(5):721–32. doi:10.1083/jcb.200502073.

42. Cotsarelis G, Sun TT, Lavker RM. Label-retaining cells reside in the bulge area of pilosebaceous unit: implications for follicular stem cells, hair cycle, and skin carcinogenesis. Cell. 1990;61(7):1329–37.

43. Grompe M. Tissue stem cells: new tools and functional diversity. Cell Stem Cell. 2012;10(6):685–9. doi:10.1016/j.stem.2012.04.006.

44. Anversa P, Leri A, Kajstura J. Biased DNA segregation during stem cell division. Circ Res. 2012;110(11):1403–7. doi:10.1161/circresaha.112.268961.

45. Oliver JA, Maarouf O, Cheema FH, Liu C, Zhang QY, Kraus C, et al. SDF-1 activates papillary label-retaining cells during kidney repair from injury. Am J Physiol Renal Physiol. 2012;302(11):F1362–73. doi:10.1152/ajprenal.00202.2011.

46. Kusaba T, Lalli M, Kramann R, Kobayashi A, Humphreys BD. Differentiated kidney epithelial cells repair injured proximal tubule. Proc Natl Acad Sci U S A. 2014;111(4):1527–32. doi:10.1073/pnas.1310653110.

3

A novel method for banking stem cells from human exfoliated deciduous teeth: lentiviral TERT immortalization and phenotypical analysis

Zhanhai Yin[1†], Qi Wang[2†], Ye Li[2], Hong Wei[3], Jianfeng Shi[3] and Ang Li[2,3*]

Abstract

Background: Stem cells from human exfoliated deciduous teeth (SHED) have recently attracted attention as novel multipotential stem cell sources. However, their application is limited due to in vitro replicative senescence. Ectopic expression of telomerase reverse transcriptase (TERT) is a promising strategy for overcoming this replicative senescence. Nevertheless, its potential application and the phenotype as well as tumorigenicity have never been assessed in SHED.

Methods: TERT expression was stably restored in SHED (TERT-SHED) isolated from healthy children aged 6–8 years using lentiviral transduction with a puromycin selection marker. The expression of TERT was detected using reverse transcription polymerase chain reaction, Western blot and immunofluorescence. Surface markers of SHED were detected by flow cytometry. Enzyme-linked immunosorbent assay was used to assess senescence-associated β-galactosidase, while CCK-8 methods were used to examine the proliferation capacity of SHED and TERT-SHED at different passages. Moreover, multilineage differentiation, karyotype, colony formation in soft agar, and tumor formation in nude mice of SHED and TERT-SHED were also examined.

Results: Lentiviral transduction induced stable TERT expression even in SHED at the 40th passage. TERT-SHED showed robust proliferation capacity and low concentration of β-galactosidase. Although they had some different biomarkers than early passage SHED, TERT-SHED at late passage showed similar mutilineage differentiation as TERT at early passage. Moreover, TERT-SHED at late passage showed normal karyotype, no soft agar colony formation, and no tumor formation in nude mice.

Conclusions: TERT-immortalized SHED may be a promising resource for stem-cell therapy, although attention should be paid to the biological behavior of the cells.

Keywords: Stem cells from human exfoliated deciduous teeth, Telomerase, TERT, Immortalization, Tumorigenicity

Background

Tissue engineering depends on the association of stem cells, growth factors, organ tissue culture, and tissue engineering materials [1]. In dentistry, regenerative strategies are of great relevance because of hard dental tissue damage, especially as a result of caries lesions, trauma, or iatrogenic procedures. The principles of regenerative medicine can be applied to endodontic tissue engineering. Regeneration of the pulp-dentin complex will allow natural replacement of damaged or missing tooth structures through the activation of tissue-specific stem cells of animal origin or transplantation of stem cells isolated and ex vivo expanded [2]. Stem cells can be found to be quiescent in their niche in all tissues of an adult organism. In response to organ injury, these cells initiate their proliferation and differentiation with the aim of healing

* Correspondence: drliang234@163.com

†Equal contributors

²Department of Periodontology, Stomatological Hospital, College of Medicine, Xi'an Jiaotong University, Xi'an 710004, P. R. China

³Research Center for Stomatology, Stomatological Hospital, College of Medicine, Xi'an Jiaotong University, Xi'an 710004, P. R. China

Full list of author information is available at the end of the article

injured tissue. Among various dental pulp stem cells (DPSCs), stem cells from human exfoliated deciduous teeth (SHED) have recently attracted attention as novel multipotential stem cell sources [3]. The loss of primary teeth creates the perfect opportunity to recover and store this convenient source of stem cells. Isolating SHED is simple, painless and convenient, and involves little or no trauma. These immature stem cells are important in the regeneration and repair of craniofacial defects, tooth loss, and bones because of their capability to proliferate and differentiate [4]. SHED generate rapidly and grow faster than adult stem cells, thus suggesting that they are less mature. SHED are postnatal stem cells capable of differentiating into osteogenic, odontogenic, adipogenic, and neural cells [5].

Although they have high self-renewal capacity, dental stem cells such as SHED will ultimately enter an irreversible proliferation-arrested state referred to as replicative senescence after long term in vitro culture under culture stresses, such as hyperoxia and elevated temperature [6]. Several strategies have been explored to overcome this replicative senescence, such as treatment of SHED with benzopyrene (BaP), radiation, or ectopic expression of viral oncogenes [7]. Long-term treatment with carcinogenic agents such as BaP results in transformation of cells as evidenced by chromosomal abnormalities, anchorage-independent growth in soft agar, and tumorigenicity in nude mice [8]. Radiation has been shown to be sufficient for immortalization of breast epithelial cells [9]. However, immortalization by radiation occurs relatively infrequently and results in morphological transformation of cells [9] and formation of tumors in nude mice [10]. A number of viral oncogenes, including simian virus-40 (SV40) large T-antigen, adenovirus E1A and E1B, polyoma T-antigen, and papillomavirus E6 and E7, have also been used to immortalize human cells [11–13]. Although immortalized cell lines have been successfully established by transfecting cells with viral oncogenes, inactivation of protein products of the tumor suppressor p53 and retinoblastoma (Rb) [14, 15], introduction of karyotypic instability, and transformation of phenotype have been reported in many studies [16, 17].

The cellular senescence and the lifespan depend on the loss rate of telomeres during each cell division and the primary length of the telomere [18]. Telomerase reverse transcriptase (TERT; catalysis subunit of telomerase) plays critical roles in the maintenance of telomere length during cell division [19–21]. It has been demonstrated that telomerase reconstitution via TERT expression could extend the telomere, prolong the lifespan of cells, and even immortalize cells [22, 23]. It has been established that the expression of TERT is a key step in human cellular proliferation, differentiation, and apoptosis. Moreover, recent findings indicate that TERT regulates stem cell properties in stemness sustaining and self-renew characterizations [24, 25]. While ectopic expression of TERT does significantly lengthen the lifespan of cells, enhanced telomerase activity is also a feature of many types of tumors and malignancies [26, 27]. The potential tumorigenicity of TERT-expressed stem cells remains controversial [27–29]. Therefore, the tumorigenicity of TERT expression in human stem cells needs to be further validated.

Our primary goal in this study was to create an immortalized SHED cell line by stable expression of TERT. Moreover, we assessed the multipotency and the potential tumorigenicity of our immortalized SHED cell line.

Methods

Subjects and cell culture

The SHED were obtained from the deciduous teeth of children aged 6–8 years. Every patient involved in the study consented to participate in the study and signed the paper consent. This study was approved according to guidelines set by the Ethic Committee of the Dental Hospital, Xi'an Jiaotong University. The deciduous anterior teeth used in this study were near natural exfoliation, with less than one third of the root remaining, and without any deep caries, restoration, periapical lesions, or internal resorption. After extraction, pulp tissues from the deciduous teeth were extirpated using a barbed broach (Mani, Utsunomiya Toshi-ken, Japan), washed with phosphate-buffered saline (PBS; Invitrogen, Carlsbad, CA, USA), and then treated with collagenase type I (3 mg/ml; Invitrogen) and dispase (4 mg/ml; Invitrogen) for 30 min at 37 °C; they were then filtered through a 70-μm cell strainer. The SHED were cultured in a DMEM/F12 medium supplemented with 15 % fetal bovine serum, 2 mmol/l L-glutamine, 100 μmol/l L-ascorbic acid-2-phosphate, 100 U/ml penicillin, 100 μg/ml streptomycin, and 0.25 μg/ml amphotericin B. After 7 days, cell colonies were observed. Individual cell colonies were collected by the filter paper enzyme digestion method. The derived cells were SHED.

Cloning of TERT in lentiviral expression plasmid and lentiviral production

pCMV6-XL5 plasmid (OriGene Technologies, Beijing, China) containing full-length cDNA of human TERT (3.6 kb) [Genbank:NM_198253.2] was amplified in DH5α E. coli strain. The cDNA clone of TERT and GV166 lentiviral vector (GeneChem Co., Ltd., Shanghai, China) were digested by a cocktail of EcoR I and Sal I (New England Biolabs, Ipswich, USA). The subsequent fragments were purified and recombined by T4 ligase (New England Biolabs) and then transformed into DH5α E. coli selecting for ampicillin resistance. The transformants were screened for correct insertion/orientation of the TERT fragment by restriction analysis. GV166 vector not recombined with

TERT was used as the control vector. For lentiviral production, the GV166-TERT or control plasmid was co-transfected into 293FT cells with Lenti-Easy Packaging Mix (GeneChem Co., Ltd.) at a 1:3 ratio using Lipofectamine™ reagent (Invitrogen). Forty-eight hours after transfection, the virus-containing supernatant was harvested and stored in aliquots at −80 °C. All cell culture procedures were performed under biosafety level 2 conditions.

Transduction of SHED with lentiviral vectors

Cells were plated 24 h before transduction at a density of 5×10^4 cells per well in six-well plates in the presence of 5 µg/ml polybrene. Transduction of SHED was carried out with TERT or control lentivirus at a multiplicity of infection (MOI) of 65. Transduced cells were passaged, and selected with puromycin (1.5 mg/ml) for 5 days.

Extraction of total RNA and RT-PCR

Total RNA was extracted using TRIzol reagent (Invitrogen, Carlsbad, CA, USA) according to the manufacturer's protocol, and RNase-free DNase I was used to remove DNA contamination. Reverse transcription (RT) was performed with 2 µg total RNA using M-MLV Reverse transcriptase (Promega, Madison, WI, USA) to synthesize first-strand cDNA according to the manufacturer's recommendation, followed by cDNA amplification using the specific primers for *TERT* and the β-actin primer. Primers used in this study were as follows: 5′-AGAGTGTCTGGAGCAAGTTG-3′ (forward) and 5′-GGATGAAGCGGAGTCTGG-3′ (reverse) for *TERT*; 5′-ATCGTGCGTGACATTAAGGAG AAG-3′ (forward) and 5′-GAGGAAGGAAGGCTGG AAGAGTG-3′ (reverse) for β-actin; and the corresponding polymerase chain reaction (PCR) products were 140 bp and 179 bp, respectively.

Western blot

Cells were lysed in RIPA lysis buffer, and the lysates were harvested by centrifugation ($13,523 \times g$) at 4 °C for 30 min. Approximately 20 µg protein samples were then separated by electrophoresis in a 12 % sodium dodecyl sulfate polyacrylamide gel and transferred onto a polyvinylidene fluoride membrane. After blocking the non-specific binding sites for 60 min with 5 % non-fat milk, the membranes were incubated overnight at 4 °C with a mouse monoclonal antibody against human TERT (Abgent, USA, at a 1:1000 dilution). The membranes were then washed three times with TBST (tris-buffered saline with tween-20) for 10 min and probed with the horseradish peroxidase (HRP)-conjugated rabbit anti-mouse IgG antibody (Immunology Consultants Laboratory, USA, at a 1:2000 dilution) at 37 °C for 1 h. After three washes, the membranes were developed by an enhanced chemiluminescence system (Cell Signaling

Technology, Danvers, MA, USA). A rabbit polyclonal antibody against human β-actin (Abcam, UK, at a 1:10000) was set as the inner control.

Immunofluorescence staining

Cells were fixed in 4 % paraformaldehyde, permeabilized with 0.1 % Triton1-X100, and blocked with 10 % horse serum in PBS for 1 h. The TERT antibody was prepared at 1:100 dilution and further incubated with the samples for 18 h at 4 °C. After washing with PBS, the cell was incubated with FITC-labeled goat antimouse antibody (Invitrogen) at 1:500 dilution for 45 min. Cells were also counterstained with DAPI. The fluorescence was evaluated by fluorescence microscope (Apotome).

Detection of surface markers by flow cytometry

The stem cell nature of SHED was analyzed using flow cytometry. Cells cultured with basal medium before cell differentiation were harvested using trypsin and washed twice with PBS. For cell surface staining, cells were fixed, washed and incubated with FITC-conjugated monoclonal or polyclonal antibodies against CD34, CD45, CD146 or STRO-1 (all Biolegend, CA, USA). For intracellular staining, cells were fixed and permeabilized using the Fix & Perm kit (Invitrogen), then washed and incubated with FITC-conjugated monoclonal or polyclonal antibodies against Oct-4 or Nanog (both Biolegend). FITC-conjugated IgG (Biolegend) was used as a negative control. The cells were analyzed using a FACScan flow cytometer (Becton Dickinson, Franklin Lakes, NJ, USA).

Multilineage differentiation assays

In vitro osteogenic differentiation of SHED was performed as previously published [30] using 100 nM dexamethasone, 10 mM β-glycerophosphate, and 50 M L-ascorbic acid-2-phosphate (Sigma). A total of 5×10^3 cells/well were seeded on to a six-well plate. After 16 days cells were assayed by von Kossa staining using a standard protocol. Adipogenic differentiation was accomplished as previously published [30] by 1 M dexamethasone, 0.2 mM indomethacin, 0.1 mg/ml insulin, and 1 mM 3-isobutyl-1-methylxanthin (IBMX) (Sigma). The maintenance medium consisted of 0.1 mg/ml insulin in standard medium. A total of 4×10^3 cells/well were seeded on to a 12-well plate. Stimulation was started when cells reached full confluency. Cells were grown for 5 days in induction medium and thereafter for 2 days in maintenance medium, and were then switched to induction medium again. After 16 days of stimulation, the cells were assayed by oil red staining using a standard protocol. Chondrogenic differentiation was achieved in aggregate cultures as previously published [30] with 100 nM dexamethasone, 1 mM pyruvate, 195 M L-ascorbic-acid-2-phosphate, 350 M L-proline, 1.25 % (v/v) insulin-transferrin-selenious acid mix (ITS, 100×), 5.35 g/ml linolic acid, 1.25 mg/ml bovine serum albumin (BSA; Sigma), and transforming growth factor-3 (TGF-3, 10 ng/ml; R&D Systems, Minneapolis, MN, USA). A total of 2.5×10^5 cells were used per pellet. Sections (12 μm) were cut with a cryostat vacutome HM 200 OM (Microm, Walldorf, Germany). Anionic sulfated proteoglycans were detected by toluidine blue metachromasia. Slices were stained in 1 % toluidine blue solution (Sigma, Munich, Germany) and 1 % sodium tetraborate (Sigma).

Real-time PCR

Total RNA and cDNA of differentiation-induced cells were prepared according to the above-mentioned protocols. Differentiation markers and GAPDH were amplified by quantitative real-time PCR using the following primers: ALP forward 5′-CATGCTGAGTGACACAGACAAGAA-3′, reverse: 5′-ACAGCAGACTGCGCCTGGTA-3′; BSP forward 5′-CTGGCACAGGGTATACAGGGTTAG-3′, reverse: 5′-GCCTCTGTGCTGTTGGTACTGGT-3′; LPL forward: 5′-GTCACGGGCTCAGGAGCATTA-3′, reverse: 5′-GCTCCAAGGCTGTATCCCAAGA-3′; PPAγ-2 forward: 5′-GCTCTGCAGGAGATCACAGA-3′, reverse: 5′-GGGCTCCATAAAGTCACCAA-3′; ACAN forward: 5′-ACGAAGACGGCTTCCACCAG-3′, reverse: 5′-TCGG ATGCCATACGTCCTCA-3′; COL2A1 forward: 5′-CCAG TTGGGAGTAATGCAAGGA-3′, reverse: 5′-ACACCAG GTTCACCAGGTTCA-3′; GAPDH forward: 5′-CTCCTC CTGTTCGACAGTCAGC-3′, reverse: 5′-CCCAATACGA CCAAATCCGTT-3′. Gene-specific amplification was performed in an ABI 7900HT real-time PCR system (Life Technologies, Carlsbad, CA) with a 15-μl PCR mix containing 0.5 μl cDNA, 7.5 μl 2×SYBR Green master mix (Invitrogen), and 200 nM of the appropriate primers. The mix was preheated at 95 °C for 10 min and then amplified in 45 cycles of 95 °C for 30 s and 60 °C for 1 min. The resolution curve was measured at 95 °C for 15 s, 60 °C for 15 s, and 95 °C for 15 s. The Ct (threshold cycle) value of each sample was calculated, and the relative mRNA expression was normalized to the GAPDH value ($2^{-\Delta\Delta Ct}$ method). The final expression value of differentiation markers was standardized according to that of control cultures.

Senescence-associated β-galactosidase assay by ELISA

Cells (1×10^6) were lysed and the supernatant was collected by centrifuge. The activity of β-galactosidase (β-GAL) in SHED was assessed using the human β-GAL enzyme-linked immunosorbent assay (ELISA) Kit (CSB-E09463h, Cusabio, China) according to the manufacturer's recommendations.

Proliferation assay

Cells were plated at a density of 1×10^3/well in 96-well plates and cultured in basal medium. A CCK-8 assay

was performed twice a day according to the cell counting kit protocol (Keygen Biotech, Nanjing, China) for 12 consecutive days. The values for each well were spectrophotometrically measured at 450 nm.

Cytogenetic analysis

Metaphase spreads were prepared from exponentially growing TERT-SHED at various passages. Cells were harvested and fixed following standard protocols [31]. Chromosome analysis was performed using the GTG-banding technique [31]. Fifteen metaphases captured by a CCD camera were analyzed and karyotyped using the CytoVision system (Leica Biosystems, Nussloch, Germany). Chromosome identification and karyotype description were made in accordance with the International System for Chromosome Nomenclature [32].

DNA isolation and PCR analysis of CDKN2A

Genomic DNA from SHED P4, and TERT-SHED P20 and P40 was extracted using the QIAamp DNA Mini Kit (QIAGEN, Duesseldorf, Germany) according to the manufacturer's recommendations. The exon 2 of the CDKN2A gene was amplified in an ABI GeneAmp 9600 PCR system using the following primers: forward: 5′-CC TGGCTCTGACCATTCTGTTC-3′, reverse: 5′-GCTTTG GAAGCTCTCAGGGTAC-3′. The PCR mixture contained 100 ng genomic DNA, 10 μl 2×Taq PCR Master Mix (TIANGEN Biotech, Beijing, China), and 0.1 pmol/μl of each primer in a total 20 μl volume. The PCR cycling conditions were 94 °C (2 min) for 1 cycle, 94 °C (30 s), 55 °C (40 s), 72 °C (2 min) for 36 cycles, and a final extension of 72 °C (10 min). The PCR products were 386 bp in length.

Soft agar assay

TERT-SHED or tongue cancer cells (Tca-8113; 2500 cells) in logarithmic growth phase were trypsinized and suspended into a single cell suspension in 0.5 ml 0.8 % top agar solution (37 °C). The cells were aliquoted on the top of a pre-prepared 1 ml 1.2 % base agar layer, and then incubated for 28 days at 37 °C with 5 % CO_2. The well plates were stained with 0.005 % crystal violet for 2 h before a photograph was taken under a microscope [33].

Tumorigenicity in nude mice

Twenty athymic nude mice were divided into four groups: SHED P2, TERT-SHED P20, Tca-8113, and PBS control. Cells (2×10^6) of each cell type were suspended in 200 μl PBS and injected subcutaneously into the fore and hind limb armpit under general anesthesia. Mice were sacrificed after 8 weeks by CO_2 overdose. All procedures were performed according to animal protection legislation and approved by the Ethics Committee of the Dental Hospital, Xi'an Jiaotong University. Photographs were taken every week for macroscopic evaluation. Skin and underlying soft tissue of the relevant area were dissected 8 weeks after cell implantation. Slides were prepared and stained with hematoxylin and eosin dye, and investigated for possible tumor growth [34].

Statistical analysis
The data for β-GAL concentration and proliferation were expressed as mean ± standard deviation (SD). One-way proliferation data were used to compare the difference in β-GAL and followed by Fisher's LSD post hoc test. Repeated measurement analysis of variance was used to compare the differences between proliferation curves. All statistical analyses were performed using IBM SPSS Statistics 19.0 software and a P value <0.05 was considered significant.

Results
SHED morphology
The morphology of SHED was analyzed under phase contrast microscope (Fig. 1). SHED spread along the surface of the culture plates, showing rapid growth within the first week. Cell colonies were seen from the fifth day, as the cells grew to confluence at 10–14 days.

Expression of TERT in SHED after lentiviral transduction
We first analyzed the expression of TERT in SHED after stable lentiviral transduction using Western blot and immunofluorescence. As shown in Fig. 2, TERT transduction remarkably restored the expression of TERT in SHED (TERT-SHED), while the control virus did not show TERT expression (SHED).

Expression of TERT in TERT-SHED during passage
We further analyzed the expression of TERT during the passage of TERT-SHED using RT-PCR and Western blot.

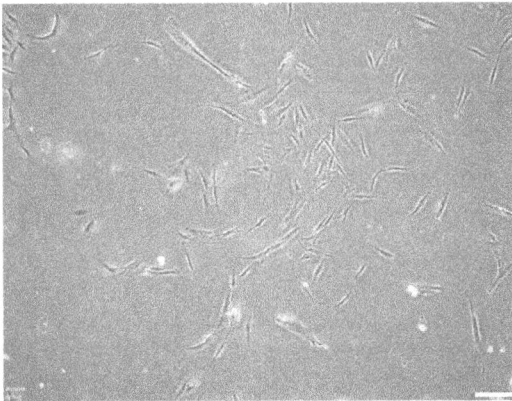

Fig. 1 Typical cell morphology of SHED under phase contrast microscope. SHED revealed a typical fibroblast-like morphology. *Scale bar* = 50 μm

As shown in Fig. 3, TERT expression remained stable even in the 40th passage.

Surface markers of SHED
The flow cytometry analysis was applied to quantify the expression ratios of specific surface antigens in SHED and TERT-SHED. The SHED at the fourth passage (SHED P4) showed robust expression of CD146 (90.45 %), STRO-1 (72.10 %), CD34 (65.76 %), and Oct-4 (85.16 %), weak expression of Nanog (13.84 %), and nearly negative expression of CD45 (3.79 %) (Fig. 4a). The TERT-SHED at the 20th passage (SHED P20) showed decreased expression of CD146 (48.51 %), STRO-1 (58.47 %), Oct-4 (10.48 %), Nanog (7.64 %), and CD34 (3.71 %) (Fig. 4b).

Multilineage differentiation of SHED
The multilineage differentiation potential of SHED was measured by differentiation induction assay. SHED P4 could different into osteogenic, adipogenic, and chondral cell lineages, as revealed by positive staining for Alizarin Red S, Oil Red O and Toluidine blue, respectively (Fig. 5a–c). Moreover, TERT-SHED P20 showed similar differentiation capacity to SHED P4 (Fig. 5e–g). Furthermore, real-time PCR analysis confirmed that TERT-SHED P20 and SHED P4 had similar expression of osteogenic (ALP and BSP), adipogenic (LPL and PPAR-γ), and chondrogenic (ACAN and COL2A1) differentiation markers (Fig. 5d and h).

Cell senescence and proliferation capacity of SHED
β-GAL activity at pH 6 is a known characteristic of senescent cells which is not found in presenescent, quiescent or immortal cells. We first examined the senescence marker, β-GAL, in SHED and TERT-SHED at different passages. As shown in Fig. 6a, the concentration of β-GAL in the 20th passage of SHED (SHED P20) was as 120 times that in SHED P4. However, β-GAL concentration remained at a very low level in the 40th passage of TERT-SHED (TERT-SHED P40). The proliferation capacity of SHED was detected by CCK-8 assay. As shown in Fig. 6b, the proliferation capacity of SHED at late passage (P20) significantly decreased. However, TERT-SHED P40 had similar proliferation potential to the early passage (P4) of SHED.

Cytogenetic of TERT-SHED
Atypia is one of the major characteristics of cancer. We therefore analyzed the karyotype of TERT-SHED at late passage (P20). As shown in Fig. 7a, no polyploid mutation or chromosomal deletion was found in TERT-SHED P40.

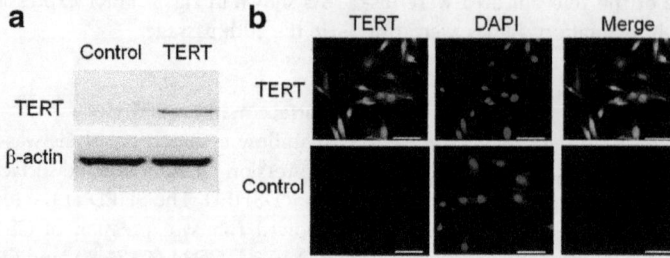

Fig. 2 Expression of TERT assessed by Western blot (**a**) and immunofluorescence (**b**) in SHED at the 20th passage 60 h after control and lentiviral transduction. Transduction of TERT-recombined lentivirus restored the expression of TERT in SHED and the optimal MOI was 65. *Scale bar* = 50 μm. *TERT* telomerase reverse transcriptase

Integrity of genomic *CDKN2A*

Since a deletion of the *CDKN2A* gene locus has been described after ectopic TERT expression using retroviral vectors [35], we analyzed the integrity of *CDKN2A*. PCR amplification yielded a band of the expected size for SHED P4, TERT-SHED P20 and TERT-SHED P40 (Fig. 7b).

Colony formation of TERT-SHED

The soft agar colony formation assay is a common method to monitor anchorage-independent growth. We thus examined the colony formation of TERT-SHED P20. As shown in Fig. 7c, only a single cell was noticed growing in soft agar in the culture of TERT-SHED P20, while cell aggregates were formed in the culture of tongue cancer cells (Tca-8113).

Tumorigenicity of TERT-SHED in nude mice

We further assessed tumor formation of TERT-SHED P20 in nude mice. As shown in Fig. 7d, no tumor formation was seen in TERT-SHED P20. However, tumor formation was noticed after Tca-8113 cell inoculation.

Discussion

In this study, we established a method to immortalize SHED using ectopic stable expression of TERT by lentiviral vector. We found that TERT-SHED showed a robust proliferation capacity even in late passages without cell senescence as indicated by low activity of β-GAL. Although they had some different biomarkers compared to early-passage SHED, TERT-SHED at late passage showed similar mutilineage differentiation to TERT at early passage. We also assessed the potential tumorigenicity of TERT-SHED, and found that TERT-SHED at late passage showed low tumorigenicity, as indicated by normal karyotype, no soft agar colony formation, and no tumor formation in nude mice. These data suggest that TERT expression may be a safe technique for banking SHED for tissue repair.

SHED are mesenchymal-like cells and are an attractive candidate for use in tissue repair thanks to their multi-potentiality, easy availability, and immunoprivileged status [36]. They do not induce an allogenic reaction and may even suppress host T-cell proliferation [37], suggesting that cells cultured from a single donor may be expanded in vitro to form a reserve pool that could be used for multiple recipients. However, during in vitro culturing, SHED undergoes replicative senescence and

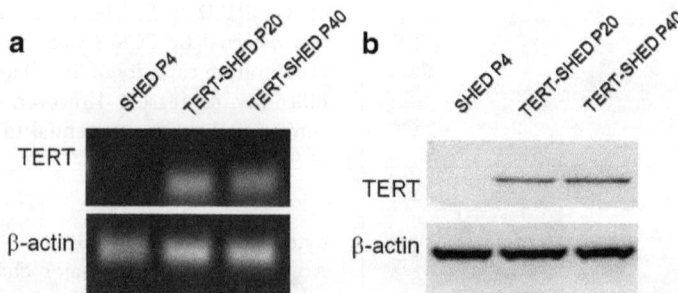

Fig. 3 Expression of TERT assessed by RT-PCR (**a**) and Western blot (**b**) in SHED and TERT-SHED at different passages. SHED showed robust mRNA and protein expression of TERT even at the 40th passage after lentiviral transduction. *P* passage, *SHED* stem cells from human exfoliated deciduous teeth, *TERT* telomerase reverse transcriptase

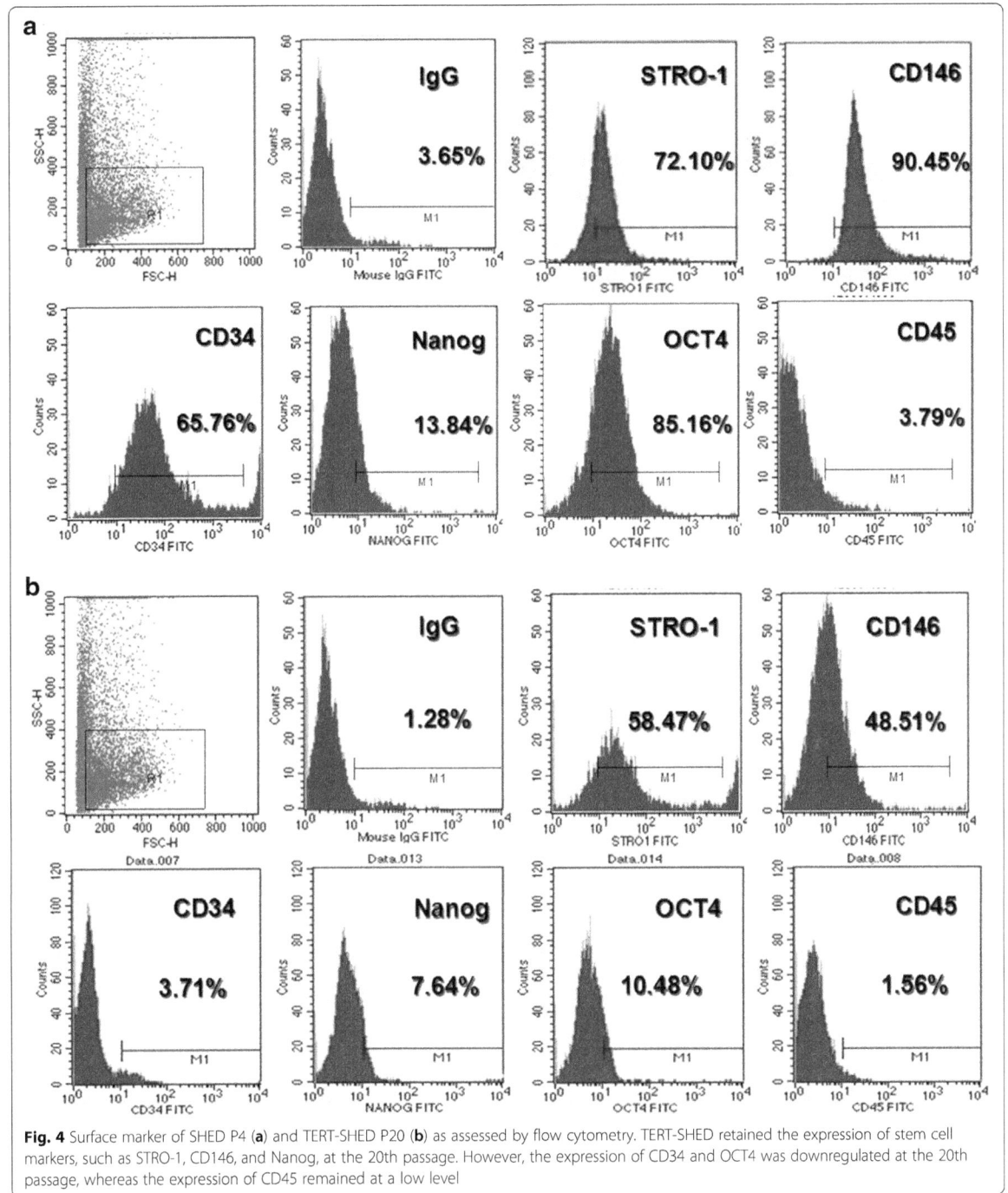

Fig. 4 Surface marker of SHED P4 (**a**) and TERT-SHED P20 (**b**) as assessed by flow cytometry. TERT-SHED retained the expression of stem cell markers, such as STRO-1, CD146, and Nanog, at the 20th passage. However, the expression of CD34 and OCT4 was downregulated at the 20th passage, whereas the expression of CD45 remained at a low level

loses its ability to differentiate over time [38]. Thus, immortalization of dental stem cells (DSCs) and establishment of a dental stem cell line are important for DSC research and regenerative dentistry.

It is generally thought that replicative senescence of stem cells is a result of genetic instability after critical shortening of telomeres [39]. Telomerase had the enzymatic activity to maintain and elongate telomere length during cell division. Ectopic expression of TERT has been proven in many studies to maintain the telomere length in different types of cells, thus immortalize cells and prevent cells from loss of function. Using this approach, stable

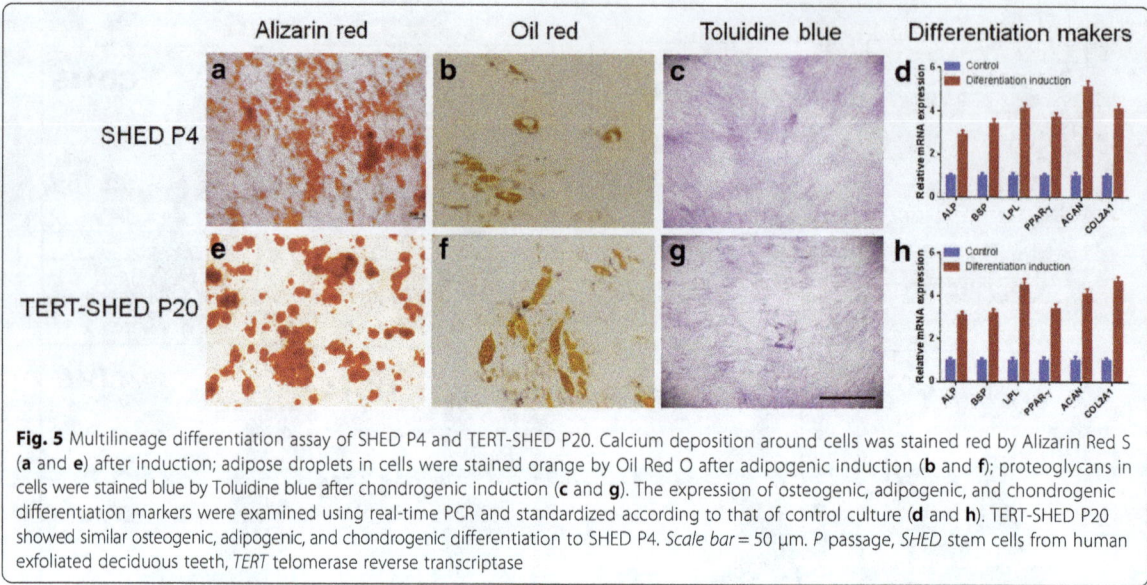

Fig. 5 Multilineage differentiation assay of SHED P4 and TERT-SHED P20. Calcium deposition around cells was stained red by Alizarin Red S (**a** and **e**) after induction; adipose droplets in cells were stained orange by Oil Red O after adipogenic induction (**b** and **f**); proteoglycans in cells were stained blue by Toluidine blue after chondrogenic induction (**c** and **g**). The expression of osteogenic, adipogenic, and chondrogenic differentiation markers were examined using real-time PCR and standardized according to that of control culture (**d** and **h**). TERT-SHED P20 showed similar osteogenic, adipogenic, and chondrogenic differentiation to SHED P4. *Scale bar* = 50 μm. *P* passage, *SHED* stem cells from human exfoliated deciduous teeth, *TERT* telomerase reverse transcriptase

expression of TERT prevents replicative senescence in human mesenchymal stem cells (hMSCs) [40–43], with a lifespan extension of more than 3 years [44]. These findings are consistent with our results that stable TERT expression causes a continuous proliferation of SHED with a lack of senescence-associated β-GAL staining in robust cells even in the 40th passage, whereas untransduced cells went into senescence in the 20th passage, indicating that TERT expression may be a useful strategy for immortalizing stem cells.

Usually, long-term in vitro culture of stem cells results in impaired differentiation capacity [45]. Previous studies have demonstrated that MSCs overexpressing TERT

exhibit an increased osteogenic differentiation potential [45], while telomerase deficiency impairs differentiation of hMSCs [46]. In our study, although TERT-SHED at late passage showed some decrease in the biomarkers, its differentiation into osteal, adipic, and chondric cells was similar to that in SHED at early passage, suggesting that telomere length maintenance plays an important role in the differentiation of stem cells.

With the success of immortalization of stem cells by TERT, concerns about potential malignant transformation by viral TERT transduction were raised. It has been reported that long-term culture of TERT-transduced adult MSCs using a retrovirus resulted in neoplastic transformation

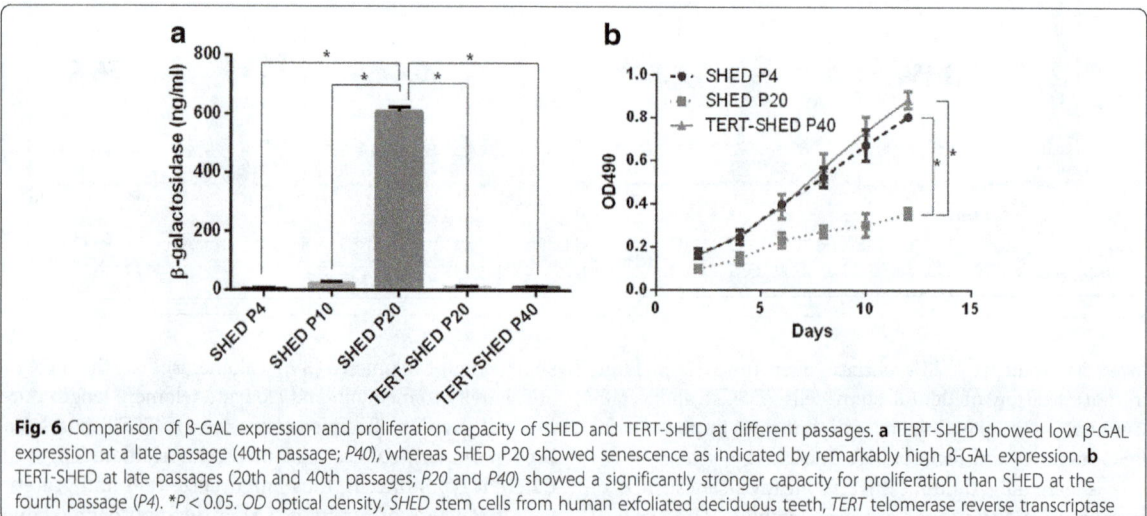

Fig. 6 Comparison of β-GAL expression and proliferation capacity of SHED and TERT-SHED at different passages. **a** TERT-SHED showed low β-GAL expression at a late passage (40th passage; *P40*), whereas SHED P20 showed senescence as indicated by remarkably high β-GAL expression. **b** TERT-SHED at late passages (20th and 40th passages; *P20* and *P40*) showed a significantly stronger capacity for proliferation than SHED at the fourth passage (*P4*). *$*P < 0.05$. *OD* optical density, *SHED* stem cells from human exfoliated deciduous teeth, *TERT* telomerase reverse transcriptase

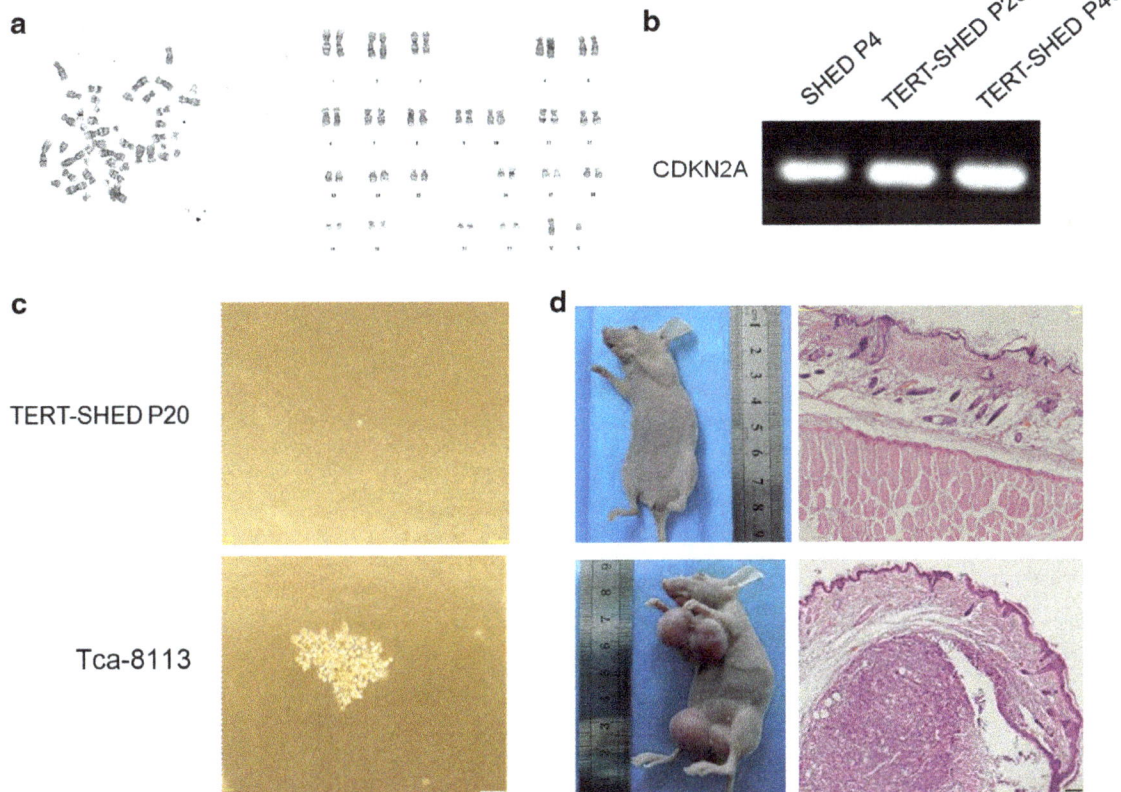

Fig. 7 Assessment of potential tumorigenicity of TERT-SHED. **a** TERT-SHED P40 showed no abnormality of karyotype. **b** PCR application showed that there was no deletion of CDKN2A in SHED P4, TERT-SHED P20 or TERT-SHED P40. **c** TERT-SHED P20 showed no colony formation in soft agar, with Tca-8113 cancer cells as a positive control. *Scale bar = 50 μm.* **d** TERT-SHED P20 showed no tumorigenicity in nude mice, with Tca-8113 cells as a positive control. *Scale bar = 50 μm. P passage, SHED stem cells from human exfoliated deciduous teeth, TERT telomerase reverse transcriptase*

[26, 35]. Insertional mutagenesis by long terminal repeat (LTR) has been a limitation of retroviral gene transfer [35] since oncogenesis occurred at an unexpected high frequency in the X-SCID gene therapy trial. However, so far, all available data suggest that lentiviral vectors are safe vehicles for ex vivo gene therapy and no adverse events have been reported upon transplantation of lentivirus vector-transduced cells [47]. The major reason for the low genotoxicity of lentiviral vectors may be the lack of transcriptionally active LTR. On the other hand, telomere length maintenance plays critical roles in preventing chromosomal instability and subsequent carcinogenesis. Markedly elevated risks of tumors (about 11 times that of the general population) are observed in patients with dyskeratosis congenita, a disease with very short telomeres caused by germline mutations in the components of the telomerase complex [18]. Mouse models also support the notion that abnormally short telomere length increases the risk of cancers [48]. Recent prospective epidemiological studies have demonstrated

that a short telomere is significantly associated with increased cancer incidence and death [49, 50]. Therefore, it is reasonable that lentiviral vector-mediated TERT expression had low tumorigenicity in SHED at late passage, as indicated by no abnormal karyotype, no colony formation in soft agar, and no tumor formation in nude mice. However, our findings need to be extended in SHED over hundreds of passages, and emphasize the caution in the use of TERT-immortalized cells in studies of normal cell biology and in tissue engineering.

There are several limitations in our study. Like most laboratory studies [5], our SHED were maintained and expanded in bovine serum-containing medium, which raises the concern about its clinical application due to the high lot-to-lot variability, risk of contamination, and immune response against xenogenic proteins in bovine serum [51]. Because of the numerous constituents of bovine serum, the development of chemically defined serum-free media with an optimal composition of the few essential factors is only beginning. Thus, bovine serum remains the gold standard medium supplement for laboratory-scale MSC culture and has been used in clinical trials approved by the US Food and Drug Administration [52]. Recent studies have demonstrated that human blood-derived components may be an ideal substitute for bovine serum in the therapeutic application of stem cells. Therefore, SHED expanded in xeno-free media are needed for clinical therapy.

Conclusions

In this study, we show that a lentiviral TERT gene transduction could establish a stable SHED cell line that is completely multipotential; even after long-term in vitro passaging, no evidence of genetic instability or malignant biological behavior of these cells was observed. These findings provide novel strategies to prevent the senescence and maintain the stemness of ex vivo-maintained SHED for potential clinical therapies, although attention should be paid to the biological behavior of these cells.

Abbreviations
BaP: benzopyrene; β-GAL: β-galatosidase; Ct: threshold cycle; DSC: dental stem cell; ELISA: enzyme-linked immunosorbent assay; hMSC: human mesenchymal stem cell; LTR: long terminal repeat; MOI: multiplicity of infection; MSC: mesenchymal stem cell; P: passage; PBS: phosphate-buffered saline; PCR: polymerase chain reaction; RT: reverse transcription; SHED: stem cells from human exfoliated deciduous teeth; TERT: telomerase reverse transcriptase.

Competing interests
The authors declare that they have no competing interests.

Authors' contributions
AL and ZY designed the study. QW and YL were responsible for collection and/or assembly of experimental data. HW and JS were responsible for conception and design, data analysis and interpretation. AL and ZY drafted the manuscript and revised the manuscript. All authors read and approved the final manuscript.

Acknowledgments
This study was supported by grants from the Shaanxi Provincial Science and Technology Innovation Project co-ordination of the Resources Oriented Industries of Key Technologies Project (2011KTCL03-24), Shaanxi Province Natural Science Basic Research (2013JM4042), Fundamental Research Funds for the Central Universities of China, and National Natural Science Foundation of China (30801173, 81371943).

Author details
[1]Department of Orthopedics, First Affiliated Hospital, College of Medicine, Xi'an Jiaotong University, Xi'an 710061, P. R. China. [2]Department of Periodontology, Stomatological Hospital, College of Medicine, Xi'an Jiaotong University, Xi'an 710004, P. R. China. [3]Research Center for Stomatology, Stomatological Hospital, College of Medicine, Xi'an Jiaotong University, Xi'an 710004, P. R. China.

References
1. Murray PE, Garcia-Godoy F, Hargreaves KM. Regenerative endodontics: a review of current status and a call for action. J Endod. 2007;33(4):377–90. doi:10.1016/j.joen.2006.09.013.
2. Pfammatter JP, Gertsch M, Weber JW, Stocker FP, Moser H, Kappenberger L. Stress-induced polymorphous ventricular tachyarrhythmias in two brothers:

unusual pattern of inheritance in the long QT syndrome. Clin Cardiol. 1993; 16(6):517–20.
3. Telles PD, Machado MA, Sakai VT, Nor JE. Pulp tissue from primary teeth: new source of stem cells. J Appl Oral Sci. 2011;19(3):189–94.
4. Nishino Y, Yamada Y, Ebisawa K, Nakamura S, Okabe K, Umemura E, et al. Stem cells from human exfoliated deciduous teeth (SHED) enhance wound healing and the possibility of novel cell therapy. Cytotherapy. 2011;13(5): 598–605. doi:10.3109/14653249.2010.542462.
5. Miura M, Gronthos S, Zhao M, Lu B, Fisher LW, Robey PG, et al. SHED: stem cells from human exfoliated deciduous teeth. Proc Natl Acad Sci U S A. 2003;100(10):5807–12. doi:10.1073/pnas.0937635100.
6. Parchment RE, Natarajan K. A free-radical hypothesis for the instability and evolution of genotype and phenotype in vitro. Cytotechnology. 1992;10(2):93–124.
7. Gudjonsson T, Villadsen R, Ronnov-Jessen L, Petersen OW. Immortalization protocols used in cell culture models of human breast morphogenesis. Cell Mol Life Sci. 2004;61(19–20):2523–34. doi:10.1007/s00018-004-4167-z.
8. Russo J, Tahin Q, Lareef MH, Hu YF, Russo IH. Neoplastic transformation of human breast epithelial cells by estrogens and chemical carcinogens. Environ Mol Mutagen. 2002;39(2–3):254–63. doi:10.1002/em.10052.
9. Band V. Preneoplastic transformation of human mammary epithelial cells. Semin Cancer Biol. 1995;6(3):185–92. doi:10.1006/scbi.1995.0015.
10. Wazer DE, Chu Q, Liu XL, Gao Q, Safaii H, Band V. Loss of p53 protein during radiation transformation of primary human mammary epithelial cells. Mol Cell Biol. 1994;14(4):2468–78.
11. Band V. In vitro models of early neoplastic transformation of human mammary epithelial cells. Methods Mol Biol. 2003;223:237–48. doi:10.1385/1-59259-329-1:237.
12. DiPaolo JA. Relative difficulties in transforming human and animal cells in vitro. J Natl Cancer Inst. 1983;70(1):3–8.
13. Shay JW, Van Der Haegen BA, Ying Y, Wright WE. The frequency of immortalization of human fibroblasts and mammary epithelial cells transfected with SV40 large T-antigen. Exp Cell Res. 1993;209(1):45–52. doi:10.1006/excr.1993.1283.
14. Bryan TM, Reddel RR. SV40-induced immortalization of human cells. Crit Rev Oncog. 1994;5(4):331–57.
15. Dimova S, Brewster ME, Noppe M, Jorissen M, Augustijns P. The use of human nasal in vitro cell systems during drug discovery and development. Toxicol In Vitro. 2005;19(1):107–22. doi:10.1016/j.tiv.2004.07.003.
16. Egbuniwe O, Idowu BD, Funes JM, Grant AD, Renton T, Di Silvio L. P16/p53 expression and telomerase activity in immortalized human dental pulp cells. Cell Cycle. 2011;10(22):3912–9. doi:10.4161/cc.10.22.18093.
17. vom Brocke J, Schmeiser HH, Reinbold M, Hollstein M. MEF immortalization to investigate the ins and outs of mutagenesis. Carcinogenesis. 2006;27(11): 2141–7. doi:10.1093/carcin/bgl101.
18. Calado RT, Young NS. Telomere diseases. N Engl J Med. 2009;361(24):2353–65. doi:10.1056/NEJMra0903373.
19. Bayne S, Jones ME, Li H, Pinto AR, Simpson ER, Liu JP. Estrogen deficiency leads to telomerase inhibition, telomere shortening and reduced cell proliferation in the adrenal gland of mice. Cell Res. 2008;18(11):1141–50. doi:10.1038/cr.2008.291.
20. Borssen M, Cullman I, Noren-Nystrom U, Sundstrom C, Porwit A, Forestier E, et al. hTERT promoter methylation and telomere length in childhood acute lymphoblastic leukemia: associations with immunophenotype and cytogenetic subgroup. Exp Hematol. 2011;39(12):1144–51. doi:10.1016/j.exphem.2011.08.014.
21. Venturini L, Daidone MG, Motta R, Collini P, Spreafico F, Terenziani M, et al. Telomere maintenance in Wilms tumors: first evidence for the presence of alternative lengthening of telomeres mechanism. Genes Chromosomes Cancer. 2011;50(10):823–9. doi:10.1002/gcc.20903.
22. Soares J, Lowe MM, Jarstfer MB. The catalytic subunit of human telomerase is a unique caspase-6 and caspase-7 substrate. Biochemistry. 2011;50(42): 9046–55. doi:10.1021/bi2010398.
23. Hao LY, Armanios M, Strong MA, Karim B, Feldser DM, Huso D, et al. Short telomeres, even in the presence of telomerase, limit tissue renewal capacity. Cell. 2005;123(6):1121–31. doi:10.1016/j.cell.2005.11.020.
24. Cohen S, Jacob E, Manor H. Effects of single-stranded DNA binding proteins on primer extension by telomerase. Biochim Biophys Acta. 2004;1679(2): 129–40. doi:10.1016/j.bbaexp.2004.06.002.
25. Bianchi A, Shore D. How telomerase reaches its end: mechanism of telomerase regulation by the telomeric complex. Mol Cell. 2008;31(2):153–65. doi:10.1016/j.molcel.2008.06.013.

26. Burns JS, Abdallah BM, Guldberg P, Rygaard J, Schroder HD, Kassem M. Tumorigenic heterogeneity in cancer stem cells evolved from long-term cultures of telomerase-immortalized human mesenchymal stem cells. Cancer Res. 2005;65(8):3126–35. doi:10.1158/0008-5472.CAN-04-2218.

27. Li N, Yang R, Zhang W, Dorfman H, Rao P, Gorlick R. Genetically transforming human mesenchymal stem cells to sarcomas: changes in cellular phenotype and multilineage differentiation potential. Cancer. 2009;115(20):4795–806. doi:10.1002/cncr.24519.

28. Huang G, Zheng Q, Sun J, Guo C, Yang J, Chen R, et al. Stabilization of cellular properties and differentiation multipotential of human mesenchymal stem cells transduced with hTERT gene in a long-term culture. J Cell Biochem. 2008;103(4):1256–69. doi:10.1002/jcb.21502.

29. Kelland LR. Overcoming the immortality of tumour cells by telomere and telomerase based cancer therapeutics—current status and future prospects. Eur J Cancer. 2005;41(7):971–9. doi:10.1016/j.ejca.2004.11.024.

30. Bocker W, Rossmann O, Docheva D, Malterer G, Mutschler W, Schieker M. Quantitative polymerase chain reaction as a reliable method to determine functional lentiviral titer after ex vivo gene transfer in human mesenchymal stem cells. J Gene Med. 2007;9(7):585–95. doi:10.1002/jgm.1049.

31. Vorsanova SG, Yurov YB, Iourov IY. Human interphase chromosomes: a review of available molecular cytogenetic technologies. Mol Cytogenet. 2010;3:1. doi:10.1186/1755-8166-3-1.

32. Brothman AR, Persons DL, Shaffer LG. Nomenclature evolution: changes in the ISCN from the 2005 to the 2009 edition. Cytogenet Genome Res. 2009; 127(1):1–4. doi:10.1159/000279442.

33. Magaye R, Zhou Q, Bowman L, Zou B, Mao G, Xu J, et al. Metallic nickel nanoparticles may exhibit higher carcinogenic potential than fine particles in JB6 cells. PLoS One. 2014;9(4):e92418. doi:10.1371/journal.pone.0092418.

34. Bocker W, Yin Z, Drosse I, Haasters F, Rossmann O, Wierer M, et al. Introducing a single-cell-derived human mesenchymal stem cell line expressing hTERT after lentiviral gene transfer. J Cell Mol Med. 2008;12(4):1347–59. doi:10.1111/j.1582-4934.2008.00299.x.

35. Serakinci N, Guldberg P, Burns JS, Abdallah B, Schrodder H, Jensen T, et al. Adult human mesenchymal stem cell as a target for neoplastic transformation. Oncogene. 2004;23(29):5095–8. doi:10.1038/sj.onc.1207651.

36. Neoptolemos JP, Stocken DD, Bassi C, Ghaneh P, Cunningham D, Goldstein D, et al. Adjuvant chemotherapy with fluorouracil plus folinic acid vs gemcitabine following pancreatic cancer resection: a randomized controlled trial. JAMA. 2010;304(10):1073–81. doi:10.1001/jama.2010.1275.

37. Le Blanc K, Ringden O. Immunomodulation by mesenchymal stem cells and clinical experience. J Intern Med. 2007;262(5):509–25. doi:10.1111/j.1365-2796.2007.01844.x.

38. Mimeault M, Batra SK. Recent insights into the molecular mechanisms involved in aging and the malignant transformation of adult stem/progenitor cells and their therapeutic implications. Ageing Res Rev. 2009;8(2):94–112. doi:10.1016/j.arr.2008.12.001.

39. Herbig U, Jobling WA, Chen BP, Chen DJ, Sedivy JM. Telomere shortening triggers senescence of human cells through a pathway involving ATM, p53, and p21(CIP1), but not p16(INK4a). Mol Cell. 2004;14(4):501–13.

40. Simonsen JL, Rosada C, Serakinci N, Justesen J, Stenderup K, Rattan SI, et al. Telomerase expression extends the proliferative life-span and maintains the osteogenic potential of human bone marrow stromal cells. Nat Biotechnol. 2002;20(6):592–6. doi:10.1038/nbt0602-592.

41. Jun ES, Lee TH, Cho HH, Suh SY, Jung JS. Expression of telomerase extends longevity and enhances differentiation in human adipose tissue-derived stromal cells. Cell Physiol Biochem. 2004;14(4–6):261–8. doi:10.1159/000080335.

42. Shi S, Gronthos S, Chen S, Reddi A, Counter CM, Robey PG, et al. Bone formation by human postnatal bone marrow stromal stem cells is enhanced by telomerase expression. Nat Biotechnol. 2002;20(6):587–91. doi:10.1038/nbt0602-587.

43. Xu C, Jiang J, Sottile V, McWhir J, Lebkowski J, Carpenter MK. Immortalized fibroblast-like cells derived from human embryonic stem cells support undifferentiated cell growth. Stem Cells. 2004;22(6):972–80. doi:10.1634/stemcells.22-6-972.

44. Abdallah BM, Haack-Sorensen M, Burns JS, Elsnab B, Jakob F, Hokland P, et al. Maintenance of differentiation potential of human bone marrow mesenchymal stem cells immortalized by human telomerase reverse transcriptase gene despite [corrected] extensive proliferation. Biochem Biophys Res Commun. 2005;326(3):527–38. doi:10.1016/j.bbrc.2004.11.059.

45. Christiansen M, Kveiborg M, Kassem M, Clark BF, Rattan SI. CBFA1 and topoisomerase I mRNA levels decline during cellular aging of human trabecular osteoblasts. J Gerontol A Biol Sci Med Sci. 2000;55(4):B194–200.

46. Liu L, DiGirolamo CM, Navarro PA, Blasco MA, Keefe DL. Telomerase deficiency impairs differentiation of mesenchymal stem cells. Exp Cell Res. 2004;294(1):1–8. doi:10.1016/j.yexcr.2003.10.031.

47. Levine BL, Humeau LM, Boyer J, MacGregor RR, Rebello T, Lu X, et al. Gene transfer in humans using a conditionally replicating lentiviral vector. Proc Natl Acad Sci U S A. 2006;103(46):17372–7. doi:10.1073/pnas.0608138103.

48. Artandi SE, Chang S, Lee SL, Alson S, Gottlieb GJ, Chin L, et al. Telomere dysfunction promotes non-reciprocal translocations and epithelial cancers in mice. Nature. 2000;406(6796):641–5. doi:10.1038/35020592.

49. Willeit P, Willeit J, Mayr A, Weger S, Oberhollenzer F, Brandstatter A, et al. Telomere length and risk of incident cancer and cancer mortality. JAMA. 2010;304(1):69–75. doi:10.1001/jama.2010.897.

50. Weischer M, Nordestgaard BG, Cawthon RM, Freiberg JJ, Tybjaerg-Hansen A, Bojesen SE. Short telomere length, cancer survival, and cancer risk in 47102 individuals. J Natl Cancer Inst. 2013;105(7):459–68. doi:10.1093/jnci/djt016.

51. Kinzebach S, Bieback K. Expansion of mesenchymal stem/stromal cells under xenogenic-free culture conditions. Adv Biochem Eng Biotechnol. 2013;129:33–57. doi:10.1007/10_2012_134.

52. Haque N, Kasim NH, Rahman MT. Optimization of pre-transplantation conditions to enhance the efficacy of mesenchymal stem cells. Int J Biol Sci. 2015;11(3):324–34. doi:10.7150/ijbs.10567.

4

Functional imaging for regenerative medicine

Martin Leahy[1,5*], Kerry Thompson[2], Haroon Zafar[1], Sergey Alexandrov[1], Mark Foley[3], Cathal O'Flatharta[4] and Peter Dockery[2]

Abstract

In vivo imaging is a platform technology with the power to put function in its natural structural context. With the drive to translate stem cell therapies into pre-clinical and clinical trials, early selection of the right imaging techniques is paramount to success. There are many instances in regenerative medicine where the biological, biochemical, and biomechanical mechanisms behind the proposed function of stem cell therapies can be elucidated by appropriate imaging. Imaging techniques can be divided according to whether labels are used and as to whether the imaging can be done in vivo. In vivo human imaging places additional restrictions on the imaging tools that can be used. Microscopies and nanoscopies, especially those requiring fluorescent markers, have made an extraordinary impact on discovery at the molecular and cellular level, but due to their very limited ability to focus in the scattering tissues encountered for in vivo applications they are largely confined to superficial imaging applications in research laboratories. Nanoscopy, which has tremendous benefits in resolution, is limited to the near-field (e.g. near-field scanning optical microscope (NSNOM)) or to very high light intensity (e.g. stimulated emission depletion (STED)) or to slow stochastic events (photo-activated localization microscopy (PALM) and stochastic optical reconstruction microscopy (STORM)). In all cases, nanoscopy is limited to very superficial applications. Imaging depth may be increased using multiphoton or coherence gating tricks. Scattering dominates the limitation on imaging depth in most tissues and this can be mitigated by the application of optical clearing techniques that can impose mild (e.g. topical application of glycerol) or severe (e.g. CLARITY) changes to the tissue to be imaged. Progression of therapies through to clinical trials requires some thought as to the imaging and sensing modalities that should be used. Smoother progression is facilitated by the use of comparable imaging modalities throughout the discovery and trial phases, giving label-free techniques an advantage wherever they can be used, although this is seldom considered in the early stages. In this paper, we will explore the techniques that have found success in aiding discovery in stem cell therapies and try to predict the likely technologies best suited to translation and future directions.

Keywords: Microscopy, Imaging, Stem cells, Label-free, Optical coherence tomography, Photoacoustic imaging, Functional

Background

A well-chosen imaging technique provides a means to produce high-impact discovery and validation data for the translation of novel regenerative therapies, but choosing the right imaging tool can be tricky and is too often biased by familiarity. Hence we try to provide, in this paper, a means to compare the best known imaging technologies in terms of their capabilities and limitations for stem cell research. Table 1 provides an overview of the optimal stem cell tracking characteristics, the probes used to achieve this, and the appropriate imaging modalities with their advantages and disadvantages. The techniques are discussed in more detail in the following paragraphs.

Main text

Overview of functional imaging for regenerative medicine

Functional imaging, especially when provided in its structural context, provides a platform for all branches

* Correspondence: martin.leahy@nuigalway.ie
[1]Tissue Optics & Microcirculation Imaging Group, School of Physics, National University of Ireland (NUI), Galway, Ireland
[5]Chair of Applied Physics, National University of Ireland (NUI), Galway, Ireland
Full list of author information is available at the end of the article

Table 1 An overview of the optimal stem cell tracking characteristics, the probes used to achieve this, and the appropriate imaging modalities with their advantages and disadvantages

Optimal stem cell tracking probe characteristic	Optimal cellular probe	Examples	Probe disadvantages	Imaging modality
Absorbance/emission spectra within "optical window"	Fluorescence	Reporter genes (e.g. iRFP), quantum dots, exogenous probes (e.g. PKH26)	Requires genetic modification and excitation light, high background due to autofluorescence, signal loss with cell division, low depth of imaging, limited spatial resolution	FLI
	Bioluminescence	Reporter genes (e.g. fLuc)	Requires genetic modification and exogenous substrate administration	BLI
	Photoacoustic	Reporter genes (e.g. LacZ, iRFP), endogenous labels (e.g. Hb, melanin)	Requires excitation light and may require genetic modification, expensive equipment	PAI
High signal sensitivity/intensity	Radionuclide	Reporter genes (e.g. hNIS), ^{99m}Tc, ^{111}In, ^{18}F FDG	Ionizing radiation, poor anatomical detail (but can be combined with magnetic resonance or x-ray), radioactive decay limits imaging time, cellular toxicity, may require genetic modification, expensive	SPECT, PET
	Electron density	Gold nanoparticles	Limited spatial/soft tissue resolution, ionizing, not indicative of cell viability, expensive	x-ray, CT
	Fluorescence	As described above	As described above	FLI
	Bioluminescence	As described above	As described above	BLI
	Photoacoustic	As described above	As described above	PAI
High spatial resolution	Magnetic resonance	Iron oxides, microcapsules	Low signal intensity, not indicative of cell viability, expensive	MRI
High temporal resolution/real time tracking	Echography	Microbubbles, perfluorocarbons	Low resolution, acoustic artefacts, subject to user bias	US
	Fluorescence	As described above	As described above	FLI
	Bioluminescence	As described above	As described above	BLI
	Photoacoustic	As described above	As described above	PAI
	Radionuclide	As described above	As described above	SPECT, PET
High imaging depth	Photoacoustic	As described above	As described above	PAI
	Echography	As described above	As described above	US
	Radionuclide	As described above	As described above	SPECT, PET
High cellular retention/signal retention upon cell division	Fluorescence	Reporter genes (e.g. iRFP)	As described above	FLI
	Bioluminescence	As described above	As described above	BLI
	Photoacoustic	As described above	As described above	PAI
High anatomical detail	Magnetic resonance	As described above	As described above	MRI
	Electron density	As described above	As described above	x-ray, CT
	Multimodal systems which include MRI or x-ray			
Low cellular toxicity/non-ionizing	Echography	As described above	As described above	US
	Magnetic resonance	As described above	As described above	MRI
	Fluorescence	As described above	As described above	FLI
	Bioluminescence	As described above	As described above	BLI
Quantifiable signal	Fluorescence	As described above	As described above	FLI
	Bioluminescence	As described above	As described above	BLI
No cellular genetic modification	Echography	As described above	As described above	US
	Radionuclide	^{99m}Tc, ^{111}In, ^{18}F FDG	As described above	SPECT, PET

BLI bioluminescence imaging, *CT* computed tomography, *FLI* fluorescence imaging, *^{18}F FDG* fluoro-2-deoxy-D-glucose, *Hb* haemoglobin, *^{111}In* indium, *MRI* magnetic resonance imaging, *PAI* photoacoustic imaging, *PET* positron emission tomography, *SPECT* single photon emission computed tomography, *^{99m}Tc* technetium, *US* ultrasound

of regenerative medicine research. The technology is constantly being advanced to image faster, deeper, less invasively, and more quantitatively, driving discovery of both biological and clinical mechanisms. This article will review some of the plethora of advances that have been made in recent years in technologies that have enabled discovery in the field of stem cell research. Topics such as in vivo fluorescence imaging and the benefits of label-free techniques such as optical coherence tomography (OCT) and photoacoustic imaging (PAI) will be discussed, along with super resolution microscopy and radionuclide imaging.

Stem cell imaging in regenerative medicine

Stem cells have the ability to undergo clonal expansion and to differentiate into multiple cell types; adult stem cells offer advantages over embryonic stem cells due to their ease of isolation and lack of ethical issues [1]. Regenerative medicine, or the use of stem cells as therapies, consists of multi-disciplinary approaches with the aim of restoring function to diseased tissues and organs. Such cell-based therapies have been extensively investigated as promising avenues of treatment for a host of disease types, including, but not limited to, cardiac disease, diabetes and orthopaedics. For the current rate of progress to be maintained, non-invasive and reproducible methods to monitor and assess stem cell integration and survival in disease models are of paramount importance. Imaging techniques with high spatial and temporal resolution will enable accurate tracking of transplanted stem cells to disease loci in vivo over a long period of time in pre-clinical (animal) models and, ultimately, in clinical trials. Information obtained from such studies will also allow scientists and clinicians to optimise stem cell administration regimens (e.g. dose, route of administration, timing) and to assess the efficacy of a cell-based treatment.

Currently, tracking stem cell migration and engraftment is achieved using appropriate imaging systems in parallel with endogenous and exogenous cell-labelling methods. An ideal cellular label should:

- be biocompatible and non-toxic to cells;
- be quantifiable;
- be inexpensive;
- remain undiluted following cell division;
- not leak into adjacent non-transplanted cells;
- remain stable over long periods of time in vivo;
- not interfere with normal cell function;
- not require genetic modification or the injection of a contrast agent.

Stem cells can be genetically modified to express reporter genes or proteins that can emit fluorescence/ bioluminescence (or other useful proteins such as lacZ or NIS) or be treated to uptake exogenous contrast agents, such as organic dyes, nanoparticles, radionuclides, or magnetic compounds [2].

In vivo fluorescence imaging

The collection of data from an innate biological site is one of the largest advantages of in vivo imaging of any form. Macroscopic imaging of either animal or human sources, as opposed to the imaging of tissue explants or cells from culture, encounters an array of complications. In vivo fluorescence imaging is similar to conventional fluorescence microscopy in that high-end low-light cameras are used to detect an emission signal generated from a fluorophore or probe [3, 4]. In recent years, the development of stem cell therapies for treatment of a vast array of diseases has progressed rapidly [5]. Molecular tagging and the addition of probes to monitor, track, and assess the administered cells in a non-invasive manner in vivo, in both animal and human clinical studies, will be discussed in this section. Further to this, the use of multimodal approaches (fluorescence in conjunction with bioluminescence and high-resolution imaging techniques) will be briefly highlighted.

Ex vivo histopathological analysis of modified stem cell behaviour was traditionally carried out, using fluorescent probes, on excised biopsies from animal model studies. These examinations were incapable of providing real-time information about alterations to the tissues under study. Despite this limitation, these probes provided the framework for many of the newer generations of markers currently in use today to be developed and refined. The incorporation of reporter genes into cellular machinery has provided scientists with a method to visualise cells, via fluorescent modifications, to a depth of about 2 mm into the tissue. The incorporation of these genes into a cell is referred to as indirect labelling. Reporter genes allow the monitoring of physiologically relevant biological processes as they occur in situ. Traditionally, green fluorescent protein (GFP) tags were used in fluorescence imaging to identify cells [6]. The main advantage of this form of labelling is that expression of the functional reporter probe only occurs after the cell has transcribed the gene of interest and the mRNA is translated into the modified version of the protein and a biosensor is created. This allows direct correlations to be drawn between the levels of expression of the probe and cell viability. The expression of the modified gene is propagated to future generations of cells and, in this way, the longevity of this method is preferable in an in vivo scenario as it would potentially create a long-term reporter of cell stem functionality and enable tracking/ tracing over a lengthier period of time. Genetic modification of cells, via transfection (non-viral vectors) or transduction (viral vectors), that are employed in order to allow the incorporation of these reporter genes is, at present, the major limiting factor of this technique [7]. The long-term safety of incorporating transformed genetic material and the potential for immune responses or tumour development in recipients of these therapies requires further investigation and regulation at a clinical trial level. With a strong focus on safety and therapeutic efficacy for stem cell delivery, many laboratories are developing alternative methods to allow the integration of reporters into the cellular genome [8]. Recent work has focused on the development of fluorescent probes for incorporation in reporter genes amongst other uses. Fluorescent probes whose spectra are in the far red, towards the near infrared (NIR) portions of the spectrum of light (650–900 nm), are experimentally the most desirable for scientists wishing to carry out in vivo imaging. The potential for alterations to the physiological state of the cell under study must be monitored when utilising any type of fluorescence imaging technique. The benefits of imaging in this portion of the spectrum will be discussed in later sections. Earlier probe variants including mKate, with excitation and emission at 588 and 635 nm and synthesised from the sea anemone *Entacmaea quadricolor*, were developed for whole body imaging, and more recently phytochrome (photosensor) from the bacteria *Deinococcus radiodurans* has allowed production of the IFP 1.4 marker [9, 10]. Despite these advances, quantum yield for these probes remained poor. Newer probes including iRFP (near-infrared fluorescent protein) are aimed at increasing the fluorescence output and signal intensity through modifications of these phytochromes, and display improved pH and photo-stability in vivo [11]. The use of optogenetics, or the control of biological processes in mammals (both cells and tissues) by light, is emerging as a very powerful manipulation technique. This method combines the genetic modifications discussed above, with the possible inclusion of NIR probes, and the potential to act as a therapy mediator for stem cell treatments [12, 13]. Work to date has concentrated on mainly neural stem cells in animal models [14, 15].

The combination of fluorescence, bioluminescence, and high-resolution probes are referred to as multimodal reporter probes. The combination of the best aspects of all probes and techniques allows a much great amount of data to be collected from one source. Recent work from Roger Tsien's group has shown that one of these triple modality reporters has been implemented in an in vivo animal study for qualitative tumour therapy and efficacy of drug delivery [16]. The development and advancement in the engineering and construction of these fluorescent and multimodal probes holds most hope for successful deep tissue in vivo fluorescence imaging.

In summary, fluorescent imaging modalities are simpler, cheaper, more user friendly, and convenient to carry out than their higher resolution counterparts. The development of high-sensitivity cameras, which are capable of detecting very low levels of gene expression, and the quantitatively close relationship between cell number and fluorescence detection signals are all major benefits of these techniques.

The advantages of label-free optical imaging techniques

Appropriate imaging modalities are needed for the tracking of stem cells to investigate various biological processes such as cell migration, engraftment, homing, differentiation, and functions. The ideal modality for tracking stem cells requires high sensitivity and high spatial resolution, non-toxic imaging. Contrast agents should be biocompatible and highly specific to reduce perturbation of the target cells. The ideal modality should provide non-invasive, depth-resolved imaging in situ and be able to detect single cells, and should show a difference between cell loss and cell proliferation. Currently none of the known imaging modalities has all of these characteristics [17, 18].

In contrast to the above-mentioned modalities, this section will focus on those techniques which do not employ the use of an endogenous/exogenous contrasting agent. Label-free imaging techniques provide the unique possibility to image and study cells in their natural environment.

For example, such techniques can be used for the isolation of human pluripotent stem cells (hPSCs), enriched to 95–99 % purity with >80 % survival, and to keep normal transcriptional profiles, differentiation potential, and karyotypes [19]. Well-known label-free imaging modalities, such as quantitative phase microscopy (QPM), are used to reconstruct nanoscale phase information within cells, including living cells [20]. Interference reflection microscopy (IRM), also sometimes referred to as Interference Reflection Contrast, or Surface Contrast Microscopy, is often used in conjunction with QPM [21]. This non-invasive label-free technique is employed in the study of cellular adhesions, migration, cell mitosis, and cytotoxicity amongst other parameters in stem cell cultures such as human induced pluripotent stem cells (hIPSCs). Greyscale images are created from the slight variations generated in optical path differences where reflected light is used to visualise structures which are at, or near, a glass coverslip surface [22]. This technique can provide quantitative information on the intracellular cytoplasmic and nuclear alterations often required by scientists whilst assessing stem cells and their differentiation state in culture, and therefore assist in the screening selection of hIPSC colonies [21]. Optical diffraction tomography permits three-dimensional (3D) image reconstruction of a single cell [23–25]. The oblique-incidence reflectivity difference (OI-RD) microscope was proposed for label-free, real-time detection of cell surface markers and applied to analyse stage-specific embryonic antigen 1 (SSEA1) on stem cells in the native state [26]. Another imaging modality, digital holographic microscopy (DHM), provides the possibility for imaging of a 3D volume with a single exposure which is very useful for imaging of living cells. DHM was combined with light scattering angular spectroscopy to provide spatially resolved quantitative morphological information [27–29], improved resolution via a synthetic aperture approach [30–32], and used for 3D tomographic imaging [33]. The disadvantages of these techniques are that they are not depth-resolved and cannot be applied to highly scattered media like tissue, or they are too slow and not suitable for in vivo applications.

The recently developed spectral encoding of the spatial frequency (SESF) approach provides the means for label-free visualization of the internal submicron structure in real time with nanoscale sensitivity [34, 35], which could be a good alternative for in vivo stem cell investigation. Precise characterisation of the internal structure with nanoscale accuracy and sensitivity can be performed using the spectral distribution of scattered light to reconstruct the nanoscale structural characteristics for each pixel [36]. The theoretical basis for tomographic imaging with increased spatial resolution and depth-resolved characterization of the 3D structure has been established [37]. Label-free, depth-resolved structural characterization of highly scattering media (tissue, skin) with nanoscale sensitivity, based on the SESF approach, has been proposed [38, 39]. Label-free, super-resolution imaging using the SESF approach has been demonstrated recently [40]. The parallel development of label-free imaging techniques and the use of new non-toxic contrast agents are very encouraging.

Optical coherence tomography for study of the stem cells
OCT is one of the promising techniques for depth-resolved imaging of biomedical objects. OCT, developed in 1991 by Fujimoto and co-workers at Massachusetts Institute of Technology [41], can be considered as an optical analogue of the ultrasound technique. In comparison with ultrasound, OCT provides improved resolution of depth-resolved images to microscale, but the penetration depth is limited. OCT can provide unique depth-resolved morphologic and functional information. For example, OCT facilitates cellular level structural and functional imaging of living animals and human tissues [42–44], performs vibration measurements in the retina and ear at the nanoscale [45, 46], and depth-resolved imaging of the cornea and mapping of vasculature networks within human skin [47–51]. OCT has also

received much attention in the field of tissue engineering [52–54]. In contrast to confocal microscopy, two-photon microscopy, and other optical depth-resolved imaging techniques, OCT provides a much better penetration depth: about 2 mm in tissue instead of 100–500 microns. Recently, OCT (the standard spectral radar-OCT (SR-OCT) system (Model OCP930SR; Thorlabs Inc., Newton, NJ, USA)) has been applied as a new imaging strategy to investigate planarian regeneration in vivo in real time [55]. The signal attenuation rates, intensity ratios, and image texture features of the OCT images were analysed to compare the primitive and regenerated tissues, showing that they might provide useful biological information regarding cell apoptosis and the formation of a mass of new cells during planarian regeneration.

The spatial resolution of conventional OCT systems is limited to about 10 microns and is insufficient for cell imaging. Only some specific complicated systems—optical coherence microscopes (OCMs; http://www.rle.mit.edu/boib/research/optical-coherence-microscopy), such as high-definition OCT (HD-OCT) and micro-OCT—provide micrometre resolution in both transverse and axial directions in order to visualise individual cells (Skintell; Agfa Healthcare, Mortsel, Belgium) [56]. This system uses a two-dimensional, infrared-sensitive (1000–1700 nm) imaging array for light detection and enables focus tracking along the depth of the sample. The movements of the focal plane and the reference mirror are synchronised. As a result, the lateral resolution is 3 μm at all depths of the sample. Together with limited resolution, OCT provides only limited molecular sensitivity. To solve the problem, application of OCT for stem cell research is based on using extrinsic contrast agents such as magnetic and iron oxide particles, proteins, dyes, various types of gold nanoparticles, carbon nanotubes, and so forth. For example, the first report to demonstrate the feasibility of photothermal optical coherence tomography (PT-OCT) to image human mesenchymal stem cells (hMSCs) labelled with single-walled carbon nanotubes (SWNTs) for in vitro cell tracking in 3D scaffolds has been presented recently [57]. A photothermal BMmode scan was performed with excitation laser driving with a frequency of 800 Hz. Figure 1a shows the cross-sectional image of the combined structural and photothermal signal of the scaffold seeded with SWNT-loaded MSCs with the photothermal excitation laser turned on. Figure 1b shows the corresponding image with the excitation laser turned off. It was shown that PT-OCT imaging together with the SWNT nanoprobes looks promising for visualising and tracking of MSCs in vitro and in vivo.

Another possibility is multimodal imaging, which may minimise the potential drawbacks of using each imaging modality alone [17], such as the combination of OCT and other imaging techniques (confocal microscopy,

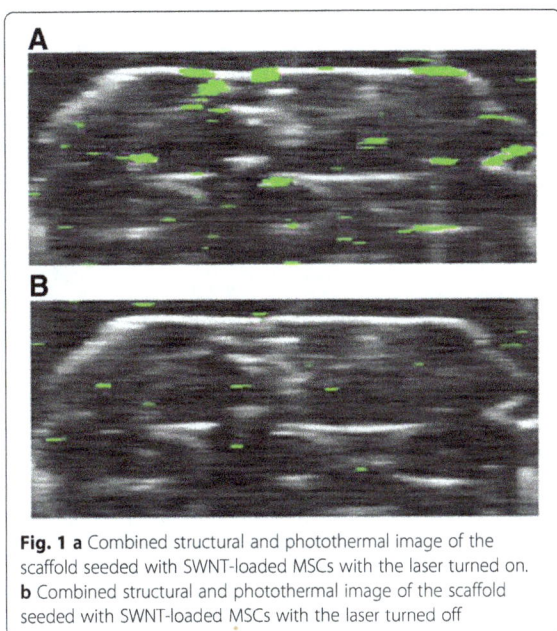

Fig. 1 a Combined structural and photothermal image of the scaffold seeded with SWNT-loaded MSCs with the laser turned on. **b** Combined structural and photothermal image of the scaffold seeded with SWNT-loaded MSCs with the laser turned off

dielectric spectroscopy (DS), fluorescence microscopy, and so forth) [56–60]. Bagnaninchi [58] used a spectral domain optical coherence tomography (SDOCT) combined with DS to qualitatively assess adipose-derived stem cells loaded in 3D carriers. The broadband (from 20 MHz to 1 GHz) DS spectra were acquired at high cell concentration simultaneously with 3D OCT imaging. Chen et al. [59] used high-resolution OCT to visualise the microstructures of the engineered tissue scaffolds in 3D and to investigate the key morphological parameters for macroporous scaffolds, while fluorescence imaging was conducted to monitor the population of labelled hMSCs loaded on to the surface of the scaffolds. Ksander et al. [60] used confocal microscopy, multiphoton microscopy and OCT to study the conditions for limbal stem cell maintenance, and corneal development and repair. Lathrop et al. [61] showed, using a combination of OCT and confocal microscopy, that OCT successfully identified the limbal palisades of Vogt that constitute the corneal epithelial stem cell niche, and offered the potential to assess and intervene in the progression of stem cell depletion by monitoring changes in the structure of the palisades. Schwartz et al. [62] used SDOCT together with visual field testing, slit-lamp biomicroscopy, ophthalmoscopy, fluorescein angiography, autofluorescence imaging, fundus photography, and electroretinography to study human embryonic stem cell-derived retinal pigment epithelium in patients with age-related macular degeneration and Stargardt's macular dystrophy. The results provide evidence of the medium- to long-term safety, graft survival, and possible biological activity of

pluripotent stem cell progeny in individuals with any disease, and suggest that human embryonic stem-derived cells could provide a potentially safe new source of cells for the treatment of various unmet medical disorders requiring tissue repair or replacement.

A potential alternative to using contrast agents is the recently developed nano-sensitive OCT which increases sensitivity to structural alterations in space and in time by more than 100 times [38, 39].

Optical coherence phase microscope

In 2011, Bagnaninchi's group demonstrated that live stem cells could be differentiated from their surrounding environment by mapping the optical phase fluctuations resulting from cellular viability and associated cellular and intracellular motility with an optical coherence phase microscope (OCPM) [63], an OCT modality that has been shown to be sensitive to nanometer-level fluctuations. In subsequent studies [64, 65], they examined murine pre-osteoblasts and human adipose-derived stem cells growing within two distinct polymeric constructs: 1) a 3D printed poly(D,L-lactic-co-glycolic acid) fibrous scaffold; and 2) hydrogel sponges (alginate). In addition to providing cell viability information, the endogenous contrast between cells and scaffolds generated by cellular motility enabled real-time, label-free monitoring of 3D engineered tissue development [65].

Photoacoustic imaging

PAI (less often called optoacoustic imaging) is an emerging biomedical imaging technique that exploits laser generated ultrasound (US) waves to generate 3D images of soft tissues. Tissue is exposed to pulsed nanosecond laser light, resulting in localised heating of the tissue. The increase in temperature of few degrees milliKelvin causes transient thermoelastic tissue expansion which generates broadband (MHz) pressure waves. The ultrasonic waves created are then detected using wideband transducers and further converted into images. PAI is a hybrid imaging modality that combines the high contrast and spectroscopic-based specificity of optical imaging with the high spatial resolution of US imaging [66]. It provides an integrated platform for functional and structural imaging, which is suitable for clinical translation.

PAI breaks through the optical diffusion limit [67] and provides real-time images with relatively high spatial resolution, without ionizing radiation being involved. The key advantages of the PAI technique over other imaging modalities include:

- the detection of haemoglobin, lipids, water, and other light absorbing molecules with higher penetration depth than pure optical imaging techniques;

- the ability to provide tissue information using an endogenous contrast alone [68];
- the imaging of optical absorption with 100 % sensitivity, which is two times greater than those of OCT and confocal microscopy;
- unlike ultrasonography and OCT, it is speckle-free [69] and provides inherently background-free detection.

The development of PAI techniques continues to be of substantial interest for clinical imaging applications in oncology, including screening, diagnosis, treatment planning, and therapy monitoring [70, 71]. PAI-based routines have also been extensively used in accurate determination of metabolic rate during early diagnosis and treatment of various skin and subcutaneous tissue disorders. The other potential implications of PAI encompass the domains of dermatology [72, 73], cardiology [74, 75], vascular biology [76, 77], gastroenterology [78, 79], neurology [80–82], and ophthalmology [83, 84]. Figure 2 summarises the potential clinical applications of PAI.

In PAI, stem cells are typically labelled using biocompatible materials with optical properties such as gold (Au) nanoparticles (NPs) or Au nanorods (NRs). In a recent study, hMSCs were labelled with 20-nm Au NPs before their incorporation into PEGylated fibrin gel [85]. After injecting the fibrin gel intramuscularly into the lateral gastrocnemius (lower limb) of an anaesthetised Lewis rat, PAI was performed to visualise the in vivo neovascularisation and differentiation of hMSCs.

Au NRs have plasmon resonance absorption and scattering in the NIR region, which makes them attractive probes for PAI [86]. In another study, hMSCs were labelled and imaged by silica-coated Au NRs (SiGNRs) [87]. The researchers found that the cellular uptake of SiGNRs can be dramatically increased (fivefold) by silica coating without changing function and viability of hMSCs.

Microcirculation imaging
Several techniques, including OCT and PAI, can be used to image microcirculatory function. The microcirculation is the usual route for delivery of stem cells by systemic or local intravascular injection. It is also affected by the stem cell therapies which may stimulate or suppress angiogenesis and will often have a major role in regeneration. In addition to the 3D techniques discussed in detail here, several other techniques are available to investigate the microcirculatory response to stem cell therapy, e.g. laser doppler, laser speckle, tissue viability imaging (TiVi), and side stream dark field microscopy [88].

Confocal reflectance microscopy
Confocal reflectance microscopy employs innate alterations in the refractive index of biological samples to create contrast within an image. Intracellular organelles and protein-protein interactions between these components, or even the interface between two different cell types as would be evident in an epithelial stromal interface, would contribute to contrast variation [89]. In recent years this technique has been used to non-invasively study skin biopsies, myelinated axons, and gather information from the excised bone marrow stem cell niche [90–92]. A combination of both fluorescent and reflectance images can be captured through the installation of a beam splitter into the light path, which allows reflected light from the sample to pass into the detection unit. In highly scattering tissues, like skin, the advantages of confocal microscopy can be combined with OCT techniques to produce the optical coherence microscope (OCM). In this way, higher numerical aperture lenses and coherence gating allows the collection of clearer images through a greater depth in tissues, when compared to either OCT or reflectance confocal modalities alone [93].

Super-resolution microscopy (nanoscopy)
Sub-cellular imaging, for example of organelles, requires diffraction-unlimited 'super-resolution' techniques. True super-resolution is only achievable with near-field optical techniques such as near-field scanning optical microscopy and 4π microscopy. However, mainstream

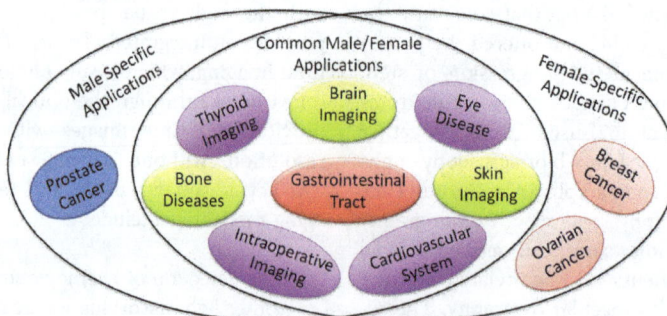

Fig. 2 An overview of potential clinical applications of PAI

functional super-resolution microscopy or nanoscopy uses the ability to switch fluorescent molecules on and off in a spot size smaller than the Abbé limit to overcome the diffraction limit for image resolution. Fluorescent molecules become "bleached" for some period of time once they have emitted a fluorescent photon. In stimulated emission depletion (STED), the illumination (excitation) spot remains diffraction-limited, but a concentric de-excitation doughnut-shaped beam turns off fluorescence in most of that spot [94]. Since the illumination wavelength is filtered out, only the longer fluorescent wavelength is detected or visible in the microscope. Hence, the smaller the spot at the centre of the doughnut which is allowed to fluoresce, the smaller the spot which can be imaged. Thus, the technique gets around the Abbé limit rather than breaks it. The size of the spot which can be imaged is only limited by the intensity of the doughnut-shaped beam. As this intensity gets larger (GW/cm^2 have been used), the size of the spot from whence fluorescence can be emitted gets smaller. STED and reversible saturable optical linear fluorescence transitions (RESOLFT) nanoscopy has been found especially useful for neurons or fixed cells and can be used in fast processes [95].

Some other techniques like photo-activated localization microscopy (PALM) and stochastic optical reconstruction microscopy (STORM) tackle this problem statistically [95]. These techniques find the locus of a molecule by fitting a Gaussian profile to the emission. If enough photons are collected, the locus can be identified with an uncertainty less than the diffraction limit. Conversely, two molecules within the lateral optical resolution can only be localised if the emitted photons occur at different times. Thus, these techniques are more suited to slower processes.

PALM, STORM, and STED share the need to switch off molecules and are essentially limited to imaging fluorophores or objects which are labelled with fluorophores which are generally toxic. Nonetheless, there are now well-established methods for labelling almost anything (typically cells or cell components) with fluorescent molecules. They also share the further steps of identification and localization [96]. Ultimately, of course, they are limited by the size of the fluorescent molecule and practical considerations such as the integrity, viability, and drift of the sample. With samples bigger than an individual cell, refractive index variations will cause distortions which are significant on the nanoscale.

Microcomputed tomography

We are all familiar with the extraordinary imaging capabilities of x-ray computed tomography (CT) in the hospital. However, the resolution is limited to approximately 1 mm in favour of penetration depth of tens of centimetres. With higher x-ray dose per voxel, the signal to noise ratio can be sufficient to achieve sub-micron resolution in engineering materials after several hours, although this dose would be too great for living cells and tissues. In vivo microCT uses a small sample bore typically sufficient for a mouse and can generate exquisite structural images with approximately 100-μm resolution in all directions. MicroCT application to stem cell research has already been reviewed by Boerckel et al. in this series [97].

Radionuclide imaging

Adding the functional capabilities provided by positron emission tomography (PET), PET-CT, and single-photon emission computed tomography (SPECT) imaging allows the stem cell functions to be put in their proper structural context. The earliest studies utilising the tracer principle [98], the use of small amounts of radionuclides in subjects, can be traced back to the 1920s [99]. However, it was development of the sodium iodide (NaI(Tl)) scintillation camera in the 1950s by Hal Anger [100] which was the bedrock of clinical nuclear medicine imaging systems for many decades. In the last decade there has been significant progress made in the development of various pre-clinical imaging systems across many modalities, and SPECT has become one of the principle tools [101, 102]. Several groups, including our own, have been demonstrating the capabilities of new SPECT system configurations [103–107]. Research innovation in this field has been significant with developments in aspects such as image reconstruction, collimation, detection, dual isotope imaging, and multimodality systems. Small animal SPECT (and PET) systems are exquisitely sensitive, capable of measuring picomolar concentrations of radiolabelled biomolecules in vivo with sub-millimetre resolution.

In terms of applications, there is considerable interest in methods where the radiation source is inside the subject and therapeutic applications are mediated by the human sodium iodide symporter (NIS). Several groups have evaluated the potential for the introduction of NIS expression to support imaging and treatment for various cancer types. For example, MSCs can be engineered to express NIS and then home to the tumour site for delivery of therapy [108]. SPECT imaging using ^{123}I or ^{99m}Tc can be used to confirm the migration of the MSCs to the tumour site, and then ^{131}I can be used for therapy.

During the last 10–15 years, small animal radionuclide imaging has undergone rapid technological development and improvement in image performance metrics. Innovations in several areas currently under investigation by several groups will lead to further improvements in the future, and radionuclide imaging will continue to play a vital role in future molecular imaging applications. The development of hybrid imaging with modalities such as

Table 2 Advantages and disadvantages of techniques listed in the manuscript

Technique	Advantages	Disadvantages
In vivo fluorescence imaging	• Simple, cheap, user friendly techniques • High spatial resolution (~200 nm in x,y,) with high sensitivity cameras • Development of FarRed and NIR probes allow greater tissue visualization with much less damage whilst imaging • High sensitivity (10^{-12} to 10^{-15} µm/L)	• Use of a probe generally required which may have repercussions on stem cell physiology • Photo-toxicity to tissue and depth resolved imaging still an issue • Vectors employed to introduce reporter genes are still under scrutiny for safety and efficacy of use in clinical trials
QPM	• Accurate quantitative visualisation of phase changes within cells	• No depth-resolving capabilities
ODT	• Depth-resolving capabilities, resolution of up to 1 µm	• Low penetration depth (a few hundred microns), not suitable for real-time imaging (slow techniques)
DHM	• Imaging of a 3D volume with a single exposure, structural and phase imaging, and also flexibility for image processing. Resolution almost as in conventional microscopy	• Relative complexity (more complicated optical set up), limitation on coherent properties of the light source, on environmental conditions (vibrations, etc.)
SESF and srSESF	• High (nano-scale, ~10 nm demonstrated) sensitivity to structural alterations within object and super- resolution imaging	• More complicated optical set up, for example for detailed quantitative analysis of the structure an imaging spectrometer or swept light source is needed
OCT	• Improved image resolution (morphological and functional information) of depth-resolved images • Can be combined with other imaging techniques for multimodal imaging • Suitable for clinical translation	• Penetration depth is limited ~2 mm into tissue • Spatial resolution is typically limited to ~10 µm, making this technique unsuitable for cell imaging • Limited molecular sensitivity of tissue
OCM	• Enhanced penetration depth compared to standard confocal microscopy; dramatically improved resolution over OCT imaging (up to 1 micron)	• Small penetration depth (compared with OCT)
nsOCT	• Depth-resolved images with high sensitivity (~30 nm demonstrated experimentally)	• Resolution and penetration depth are approximately the same as conventional OCT
OCPM	• Quantitative phase information with high sensitivity, useful for 3D intracellular imaging	• Small depth of field
PAI	• Capable of collecting molecular and spatial information from the tissue using endogenous contrast alone • Greater sensitivity than OCT and confocal imaging • Suitable for clinical translation • The ratio of the imaging depth to the best spatial resolution is roughly a constant of 200	• Sometimes requires the use of biocompatible labelling materials such as gold or silver nanoparticles
Confocal reflectance microscopy	• High spatial resolution images achievable (diffraction limited ~200 nm) • Can work in combination with other modes of microscopy including fluorescence and OCT	• Lack of specific light reflecting probes for confocal microscopy when used in reflectance mode
Super-resolution microscopy (nanoscopy)	• Images created have a higher spatial resolution that normal diffraction limited techniques. (STED x.y resolution ~20–100 nm, PALM and STORM x.y ~20–50 nm) • Increased localization and clarity of intracellular structures due to increased resolution	• Fluorophores or fluorescent markers must be used. Potential for photo bleaching of the sample under study • Expensive equipment • Currently most super resolution techniques are not suitable for live cell imaging • Refractive index variations in the substrate can cause distortions which when translated to the nanoscale can be significant
Microcomputed tomography	• Can generate defined structural images with increased all round resolution (100 µm in x,y and z dimensions) • Suitable for clinical translation	• Exposure to ionizing radiation which can cause DNA damage • Not suitable for soft tissues
Radionuclide imaging	• Only low doses of labels need to be employed due to the high sensitivity of the probes • Good tissue penetration of the probe • Suitable for clinical translation • Fair sensitivity (10^{-8} to 10^{-9} µm/L)	• Exposure to ionizing radiation which can cause DNA damage • Half-life of the probe must be considered

3D three-dimensional, *DHM* digital holographic microscopy, *NIR* near infrared, *nsOCT, OCPM, OCT* optical coherence microscope, *ODT, PAI* photoacoustic imaging, *PALM* photo-activated localization microscopy, *QPM* quantitative phase microscopy, *SESF* spectral encoding of the spatial frequency, *srSESF, STED* stimulated emission depletion, *STORM* stochastic optical reconstruction microscopy, *nsOCT* nano-sensitive optical coherence tomography, *OCPM* optical coherence phase microscopy, *ODT* optical doppler tomography, *srSESF* super-resolution spectral encoding of spatial frequency

PET/CT, PET/MR, SPECT/CT, and, possibly in the near future, SPECT/MR will enable biologists to observe processes in varying time windows from minutes to weeks.

Stem cell tracking requires high spatial resolution and sensitivity. Given that each imaging technique presents its unique set of advantages and disadvantages, the selection of an appropriate imaging modality depends on the application, the goal of the experiment, the subject under study, and so forth. No imaging technique is perfect in all aspects. Optical imaging techniques offer many distinctive advantages such as non-invasiveness, resolution, high spatial and temporal sensitivity, and adaptability, but these techniques are limited by relatively poor tissue depth. Radionuclide imaging has a fair sensitivity (10^{-8} to 10^{-9} μm/L), but it is not suitable for long-term cell tracking due to radioisotope decay. Fluorescence imaging has very high sensitivity (10^{-12} to 10^{-15} μm/L), but this technique is constrained by relatively shallow tissue depth [17]. An overview of the advantages and disadvantages of each technique is presented in Table 2.

Future directions should focus on multimodality imaging approaches that can combine the strength of each modality for a comprehensive detection and minimise potential drawbacks of using the imaging technique alone. Developing biodegradable contrast agents and multimodal contrast agents is another future development direction. The cytotoxicity and potential toxicity can be effectively reduced using degradable contrast agents by facilitating the clearance of the contrast materials [109]. Future directions of microscopic-related technologies will more than likely be in parallel with the development of advanced label-free imaging techniques and those which employ non-toxic cellular contrasting agents. Future development of imaging modalities for stem cell study should be focused on specific needs for different applications, but all applications would benefit from increased resolution, sensitivity, and reduced toxicity.

Conclusions

The vast array of technologies discussed above that are available to clinical and scientific researchers in the field of regenerative medicine allow multiple different elucidating conclusions to be drawn from imaging or analysing the tissue under study. The development of multimodal techniques which have the capacity to employ more sensitive, accurate, and less toxic labels to image deeper into the innate tissue in vivo will in time greatly further discoveries in this field. In relation to stem cell tracking for regenerative medicine, the availability of imaging systems (combination of hardware and cell-labelling strategy) will determine the cell-labelling strategy, with each approach having advantages and disadvantages. In general, the ideal system should have high spatial (ability to resolve single cells) and temporal resolution, contrast, sensitivity (detect small numbers of cells), be relatively easy of use, and be inexpensive. No imaging strategy will tick all the boxes; however, the current trend towards multimodal imaging can exploit one system's advantages while negating the disadvantages of another.

Abbreviations
3D: Three-dimensional; Au: gold; CLARITY: clear lipid-exchanged acrylamide-hybridized rigid imaging/immunostaining/in situ-hybridization-compatible tissue hydrogel; CT: computed tomography; DHM: digital holographic microscopy; DS: dielectric spectroscopy; GFP: green fluorescent protein; HD-OCT: high-definition optical coherence tomography; hIPSC: human induced pluripotent stem cell; hMSC: human mesenchymal stem cell; hPSC: human pluripotent stem cell; iRFP: near-infrared fluorescent protein; IRM: interference reflection microscopy; MR: magnetic resonance; MSC: mesenchymal stem cell; NIR: near infrared; NIS: sodium iodide symporter; NP: nanoparticle; NR: nanorod; OI-RD: oblique-incidence reflectivity difference; OCM: optical coherence microscope; OCT: optical coherence tomography; PAI: photoacoustic imaging; PALM: photo-activated localization microscopy; PET: positron emission tomography; PT-OCT: photothermal optical coherence tomography; QPM: quantitative phase microscopy; SESF: spectral encoding of the spatial frequency; SiGNR: silica-coated gold nanorod; SR-OCT: spectral radar optical coherence tomography; SDOCT: spectral domain optical coherence tomography; SPECT: single-photon emission computed tomography; SSEA1: stage-specific embryonic antigen 1; STED: stimulated emission depletion; STORM: stochastic optical reconstruction microscopy; SWNT: single-walled carbon nanotube; US: ultrasound.

Competing interests
The authors declare that they have no competing interests.

Authors' contributions
ML provided overall guidance for the manuscript and wrote the paper. KT wrote the in vivo fluorescence imaging section and edited the whole manuscript. HZ wrote the photoacoustic imaging section and edited the whole manuscript. SA wrote the optical coherence tomography and phase microscopy sections. MF wrote the radionuclide imaging section. CO'F wrote the section on requirements for stem cell imaging and compiled Table 1. PD contributed to the microscopy sections and edited the whole manuscript. All authors read and approved the final manuscript.

Acknowledgements
This work was supported by the National Biophotonics Imaging Platform (NBIP) Ireland funded under the Higher Education Authority PRTLI Cycle 4, co-funded by the Irish Government and the European Union—Investing in your future. The photoacoustic research was supported by the Science Foundation Ireland (SFI). HZ is supported by a Hardiman Fellowship from NUI Galway.

Author details
[1]Tissue Optics & Microcirculation Imaging Group, School of Physics, National University of Ireland (NUI), Galway, Ireland. [2]Centre for Microscopy and Imaging, Anatomy, School of Medicine, National University of Ireland (NUI), Galway, Ireland. [3]Medical Physics Research Cluster, School of Physics, National University of Ireland (NUI), Galway, Ireland. [4]Regenerative Medicine Institute (REMEDI), National University of Ireland (NUI), Galway, Ireland. [5]Chair of Applied Physics, National University of Ireland (NUI), Galway, Ireland.

References
1. Azene N, Fu Y, Maurer J, Kraitchman DL. Tracking of stem cells in vivo for cardiovascular applications. J Cardiovasc Magn Reson. 2014;16(1):7.
2. Lacroix LM, Delpech F, Nayral C, Lachaize S, Chaudret B. New generation of magnetic and luminescent nanoparticles for in vivo real-time imaging. Interface Focus. 2013;3(3):20120103.

3.   Rao J, Dragulescu-Andrasi A, Yao H. Fluorescence imaging in vivo: recent advances. Curr Opin Biotechnol. 2007;18:17–25.

4.   Merian J, Gravier J, Navarro F, Texier I. Fluorescent nanoprobes dedicated to in vivo imaging: from preclinical validations to clinical translation. Molecules. 2012;17:5564–91.

5.   Nguyen PK, Riegler J, Wu JC. Stem cell imaging: from bench to bedside. Cell Stem Cell. 2014;14:431–44.

6.   Mehta S, Zhang J. Reporting from the field: genetically encoded fluorescent reporters uncover signaling dynamics in living biological systems. Annu Rev Biochem. 2011;80:375–401.

7.   Nowakowski A, Andrzejewska A, Janowski M, Walczak P, Lukomska B. Genetic engineering of stem cells for enhanced therapy. Acta Neurobiol Exp. 2013;73:1–18.

8.   Ronald JA, Cusso L, Chuang H-Y, Yan X, Dragulescu-Andrasi A, Gambhir SS. Development and validation of non-integrative, self-limited, and replicating minicircles for safe reporter gene imaging of cell-based therapies. PLoS One. 2013;8:e73138.

9.   Shu X, Royant A, Lin MZ, Aguilera TA, Lev-Ram V, Steinbach PA, et al. Mammalian expression of infrared fluorescent proteins engineered from a bacterial phytochrome. Science. 2009;324:804–7.

10. Shcherbo D, Merzlyak EM, Chepurnykh TV, Fradkov AF, Ermakova GV, Solovieva EA, et al. Bright far-red fluorescent protein for whole-body imaging. Nat Meth. 2007;4:741–6.

11. Filonov GS, Piatkevich KD, Ting L-M, Zhang J, Kim K, Verkhusha VV. Bright and stable near-infrared fluorescent protein for in vivo imaging. Nat Biotech. 2011;29:757–61.

12. Iyer SM, Delp SL. Optogenetic regeneration. Science. 2014;344:44–5.

13. Piatkevich KD, Subach FV, Verkhusha VV. Engineering of bacterial phytochromes for near-infrared imaging, sensing, and light-control in mammals. Chem Soc Rev. 2013;42:3441–52.

14. Tu J, Yang F, Wan J, Liu Y, Zhang J, Wu B, et al. Light-controlled astrocytes promote human mesenchymal stem cells toward neuronal differentiation and improve the neurological deficit in stroke rats. Glia. 2014;62:106–21.

15. Bryson JB, Machado CB, Crossley M, Stevenson D, Bros-Facer V, Burrone J, et al. Optical control of muscle function by transplantation of stem cell-derived motor neurons in mice. Science. 2014;344:94–7.

16. Levin RA, Felsen CN, Yang J, Lin JY, Whitney MA, Nguyen QT, et al. An optimized triple modality reporter for quantitative in vivo tumor imaging and therapy evaluation. PLoS One. 2014;9:e97415.

17. Zhang SJ, Wu JC. Comparison of imaging techniques for tracking cardiac stem cell therapy. J Nucl Med. 2007;48(12):1916–9.

18. Frangioni JV, Hajjar RJ. In vivo tracking of stem cells for clinical trials in cardiovascular disease. Circulation. 2004;110(21):3378–83.

19. Singh A, Suri S, Lee T, Chilton JM, Cooke MT, Chen WQ, et al. Adhesion strength-based, label-free isolation of human pluripotent stem cells. Nat Methods. 2013;10(5):438.

20. Zuo C, Chen Q, Qu WJ, Asundi A. Noninterferometric single-shot quantitative phase microscopy. Opt Lett. 2013;38(18):3538–41.

21. Sugiyama N, Asai Y, Yamauchi T, Kataoka T, Ikeda T, Iwai H, et al. Label-free characterization of living human induced pluripotent stem cells by subcellular topographic imaging technique using full-field quantitative phase microscopy coupled with interference reflection microscopy. Biomed Opt Express. 2012;3(9):2175–83.

22. Barr VA, Bunnell SC. Interference reflection microscopy. Curr Protoc Cell Biol. 2009;Chapter 4:Unit 4.23. doi:10.1002/0471143030.cb0423s45.

23. Cotte Y, Toy F, Jourdain P, Pavillon N, Boss D, Magistretti P, et al. Marker-free phase nanoscopy. Nat Photonics. 2013;7(2):113–7.

24. Choi W, Fang-Yen C, Badizadegan K, Oh S, Lue N, Dasari RR, et al. Tomographic phase microscopy. Nat Methods. 2007;4(9):717–9.

25. Kim T, Zhou RJ, Mir M, Babacan SD, Carney PS, Goddard LL, et al. White-light diffraction tomography of unlabelled live cells. Nat Photonics. 2014;8(3):256–63.

26. Zhu XD, Landry JP, Sun YS, Gregg JP, Lam KS, Guo XW. Oblique-incidence reflectivity difference microscope for label-free high-throughput detection of biochemical reactions in a microarray format. Appl Optics. 2007;46(10):1890–5.

27. Alexandrov SA, Hillman TR, Sampson DD. Spatially resolved Fourier holographic light scattering angular spectroscopy. Opt Lett. 2005;30(24):3305–7.

28. Hillman TR, Alexandrov SA, Gutzler T, Sampson DD. Microscopic particle discrimination using spatially-resolved Fourier-holographic light scattering angular spectroscopy. Opt Express. 2006;14(23):11088–102.

29. Gutzler T, Hillman TR, Alexandrov SA, Sampson DD. Three-dimensional depth-resolved and extended-resolution micro-particle characterization by holographic light scattering spectroscopy. Opt Express. 2010;18(24):25116–26.

30. Alexandrov SA, Hillman TR, Gutzler T, Sampson DD. Synthetic aperture Fourier holographic optical microscopy. Phys Rev Lett. 2006;97(16):168102.

31. Hillman TR, Gutzler T, Alexandrov SA, Sampson DD. High-resolution, wide-field object reconstruction with synthetic aperture Fourier holographic optical microscopy. Opt Express. 2009;17(10):7873–92.

32. Mico V, Zalevsky Z. Superresolved digital in-line holographic microscopy for high-resolution lens-less biological imaging. J Biomed Opt. 2010;15(4):046027.

33. Kim MK. Tomographic three-dimensional imaging of a biological specimen using wavelength-scanning digital interference holography. Opt Express. 2000;7(9):305–10.

34. Alexandrov SA, Uttam S, Bista RK, Zhao CQ, Liu Y. Real-time quantitative visualization of 3D structural information. Opt Express. 2012;20(8):9203–14.

35. Alexandrov SA, Uttam S, Bista RK, Liu Y. Spectral contrast imaging microscopy. Opt Lett. 2011;36(17):3323–5.

36. Alexandrov SA, Uttam S, Bista RK, Staton K, Liu Y. Spectral encoding of spatial frequency approach for characterization of nanoscale structures. Appl Phys Lett. 2012;101(3):033702.

37. Uttam S, Alexandrov SA, Bista RK, Liu Y. Tomographic imaging via spectral encoding of spatial frequency. Opt Express. 2013;21(6):7488–504.

38. Alexandrov SA, Subhash HM, Zam a, Leahy M. Nano-sensitive optical coherence tomography. Nanoscale. 2014;6(7):3545–9.

39. Alexandrov S, Subhash HM, Leahy M. Nanosensitive optical coherence tomography for the study of changes in static and dynamic structures. Quantum Electronics. 2014;44(7):657–3.

40. Alexandrov S, McGrath J, Subhash H, Boccafoschi F, Giannini C, Leahy M. Novel approach for label free super-resolution imaging in far field. Nat Sci Rep. 2015;5:13274.

41. Huang D, Swanson EA, Lin CP, Schuman JS, Stinson WG, Chang W, et al. Optical coherence tomography. Science. 1991;254(5035):1178–81.

42. Adler DC, Chen Y, Huber R, Schmitt J, Connolly J, Fujimoto JG. Three-dimensional endomicroscopy using optical coherence tomography. Nat Photonics. 2007;1(12):709–16.

43. Liu LB, Gardecki JA, Nadkarni SK, Toussaint JD, Yagi Y, Bouma BE, et al. Imaging the subcellular structure of human coronary atherosclerosis using micro-optical coherence tomography. Nat Med. 2011;17(8):1010–U132.

44. Vakoc BJ, Lanning RM, Tyrrell JA, Padera TP, Bartlett LA, Stylianopoulos T, et al. Three-dimensional microscopy of the tumor microenvironment in vivo using optical frequency domain imaging. Nat Med. 2009;15(10):1219–U151.

45. Choi W, Potsaid B, Jayaraman V, Baumann B, Grulkowski I, Liu JJ, et al. Phase-sensitive swept-source optical coherence tomography imaging of the human retina with a vertical cavity surface-emitting laser light source. Opt Lett. 2013;38(3):338–40.

46. Subhash HM, Anh NH, Wang RKK, Jacques SL, Choudhury N, Nuttall AL. Feasibility of spectral-domain phase-sensitive optical coherence tomography for middle ear vibrometry. J Biomed Opt. 2012;17(6):060505.

47. Maenz M, Morcos M, Ritter T. A comprehensive flow-cytometric analysis of graft infiltrating lymphocytes, draining lymph nodes and serum during the rejection phase in a fully allogeneic rat cornea transplant model. Mol Vis. 2011;17:420–9.

48. Enfield J, Jonathan E, Leahy M. In vivo imaging of the microcirculation of the volar forearm using correlation mapping optical coherence tomography (cmOCT). Biomed Opt Express. 2011;2(5):1184–93.

49. Jonathan E, Enfield J, Leahy MJ. Correlation mapping method for generating microcirculation morphology from optical coherence tomography (OCT) intensity images. J Biophotonics. 2011;4(9):583–7.

50. Zafar H, Enfield J, O'Connell ML, Ramsay B, Lynch M, Leahy MJ. Assessment of psoriatic plaque in vivo with correlation mapping optical coherence tomography. Skin Res Technol. 2014;20(2):141–6.

51. Subhash HM, Leahy MJ. Microcirculation imaging based on full-range high-speed spectral domain correlation mapping optical coherence tomography. J Biomed Opt. 2014;19(2):21103.

52. Yang Y, Dubois A, Qin XP, Li J, El Haj A, Wang RK. Investigation of optical coherence tomography as an imaging modality in tissue engineering. Phys Med Biol. 2006;51(7):1649–59.

53. Bagnaninchi PO, Yang Y, Zghoul N, Maffulli N, Wang RK, El Haj AJ. Chitosan microchannel scaffolds for tendon tissue engineering characterized using optical coherence tomography. Tissue Eng. 2007;13(2):323–31.

54. Tan W, Sendemir-Urkmez A, Fahrner LJ, Jamison R, Leckband D, Boppart SA. Structural and functional optical imaging of three-dimensional engineered tissue development. Tissue Eng. 2004;10(11-12):1747–56.

55. Lin YS, Chu CC, Lin JJ, Chang CC, Wang CC, Wang CY, et al. Optical coherence tomography: a new strategy to image planarian regeneration. Sci Rep-Uk. 2014;4:6316.

56. Boone MALM, Norrenberg S, Jemec GBE, Del Marmol V. Imaging of basal cell carcinoma by high-definition optical coherence tomography: histomorphological correlation. A pilot study. Br J Dermatol. 2012;167(4):856–64.

57. Subhash HM, Connolly E, Murphy M, Barron V, Leahy M, editors. Photothermal optical coherence tomography for depth-resolved imaging of mesenchymal stem cells via single wall carbon nanotubes. Proc. SPIE 8954, Nanoscale Imaging, Sensing, and Actuation for Biomedical Applications XI, 89540C (March 4, 2014); doi:10.1117/12.2038535.

58. Bagnaninchi PO. Monitoring adipose-derived stem cells within 3D carrier by combined dielectric spectroscopy and spectral domain optical coherence topography. Proc. SPIE 7566, Optics in Tissue Engineering and Regenerative Medicine IV, 756602.

59. Chen C-W, Betz MW, Fisher JP, Paek A, Jiang J, Ma H, et al. Investigation of pore structure and cell distribution in EH-PEG hydrogel scaffold using optical coherence tomography and fluorescence microscopy. Proc. SPIE 7566, Optics in Tissue Engineering and Regenerative Medicine IV, 756603.

60. Ksander BR, Kolovou PE, Wilson BJ, Saab KR, Guo Q, Ma J, et al. ABCB5 is a limbal stem cell gene required for corneal development and repair. Nature. 2014;511(7509):353.

61. Lathrop KL, Gupta D, Kagemann L, Schuman JS, SundarRaj N. Optical coherence tomography as a rapid, accurate, noncontact method of visualizing the palisades of Vogt. Invest Ophth Vis Sci. 2012;53(3):1381–7.

62. Schwartz SD, Regillo CD, Lam BL, Eliott D, Rosenfeld PJ, Gregori NZ, et al. Human embryonic stem cell-derived retinal pigment epithelium in patients with age-related macular degeneration and Stargardt's macular dystrophy: follow-up of two open-label phase 1/2 studies. Lancet. 2015;385(9967): 509-16.

63. Bagnaninchi PO, Holmes C, Drummond N, Daoud J, Tabrizian M. Two-dimensional and three-dimensional viability measurements of adult stem cells with optical coherence phase microscopy. J Biomed Opt. 2011;16(8):086003.

64. Holmes C, Bagnaninchi P, Daoud J, Tabrizian M. Polyelectrolyte multilayer coating of 3D scaffolds enhances tissue growth and gene delivery: non-invasive and label-free assessment. Adv Healthc Mater. 2014;3(4):572–80.

65. Holmes C, Tabrizian M, Bagnaninchi P. Motility imaging via optical coherence phase microscopy enables label-free monitoring of tissue growth and viability in 3D tissue engineering scaffolds. J Tissue Eng Regen Med. 2015. doi:10.1002/term.1687.

66. Beard P. Biomedical photoacoustic imaging. Interface Focus. 2011;1(4):602–31.

67. Wang LV. Multiscale photoacoustic microscopy and computed tomography. Nat Photonics. 2009;3:503–9.

68. Zackrisson S, van de Ven SM, Gambhir SS. Light in and sound out: emerging translational strategies for photoacoustic imaging. Cancer Res. 2014;74(4):979–1004.

69. Guo Z, Li L, Wang LV. On the speckle-free nature of photoacoustic tomography. Med Phys. 2009;36:4084–8.

70. Erpelding TN et al. Sentinel lymph nodes in the rat: noninvasive photoacoustic and US imaging with a clinical US System. Radiology. 2010;256:102–10.

71. Kim C et al. In vivo molecular photoacoustic tomography of melanomas targeted by bioconjugated gold nanocages. ACS Nano. 2010;4:4559–64.

72. Zhang EZ et al. Multimodal photoacoustic and optical coherence tomography scanner using an all optical detection scheme for 3D morphological skin imaging. Biomed Opt Express. 2011;2:2202–15.

73. Zafar H, Breathnach A, Subhash HM, Leahy MJ. Linear array-based photoacoustic imaging of human microcirculation with a range of high frequency transducer probes. J Biomed Opt. 2015;20(5):051021.

74. Jansen K, van der Steen AFW, van Beusekom HMM, Oosterhuis JW, van Soest G. Intravascular photoacoustic imaging of human coronary atherosclerosis. Opt Lett. 2011;36:597–9.

75. Wang B et al. Plasmonic intravascular photoacoustic imaging for detection of macrophages in atherosclerotic plaques. Nano Lett. 2008;9:2212–7.

76. Oladipupo S et al. VEGF is essential for hypoxia-inducible factor-mediated neovascularization but dispensable for endothelial sprouting. Proc Natl Acad Sci U S A. 2011;108:13264–9.

77. Oladipupo SS et al. Conditional HIF-1 induction produces multistage neovascularization with stage-specific sensitivity to VEGFR inhibitors and myeloid cell independence. Blood. 2011;117:4142–53.

78. Yang J-M et al. Photoacoustic endoscopy. Opt Lett. 2009;34:1591–3.

79. Yang J-M et al. Toward dual-wavelength functional photoacoustic endoscopy: laser and peripheral optical systems development. Proc SPIE. 2012;8223:822316.

80. Hu S, Wang LV. Neurovascular photoacoustic tomography. Front Neuroenerg. 2010;2:10.

81. Hu S, Yan P, Maslov K, Lee J-M, Wang LV. Intravital imaging of amyloid plaques in a transgenic mouse model using optical-resolution photoacoustic microscopy. Opt Lett. 2009;34:3899–901.

82. Wang X et al. Noninvasive laser-induced photoacoustic tomography for structural and functional in vivo imaging of the brain. Nat Biotechnol. 2003;21:803–6.

83. Hu S, Rao B, Maslov K, Wang LV. Label-free photoacoustic ophthalmic angiography. Opt Lett. 2010;35:1–3.

84. Jiao S et al. Photoacoustic ophthalmoscopy for in vivo retinal imaging. Opt Express. 2010;18:3967–72.

85. Nam SY, Ricles LM, Suggs LJ, Emelianov SY. In vivo ultrasound and photoacoustic monitoring of mesenchymal stem cells labeled with gold nanotracers. PLoS One. 2012;7:e37267.

86. Tong L, Wei QS, Wei A, Cheng JX. Gold nanorods as contrast agents for biological imaging: optical properties, surface conjugation and photothermal effects. Photochem Photoboil. 2009;85:21–32.

87. Jokerst JV, Thangaraj M, Kempen PJ, Sinclair R, Gambhir SS. Photoacoustic imaging of mesenchymal stem cells in living mice via silica-coated gold nanorods. ACS Nano. 2012;6:5920–30.

88. Daly SM, Leahy MJ. 'Go with the flow': a review of methods and advancements in blood flow imaging. J Biophotonics. 2013;6(3):217–55.

89. Wang Z, Glazowski CE, Zavislan JM. Modulation transfer function measurement of scanning reflectance microscopes. J Biomed Opt. 2007;12:051802.

90. Hofmann-Wellenhof R, Wurm EM, Ahlgrimm-Siess V, Richtig E, Koller S, Smolle J, et al. Reflectance confocal microscopy—state-of-art and research overview. Semin Cutan Med Surg. 2009;28:172–9.

91. Takaku T, Malide D, Chen J, Calado RT, Kajigaya S, Young NS. Hematopoiesis in 3 dimensions: human and murine bone marrow architecture visualized by confocal microscopy. Blood. 2010;116:e41–55.

92. Schain AJ, Hill RA, Grutzendler J. Label-free in vivo imaging of myelinated axons in health and disease with spectral confocal reflectance microscopy. Nat Med. 2014;20:443–9.

93. Zhao Y, Bower AJ, Graf BW, Boppart MD, Boppart SA. Imaging and tracking of bone marrow-derived immune and stem cells. Methods Mol Biol. 2013;1052:57–76.

94. Hell SW, Wichmann J. Breaking the diffraction resolution limit by stimulated emission: stimulated-emission-depletion fluorescence microscopy. Opt Lett. 1994;19(11):780–2.

95. MacDonald L, Baldini G, Storrie B. Does super resolution fluorescence microscopy obsolete previous microscopic approaches to protein co-localization? Methods Mol Biol (Clifton, NJ). 2015;1270:255–75. doi:10.1007/978-1-4939-2309-0_19.

96. Requejo-Isidro J. Fluorescence nanoscopy: methods and applications. J Chem Biol. 2013;6(3):97–120.

97. Boerckel JD, Mason DE, McDermott AM, Alsberg E. Microcomputed tomography: approaches and applications in bioengineering. Stem Cell Res Ther. 2014;5:144.

98. Chiewitz O, Hevesy G. Radioactive indicators in the study of phosphorous metabolism in rats. Nature. 1935;136:754–5.

99. Blumgart HL, Weiss S. Studies on the velocity of blood flow: VII. The pulmonary circulation time in normal resting individuals. J Clin Invest. 1927;4:399–425.

100. Anger HO. Use of a gamma-ray pinhole camera for in vivo studies. Nature. 1952;170:200–1.

101. Meikle SR, Kench P, Kassiou M, Banati RB. Small animal SPECT and its place in the matrix of molecular imaging technologies. Phys Med Biol. 2005;50:R45–61.

102. Franc BL, Acton PD, Mari C, Hasegawa BH. Small-animal SPECT and SPECT/CT: important tools for preclinical investigation. J Nucl Med. 2008;49:1651–63.

103. Beekman FJ, van der Have F, Vastenhouw B, van der Linden AJ, van Rijk PP, Burbach JP, et al. U-SPECT-I: a novel system for submillimeter-resolution tomography with radiolabeled molecules in mice. J Nucl Med. 2005;46:1194–200.

104. Havelin RJ, Miller BW, Barrett HH, Furenlid LR, Murphy JM, Dwyer RM, et al. Design and performance of a small-animal imaging system using synthetic collimation. Phys Med Biol. 2013;58:3397–412.

105. Havelin RJ, Miller BW, Barrett HH, Furenlid LR, Murphy JM, Foley MJ. A SPECT imager with synthetic collimation. Proc. SPIE 8853, Medical Applications of Radiation Detectors III, 885309 (September 26, 2013).

106. Furenlid LR, Wilson DW, Chen YC, Kim H, Pietraski PJ, Crawford MJ, et al. FastSPECT II: a second-generation high-resolution dynamic SPECT imager. IEEE Trans Nucl Sci. 2004;51:631–5.

107. Hasegawa BH, Barber WC, Funk T, Hwang AB, Taylor C, Sun M, et al. Implementation and applications of dual-modality imaging. Nucl Instrum Methods Phys Res, Sect A. 2004;525:236–41.

108. Dwyer RM, Ryan J, Havelin RJ, Morris JC, Miller BW, Liu Z, et al. Mesenchymal stem cell-mediated delivery of the sodium iodide symporter supports radionuclide imaging and treatment of breast cancer. Stem cells (Dayton, Ohio). 2011;29:1149–57.

109. Wang J, Jokerst JV. Stem cell imaging: tools to improve cell delivery and viability. Stem Cells Int. 2016;2016:9240652.

microRNA-206 is involved in survival of hypoxia preconditioned mesenchymal stem cells through targeting Pim-1 kinase

You Zhang[1†], Wei Lei[2†], Weiya Yan[2], Xizhe Li[3], Xiaolin Wang[4], Zhenao Zhao[2], Jie Hui[1*], Zhenya Shen[2*] and Junjie Yang[2*]

Abstract

Background: Overexpression of Pim-1 in stem/progenitor cells stimulated cell cycling and enhanced cardiac regeneration in vivo. We proposed that hypoxic preconditioning could increase survival of bone marrow mesenchymal stem cells (MSCs) via upregulation of Pim-1 and aimed to determine the microRNAs that modulate the expression of Pim-1.

Methods and results: MSCs were subjected to hypoxia exposure. The expression of Pim-1 in MSCs was enhanced in a time-dependent manner, detected by quantitative PCR and western blot. miR-206 is predicted as one of the potential microRNAs that target Pim-1. The expression of miR-206 was decreased in hypoxic MSCs and reversely correlated with Pim-1 expression. Luciferase activity assay further confirmed Pim-1 as a putative target of miR-206. In addition, gain and loss-of-function studies with miR-206 mimics and inhibitors showed that inhibition of miR-206 in hypoxic MSCs promoted the migration ability of the cells, prevented cell apoptosis, and protected membrane potential of mitochondria, while the benefits were all blocked by Pim-1 inhibitor. In an acute model of myocardial infarction, transplanted hypoxic MSCs showed a significantly improved survival as compared with hypoxic MSCs overexpressing miR-206.

Conclusions: Hypoxic preconditioning could increase short-term survival of bone marrow MSCs via upregulation of Pim-1, and miR-206 was one of the critical regulators in this process.

Keywords: Hypoxia, Mesenchymal stem cells, Pim-1, miR-206, Apoptosis

Background

Pim-1, a proto-oncogenic serine–threonine kinase, was originally discovered as the proviral integration site for Moloney murine leukemia virus [1]. Pim-1 plays a role in the proliferation and survival of hematopoietic cells [2]. However, the role of Pim-1 in cardiac development has been overlooked for a long time. In 2007, Dr Mark A Sussman reported that Pim-1 could regulate cardiomyocyte survival downstream of Akt using Pim-1-deficient mice [3]. Also, overexpression of Pim-1 inhibited cardiomyocyte apoptosis and protected mice from infarction injury, which revealed the potential cardioprotective role of Pim-1. Successively, Sussman's group showed that overexpression of Pim-1 in cardiac progenitor cells enhanced cardiac regeneration [4] and stimulated cell cycling [5]. Rejuvenation of stem cells by Pim-1 overexpression thus endued the cells with increased proliferative and regenerative potentials [6–8]. However, genetic modification of stem cells is not the first promising option in their utilization for tissue regeneration. In 2014, Hu et al. [9] reported that hypoxic preconditioning increases survival of cardiac progenitor cells via upregulation of Pim-1, discovering a method for nongenetic modification of Pim-1 in stem cells. In our study, we proposed that hypoxic preconditioning could increase survival of bone marrow mesenchymal stem cells (MSCs) via upregulation of Pim-1 and aimed to determine some microRNAs (miRNAs) that modulate the expression of Pim-1.

* Correspondence: huijie92@163.com; uuzyshen@aliyun.com; jjyang@suda.edu.cn
†Equal contributors
[1]Department of Cardiology of the First Affiliated Hospital, Soochow University, Suzhou, China
[2]Institute for Cardiovascular Science & Department of Cardiovascular Surgery of The First Affiliated Hospital, Soochow University, Suzhou, China
Full list of author information is available at the end of the article

We demonstrated that hypoxia preconditioning enhanced the expression of Pim-1 kinase in MSCs in a time-dependent manner, as detected by quantitative PCR (qPCR) and western blot. We further predicted that microRNA-206 (miR-206) is one of the potential miRNAs that targets Pim-1 through TargetScan, microRNA, and miRDB software. In addition, miR-206 is a potential regulator of proliferation, apoptosis, and differentiation of pulmonary artery smooth muscle cells [10].

We confirmed that the expression of miR-206 was decreased in hypoxic MSCs. Next, qPCR analysis revealed that miR-206 reversely regulated the expression of Pim-1. Luciferase activity assay further confirmed Pim-1 as a putative target of miR-206. In addition, gain and loss-of-function studies with miR-206 mimics and inhibitors showed that inhibition of miR-206 in HP-MSCs promoted the migration ability of the cells, prevented cell apoptosis, and protected membrane potential of mitochondria, while the benefits were all blocked by Pim-1 inhibitor. In an acute model of myocardial infarction, transplanted HP-MSCs showed significantly improved survival as compared with hypoxic MSCs overexpressing miR-206. In conclusion, hypoxic preconditioning could increase short-term survival of bone marrow MSCs via upregulation of Pim-1, and miR-206 was one of the critical regulators in this process.

Methods

Animals

All animal procedures were approved by the Ethic Committee of Soochow University, Suzhou, China and were carried out in accordance with the Guidelines for the Care and Use of Research Animals established by Soochow University. Rats were housed at the Animal Facility of Soochow University on a 12-h light/dark cycle with free access to water and standard mouse food.

Isolation and culture of bone marrow MSCs

Bone marrow MSCs were isolated from 4-week-old Sprague–Dawley rats. Briefly, the bone marrow was flushed from the femora of rats with low-glucose DMEM (Gibco, Gaithersburg, MD, USA), 10 % FBS (Gibco), and 1 % penicillin–streptomycin (HyClone, Logan, Utah, USA). Cells were collected and seeded onto a 10-cm diameter plate, and incubated in 5 % CO_2 at 37 °C. After 24 h of incubation, the medium was changed first and replaced every 3 days thereafter. When at 80–90 % confluence, MSCs were dissociated with 0.25 % trypsin (Sangon biotech, Shanghai, China) and expanded at a 1:2 dilution. MSCs at passage 4 were used in all experiments.

Hypoxia preconditioning of MSCs in vitro

For hypoxic culture, cells were cultured in a tri-gas incubator (Thermo Fisher Scientific, Marietta, OH, USA cell,

Germany) composed of 94 % N_2, 5 % CO_2, and 1 % O_2. To determine the optimal time length for hypoxic treatment, cells were cultured for 6, 12, and 24 h in hypoxia. MSCs were thus divided into four groups: normoxia, hypoxia for 6 h, hypoxia for 12 h, and hypoxia for 24 h. During the hypoxic preconditioning, MSCs were incubated with Quercetagetin (Calbiochem, San Diego, CA, USA), a specific Pim-1 activity inhibitor, at 10 µmol/l.

Western blot

MSCs were lysed with lysis buffer including 1 % phenylmethane sulfonylfluoride (PMSF) and 1 % phosphatase inhibitors and the protein was extracted. The protein concentration was determined using a BCA Protein Assay Kit (Beyotime Biotechnology, Shanghai, China). The protein sample extracted from MSCs was separated by 10 % SDS-PAGE (for Pim-1 and β-actin) and transferred to a 0.45 mm nitrocellulose membrane (Millipore, Billerica, MA, USA) using a Trans-Blot SD Semi-Dry Electrophoretic Transfer Cell (Bio-Rad, Hercules, California, USA). Milk 5 % (w/v) in 0.1 % Tween 20/PBS was used to block the membranes for 1 h. The membranes were then incubated with rabbit polyclonal antibody against Pim-1 (1:100; Santa Cruz, Dallas, Texas, USA) and β-actin (1:1000; Beyotime Biotechnology) overnight and blotted with a HRP-linked secondary antibody (1:1000; Beyotime Biotechnology) for 2 h. Afterwards, the protein band could be visualized via enhanced chemiluminescence and analyzed by the Scion Image Software (Scion, Frederich, MD, USA).

Target gene prediction

Three target prediction algorithms—TargetScan 6.2 (http://www.targetscan.org/), mirWALK (http://zmf.umm.uni-heidelberg.de/apps/zmf/mirwalk2/), and mircroRNA.org (http://www.microrna.org/microrna/)—were used to predict the potential target relationship between miR-206 and Pim-1.

microRNA expression level in MSCs

Total RNA was extracted from rat MSCs using the PureLink™ RNA Mini Kit (Ambion, Foster City, CA, USA) and quantified using an ND2000 spectrophotometer (NanoDrop Technologies, Wilmington, DE, USA). The extracted RNA was subjected to RT-PCR using the Hairpin-it™ miRNAs RT-PCR Quantitation Kit (GenePharma, Shanghai, China). miRNA expression levels were measured using Step One-Plus Real-Time PCR System (Applied Biosystems, Foster City, CA, USA), according to the manufacturer's instructions. miRNA expression was normalized to U6 snRNA.

Dual-luciferase reporter assays

Using standard procedures, wild-type (WT) or mutant 3′-untranslated regions (UTRs) of Pim-1 were subcloned into the pmiR-RB-REPORT™ vector (Ribobio, Guangzhou,

China) downstream of the luciferase gene. Pim-1-3′-UTR-WT or Pim-1-3′-UTR-Mut vectors were cotransfected with miR-206 mimics or negative control into MSCs, using Lipofectamine 2000 for 48 h. Firefly luciferase activity was measured at 48 h post transfection, using a Dual-Luciferase Reporter Assay System (Promega Corporation, Fitchburg, WI, USA), according to the manufacturer's instructions. Each reporter plasmid was transfected at least three times, and each sample was assayed in triplicate.

Transfection and groups in vitro

The miR-206 oligonucleotides including mimics, inhibitors, and negative control were purchased from GenePharma Co., Ltd (Shanghai, China). We dissolved the oligonucleotides with diethylpyrocarbonate (DEPC)-treated water to 50 nM. Cell transfection was performed with Lipofectamine 2000 (Invitrogen, Carlsbad, CA, USA), according to the manufacturer's protocols. The efficiency of transfection was confirmed using RT-qPCR. Cells were then subjected to hypoxic treatment as already described. To provide a positive control on the blocking of Pim-1 activity, Quercetagetin (Calbiochem) was also added at 10 μmol/l when cells were under hypoxic exposure. The experimental groups are thus illustrated as follows: MSCs, HP-MSCs, HP-MSCs$^{miR-206}$, HP-MSCs$^{anti-miR-206}$, and HP-MSCs$^{anti-miR-206 + Pim-1\ inhibitor}$ (see Table 1).

Transwell migration assay

Cell migration assays were performed using transwell filters with 8-μm pores (Fisher Scientific, Pittsburgh, PA, USA). Briefly, cells (5×10^4 cells/200 μl) suspended in serum-free DMEM were plated into the upper compartment of a Transwell chamber in triplicate. Lower chambers were filled with 500 μl of DMEM containing 10 % FBS. After 4 h, cells were fixed in 4 % methanal for 20 min, stained with DAPI (Invitrogen) and counted under a fluorescent microscope (Olympus, Tokyo, Japan).

Apoptosis assay

An Annexin V-fluorescein isothiocyanate (FITC) apoptosis detection kit (BD Pharmingen, San Diego, CA, USA) was used to perform the apoptosis assay. Briefly, 1×10^6 cells were collected by trypsinization and

resuspended in binding buffer containing Annexin V-FITC and propidium iodide. After incubation in the dark for 15 min, MSCs were analyzed using a BD FACS Aria flow cytometer.

Measurement of the mitochondrial membrane potential

The mitochondrial membrane potential ($\Delta\psi$m) was measured using the lipophilic cationic probe JC-1 dye (Beyotime Biotechnology), according to the manufacturer's instructions. Briefly, MSCs were stained with JC-1 (5 mol/l) at 37 °C for 20 min in the dark and rinsed three times with ice-cold working solution. The fluorescence was then monitored by the inversion fluorescence microscope (Olympus). The red fluorescence is caused by a potentially dependent aggregation in the mitochondria, reflecting $\Delta\psi$m. Green emission of the dye represents the monomeric form of JC-1. Mitochondrial depolarization is indicated by a decrease in the red/green fluorescence intensity ratio.

Acute myocardial infarction model of rat hearts and cell transplantation

Acute myocardial infarction (AMI) was made according to the method described previously [11]. Thirty female young (6–8 weeks) Sprague–Dawley rats (200–250 g) were studied. They were anesthetized once with by Avertin™ (300 mg/kg; Sigma, St. Louis, MO, USA) intraperitoneally. After a left lateral thoracotomy, the proximal portion of the left anterior descending artery was ligated with a 6–0 Prolene suture between the pulmonary artery outflow tract and the left atrium. Then, 1 million MSCs from male rats prepared in 70 μl suspension were transplanted by myocardial injection at multiple sites (3–4 sites/heart) in the free wall of the left ventricle during the acute phase of AMI. After injection, we closed the chests of the animals and the animals were allowed to recover. Four days later, the animals were euthanized for collection of the heart tissue samples for molecular studies.

Detection of cell survival in vivo

Cell survival was determined by detection of *Sry* gene using qPCR in the myocardial tissue samples. Tissue samples were snap-frozen in liquid nitrogen and

Table 1 Groups used for the in-vitro and in-vivo experiments

Group[a]	Step 1	Time (h)	Step 2	Time (h)
MSCs	Untreated			
HP-MSCs	Untreated		Hypoxia exposure	12
HP-MSCs$^{miR-206}$	Transfected with miR-206 mimics	48	Hypoxia exposure	12
HP-MSCs$^{anti-miR-206}$	Transfected with miR-206 inhibitor	48	Hypoxia exposure	12
HP-MSCs$^{anti-miR-206 + Pim-1\ inhibitor}$	Transfected with miR-206 inhibitor	48	Hypoxia with Pim-1 inhibitor treatment	12

[a]MSCs subjected to different combinations of treatment

Step 1 first step of treatment, *time* length of treatment, *Step 2* second step of treatment, *MSC* mesenchymal stem cell, *HP-MSC* mesenchymal stem cell subjected to hypoxic preconditioning, *miR-206* microRNA-206

powdered. DNA purification was performed using the Genomic DNA Isolation kit (Qiagen, Germantown, MD, USA), and the concentration of the purified DNA was determined by spectrophotometry. The primer sequences for *sry* gene and β-actin were as follows: *sry* gene, forward 5′-GAGGCACAAGTTGGCTCAACA-3′ and reverse 5′-CTCCTGCAAAAAGGGCCTTT-3′; β-actin, forward 5′-CCACCATGTACCCAGGCATT-3′ and reverse 5′-ACTCCTGCTTGCTGATCCAC-3′.

Statistical analysis

All data are shown as the mean ± standard error (SE). Differences between two mean values were evaluated by an unpaired Student two-tailed t test, and between three

or more groups were analyzed using one-way analysis of variance by GraphPad Prism software (GraphPad Software Inc., San Diego, CA, USA). $P < 0.05$ was considered significant.

Results

Pim-1 expression was significantly increased in HP-MSCs

We first performed qPCR to detect the effect of hypoxic preconditioning on the expression of Pim-1 in MSCs. A time-course study was conducted at 0, 6, 12, and 24 h to quantify the Pim-1 expression in MSCs under hypoxic conditions (Fig. 1a). We observed a significant change of Pim-1 expression under low-oxygen conditions in a time-dependent manner. Although mildly decreased at

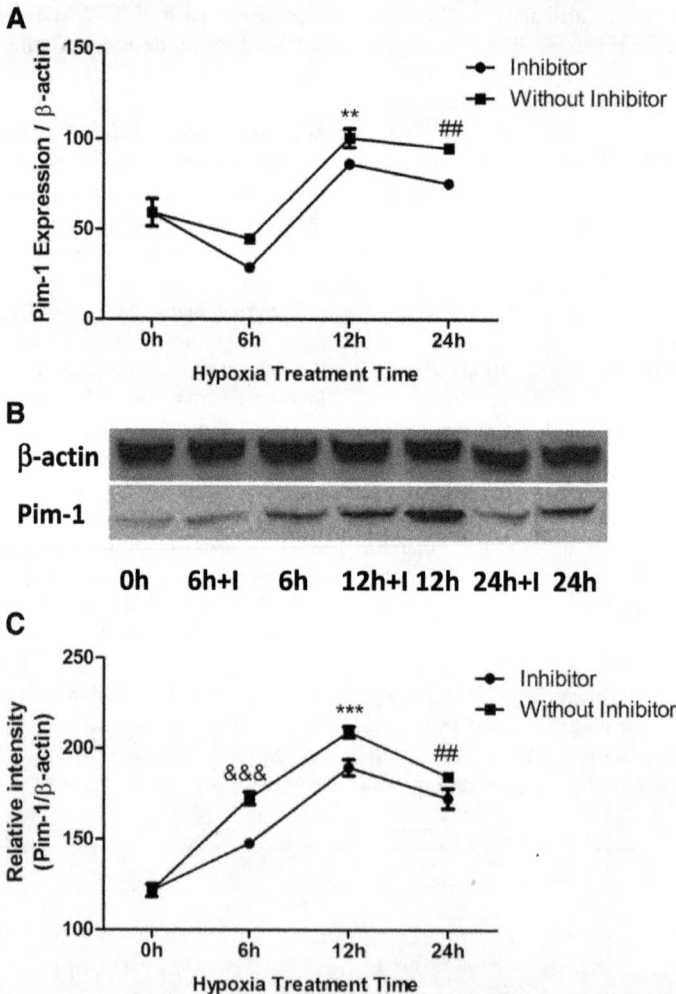

Fig. 1 Hypoxia preconditioning changes the expression of Pim-1 time-dependently. **a** Expression of Pim-1 in MSCs at 0, 6, 12, and 24 h of hypoxia with or without Pim-1 inhibitor detected by qPCR ($n = 4$). **$P < 0.01$ vs 0 h and 6 h; ##$P < 0.01$ vs 0 h and 6 h. **b** Western blot analysis of Pim-1 in MSCs with varied pretreatment. **c** Statistical analysis of Pim-1 intensity ($n = 6$). &&&$P < 0.001$ vs 0 h; ***$P < 0.01$ vs 0 h and 6 h; ##$P < 0.01$ vs 0 h and 12 h

6 h, the level of Pim-1 expression in hypoxic MSCs was increased to about twofold at 12 h ($P < 0.01$ vs 0 h and 6 h), starting to decrease afterwards. It is worthy of note that the expression of Pim-1 was slightly inhibited by Quercetagetin, a specific inhibitor of Pim-1 activity (Fig. 1a). Furthermore, western blot analysis confirmed the tendency of Pim-1 expression (Fig. 1b, c). In contrast with its mRNA expression, the protein expression of Pim-1 significantly increased at 6 h under hypoxic conditions, suggesting a posttranscriptional regulation of the induction of Pim-1.

Hypoxia pretreatment suppressed miR-206 expression and Pim-1 is a putative target of miR-206

miRNAs play a critical role in the posttranscriptional regulation of gene expression, and thus we could determine the miRNAs modulating Pim-1. Three databases (TargetScan, microRNA, and miRDB) were used to predict the putative miRNAs that target Pim-1, and miR-206 was predicted as one of the potential miRNAs

binding on the 3′-UTR of the Pim-1 gene (Fig. 2a). To further confirm the targeting relationship between miR-206 and Pim-1, we first analyzed miR-206 expression using qPCR in HP-MSCs and found it was decreased to 50 ± 13 % under hypoxia vs normoxia (Fig. 2b). Furthermore, we constructed the luciferase reporter vector (Fig. 2c) to perform the luciferase reporter assay. As shown in Fig. 2d, miR-206 significantly decreased the relative luciferase reporter activity of the wild-type Pim-1 3′-UTR, whereas that of the mutant Pim-1 3′-UTR did not change significantly, which suggests that miR-206 could directly bind to the 3′-UTR of Pim-1. We then transfected MSCs with miR-206 mimics or miR-206 inhibitors, and over 90 % of the cells were successfully transfected. The expression of miR-206 was significantly upregulated or downregulated after transfection with mimics and inhibitors, respectively (Fig. 2e). qPCR showed that the inhibition of miR-206 in MSCs was concurrent with the increased expression of Pim-1 and vice versa (Fig. 2f). These results provided clear evidence that hypoxia pretreatment

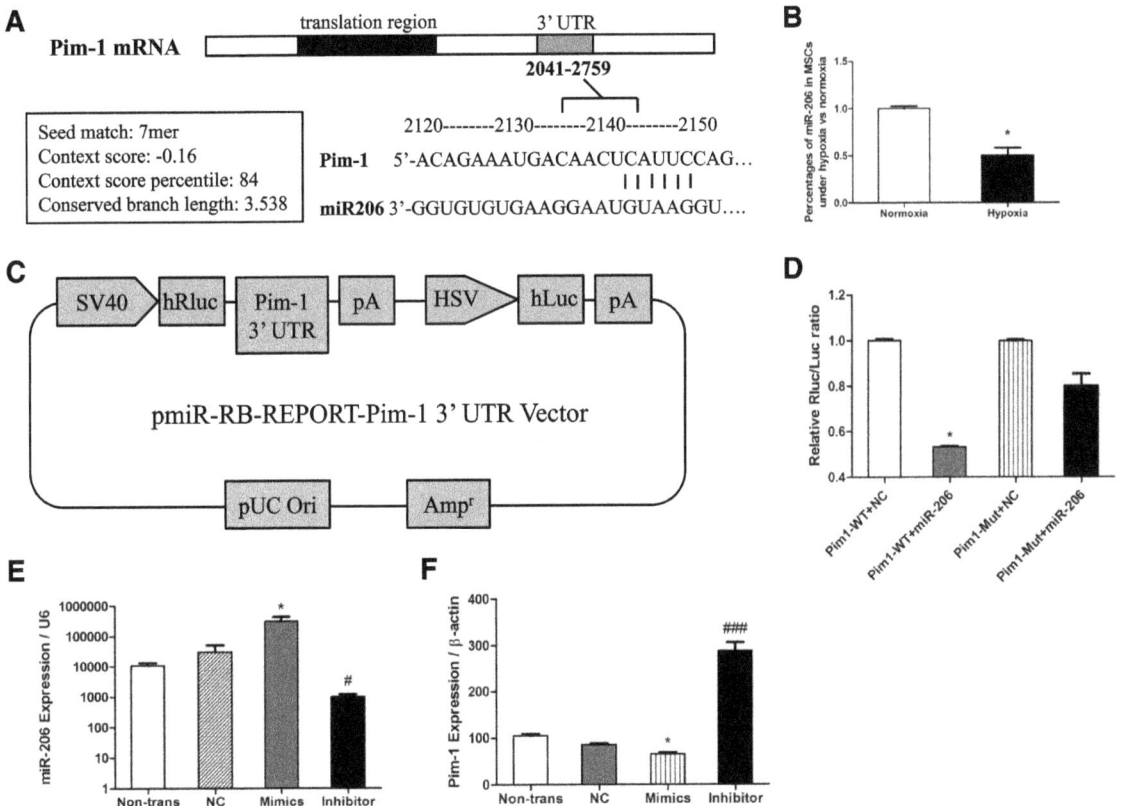

Fig. 2 Pim-1 is a putative target of miR-206 in MSCs. **a** Binding motif (2136–2142) of miR-206 on Pim-1 3′-UTR (2041–2759). **b** Percentages of miR-206 expression in MSCs under hypoxia vs normoxia ($n = 3$). *$P < 0.05$ vs normoxia. **c** Construction of luciferase construct. **d** Relative Rluc/Luc ratio ($n = 3$). *$P < 0.05$ vs Pim-1-WT + NC. **e** Expression of miR-206 in MSCs after transfection of mimics and inhibitors ($n = 6$). *$P < 0.05$ vs non-trans, NC, and inhibitor; #$P < 0.05$ vs non-trans and NC. **f** Expression of Pim-1 in MSCs after transfection of miR-206 mimics and inhibitors ($n = 6$). *$P < 0.05$ vs non-trans; ###$P < 0.001$ vs non-trans, NC negative control and mimics. *miR-206* microRNA-206, *MSC* mesenchymal stem cell, *UTR* untranslated region, *WT* wild type

suppressed the expression of miR-206 and Pim-1 is a putative target of miR-206.

Abrogation of miR-206 in HP-MSCs promotes the migration ability of the cells

The transwell assay was conducted to examine the migration abilities of MSCs under different conditions. MSCs without any pretreatment were used as a baseline control for all other experimental groups. As shown in Fig. 3a, the migration ability of HP-MSCs had a significant increase ($P < 0.01$ vs MSCs), but was inhibited when overexpressing miR-206 in HP-MSCs ($P < 0.001$ vs HP-MSCs). Besides, HP-MSCs with the inhibition of miR-206 had much more migrated cells, but when the cells were treated with Quercetagetin at the same time, migrated cells were reduced by about 70 % ($P < 0.001$ vs HP-MSCs$^{anti-miR-206}$) (Fig. 3b). This suggests that hypoxic preconditioning could promote the migration ability of MSCs by inhibiting miR-206 expression, and Pim-1 is a downstream target of miR-206.

Cytoprotective effects of Pim-1/miR-206 in HP-MSCs

To further study the role of miR-206 and Pim-1 in cytoprotection of HP-MSCs, we examined the apoptosis of MSCs in different groups by flow cytometry analysis (Fig. 4a). Hypoxic pretreatment slightly delayed the early apoptosis of the cells ($P > 0.05$ vs MSCs). However, HP-MSCs overexpressing miR-206 showed a higher apoptosis rate ($P < 0.001$ vs HP-MSCs) (Fig. 4b). In addition, Pim-1 inhibition in HP-MSCs after anti-miR-206 transfection significantly accelerated the early apoptosis of the cells ($P < 0.05$ vs HP-MSCs$^{anti-miR-206}$), which further confirmed the importance of miR-206 and Pim-1 in cytoprotection of HP-MSCs and the biological relationship between them.

Inhibition of miR-206 in HP-MSCs protects the membrane potential of mitochondria

The protective effect of Pim-1 in apoptosis via the mitochondrial pathway has been widely studied [12]. Therefore, we performed JC-1 staining to examine the protective effects of miR-206 and Pim-1 on mitochondrial integrity. The red fluorescence of JC-1 is caused by a potentially dependent aggregation in the mitochondria, reflecting $\Delta\psi$m. Green emission of the dye represents the monomeric form of JC-1. Mitochondrial depolarization is indicated by a decrease in the red/green fluorescence intensity ratio. As shown in Fig. 5a and 5b, hypoxia exposure of MSCs significantly decreased mitochondria depolarization compared with untreated MSCs ($P < 0.05$ vs MSCs). However, the red/green ratio of JC-1 was remarkably decreased when HP-MSCs were transfected with miR-

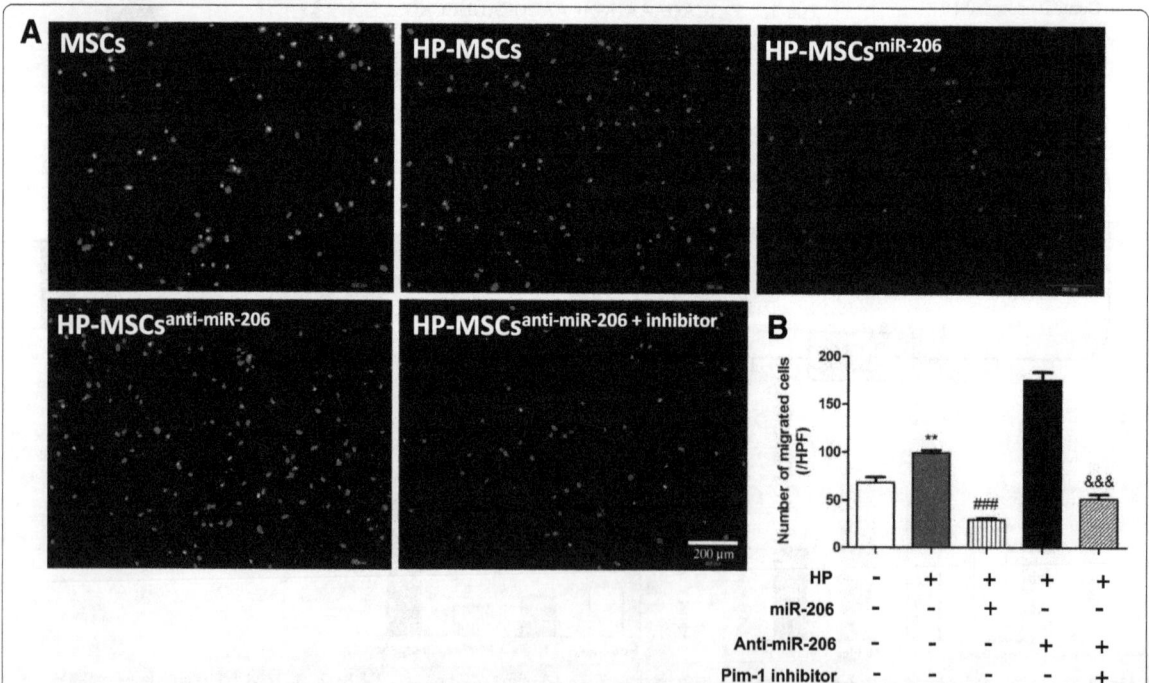

Fig. 3 Hypoxia exposure of MSCs promotes the migration ability of the cells by modulation of miR-206/Pim-1. **a** Representative images of migrated cells stained with DAPI. *Bar*, 200 μm. **b** Numbers of migrated cells per high-power field (*HPF*) were statistically analyzed ($n = 4$). **$P < 0.01$ vs MSCs; $^{###}P < 0.001$ vs HP-MSCs; $^{&&&}P < 0.001$ vs HP-MSCs$^{anti-miR-206}$. *miR-206* microRNA-206, *MSC* mesenchymal stem cell, *HP-MSC* MSC subjected to hypoxic preconditioning

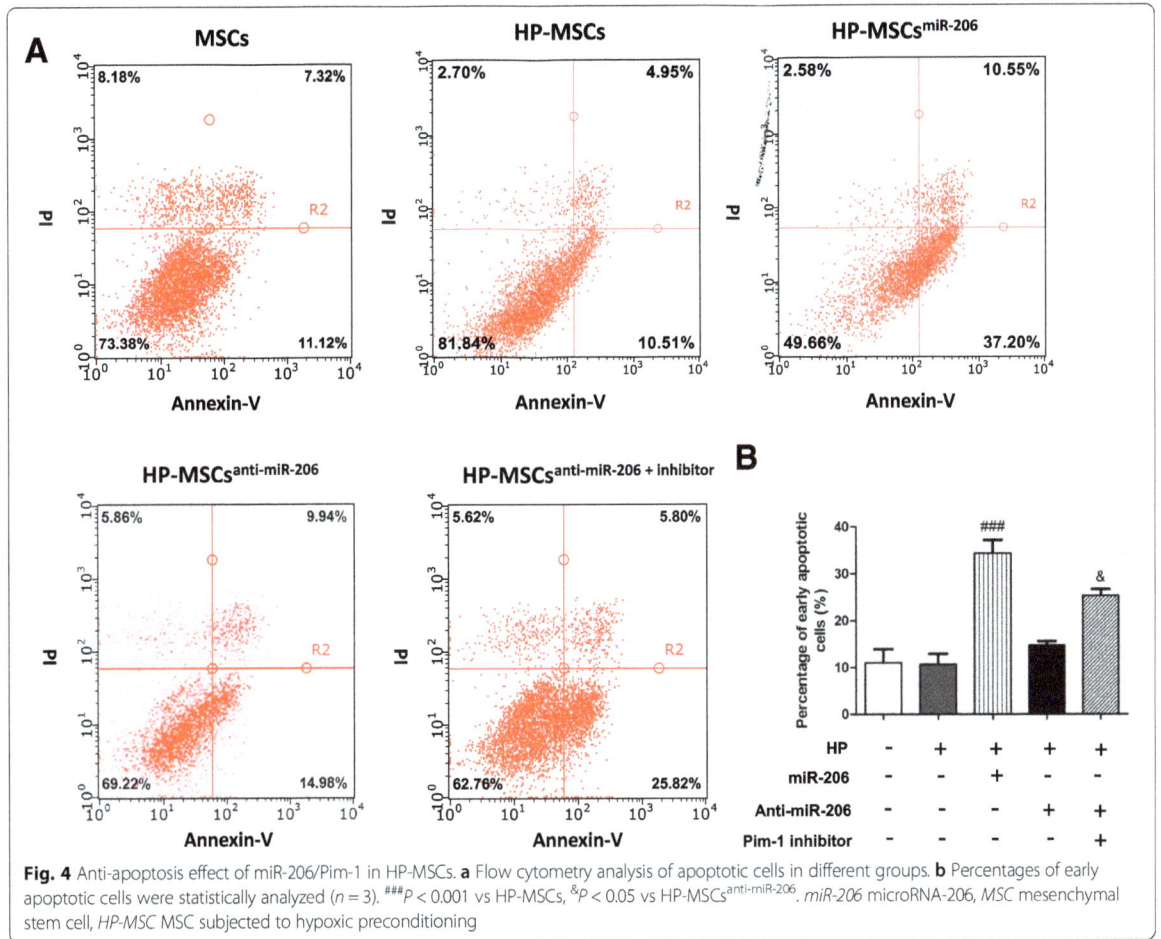

Fig. 4 Anti-apoptosis effect of miR-206/Pim-1 in HP-MSCs. **a** Flow cytometry analysis of apoptotic cells in different groups. **b** Percentages of early apoptotic cells were statistically analyzed ($n = 3$). $^{###}P < 0.001$ vs HP-MSCs, $^{&}P < 0.05$ vs HP-MSCs$^{anti-miR-206}$. *miR-206* microRNA-206, *MSC* mesenchymal stem cell, *HP-MSC* MSC subjected to hypoxic preconditioning

206 mimics ($P < 0.01$ vs HP-MSCs). In addition, Pim-1 inhibition in HP-MSCs after anti-miR-206 transfection decreased the red/green ratio of JC-1 from 20 ± 7 % (HP-MSCs$^{anti-miR-206}$) to 3 ± 0.6 % (HP-MSCs$^{anti-miR-206 + inhibitor}$) ($P < 0.001$).

HP-MSCs with lower miR-206 showed improved survival in the infarcted heart

MSCs from male donors were transplanted to female rats subjected to myocardial infarction. The survival of transplanted cells was evaluated by detection of *Sry* gene in the ischemic hearts as described previously [13]. Hearts were collected 4 days after cell transplantation, and cell survival was examined (Fig. 6a). In vivo, MSCs survived more by hypoxic preconditioning ($P < 0.05$ vs MSCs), which was prevented by overexpression of miR-206 ($P < 0.01$ vs HP-MSCs) (Fig. 6b). As it was observed in vitro, inhibition of miR-206 in HP-MSCs showed improved survival, which was prevented by Pim-1 inhibitor ($P < 0.001$ vs HP-MSCs$^{anti-miR-206}$).

Discussion

The important findings of our study include the following: Pim-1 kinase is upregulated in MSCs under hypoxic conditions; miR-206 plays a mechanistic role in the migration and survival of MSCs via its putative target Pim-1; the prosurvival effect of miR-206/Pim-1 maintains the mitochondria membrane potential during hypoxic treatment of MSCs; and abrogation of miR-206 in hypoxic MSCs enhanced survival of the cells in the ischemic myocardium.

The role of Pim-1, a proto-oncogenic serine–threonine kinase, in cardiac development has been overlooked for a long time. Dr Sussman first discovered that Pim-1, downstream of Akt, regulates cardiomyocyte survival [3]. Regenerative therapies utilizing stem/progenitors cells engineered with Pim-1 enhanced regenerative potential of the cells [4, 8, 14], thus making Pim-1 an important player in the treatment of severe heart failure. Very recently, it was reported that hypoxic preconditioning increases survival of cardiac progenitor cells via

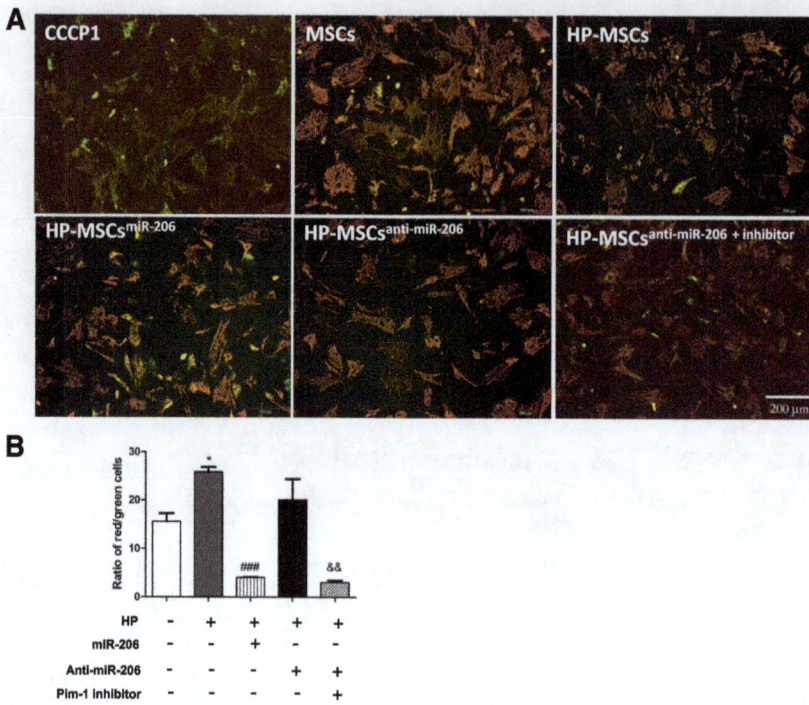

Fig. 5 Inhibition of miR-206 in HP-MSCs protects membrane potential of mitochondria. **a** Representative images of mitochondria membrane potential stained by JC-1. Cells treated with CCCP-1 served as a control. *Bar*, 200 μm. **b** Ratios of red/green cells were calculated and analyzed ($n = 4$). *$P < 0.05$ vs MSCs; ###$P < 0.01$ vs HP-MSCs; &&$P < 0.001$ vs HP-MSCs[anti-miR-206]. *miR-206* microRNA-206, *MSC* mesenchymal stem cell, *HP-MSC* MSC subjected to hypoxic preconditioning

upregulation of Pim-1, discovering a method for nongenetic modification of Pim-1 in stem cells. Another report [15] also claimed that Pim-1 could promote MSC proliferation and prevent MSC apoptosis. Thus we raised a proposal that hypoxic preconditioning could increase survival of MSCs via upregulation of Pim-1. Not surprisingly, we demonstrated that Pim-1 was gradually increased, reaching a peak at 12 h of hypoxia. Next, we aimed to determine the miRNAs that modulate the expression of Pim-1 in MSCs under hypoxic conditions. Through the three

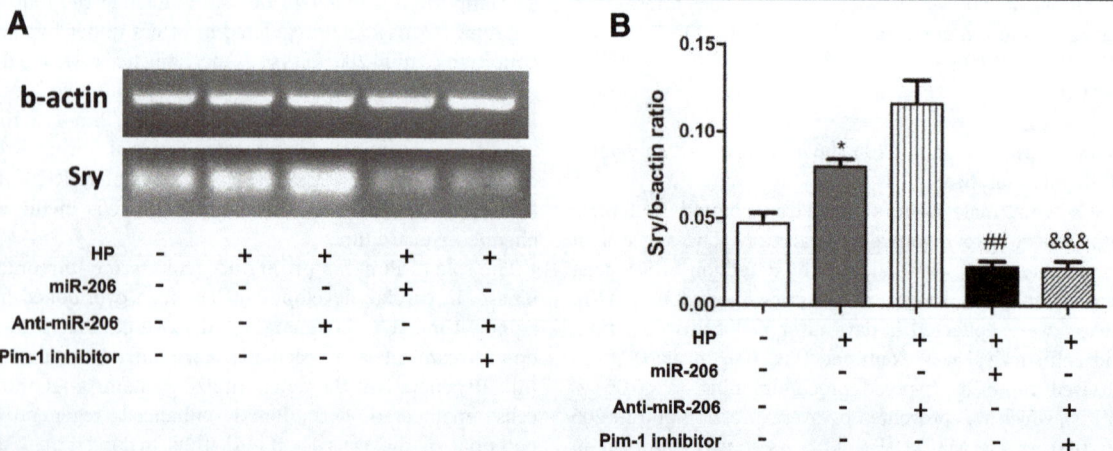

Fig. 6 HP-MSCs with lower miR-206 showed improved survival in the infarcted heart. **a** Electrophoresis analysis of *sry* gene in the infarcted hearts 4 days post transplantation. **b** Ratios of *Sry*/β-actin were calculated and analyzed ($n = 4$). *$P < 0.05$ vs MSCs; ##$P < 0.01$ vs HP-MSCs; &&&$P < 0.001$ vs HP-MSCs[anti-miR-206]. *HP* hypoxic preconditioning, *miR-206* microRNA-206

target gene prediction software packages, we predicted that miR-206, miR-328, miR-327, miR-532-3p, and miR-760-3p had putative binding sites on the 3'-UTR region of Pim-1. Through literature searching, we screened out miR-206 and miR-328 since they were reported to be downregulated in hypoxic situations [16, 17]. miR-206 was further confirmed to be downregulated in hypoxic MSCs, while miR-328 was not changed (data not shown). Next, the targeting relationship of miR-206 and Pim-1 was validated by qPCR and luciferase reporter activity.

Although little is known about miR-206 for its function during the preconditioning of stem cells, miR-206 has been mostly studied for its association with the pathogenesis of human cancers. A number of studies have shown that miR-206 is frequently downregulated in many human malignancies, including colorectal cancer [18], cervical cancer [19], lung cancer [20], gastric cancer [21], and breast cancer [22], and is associated with a malignant phenotype. These studies have validated that upregulation of miR-206 inhibited cancer cell proliferation and migration, blocked the cell cycle, and activated apoptosis. In our study, we observed that during the hypoxic preconditioning of MSCs, the inhibition of miR-206 had anti-apoptotic and migration-promoting effects on the MSCs, which were blocked either by overexpression of miR-206 or by abrogation of Pim-1 activity using the specific inhibitor, consistent with the observations in cancer research. In HeLa and C2C12 cells, miR-206 targets NOTCH3 expression to induce cell-cycle arrest and the inhibition of tumor cell migration [23]. It was also shown that through the c-Met pathway, miR-206 effectively inhibited gastric cancer progression [21]. In this study we provide the first evidence that hypoxic preconditioning regulated the expression of miR-206, which plays an essential role in mediating migration and apoptosis by targeting Pim-1 kinase in MSCs.

A loss of mitochondrial membrane potential is the initial manifestation of mitochondrial damage. In the apoptosis pathway, mitochondrial membrane potential was impaired first, and then apoptosis cascade starts [24, 25]. In this study, the protective effect of Pim-1/miR-206 in HP-MSCs maintains the membrane potential of mitochondria, slowing down the progression of apoptosis. In addition, hypoxic treatment significantly enhanced survival of MSCs after transplantation in ischemic hearts, while the pretreatment of MSCs with miR-206 and Pim-1 inhibitor significantly abolished the cytoprotective effects of preconditioning.

Conclusions

We demonstrate that miR-206 is modulated by hypoxic preconditioning of MSCs through targeting Pim-1 kinase, elucidating a new aspect of prosurvival signaling in hypoxia. Moreover, the protective effects of miR-206/

Pim-1 in HP-MSCs supported their short-term survival under ischemic stress after transplantation. miR-206/Pim-1 may serve as a molecular therapeutic strategy in stem cell translational research.

Availability of data and materials
The datasets supporting the conclusions of this article are included within the article.

Abbreviations
AMI: acute myocardial infarction; $\Delta\psi$m: mitochondrial membrane potential; HP-MSC: mesenchymal stem cell subjected to hypoxic preconditioning; miR-206: microRNA-206; miRNA: microRNA; MSC: mesenchymal stem cell; qPCR: quantitative PCR; UTR: untranslated region; WT: wild type.

Competing interests
The authors declare that they have no competing interests.

Authors' contributions
YZ, WL, WY, XL, XW, and JY were responsible for performance of experiment, data analysis, and manuscript preparation. ZZ, JH, ZS, and JY were responsible for manuscript writing and revision, and experimental design. All authors read and approved the manuscript.

Sources of funding
This work was supported by National Natural Science Foundation of China (No. 81400199 and No. 30972696), Natural Science Foundation of Jiangsu Province (BK20150321), and Suzhou Municipal Science and Technology Project of China (No. SYS201414 and No. SYS201540).

Author details
[1]Department of Cardiology of the First Affiliated Hospital, Soochow University, Suzhou, China. [2]Institute for Cardiovascular Science & Department of Cardiovascular Surgery of The First Affiliated Hospital, Soochow University, Suzhou, China. [3]Department of Cardiovascular Surgery, Affiliated Shanghai 1st People's Hospital, Shanghai Jiaotong University, Shanghai, China. [4]Department of Thoracic and Cardiovascular Surgery, Northern Jiangsu People's Hospital, Yangzhou, China.

References
1. Wang Z, Bhattacharya N, Weaver M, Petersen K, Meyer M, Gapter L, et al. Pim-1: a serine/threonine kinase with a role in cell survival, proliferation, differentiation and tumorigenesis. J Vet Sci. 2001;2:167–79.
2. Aho TL, Sandholm J, Peltola KJ, Mankonen HP, Lilly M, Koskinen PJ. Pim-1 kinase promotes inactivation of the pro-apoptotic Bad protein by phosphorylating it on the Ser112 gatekeeper site. FEBS Lett. 2004;571:43–9.
3. Muraski JA, Rota M, Misao Y, Fransioli J, Cottage C, Gude N, et al. Pim-1 regulates cardiomyocyte survival downstream of Akt. Nat Med. 2007;13:1467–75.
4. Fischer KM, Cottage CT, Wu W, Din S, Gude NA, Avitable D, et al. Enhancement of myocardial regeneration through genetic engineering of cardiac progenitor cells expressing Pim-1 kinase. Circulation. 2009;120:2077–87.
5. Cottage CT, Bailey B, Fischer KM, Avitable D, Collins B, Tuck S, et al. Cardiac progenitor cell cycling stimulated by pim-1 kinase. Circ Res. 2010;106:891–901.
6. Karantalis V, Schulman IH, Hare JM. Nitroso-redox imbalance affects cardiac structure and function. J Am Coll Cardiol. 2013;61:933–5.
7. Mohsin S, Khan M, Nguyen J, Alkatib M, Siddiqi S, Hariharan N, et al. Rejuvenation of human cardiac progenitor cells with Pim-1 kinase. Circ Res. 2013;113:1169–79.
8. Quijada P, Toko H, Fischer KM, Bailey B, Reilly P, Hunt KD, et al. Preservation of myocardial structure is enhanced by pim-1 engineering of bone marrow cells. Circ Res. 2012;111:77–86.
9. Hu S, Yan G, Xu H, He W, Liu Z, Ma G. Hypoxic preconditioning increases survival of cardiac progenitor cells via the pim-1 kinase-mediated anti-apoptotic effect. Circ J. 2014;78:724–31.

10. Jalali S, Ramanathan GK, Parthasarathy PT, Aljubran S, Galam L, Yunus A, et al. Mir-206 regulates pulmonary artery smooth muscle cell proliferation and differentiation. PLoS One. 2012;7:e46808.

11. Zhang Z, Yang J, Yan W, Li Y, Shen Z and Asahara T. Pretreatment of cardiac stem cells with exosomes derived from mesenchymal stem cells enhances myocardial repair. J Am Heart Assoc. 2016; 5:1-18. doi: 10.1161/JAHA.115.002856.

12. Sussman MA. Mitochondrial integrity: preservation through Akt/Pim-1 kinase signaling in the cardiomyocyte. Expert Rev Cardiovasc Ther. 2009;7:929–38.

13. Kim HW, Mallick F, Durrani S, Ashraf M, Jiang S, Haider KH. Concomitant activation of miR-107/PDCD10 and hypoxamir-210/Casp8ap2 and their role in cytoprotection during ischemic preconditioning of stem cells. Antioxid Redox Signal. 2012;17:1053–65.

14. Mohsin S, Khan M, Toko H, Bailey B, Cottage CT, Wallach K, et al. Human cardiac progenitor cells engineered with Pim-I kinase enhance myocardial repair. J Am Coll Cardiol. 2012;60:1278–87.

15. Zhao Y, Hu J, Buckingham B, Pittenger MF, Wu ZJ, Griffith BP. Pim-1 kinase cooperates with serum signals supporting mesenchymal stem cell propagation. Cells Tissues Organs. 2014;199:140–9.

16. Guo L, Qiu Z, Wei L, Yu X, Gao X, Jiang S, et al. The microRNA-328 regulates hypoxic pulmonary hypertension by targeting at insulin growth factor 1 receptor and L-type calcium channel-alpha1C. Hypertension. 2012;59:1006–13.

17. Yue J, Guan J, Wang X, Zhang L, Yang Z, Ao Q, et al. MicroRNA-206 is involved in hypoxia-induced pulmonary hypertension through targeting of the HIF-1alpha/Fhl-1 pathway. Lab Invest. 2013;93:748–59.

18. Wang XW, Xi XQ, Wu J, Wan YY, Hui HX, Cao XF. MicroRNA-206 attenuates tumor proliferation and migration involving the downregulation of NOTCH3 in colorectal cancer. Oncol Rep. 2015;33:1402–10.

19. Ling S, Ruiqin M, Guohong Z, Bing S, Yanshan C. Decreased microRNA-206 and its function in cervical cancer. Eur J Gynaecol Oncol. 2015;36:716–21.

20. Zhang YJ, Xu F, Li HB, Han JC, Li L. miR-206 inhibits non small cell lung cancer cell proliferation and invasion by targeting SOX9. Int J Clin Exp Med. 2015;8:9107–13.

21. Zheng Z, Yan D, Chen X, Huang H, Chen K, Li G, et al. MicroRNA-206: effective Inhibition of gastric cancer progression through the c-Met pathway. PLoS One. 2015;10:e0128751.

22. Ge X, Lyu P, Cao Z, Li J, Guo G, Xia W, et al. Overexpression of miR-206 suppresses glycolysis, proliferation and migration in breast cancer cells via PFKFB3 targeting. Biochem Biophys Res Commun. 2015;463:1115–21.

23. Song G, Zhang Y, Wang L. MicroRNA-206 targets notch3, activates apoptosis, and inhibits tumor cell migration and focus formation. J Biol Chem. 2009;284:31921–7.

24. Milani L. Mitochondrial membrane potential: a trait involved in organelle inheritance? Biol Lett 2015; 11:1-4. doi: 10.1098/rsbl.2015.0732.

25. Zhang BB, Wang DG, Guo FF, Xuan C. Mitochondrial membrane potential and reactive oxygen species in cancer stem cells. Fam Cancer. 2015;14:19–23.

6

Genome-wide gene expression analyses reveal unique cellular characteristics related to the amenability of HPC/HSCs into high-quality induced pluripotent stem cells

Shuai Gao[2,3,4], Li Tao[4], Xinfeng Hou[2], Zijian Xu[2], Wenqiang Liu[2,5], Kun Zhao[5], Mingyue Guo[5], Hong Wang[2,5], Tao Cai[2], Jianhui Tian[4], Shaorong Gao[2,5] and Gang Chang[1,2*]

Abstract

Background: Transcription factor-mediated reprogramming can efficiently convert differentiated cells into induced pluripotent stem cells (iPSCs). Furthermore, many cell types have been shown to be amenable to reprogramming into iPSCs, such as neural stem cells, hematopoietic progenitor and stem cells (HPC/HSCs). However, the mechanisms related to the amenability of these cell types to be reprogrammed are still unknown.

Methods: Herein, we attempt to elucidate the mechanisms of HPC/HSC reprogramming using the sequential reprogramming system that we have previously established.

Results: We found that HPC/HSCs were amenable to transcription factor-mediated reprogramming, which yielded a high frequency of fully reprogrammed HPC/HSC-iPSCs. Genome-wide gene expression analyses revealed select down-regulated tumor suppressor and mesenchymal genes as well as up-regulated oncogenes in HPC/HSCs compared with mouse embryonic fibroblasts (MEFs), indicating that these genes may play important roles during the reprogramming of HPC/HSCs. Additional studies provided insights into the contribution of select tumor suppressor genes (*p21*, *Ink4a* and *Arf*) and an epithelial-to-mesenchymal transition (EMT) factor (*Snail1*) to the reprogramming process of HPC/HSCs.

Conclusions: Our findings demonstrate that HPC/HSCs carry unique cellular characteristics, which determine the amenability of HPC/HSCs to be reprogrammed into high-quality iPSCs.

Keywords: Hematopoietic progenitor and stem cells, Induced pluripotent stem cells, Pluripotency, Reprogramming

Background

The ectopic expression of a set of transcription factors, including Oct4, Sox2, Klf4, and c-Myc (OSKM), has been shown to convert somatic cells into induced pluripotent stem cells (iPSCs) [1]. The generation of iPSCs has tremendous therapeutic potential for making patient-specific iPSCs available [2–4]. To date, many types of cells, including embryonic fibroblasts, hepatocytes, gastric epithelial cells, adult tail-tip fibroblasts (TTFs), pancreatic cells,

neural stem cells, B lymphocytes, T lymphocytes, and trophoblast stem cells, have been successfully converted into iPSCs [5–11]. In addition, iPSCs are indistinguishable from embryonic stem cells (ESCs) with respect to global transcription and epigenetic modifications and can achieve pluripotency. These findings have led to enthusiasm for conducting proof-of-principle safety evaluations for the therapeutic use of iPSCs [12–15]. The availability of high-quality iPSCs is a prerequisite for their potential therapeutic use.

During the reprogramming process, the epigenetic transition from starting cells to pluripotent cells can be achieved when cells reach an acquired pluripotency state [16]. Accumulating evidence suggests that the dynamics

* Correspondence: changgang@szu.edu.cn
[1]Institute of Molecular Medicine, Health Science Center, Shenzhen University, Shenzhen 518060, China
[2]National Institute of Biological Sciences, NIBS, Beijing 102206, China
Full list of author information is available at the end of the article

of nuclear reprogramming may be diverse; a typical example can be found in the reprogramming of B lymphocytes [8, 9]. Many studies have established that substitutes for the traditional four reprogramming factors, now known as "Yamanaka factors", can be used to induce pluripotency [17, 18], indicating the complexity of the versatile reprogramming process. Likewise, many developmental events, such as the mesenchymal-to-epithelial transition (MET), have also been found to coincide with the nuclear reprogramming process [19]. The $Ink4a/Arf$ locus, which encodes three tumor suppressor genes (p16^{Ink4a} and p19^{Arf} from $Cdkn2a$ and p15^{Ink4b} from $Cdkn2b$), was found to be a barrier to reprogramming [20, 21], further emphasizing the relationship between immortalization and reprogramming. Furthermore, a negative correlation has been revealed between excessive cell proliferation and reprogramming efficiency in human somatic cells [22]. Therefore, understanding the molecular events of reprogramming in different cell types to elucidate the core circuit of nuclear reprogramming may allow more effective interventions for the reprogramming of cells.

Hematopoietic cells belong to a well-defined population of cells that have been rigorously defined based on the expression of cluster of differentiation (CD) molecules. These cells have been used widely in nuclear transfer-mediated and transcription factor-mediated cell reprogramming studies [10, 23]. One of the most notable breakthroughs in nuclear transfer-mediated reprogramming was carried out in terminally differentiated B and T lymphocytes [24], supporting the notion that the presence of residual adult stem cells in tissues does not account for the observed successful reprogramming of mammalian somatic cells [25]. Compared with somatic cells, the epigenetic state of neural stem cells has been proven amenable to nuclear reprogramming [7]. Additionally, hematopoietic progenitor and stem cells (HPC/HSCs) were found to be susceptible to transcription factor-mediated reprogramming [10]. However, the mechanisms related to the amenability of HPC/HSCs to be reprogrammed remain unknown.

Herein, we attempt to determine the mechanisms related to the amenability of HPC/HSCs to be reprogrammed by testing the potential of HPC/HSCs to be reprogrammed into high-quality iPSCs using our previously established sequential reprogramming system [26] and performing genome-wide gene expression analyses to determine the underlying events correlated with HPC/HSC reprogramming. Our results demonstrated that the downregulation of select tumor suppressor genes and an epithelial-to-mesenchymal transition (EMT) factor as well as the upregulation of oncogenes in HPC/HSCs undergo a unique reprogramming process through which HPC/HSCs are amenable to be reprogrammed into iPSCs. Further studies showed that tumor suppressor genes ($p21$, $Ink4a$, and Arf) and an EMT factor ($Snail 1$) participated in the reprogramming of HPC/HSCs, in which independent ectopic activation of $p21$, $Ink4a$, Arf, and $Snail1$ along with OSKM in HPC/HSCs decreased the reprogramming efficiency.

Methods

Animal welfare

The protocols of all animal experiments were approved by the Animal Care and Use Committee of the National Institute of Biological Sciences, Beijing, China. All animal procedures were performed according to the National Institute of Biological Sciences Guide for the Care and Use of Laboratory Animals.

Isolation of HPC/HSCs

HPC/HSCs were isolated from tetraploid-complementation (4N) mice derived from mouse embryonic fibroblasts (MEFs) with a 129S2/Sv genetic background and a Rosa26-M2rtTA transgene [27]. In the isolation procedure, the 4N mice were euthanized, after which the tibia and femur were dissected from both legs and maintained in ice-cold PBE (phosphate-buffered saline (PBS) containing 0.5 % bovine serum albumin and 2 mM ethylenediamine tetraacetic acid). The muscles were removed from the bones using sharp surgical scissors; a 5 ml syringe containing ice-cold PBE was then inserted into one end of the bone, and the bone marrow was extruded into a 5 ml tube. After thorough mixing of the cell suspension, the cells were passed through a 70 μm nylon mesh filter into a fresh 5 ml tube to remove any cell clumps. The cell suspension was centrifuged at $300 \times g$ for 10 minutes at 4 °C, the supernatant was discarded, and the cell pellet was resuspended in 80 μl PBE per 10^8 total cells. Then, 20 μl of CD117 MicroBeads (Miltenyi, Bergisch Gladbach, Germany) was added to the cell suspension and incubated on ice for 15 minutes. The cells were washed twice with PBE in a final volume of 500 μl. Finally, the cell suspension was transferred to a PBE-pretreated MS column (Miltenyi, Bergisch Gladbach, Germany) under a magnetic field (MACS; Miltenyi, Bergisch Gladbach, Germany), and the magnetically labeled cells were flushed into PBE. The nucleated cells were centrifuged at $500 \times g$ for 10 minutes.

Flow cytometry

HSC/HPCs isolated by MACS were incubated with APC-CD117 (c-kit; eBioscience) and FITC-CD45.2 (eBioscience, San Diego, CA) and analyzed using LSR II (BD Biosciences, San Jose, CA) as described previously [28]. Flow cytometric analysis was performed for the cell proliferation rate using BD Pharmingen™ BrdU Flow Kits (BD Biosciences, San Jose, CA) according to the manufacturer's instructions.

Generation of HPC/HSC-iPSCs and cell culture

The generation of HPC/HSC-iPSCs was performed under the sequential reprogramming system we established [26]. In detail, 5×10^4 HPC/HSCs were transferred to 3.5 cm dishes with ES medium containing 50 ng/ml murine stem cell factor (SCF; Peprotech, Rocky Hill, NJ), 10 ng/ml murine interleukin (IL)-3 (Peprotech, Rocky Hill, NJ), and 10 ng/ml murine IL-6 (Peprotech, Rocky Hill, NJ). Twenty-four hours later, the medium was replaced with ES medium containing 1 µg/ml doxycycline (Dox; Sigma, St. Louis, MO) to induce the expression of OSKM under the regulation of tetracycline response elements (TRE). Dox was removed on day 14. Two days after the withdrawal of Dox, ESC-like colonies were picked and passaged three days later to yield HPC/HSC-iPSCs. All ESCs and iPSCs were cultured on mitomycin C-treated (Sigma, St. Louis, MO) MEFs in ES medium, which consisted of Dulbecco modified Eagle's medium (DMEM; Invitrogen, Carlsbad, CA) supplemented with 15 % fetal bovine serum (FBS; Hyclone, South Logan, Utah), 1 mM L-glutamine (Invitrogen, Carlsbad, CA), 0.1 mM β-mercaptoethanol (Invitrogen, Carlsbad, CA), 1 % nonessential amino acid (Invitrogen, Carlsbad, CA), and 1000 U/ml leukemia inhibitory factor (LIF; Millipore, Darmstadt, Germany).

Quantitative PCR

We extracted mRNA using TRIzol (Invitrogen, Carlsbad, CA) and reverse-transcribed the mRNA using M-MLV reverse transcriptase (Promega, Madison, WI). Quantitative PCR (Q-PCR) was carried out with SYBR Green-based PCR Master Mix (Takara, Shiga, Japan). A total volume of 20 µl containing 10 µl SYBR Green-based PCR Master Mix, 0.2 mM dNTP, 0.2 µl forward primer (10 mM), 0.2 µl reverse primer (10 mM), and 0.2 µl dye II was mixed and plated for gene expression analyses using the relative quantitation (RQ) of gene expression of the Applied Biosystems 7500 Fast Real-Time PCR System (Thermo Fisher Scientific, Waltham, MA) in accordance with the manufacturer's instructions. One independent experiment contained three replicates of both targeted genes and inner control. The results of three independent experiments in duplicate were averaged to calculate the mean value of every gene. Relative expression levels of target genes in each cell line were normalized to the level of their endogenous *Gapdh*. Paired Student's t tests were performed to assess the statistical difference. The significant standard was set as: fold-change >2; $P <0.05$. The primer pairs for real-time PCR are summarized in Additional file 1: Table S1.

Alkaline phosphatase and immunofluorescence staining

Alkaline phosphatase (AP) staining was performed using the Alkaline Phosphatase Detection Kit (Millipore, Darmstadt, Germany) according to the manufacturer's instructions. Immunofluorescence staining was performed as described previously [29].

Teratoma formation

For the teratoma formation assay, iPSCs suspended in PBS were injected subcutaneously into the forelimbs of SCID mice. SCID mice were sacrificed 3 or 4 weeks after the injection to collect the tumors, which were further dissected for hematoxylin and eosin (H & E) staining to identify the three germ layers.

Bisulfite genomic sequencing

Bisulfite genomic sequencing was conducted in triplicate to analyze the DNA methylation of *Pou5f1* and *Nanog* as described previously [29]. The bisulfite PCR primer pairs are summarized in Additional file 1: Table S1. The amplified PCR products were cloned into a vector using the pEASYTM-T5 Zero cloning kit (TransGen, Beijing, China) and were sequenced by Invitrogen and Sangon-Biotech (Sangon, Shanghai, China).

Generation of chimera and 4N mice

To generate chimera (2N) mice, 10–15 iPSCs were microinjected into eight-cell stage ICR embryos using a piezo-actuated microinjection pipette. The reconstructed embryonic day 2.5 embryos were then transplanted into the uteri of pseudo-pregnant mice. Tetraploid complementation was similarly performed using piezo-actuated microinjection. The two-cell stage ICR embryos were first electrofused into tetraploid embryos and cultured to blastocysts. Then, 10–15 iPSCs were injected into the cavum of the tetraploid blastocysts, which were then transplanted into the uteri of pseudo-pregnant mice. A cesarean section was performed at embryonic day 19.5, and the resultant pups were fostered by lactating ICR mothers.

Simple sequence length polymorphism

Primers for simple sequence length polymorphism (SSLP) were selected according to the Mouse Genome Informatics website (http://www.informatics.jax.org) and performed as reported previously [26, 30].

Western blot

Whole cell extracts were prepared using RIPA buffer and ultrasonic extraction, resolved on SDS-PAGE gels, and transferred to polyvinylidene fluoride membranes. Specific proteins were analyzed using anti-CDKN2A/p16INK4a (Abcam, Cambridge, UK), anti-CDKN2A/p19ARF (Abcam, Cambridge, UK), anti-SNAI1 (Millipore, Darmstadt, Germany), and anti-Gapdh (Sigma, St. Louis, MO). Enhanced chemiluminescence peroxidase-labeled anti-mouse, rabbit or goat antibodies (Amersham, Pittsburgh, PA) were used for further detection.

Gene construction and transient and stable transfection

The cDNAs of *Ink4a*, *Arf*, *Ink4b*, *p21*, and *Snail1* were cloned separately into the TetO-FUW plasmid, and the reconstructed plasmids were transfected into 293T cells along with lentivirus packaging plasmids ps-PAX-2 and pMD2G. Viral supernatants were harvested 48 hours after transfection, filtered through a 0.45 μm filter (Millipore, Darmstadt, Germany), and concentrated by centrifugation. HPC/HSCs from 4N mice were incubated with virus resuspended in ES medium containing SCF, IL-3, and IL-6 for 24 hours; the medium was then replaced with ES medium containing 1 μg/ml Dox to access the re-programming efficiency of specific candidate genes.

RNA sequencing

Total RNA was isolated from cell pellets using TRIzol (Invitrogen, Carlsbad, CA). The RNA integrity was con-firmed with a minimum RNA integrity number of 8 using 2100 Bioanalyzer (Agilent Technologies, Santa Clara, CA). The mRNA was enriched using oligo(dT) magnetic beads and sheared to create short fragments of ~200 base pairs. cDNA was then synthesized using random hexamer primers and purified using a PCR product extraction kit (Qiagen, Germany). Finally, the sequencing primers linked to the cDNA fragments were isolated by gel elec-trophoresis and enriched by PCR amplification to con-struct the library for sequencing. Single-end sequencing was applied to RNA sequencing at the Beijing Genom-ics Institute using the HiSeq™ 2000 system developed by Illumina (Illumina, San Diego, CA). The RNA se-quencing reads were mapped to the mouse genome using Tophat (v1.3.3) and the Ensembl genome annota-tion (Mus_musculus.NCBIM37.64.gtf) with the default parameters [31]. The fragments per kilobase of exon per million fragments mapped (FPKM) for each gene were calculated using Cufflinks (v1.2.0) [32]. Hierarch-ical clustering was presented as mean ± standard devi-ation (SD) based on two biological replicates to describe the relationship among samples.

Statistical analyses

The SD was used to assess biological significance.

Accession numbers

The genome-wide gene expression data reported in this article were deposited in Gene Expression Omnibus [GEO:GSE36294].

Results

Derivation of high-quality iPSCs from HPC/HSCs

To determine the reprogramming potential of HPC/HSCs, our sequential reprogramming system was used to gener-ate the genetically identical iPSCs [26, 27] (Fig. 1a). HPC/HSCs were isolated from 4N mice derived from 1^0-MEF-

iPSCs using MACS MicroBeads technology with a purity of approximately 92 % when analyzed by flow cytometry (Fig. 1b). The characteristics of the HPC/HSCs were veri-fied by analyzing the expression of HPC/HSC-specific genes (Additional file 2: Figure S1A). Exposure of the freshly sorted HPC/HSCs to Dox and the addition of SCF, IL-3, and IL-6 resulted in re-expression of the four tran-scription factors (*Oct4*, *Sox2*, *Klf4*, and *c-Myc*) and rapid proliferation. The suspended HPC/HSCs adhered to the bottom of the culture dishes 3 days later, and the AP-positive ESC-like colonies emerged approximately 9 days after induction, much sooner than the control MEFs (Fig. 1c, top). The average reprogramming efficiency of HPC/HSCs was 1.1 % when measured by AP staining, ap-proximately threefold greater than MEFs (Fig. 1c, bottom). Two days following the withdrawal of Dox, transgene-independent HPC/HSC-iPSC lines were established from day 16 ESC-like colonies that displayed typical expression patterns of pluripotency-related genes (*Oct4*, *Sox2*, and *SSEA-1*; Additional file 2: Figure S1B). In addition, the global expression patterns of HPC/HSC-iPSCs were in-distinguishable from those of ESCs (Additional file 2: Figure S1C). Bisulfite genomic sequencing results showed that demethylation of *Pou5f1* and *Nanog* occurred in HPC/HSC-iPSCs (Additional file 2: Figure S1D). Notably, Q-PCR results showed that the exogenous OSKM expres-sions were silenced in HPC/HSC-iPSCs, which were well maintained under the expression of endogenous pluripo-tency genes (Additional file 3: Figure S2A). The in vivo differentiation potential of HPC/HSC-iPSCs was further confirmed by teratoma formation assays and three typical embryonic germ layers were observed (Additional file 3: Figure S2B). The birth of reconstructed 2N mice further validated the chimeric and germline transmission poten-tial of HPC/HSC-iPSCs (Additional file 3: Figure S2C).

To stringently evaluate the pluripotency state of HPC/HSC-iPSCs, tetraploid complementation was performed on the HPC/HSC-iPSCs and viable 4N mice were ultim-ately generated (Fig. 1d). Healthy offspring were pro-duced after the adult 4N mice were mated with female ICR mice (Fig. 1d). To confirm the genetic background of the 4N mice, SSLP analyses were performed. The SSLP results indicated that the 4N mice were indeed produced from HPC/HSC-iPSCs and that they had a 129/Sv × M2rtTA genetic background (Fig. 1e). To the best of our knowledge, this is the first demonstration that viable, fertile 4N mice can be generated from HPC/HSC-derived iPSCs.

Interestingly, a large proportion (5/9) of HPC/HSC-iPSC lines gave rise to 4N mice (Fig. 1f and Additional file 3: Figure S2C). Based on published results, the average ratio of 4N competent iPSCs never exceeds 40 % [14, 33]. To exclude the possibility that the sequential reprogram-ming system utilized here resulted in a large proportion of

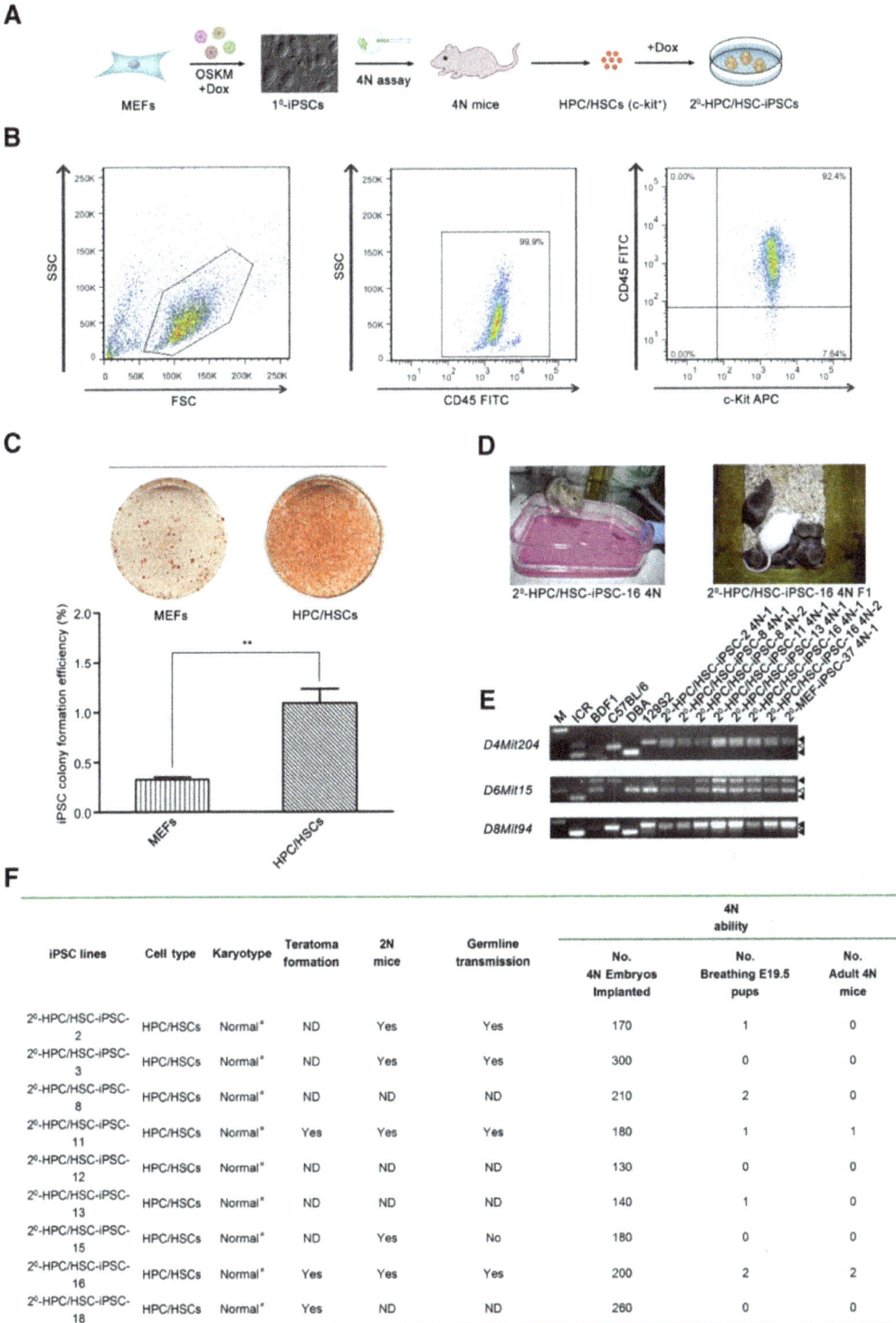

A

MEFs → OSKM +Dox → 1°-iPSCs → 4N assay → 4N mice → HPC/HSCs (c-kit⁺) → +Dox → 2°-HPC/HSC-iPSCs

B

C

MEFs   HPC/HSCs

D

2°-HPC/HSC-iPSC-16 4N   2°-HPC/HSC-iPSC-16 4N F1

E

D4Mit204

D6Mit15

D8Mit94

F

iPSC lines	Cell type	Karyotype	Teratoma formation	2N mice	Germline transmission	4N ability		
						No. 4N Embryos Implanted	No. Breathing E19.5 pups	No. Adult 4N mice
2°-HPC/HSC-iPSC-2	HPC/HSCs	Normal[a]	ND	Yes	Yes	170	1	0
2°-HPC/HSC-iPSC-3	HPC/HSCs	Normal[a]	ND	Yes	Yes	300	0	0
2°-HPC/HSC-iPSC-8	HPC/HSCs	Normal[a]	ND	ND	ND	210	2	0
2°-HPC/HSC-iPSC-11	HPC/HSCs	Normal[a]	Yes	Yes	Yes	180	1	1
2°-HPC/HSC-iPSC-12	HPC/HSCs	Normal[a]	ND	ND	ND	130	0	0
2°-HPC/HSC-iPSC-13	HPC/HSCs	Normal[a]	ND	ND	ND	140	1	0
2°-HPC/HSC-iPSC-15	HPC/HSCs	Normal[a]	ND	Yes	No	180	0	0
2°-HPC/HSC-iPSC-16	HPC/HSCs	Normal[a]	Yes	Yes	Yes	200	2	2
2°-HPC/HSC-iPSC-18	HPC/HSCs	Normal[a]	Yes	ND	ND	260	0	0

Fig. 1 (See legend on next page.)

(See figure on previous page.)
Fig. 1 Transcription factor-mediated reprogramming of HPC/HSCs results in high-quality iPSCs with full pluripotency. **a** Schematic showing the generation of iPSCs from HPC/HSCs. **b** Flow cytometry analysis of the purity of HPC/HSCs isolated by MACS. *First column*, cells were isolated based on size, as indicated by side scatter (*SSC*) and forward scatter (*FSC*). *Second column*, CD45-positive cells were further selected. *Third column*, c-Kit and CD45 double-positive cells were isolated. **c** The reprogramming efficiency of HPC/HSCs was analyzed using AP staining. MEFs were used as the control (*n* = 3 measurements). Error bars indicate the SD. **P <0.01, unpaired *t* test. **d** Viable 4N mice and the offspring (F1) of 4N mice derived from HPC/HSC-iPSCs with the indicated cell line. **e** Genetic characterization of the 4N mice using SSLP analysis. **f** Summary of the derivation of mice from HPC/HSC-iPSCs. #Karyotype was considered normal when greater than 80 %. *Dox* doxycycline, *E19.5* embryonic day 19.5, *HPC/HSC* hematopoietic progenitor and stem cell, *iPSC* induced pluripotent stem cell, *MEF* mouse embryonic fibroblast, *2N* chimera, *4N* tetraploid complementation, *ND* not determined, *OSKM* Oct4, Sox2, Klf4, and c-Myc. *APC* Allophycocyanin, *FITC* Fluorescein isothiocyanate

high-quality iPSCs, skin fibroblasts (SFs) and TTFs from the same 4N mouse as the HPC/HSCs were used to generate SF-iPSCs and TTF-iPSCs, respectively, and tetraploid complementation was performed to test the pluripotency of the derived cell lines. SF-iPSCs were found to achieve a full pluripotency state with low proportion (1/6; Additional file 4: Table S2). However, none of the TTF-iPSC lines (0/4) supported the full development of 4N mice (Additional file 4: Table S2). These results indicate that HPC/HSCs are more amenable to reprogramming into high-quality iPSCs than somatic cells.

Genome signatures of HPC/HSCs correlated with accelerated reprogramming

As already mentioned, HPC/HSCs are amenable to reprogramming into high-quality iPSCs, suggesting that the reprogramming process may be different in HPC/HSCs compared with MEFs. Concerning the morphology change of HPC/HSCs during the reprogramming process, precolonies were observed 3 days post reprogramming which are similar to the control MEFs (Additional file 5: Figure S3A). However, more compact colonies, which are an indication of naive pluripotency, were derived from HPC/HSCs in the middle and late stages of reprogramming when compared with MEFs (Fig. 2a). However, the underlying mechanisms of HPC/HSC reprogramming remain unclear.

To gain insight into the intrinsic events related to HPC/HSC reprogramming, genome-wide gene expression analyses were performed in HPC/HSCs using MEFs and pluripotent stem cells as controls. First, we compared the global gene expression patterns from HPC/HSCs, MEFs, R1(ESCs), and iPSCs (1^0-MEF-iPSC-37). Principle component analysis (PCA) showed that HPC/HSCs and MEFs demonstrated gene expression patterns that differed from those of pluripotent stem cells, indicating that dramatic cell fate transitions occurred during the process of reprogramming (Fig. 2b). Next, HPC/HSC-specific and MEF-specific genes were subjected to gene ontology (GO) analysis; the results showed that highly expressed genes in HPC/HSCs mainly correlated with functions related to hematopoiesis (Fig. 2c). In MEFs, genes that were highly expressed were associated with extracellular matrix functions (Fig. 2d).

Previous studies have shown that epithelial cells can be reprogrammed with a higher efficiency than MEFs and that the induction of MET enhances reprogramming efficiency [19, 34]. Moreover, ultrafast cycling cell populations were found to be privileged adopters of the pluripotency state, suggesting a role for the cell cycle in reprogramming [35]. To investigate whether these two events also affected the reprogramming of HPC/HSCs, mesenchymal genes, epithelial genes, and cell cycle-related genes were analyzed in HPC/HSCs, MEFs, ESCs, and iPSCs. We found that the expression levels of mesenchymal genes (*Zeb1*, *Snail1*, *Col1a1*, *Col5a1*, *Col6a1*, *Itgbl1*, and *Thy1*) were downregulated in HPC/HSCs compared with MEFs, while epithelial genes (*L1td1*, *Ocln*, *Epcam*, and *Cdh1*) demonstrated no changes (Fig. 2e), suggesting that HPC/HSCs undergoing reprogramming might require fewer steps to switch off the MET program compared with MEFs. Flow cytometric analysis of 5-bromo-2′-deoxyuridine (BrdU) incorporation was performed on HPC/HSCs to assess cell proliferation following the first 48 hours after reprogramming using MEFs as the control. We found a higher number of S-phase (BrdU⁺) cells in HPC/HSCs compared with MEFs, indicating a rapid doubling of HPC/HSCs in the first 48 hours after reprogramming (Additional file 5: Figure S3B, C). Consistently, several tumor suppressor genes among the cell cycle-related genes, including *Cdkn1a*, *Cdkn2a*, *Cdkn2b*, and *Rb1*, showed lower expression patterns in HPC/HSCs than in MEFs, whereas oncogenes, such as *Myc* and *E2f2*, were more highly expressed in HPC/HSCs (Fig. 2e). Q-PCR results confirmed differential expression patterns for *p21* (encoded by *Cdkn1a*), *Ink4b* (encoded by *Cdkn2b*), *Ink4a* (encoded by *Cdkn2b*), *Arf* (encoded by *Cdkn2b*), *E2f2*, *Myc*, *Snail1*, *Col1a*, and *Col6a* between HPC/HSCs and MEFs (*t* test, *P* <0.05) (Fig. 2f). Western blot further confirmed the differential expression patterns of *Snail1*, *Arf*, and *Ink4a* (Fig. 2g) between HPC/HSCs and MEFs. These results indicate the potential role of these genes in HPC/HSC reprogramming.

The participation of select tumor suppressor genes and an EMT factor in HPC/HSC reprogramming

To elucidate the mechanisms underlying HPC/HSC reprogramming, the functions of select cell cycle-related

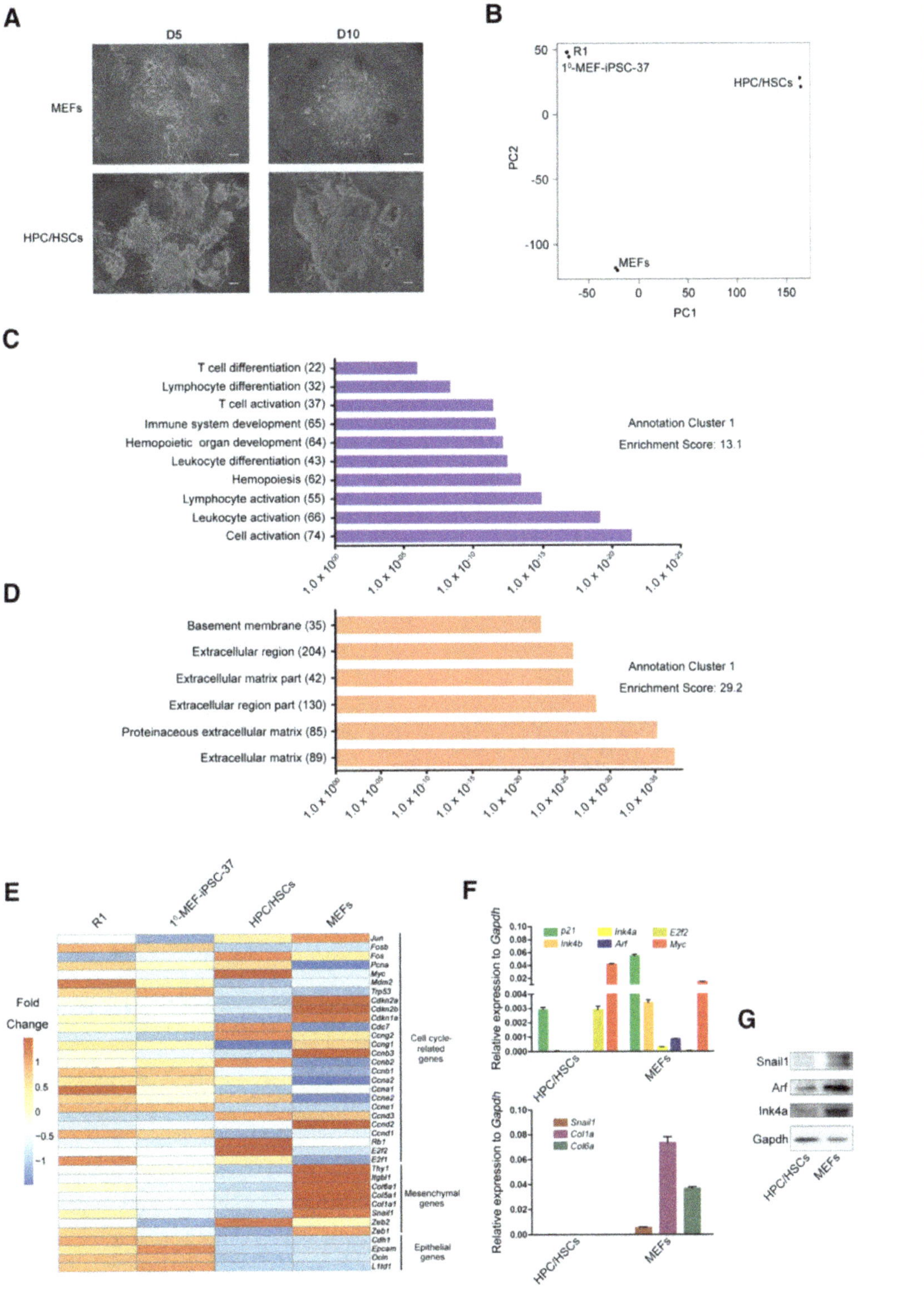

Genome-wide gene expression analyses reveal unique cellular characteristics related...

65

Fig. 2 (See legend on next page.)

(See figure on previous page.)
Fig. 2 Genome-wide signature analyses reveal the unique characteristics related to HPC/HSC reprogramming. **a** Morphology differences during the reprogramming process of HPC/HSCs and MEFs. Scale bar, 100 μm. **b** Principle component analysis of global gene expression patterns in HPC/HSCs, MEFs, R1 ESCs, and iPSCs. The analysis of each cell line was based on two biological replicates. **c** Gene ontology (GO) of highly expressed genes in HPCs/HSCs. **d** GO of highly expressed genes in MEFs. **c**, **d** *X* axis depicts the *P* value. **e** Gene expression analyses of mesenchymal genes, epithelial genes, and cell cycle-related genes in HPC/HSCs and MEFs. The cluster analysis is presented as the mean ± SD based on two biological replicates. **f** Comparison of expression analyses of indicated genes in HPC/HSCs and MEFs (*n* = 3 measurements). Error bars indicate the SD. **g** Western blot comparing the levels of the indicated genes in HPC/HSCs and in MEFs. *D* day, *HPC/HSC* hematopoietic progenitor and stem cell, *iPSC* induced pluripotent stem cell, *MEF* mouse embryonic fibroblast

genes (*p21, Ink4b, Ink4a, Arf,* and *E2f2*) and an EMT factor during reprogramming were studied. The kinetics of *p21, Ink4b, Ink4a, Arf, E2f2,* and *Snail1* expression during the reprogramming process of HPC/HSCs and MEFs were therefore analyzed for their potential functions. We found that the tumor suppressor genes *p21, Ink4b, Ink4a,* and *Arf*, as well as the EMT factor *Snail1,* showed lower expression in the intermediate HPC/HSCs than in the intermediate MEFs (*t* test, *P* <0.05), whereas the oncogene *E2f2* showed an opposite pattern (*t* test, *P* <0.05), indicating the potential participation of these genes in HPC/HSC reprogramming (Fig. 3a).

Independent ectopic activation of *p21* (*t* test, *P* <0.05), *Ink4b* (*t* test, *P* <0.05), *Ink4a, Arf,* and *Snail1,* along with OSKM, was performed to characterize the functions of these genes in HPC/HSC reprogramming (Fig. 3b, c). The results showed that the overexpression of *p21, Ink4a,* and *Arf* decreased the reprogramming efficiency of HPC/HSCs, but the overexpression of *Ink4b* exhibited no effect (Fig. 3d, e). Our observed roles of *p21, Ink4a,* and *Arf* in HPC/HSC reprogramming were consistent with previous reports [20, 21], suggesting that immortalization is indispensable for the accelerated reprogramming of HPC/HSCs. Similar to the tumor suppressor gene results, the overexpression of *Snail1* also decreased the efficiency of HPC/HSC reprogramming (Fig. 3d, e), which is in accordance with a previous report [19]. Taken together, these results indicate that select tumor suppressor genes and an EMT factor participate in the reprogramming of HPC/HSCs.

Collectively, the work presented here showed that the reprogramming of HPC/HSCs results in a high frequency of high-quality iPSCs with full pluripotency. Genome-wide analyses showed that the downregulation of select tumor suppressor genes and an EMT factor and the upregulation of select oncogenes occur in HPC/HSCs, suggesting that these factors are potentially involved in the amenability of HPC/HSCs to be reprogrammed. Further exploration demonstrated that select tumor suppressor genes (*p21, Ink4a,* and *Arf*) and an EMT factor (*Snail1*) participate in HPC/HSC reprogramming, in which independent ectopic activation of *p21, Ink4a, Arf,* and *Snail1* along with OSKM decreased the reprogramming efficiency of HPC/HSCs (Fig. 4).

Discussion

In the present study, the mechanisms of reprogramming HPC/HSCs into high-quality iPSCs were investigated by a combination of techniques, including genome-wide analysis. To our knowledge, this study is the first to demonstrate that HPC/HSCs are amenable to be reprogrammed into high-quality iPSCs with full pluripotency at a high frequency. Through genome-wide analysis, we selected several candidates and further verified their potential contribution to HSC/HPC reprogramming. Furthermore, we discovered that select tumor suppressor genes (*p21, Ink4a,* and *Arf*) and an EMT factor (*Snail1*) participated in HPC/HSC reprogramming, suggesting that both immortalization and EMT transition are indispensable for HPC/HSC reprogramming.

The quality of iPSCs determines the biosafety of cell transplantation-based regenerative medicine. Many strategies have been used to improve the quality of iPSCs [9, 14]. A previous study has shown an association between differentiation state and the reprogramming efficiency of iPSCs [10]; however, the mechanisms correlated with the amenability of HPC/HSCs to reprogram into iPSCs have not been rigorously addressed. Moreover, it is of significance to test whether increased reprogramming efficiency can result in high-quality iPSCs with full pluripotency. Our finding that a large proportion of HPC/HSC-iPSCs show full pluripotency is comparable with previous results which used the overexpression of *Zscan4* during reprogramming [36]. The presence of a large proportion of HPC/HSC-iPSCs with full pluripotency in our study demonstrates the susceptibility of HPC/HSCs to reprogramming and, reciprocally, defines the notion that different cell types face distinct barriers to achieve pluripotency. Further elucidation of the mechanisms underlying somatic stem and progenitor cell reprogramming will undoubtedly shed light on our understanding of the 4N competency of iPSCs.

The cellular states of starting cells have been shown to have a great impact on the cellular reprogramming process [7]. Furthermore, the acquisition of immortalization is found to be a crucial and rate-limiting step in the establishment of pluripotency [21]; in particular, the abrogation of Trp53 eliminates the roadblock in iPSC cellular reprogramming [37, 38]. In addition, the

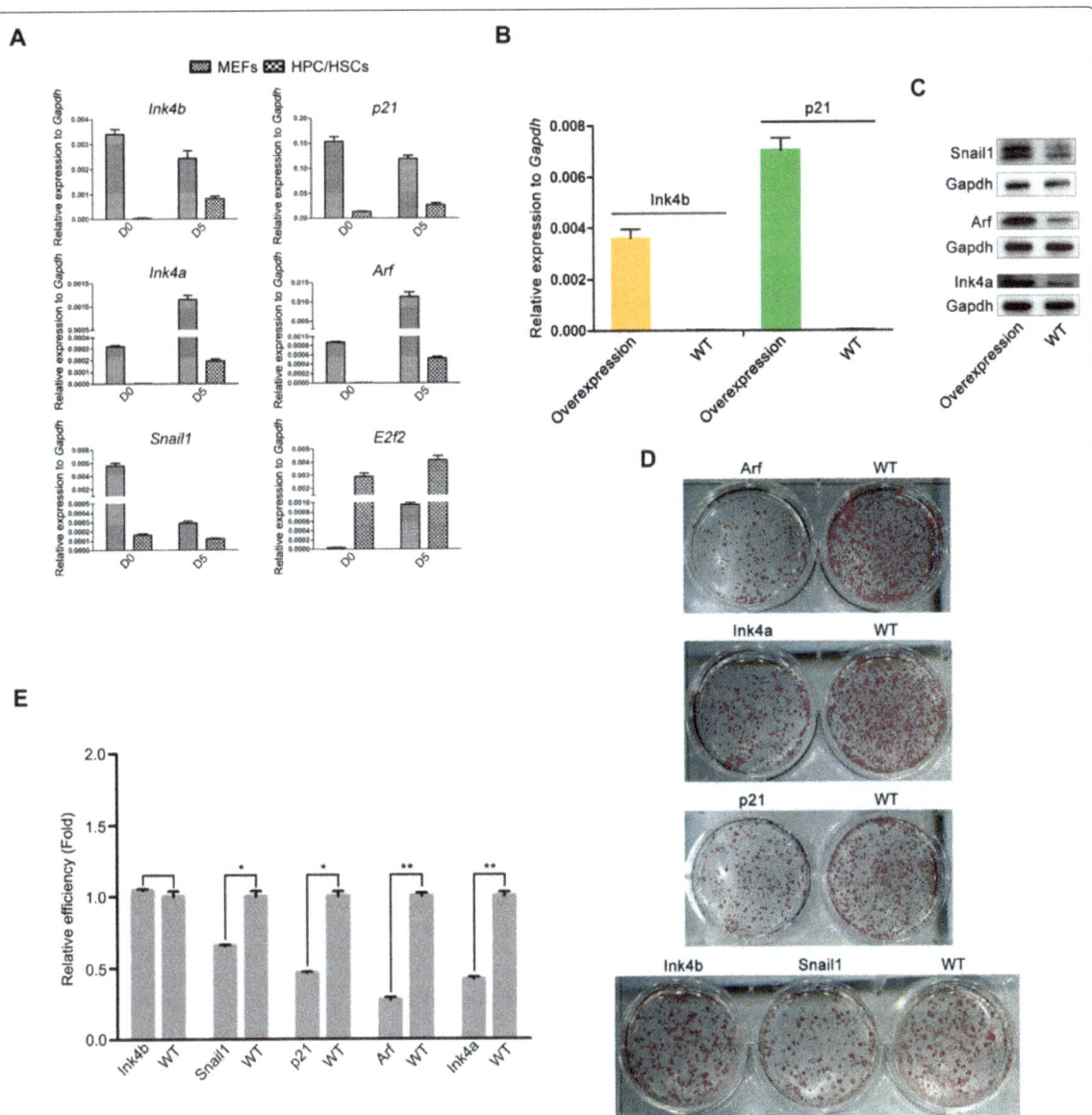

Fig. 3 Participation of select tumor suppressor genes and an EMT factor in HPC/HSC reprogramming. **a** Kinetics of expression of indicated genes in the OSKM-mediated reprogramming of HPC/HSCs and MEFs (n = 3 measurements). Error bars indicate the SD. **b, c** Ectopic activation of indicated genes during the reprogramming process of HPC/HSCs. **d** AP staining results for ectopic activation of p21, Ink4b, Ink4a, Arf, and Snail1 in the OSKM-mediated reprogramming of HPC/HSCs. **e** Reprogramming efficiencies of the indicated genotypes relative to wild-type (WT) in HPC/HSCs (n = 3 measurements). Error bars indicate the SD. *P <0.05, **P <0.01, unpaired t test. D day, HPC/HSC hematopoietic progenitor and stem cell, MEF mouse embryonic fibroblast

acquisition of immortalization through the suppression of the *Ink4a/Arf* locus can accelerate iPSC reprogramming [20, 21], which may help the cell bypass reprogramming barriers represented by the typical P53 and Rb pathways. In addition, recent data have shown that an ultrafast cycling cell population pre-existing among hematopoietic progenitors displays privileged induced reprogramming activity [35]. Furthermore, considerable evidence suggests that the EMT factor *Snail1* exerts dual effects which depend on the induction of EMT in the early stage of reprogramming and that of MET in the late stage of reprogramming [39, 40]. Similarly, we show in the present study that HPC/HSCs which demonstrate the downregulation of select tumor suppressor genes and

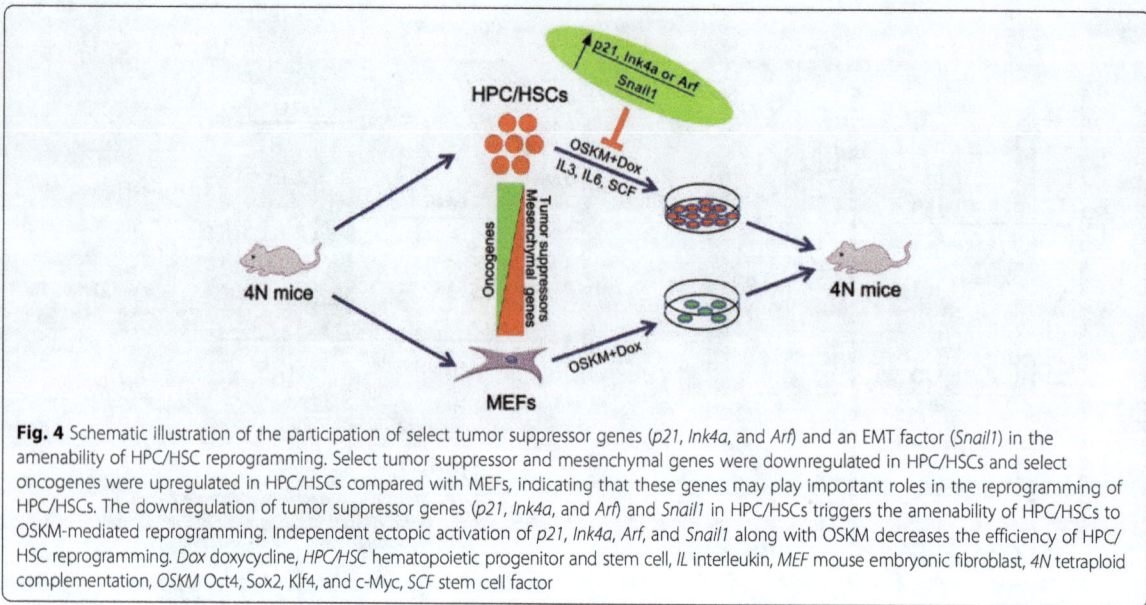

Fig. 4 Schematic illustration of the participation of select tumor suppressor genes (*p21*, *Ink4a*, and *Arf*) and an EMT factor (*Snail1*) in the amenability of HPC/HSC reprogramming. Select tumor suppressor and mesenchymal genes were downregulated in HPC/HSCs and select oncogenes were upregulated in HPC/HSCs compared with MEFs, indicating that these genes may play important roles in the reprogramming of HPC/HSCs. The downregulation of tumor suppressor genes (*p21*, *Ink4a*, and *Arf*) and *Snail1* in HPC/HSCs triggers the amenability of HPC/HSCs to OSKM-mediated reprogramming. Independent ectopic activation of *p21*, *Ink4a*, *Arf*, and *Snail1* along with OSKM decreases the efficiency of HPC/HSC reprogramming. *Dox* doxycycline, *HPC/HSC* hematopoietic progenitor and stem cell, *IL* interleukin, *MEF* mouse embryonic fibroblast, *4N* tetraploid complementation, *OSKM* Oct4, Sox2, Klf4, and c-Myc, *SCF* stem cell factor

an EMT factor and the upregulation of oncogenes could achieve pluripotency with high efficiency. Gene function analyses showed that select tumor suppressor genes and an EMT factor play important roles in HPC/HSC reprogramming, a finding that is consistent with previous reports [19, 20].

However, the efficiency of induced reprogramming observed in the present study was lower than in previous reports [10, 35]. This inconsistency may have resulted from the different origins of the cells used in these studies as well as from differences in the technical details of the reprogramming assay. The HPC/HSCs that were used in the present study included myeloid progenitors and HSCs. The major difference between these two cell types is the cycling time, with HSCs being slow cycling. In addition, the HPC/HSCs used here were cultured in ES medium with SCF, IL-3, and IL-6 for 24 hours before exposure to Dox, a method that may preferentially impact the proliferation of the ultrafast cycling population among the bulk cells. Nevertheless, under the reprogramming conditions used in the current study, high-quality iPSCs were obtained from HPC/HSCs. Further investigation of the reprogramming of HPC/HSCs will not only provide more information on the molecular barriers to reprogramming but might also provide alternatives that make it possible to overcome these inherent barriers to achieving pluripotency. Finally, it will be interesting to examine whether high-quality iPSCs can be generated from progenitor cells derived from other somatic tissues.

Conclusions

HPC/HSCs that exhibited a downregulation of select tumor suppressor and mesenchymal genes and an upregulation of select oncogenes were amenable to transcription factor-mediated reprogramming, which yielded a high frequency of fully reprogrammed HPC/HSC-iPSCs. Additional studies provided insights into the contribution of select tumor suppressor genes (*p21*, *Ink4a*, and *Arf*) and an EMT factor (*Snail1*) to the amenability of HPC/HSC reprogramming.

Additional files

Additional file 1: is Table S1 presenting the list of primers. (DOC 68 kb)

Additional file 2: is Figure S1 showing cellular characteristics and HPC/HSC reprogramming. **A** Identification of HPC/HSCs using Q-PCR to detect genes (*Ikzf1*, *Lyl1*, and *Myb*) specifically expressed by HPC/HSCs. MEFs, adipose progenitor cells (*APCs*), and R1 ESCs were used as controls (*n* = 3 measurements). Error bars indicate the SD. **B** Expression of pluripotency-related markers (Pou5f1, Sox2, and SSEA-1) detected by immunofluorescence. DNA was stained with 4′,6-diamidino-2-phenylindole (*DAPI*). **C** HPC/HSC-iPSCs were indistinguishable from MEF-iPSCs at the level of global gene expression. *Cor* Pearson correlation coefficient, *X, Y* log expression value. **D** Bisulfite genomic sequencing of *Pou5f1* and *Nanog* in HPC/HSC-iPSCs. R1 ESCs were used as the control. *Black circle* methylated CpG; *white circle* unmethylated CpG. (TIF 1727 kb)

Additional file 3: is Figure S2 showing silencing of OSKM factors and the pluripotency state of HPC/HSC-iPSCs. **A** Q-PCR was used to detect the gene expression levels of Pou5f1, Sox2, Klf4, and c-Myc in the indicated cell lines. *Gray column* total; *white column* endogenous; *red column* exogenous. Relative expression levels of these genes (*Y* axis) in each cell line were first normalized to the level of their endogenous *Gapdh*, and then the amount was normalized to R1 (*n* = 3 measurements). Error bars indicate the SD. **B**) Hematoxylin and eosin (*HE*) staining of teratomas derived from iPSCs. Ectoderm, lamellar corpuscle (×200); mesoderm, cartilage (×400); and endoderm, columnar epithelium and gland (×400). **C** viable 2N and 4N mice derived from HPC/HSC-iPSCs with the indicated cell line. The placenta of each newborn 4N mouse is displayed on the left of the 4N pup. (TIF 9159 kb)

Additional file 4: is Table S2 presenting characteristics of the SF-iPSCs and TTF-iPSCs. (DOC 28 kb)

Additional file 5: is Figure S3 showing the cell morphology change and cell proliferation rate of HPC/HSCs compared with MEFs. A The cell morphology changes in the early reprogramming process of HPC/HSCs and MEFs. Scale bar, 100 μm. B Flow cytometric analysis of BrdU/7-Aminoactinomycin D (*7-AAD*) incorporation was performed on HPC/HSCs during the first 48 hours after reprogramming to assess cell proliferation. MEFs were used as the control. C Frequencies of cell cycle phases in the early intermediate cells (48 hours after reprogramming) of HPC/HSC and MEF populations ($n = 3$ measurements). Error bars indicate the SD. *P <0.05, **P <0.01, unpaired t test. (TIF 2724 kb)

Abbreviations

AP: Alkaline phosphatase; BrdU: 5-Bromo-2'-deoxyuridine; CD: Cluster of differentiation; DMEM: Dulbecco modified Eagle's medium; Dox: Doxycycline; EMT: Epithelial-to-mesenchymal transition; ESC: Embryonic stem cell; FBS: Fetal bovine serum; FPKM: Fragments per kilobase of exon per million fragments mapped; GO: Gene ontology; H & E: Hematoxylin and eosin; HPC/HSC: Hematopoietic progenitor and stem cell; IL: Interleukin; iPSC: Induced pluripotent stem cell; LIF: Leukemia inhibitory factor; MEF: Mouse embryonic fibroblast; MET: Mesenchymal-to-epithelial transition; 2N: Chimera; 4N: Tetraploid complementation; OSKM: Oct4, Sox2, Klf4, and c-Myc; PBE: Phosphate-buffered saline containing 0.5 % bovine serum albumin and 2 mM ethylenediamine tetraacetic acid; PBS: Phosphate-buffered saline; PCA: Principle component analysis; Q-PCR: quantitative PCR; RQ: Relative quantitation; SCF: Stem cell factor; SD: Standard deviation; SF: Skin fibroblast; SSLP: Simple sequence length polymorphism; TRE: Tetracycline response elements; TTF: Tail-tip fibroblast.

Competing interests

The authors declare that they have no competing interests.

Authors' contributions

SG, LT, XH, ZX, WL, KZ, and MG performed the experiments. TC performed the bioinformatics analysis. GC, SG, HW, TC, JT, and SrG analyzed and interpreted the data. LT, XH, ZX, WL, KZ, MG, HW, and TC gave some advice for this manuscript. SG drafted the manuscript. GC, SG, JT, and SrG designed the study and reviewed and revised the manuscript. All the authors read and approved the manuscript for publication.

Acknowledgements

The authors thank Professor Rudolf Jaenisch of the Whitehead Institute, MIT (Massachusetts Institute of Technology, Cambridge, MA) for generously supplying the lentivirus vectors. They are also grateful to laboratory colleagues for their assistance with experiments and in the preparation of the manuscript. This work was supported by the Natural Science Foundation of China (31000656 to GC), the Natural Science Foundation of SZU (201407 to GC), and the Natural Science Foundation of ShanDong Province of China (ZR2014CP026 to SG).

Author details

[1]Institute of Molecular Medicine, Health Science Center, Shenzhen University, Shenzhen 518060, China. [2]National Institute of Biological Sciences, NIBS, Beijing 102206, China. [3]Translational Medical Center for Stem Cell Therapy, Shanghai East Hospital, School of Medicine, Tongji University, Shanghai 200120, China. [4]Ministry of Agriculture Key Laboratory of Animal Genetics, Breeding and Reproduction; National Engineering Laboratory for Animal Breeding; College of Animal Sciences and Technology, China Agricultural University, Beijing 100193, China. [5]School of Life Sciences and Technology, Tongji University, Shanghai 200092, China.

References

1. Takahashi K, Yamanaka S. Induction of pluripotent stem cells from mouse embryonic and adult fibroblast cultures by defined factors. Cell. 2006;126:663–76.
2. Park IH, Arora N, Huo H, Maherali N, Ahfeldt T, Shimamura A, et al. Disease-specific induced pluripotent stem cells. Cell. 2008;134:877–86.
3. Soldner F, Hockemeyer D, Beard C, Gao Q, Bell GW, Cook EG, et al. Parkinson's disease patient-derived induced pluripotent stem cells free of viral reprogramming factors. Cell. 2009;136:964–77.
4. Takebe T, Sekine K, Enomura M, Koike H, Kimura M, Ogaeri T, et al. Vascularized and functional human liver from an iPSC-derived organ bud transplant. Nature. 2013;499:481–4.
5. Aoi T, Yae K, Nakagawa M, Ichisaka T, Okita K, Takahashi K, et al. Generation of pluripotent stem cells from adult mouse liver and stomach cells. Science. 2008;321:699–702.
6. Zhao XY, Li W, Lv Z, Liu L, Tong M, Hai T, et al. Viable fertile mice generated from fully pluripotent iPS cells derived from adult somatic cells. Stem Cell Rev. 2010;6:390–7.
7. Kim JB, Zaehres H, Wu G, Gentile L, Ko K, Sebastiano V, et al. Pluripotent stem cells induced from adult neural stem cells by reprogramming with two factors. Nature. 2008;454:646–50.
8. Hanna J, Markoulaki S, Schorderet P, Carey BW, Beard C, Wernig M, et al. Direct reprogramming of terminally differentiated mature B lymphocytes to pluripotency. Cell. 2008;133:250–64.
9. Stadtfeld M, Apostolou E, Ferrari F, Choi J, Walsh RM, Chen T, et al. Ascorbic acid prevents loss of Dlk1-Dio3 imprinting and facilitates generation of all-iPS cell mice from terminally differentiated B cells. Nat Genet. 2012;44:398–405.
10. Eminli S, Foudi A, Stadtfeld M, Maherali N, Ahfeldt T, Mostoslavsky G, et al. Differentiation stage determines potential of hematopoietic cells for reprogramming into induced pluripotent stem cells. Nat Genet. 2009;41:968–76.
11. Wu T, Wang H, He J, Kang L, Jiang Y, Liu J, et al. Reprogramming of trophoblast stem cells into pluripotent stem cells by Oct4. Stem Cells. 2011;29:755–63.
12. Guenther MG, Frampton GM, Soldner F, Hockemeyer D, Mitalipova M, Jaenisch R, et al. Chromatin structure and gene expression programs of human embryonic and induced pluripotent stem cells. Cell Stem Cell. 2010;7:249–57.
13. Kang L, Wang J, Zhang Y, Kou Z, Gao S. iPS cells can support full-term development of tetraploid blastocyst-complemented embryos. Cell Stem Cell. 2009;5:135–8.
14. Zhao XY, Li W, Lv Z, Liu L, Tong M, Hai T, et al. iPS cells produce viable mice through tetraploid complementation. Nature. 2009;461:86–90.
15. Boland MJ, Hazen JL, Nazor KL, Rodriguez AR, Gifford W, Martin G, et al. Adult mice generated from induced pluripotent stem cells. Nature. 2009;461:91–4.
16. Hochedlinger K, Jaenisch R. Nuclear reprogramming and pluripotency. Nature. 2006;441:1061–7.
17. Heng JC, Feng B, Han J, Jiang J, Kraus P, Ng JH, et al. The nuclear receptor Nr5a2 can replace Oct4 in the reprogramming of murine somatic cells to pluripotent cells. Cell Stem Cell. 2010;6:167–74.
18. Gao Y, Chen J, Li K, Wu T, Huang B, Liu W, et al. Replacement of Oct4 by Tet1 during iPSC induction reveals an important role of DNA methylation and hydroxymethylation in reprogramming. Cell Stem Cell. 2013;12:453–69.
19. Li R, Liang J, Ni S, Zhou T, Qing X, Li H, et al. A mesenchymal-to-epithelial transition initiates and is required for the nuclear reprogramming of mouse fibroblasts. Cell Stem Cell. 2010;7:51–63.
20. Li H, Collado M, Villasante A, Strati K, Ortega S, Canamero M, et al. The Ink4/Arf locus is a barrier for iPS cell reprogramming. Nature. 2009;460:1136–9.
21. Utikal J, Polo JM, Stadtfeld M, Maherali N, Kulalert W, Walsh RM, et al. Immortalization eliminates a roadblock during cellular reprogramming into iPS cells. Nature. 2009;460:1145–8.
22. Gupta MK, Teo AK, Rao TN, Bhatt S, Kleinridders A, Shirakawa J, et al. Excessive cellular proliferation negatively impacts reprogramming efficiency of human fibroblasts. Stem Cells Transl Med. 2015;4:1101–8.
23. Sung LY, Gao S, Shen H, Yu H, Song Y, Smith SL, et al. Differentiated cells are more efficient than adult stem cells for cloning by somatic cell nuclear transfer. Nat Genet. 2006;38:1323–8.
24. Hochedlinger K, Jaenisch R. Monoclonal mice generated by nuclear transfer from mature B and T donor cells. Nature. 2002;415:1035–8.
25. Wilmut I, Schnieke AE, McWhir J, Kind AJ, Campbell KH. Viable offspring derived from fetal and adult mammalian cells. Nature. 1997;385:810–3.
26. Gao S, Zheng C, Chang G, Liu W, Kou Z, Tan K, et al. Unique features of mutations revealed by sequentially reprogrammed induced pluripotent stem cells. Nat Commun. 2015;6:6318.

27. Brambrink T, Foreman R, Welstead GG, Lengner CJ, Wernig M, Suh H, et al. Sequential expression of pluripotency markers during direct reprogramming of mouse somatic cells. Cell Stem Cell. 2008;2:151–9.

28. Liu Y, Cheng H, Gao S, Lu X, He F, Hu L, et al. Reprogramming of MLL-AF9 leukemia cells into pluripotent stem cells. Leukemia. 2014;28:1071–80.

29. Chang G, Miao YL, Zhang Y, Liu S, Kou Z, Ding J, et al. Linking incomplete reprogramming to the improved pluripotency of murine embryonal carcinoma cell-derived pluripotent stem cells. PLoS One. 2010;5:e10320.

30. Chang G, Gao S, Hou X, Xu Z, Liu Y, Kang L, et al. High-throughput sequencing reveals the disruption of methylation of imprinted gene in induced pluripotent stem cells. Cell Res. 2014;24:293–306.

31. Trapnell C, Pachter L, Salzberg SL. TopHat: discovering splice junctions with RNA-Seq. Bioinformatics. 2009;25:1105–11.

32. Roberts A, Pimentel H, Trapnell C, Pachter L. Identification of novel transcripts in annotated genomes using RNA-Seq. Bioinformatics. 2011;27:2325–9.

33. Kang L, Wu T, Tao Y, Yuan Y, He J, Zhang Y, et al. Viable mice produced from three-factor induced pluripotent stem (iPS) cells through tetraploid complementation. Cell Res. 2011;21:546–9.

34. Aasen T, Raya A, Barrero MJ, Garreta E, Consiglio A, Gonzalez F, et al. Efficient and rapid generation of induced pluripotent stem cells from human keratinocytes. Nat Biotechnol. 2008;26:1276–84.

35. Guo S, Zi X, Schulz VP, Cheng J, Zhong M, Koochaki SH, et al. Nonstochastic reprogramming from a privileged somatic cell state. Cell. 2014;156:649–62.

36. Jiang J, Lv W, Ye X, Wang L, Zhang M, Yang H, et al. Zscan4 promotes genomic stability during reprogramming and dramatically improves the quality of iPS cells as demonstrated by tetraploid complementation. Cell Res. 2013;23:92–106.

37. Rasmussen MA, Holst B, Tumer Z, Johnsen MG, Zhou S, Stummann TC, et al. Transient p53 suppression increases reprogramming of human fibroblasts without affecting apoptosis and DNA damage. Stem Cell Reports. 2014;3:404–13.

38. Kawamura T, Suzuki J, Wang YV, Menendez S, Morera LB, Raya A, et al. Linking the p53 tumour suppressor pathway to somatic cell reprogramming. Nature. 2009;460:1140–4.

39. Unternaehrer JJ, Zhao R, Kim K, Cesana M, Powers JT, Ratanasirintrawoot S, et al. The epithelial-mesenchymal transition factor SNAIL paradoxically enhances reprogramming. Stem Cell Reports. 2014;3:691–8.

40. Liu X, Sun H, Qi J, Wang L, He S, Liu J, et al. Sequential introduction of reprogramming factors reveals a time-sensitive requirement for individual factors and a sequential EMT-MET mechanism for optimal reprogramming. Nat Cell Biol. 2013;15:829–38.

7

Chm-1 gene-modified bone marrow mesenchymal stem cells maintain the chondrogenic phenotype of tissue-engineered cartilage

Zhuoyue Chen[1,2†], Jing Wei[1†], Jun Zhu[1], Wei Liu[1], Jihong Cui[1,2], Hongmin Li[1,2] and Fulin Chen[1,2*]

Abstract

Background: Marrow mesenchymal stem cells (MSCs) can differentiate into specific phenotypes, including chondrocytes, and have been widely used for cartilage tissue engineering. However, cartilage grafts from MSCs exhibit phenotypic alternations after implantation, including matrix calcification and vascular ingrowth.

Methods: We compared chondromodulin-1 (Chm-1) expression between chondrocytes and MSCs. We found that chondrocytes expressed a high level of Chm-1. We then adenovirally transduced MSCs with *Chm-1* and applied modified cells to engineer cartilage in vivo.

Results: A gross inspection and histological observation indicated that the chondrogenic phenotype of the tissue-engineered cartilage graft was well maintained, and the stable expression of Chm-1 was detected by immunohistological staining in the cartilage graft derived from the Chm-1 gene-modified MSCs.

Conclusions: Our findings defined an essential role for Chm-1 in maintaining chondrogenic phenotype and demonstrated that Chm-1 gene-modified MSCs may be used in cartilage tissue engineering.

Keywords: Chondromodulin-1, Mesenchymal stem cells, Chondrocytes, Cartilage tissue

Background

Cartilage regeneration and repair is often needed in orthopedic or plastic and reconstructive surgery for the treatment of cartilaginous defects and malformations. Unlike other self-repairing tissues, cartilage is an avascular tissue characterized by a low cell density and limited nutrient supply [1, 2]. Because of the limited regenerative capacity of cartilage, the treatment of various cartilaginous lesions remains a challenge to clinicians.

Tissue engineering provides an optimized alternative for cartilage regeneration and repair by combining chondrogenic cells and a scaffold [3, 4]. Chondrocytes are most commonly used for cartilage tissue engineering. However, the harvesting of a cartilage biopsy to obtain

primary chondrocytes may cause donor site morbidity, and chondrocytes will dedifferentiate and lose their chondrogenic phenotype during monolayer expansion [5]. Compared with chondrocytes, marrow mesenchymal stem cells (MSCs) could be easily isolated, expanded, and directed to differentiate into mesodermal lineages, including bone, cartilage, and adipose tissue [6, 7]. MSCs may undergo chondrogenic differentiation with the induction of transforming growth factor beta (TGF-β), especially under micropellet culture condition [8], and great efforts have been made to use MSCs for cartilage tissue engineering [9–11]. However, cartilage grafts from MSCs undergo gradual histological changes after implantation, including chondrocyte hypertrophy, extracellular matrix calcification, and vascular invasion, which may significantly influence the treatment outcome of cartilage defects [12, 13].

Coculturing with chondrocytes was usually employed to maintain the stable chondrogenic phenotype of MSCs.

* Correspondence: chenfl@nwu.edu.cn
†Equal contributors
[1]Laboratory of Tissue Engineering, Faculty of Life Science, Northwest University, 229 TaiBai North Road, Xi'an, Shaanxi Province 710069, P.R. China
[2]Provincial Key Laboratory of Biotechnology of Shaanxi, Northwest University, 229 TaiBai North Road, Xi'an, Shaanxi Province 710069, P.R. China

The approach could improve collagen type II and gly-cosaminoglycan expression as well as the deposition of MSCs. The mechanism was attributed to signaling via direct cell–cell contacts [14–16] and paracrine factors secreted by chondrocytes [17, 18]. Kang et al. [19] demonstrated that a 1:1 ratio of chondrocytes to MSCs can be used to engineer phenotypically stable cartilage, and obvious vascular invasion could not be observed 6 weeks after in-vivo implantation. However, engineering cartilage with co-seeding of chondrocytes and MSCs still requires the surgical harvesting of cartilage biopsy to obtain chondrocytes [20].

Chondromodulin-1 (Chm-1) is a glycoprotein with 25 kDa molecular weight and is found highly expressed in cartilage tissue [21]. Chm-1 could inhibit the endothe-lial cell proliferation and tube morphogenesis, induce apoptisis of vascular endothelial cells in vitro, as well as inhibit angiogenesis in the chick chorioallantoic mem-brane [22–24]. Mature cartilage contains considerable amounts of Chm-1, is avascular, and its extracellular matrix does not calcify. Meanwhile healing cartilage from MSCs within the cartilage lesions of the knee joint lacked Chm-1 expression, and exhibited excessive ossifi-cation and vascularization [25]. Based on these findings, we may deduce that the phenotypic drift of cartilage grafts from MSCs after in-vivo implantation is due to the low expression of Chm-1 in MSCs.

In the current experiment, we first compared Chm-1 expression profiles in MSCs and chondrocytes. We then engineered phenotypically stable cartilage grafts from Chm-1 gene-modified MSCs. Coral has an interconnec-tive porous structure and good osteoconductive activity which are suitable for blood vessel invasion and tissue ossification [26–28]. We chose coral as the cell-seeding scaffold to investigate the critical effect of Chm-1 on antivascularization and maintaining the chondrogenic phenotype of tissue-engineered cartilage with MSCs.

Methods

All reagents were purchased from Sigma-Aldrich (St. Louis, MO, USA) unless otherwise specified.

Isolation and culture of MSCs and chondrocytes

Rabbit MSCs were isolated and cultured as reported previously [28]. New Zealand rabbits (1 month old) were obtained from the animal holding unit of Four Military Medical University (FMMU, Xi'an, Shaanxi Province, P.R. China) and samples of bone marrow were harvested in ac-cordance with IACUC approval from Northwest Univer-sity, Xi'an, P.R. China. Briefly, the obtained marrow was suspended and cultured in Dulbecco's modified Eagle medium (DMEM; Gibco BRL, Grand Island, NY, USA) containing 10 % fetal bovine serum (FBS), 272 μg/ml L-glutamine, and 100 U/ml penicillin/streptomycin. The media were changed every 3 days. Before the cells formed a confluent monolayer, they were digested using trypsin 0.25 % and harvested by centrifugation, and cells of pas-sage 2 were used for the experiment. The cell density was adjusted to 5×10^7 cells/ml with medium before cell seeding.

Rabbit chondrocytes were isolated and cultured ac-cording to the method described by Wu et al. [29]. All New Zealand rabbits were anesthetized with ketamine (40 mg/kg, intramuscularly) and xylazine (5 mg/kg, intramuscularly). After aseptic preparation, the auricle cartilage from ear roots was dissected and cut into pieces of approximately 2 mm^3 after being rinsed three times with phosphate-buffered saline (PBS) supple-mented with 100 U/ml penicillin and 100 U/ml strepto-mycin; the cartilage samples were digested with 0.2 % collagenase type II (Gibco) in DMEM (Gibco) at 37 °C for 12 hours. The digested cell suspension was filtered through a 250 mm nylon mesh filter to remove matrix debris and was centrifuged at 1000 rpm for 5 minutes; the resulting cell pellet was washed twice with PBS and resus-pended with DMEM containing 10 % FBS, L-glutamine (272 μg/ml), and ascorbic acid (5 μg/ml). The medium was changed every 3 days. The chondrocytes were subcul-tured twice, collected by trypsin digestion, and suspended in culture medium at a density of 5×10^7 cells/ml for seeding.

RNA isolation and reverse transcription-PCR

The expression levels of *Chm-1*, *Col II*, and *AGG* in chondrocytes and MSCs were compared by reverse transcription-PCR (RT-PCR). Total RNA was isolated from chondrocytes and MSCs using TRIzol Reagent (Invitrogen, Carlsbad, CA, USA). For cDNA synthesis, RT-PCR was performed using the Takara RT-PCR Kit (Takara, Dalian, China). The reaction product cDNA was used as a template for PCR amplification. The PCR conditions were as follows: initial denaturation at 94 °C for 3 minutes; 30 cycles at 94 °C for 40 seconds, 60 °C for 40 seconds, and 72 °C for 80 seconds; and a final ex-tension at 72 °C for 5 minutes. The PCR products were visualized on 1.5 % agarose gels. *GAPDH* was used as an internal control. The primer sequences used for this analysis are presented in Table 1. Band intensity was quantified using Bandscan software. The gray values of bands were normalized relative to those of *GAPDH*. The gray values were expressed in relation to the control and presented as means ± SD from four independent experiments.

Western blot analysis

Western blot analysis was carried out for Chm-1, Col II, and AGG expression of MSCs and chondrocytes. Equal amounts of protein extracts (30 μg/lane) were separated

Table 1 Primer sequences for reverse transcription-PCR

Gene	Primers	Product size (base pairs)
GAPDH	TCACCATCTTCCAGGAGCGA	293
AGG	CACAATGCCGAAGTGGTCGT GGTCGTGGTGAAAGGTGTTGT	315
Col II	GCAGACGCATGAAGGCAAGTT AGCAGCAGCACGTGTGGTT	97
Chm-1	ATCTGGACGTTGGCAGTGTTG CCGCTCGAGCATGACCGAGAACTCGGACA	1022
	CCGGAATTCGCACCTGATACGCAAAGTGA	

by SDS-PAGE and transferred to the nitrocellulose membrane. Nonspecific binding was blocked with TBS buffer (50 mM Tris/HCl and 150 mM NaCl) containing 5 % (w/v) skimmed milk for 2 hours at room temperature. The membranes were then incubated with primary antibodies (1:1000 (v/v) for GAPDH, 1:1000 (v/v) for Chm-1, 1:1000 (v/v) for Col II, and 1:1000 (v/v) for AGG; Santa Cruz, Santa Cruz, California, USA) for 2 hours at 37 °C. After washing with TBS containing 0.05 % Tween-20 (TBST) three times, the membranes were incubated for 1 hour at 37 °C with secondary antibodies conjugated with horseradish peroxidase diluted 1:1000 in TBST. Finally, the membranes were treated with enhanced chemoluminescence (ECL) reagent (Santa Cruz) and exposed to Kodak X-ray film. GAPDH acted as internal control.

MSCs modified with Chm-1 gene

Chm-1 cDNAs were amplified from rabbit chondrocytes by PCR using the primers presented in Table 1. The PCR products were subcloned into the pDC316 expression adenovirus vector (pDC316-Chm-1) after restriction enzyme digestion. According to the manufacturer's instructions for the AdMax Kit D (Microbix Biosystems Inc.,Toronto, ON, Canada), HEK 293 producer cells were cotransfected with pBHGlox E1, 3Cre, and expression adenovirus vector (pDC316-Chm-1) to obtain the adenovirus-containing Chm-1 gene (Ad5-Chm-1).

Second-passage rabbit MSCs were transduced with an adenovirus containing either green fluorescent protein (GFP) (Ad5-GFP) or Chm-1 (Ad5-Chm-1) for 72 hours at a multiplicity of infection (MOI) of 1000 plaque-forming units (PFU)/cell. The efficiency of adenovirus gene transfer in MSCs was evaluated under a fluorescence microscope 72 hours after transfection.

Expression of Chm-1 in Ad5-Chm-1-transfected MSCs

Total RNA was isolated from MSCs before infection and 72 hours after infection using TRIzol Reagent (Invitrogen). For cDNA synthesis, the total RNA was reverse-transcribed using the Takara RT-PCR Kit for RT-PCR. The reaction product cDNA was used as a template for PCR amplification. The PCR conditions were as follows: initial denaturation at 94 °C for 3 minutes; 30 cycles at 94 °C for 40 seconds, 60 °C for 40 seconds, and 72 °C for 80 seconds; and a final extension at 72 °C for 5 minutes. PCR products were visualized on 1.5 % agarose gels. GAPDH was used as the internal control. The primer sequences used for this analysis are presented in Table 1. The gray values of bands were normalized relative to those of GAPDH. The gray values were expressed in relation to the control and presented as means ± SD from four independent experiments. To confirm the bioactivity of transgenic Chm-1, the expression levels of Chm-1, Col II, and AGG in Ad5-Chm-1-transfected MSCs (T-MSCs) were measured by western blot analysis. The western blot analysis was carried out as described previously. Furthermore, the Chm-1 and Col II proteins were detected by immunofluorescence. Briefly, the MSCs, chondrocytes, and T-MSCs were washed three times with PBS (pH 7.4). Cells were fixed for 10 minutes by incubating in 4 % formaldehyde in PBS followed by further washing and preincubation with 1 % bovine serum albumin (BSA) for 30 minutes. Incubation was with anti-Chm-1 antibody (Santa Cruz) and anti-Col II antibody (Santa Cruz) for 20 minutes at room temperature. Next, the samples were rinsed in PBS followed by incubation with Cy3-conjugated antimouse secondary antibody (Calbiochem, Darmstadt, Germany) for 20 minutes at room temperature, PBS washing, and finally staining with 5 mg/ml Hoechst 33342 for 30 minutes. The fluorescence images from stained samples were obtained using a confocal laser scanning microscope (FV1000; Olympus Corporation, Tokyo, Japan).

Construction of cell–scaffold complex

Natural coral (Gonophoresduofaciata, Hainan, China) was carefully molded into the shape of a tube that was 8 mm in diameter and 2 mm thick. The material was treated as described previously [30]. Briefly, it was immersed in 50 mg/ml sodium hypochlorite for 14 days, and the medium was changed every other day to remove foreign protein in the coral. The scaffold was then washed with distilled water and autoclaved before use.

In preparation for cell seeding, empty scaffolds were prewetted in culture medium for 10 minutes. Cells were then pipetted onto each scaffold to achieve a final seeding number of 2×10^6 cells in 40 µl suspensions. Each scaffold was seeded with cells which were

i) chondrocytes, ii) 1:1 mixture of MSCs and chondrocytes, iii) MSCs, and iv) T-MSCs. The cell–scaffold complexes were placed into dishes and moved to the incubator for 4 hours to ensure that most cells adhered to the scaffolds. Then, 2 ml of medium was carefully added around the complexes. Twelve hours later, an additional 10 ml of medium was added. The composites were then incubated for 5 days to allow for cell attachment. Prior to implantation, the scaffolds were rinsed in sterile PBS, stained with Hoechst 33342, and observed through a fluorescence microscope (Nikon, Tokyo, Japan). Twelve hours and 5 days after cell seeding, a PicoGreen DNA quantitation assay [31] was used to monitor cell-seeding efficacy and proliferation on coral scaffolds ($n = 4$). The DNA quantitation of 2×10^6 cells before seeding acted as controls.

Subcutaneous implantation in nude mice

Eight BALB/c nude mice (6 weeks old, from the animal holding unit of FMMU) were used for the experiment. The animals were acclimated for 1 week before the surgery and monitored for general appearance, activity, excretion, and weight. All procedures were approved by the IACUC of Northwest University. Before implantation, the nude mice were anesthetized by intraperitoneal injection of 10 % chloral hydrate (300 mg/kg). After aseptic preparation, the skin on the back was incised and a subcutaneous pocket was made. The coral–implant composite scaffolds loaded with cells (four kinds of composites: scaffold loaded with i) chondrocytes, ii) 1:1 chondrocytes and MSCs, iii) MSCs, and iv) T-MSCs) were implanted into eight animals (each animal received four kinds of implant). After 1 and 2 months of implantation, the animals ($n = 4$, at each time point) were sacrificed by neck dislocation and specimens were harvested. The specimens were observed by gross inspection and then fixed with 10 % phosphate-buffered formalin.

Analyses of Chm-1 distribution and vascularization in newly formed tissue

After the specimens were fixed in 10 % phosphate-buffered formalin for 24 hours and demineralized in 5 % formic acid for 5 days, they were dehydrated in graded alcohols and embedded in paraffin before preparing sections 7 µm thick. The sections were stained with hematoxylin and eosin (H & E) and toluidine blue (TB) to evaluate the cartilaginous matrix. Finally, Masson's trichrome staining (MTS) was utilized to detect the vascular structure [32].

Immunofluorescence was performed using the following primary antibodies: anti-Col II antibody (Santa Cruz) and anti-Chm-1 antibody (Santa Cruz). All incubations were performed in a humidified chamber. Next, the sections were rinsed in PBS followed by incubation with FITC-conjugated antimouse secondary antibody (Calbiochem).

Finally, the sections were examined using a fluorescence microscope.

The blood vessel density was quantitatively analyzed from MTS of representative sections from each specimen. Vessels were identified by their luminal structure and the presence of red blood cells stained yellow within their boundaries. Vessels were counted from four random fields of each section under 200× magnification as the vessel number in each specimen. The vessel density in each group was determined by the average number of blood vessels from specimens ($n = 4$).

Statistical analysis

Results are reported as the mean ± SD, and significance was determined using a probability value of $P < 0.05$. The significance of differences between groups was assessed using a two-way analysis of variance (ANOVA) with Tukey's post-hoc analysis.

Results

Different expression of *Chm-1* gene in chondrocytes and MSCs

According to the RT-PCR results, both chondrocytes and MSCs expressed genes encoding cartilage-specific matrix proteins, including *Col II* (97 base pairs (bp)) and *AGG* (315 bp) (Fig. 1a). Importantly, high *Chm-1* gene (1000 bp) expression was found in chondrocytes, whereas expression of *Chm-1* gene was low in MSCs (Fig. 1a). Furthermore, Chm-1, Col II, and AGG protein levels were also measured using western blot. As shown in Fig. 1b, Chm-1 protein was specifically expressed in chondrocytes at a level higher than that of MSCs ($P < 0.01$). Immunofluorescence tests further demonstrated that the expression of Chm-1 differed in chondrocytes and MSCs (Fig. 1c). The result demonstrated that Chm-1 was specifically expressed in chondrocytes.

Transfection of MSCs with *Chm-1* gene

As shown in Fig. 1a, the *Chm-1*, *Col II*, and *AGG* genes were detected by RT-PCR in the T-MSCs, whereas the expression of *Chm-1* gene was very low in the untransfected MSCs. To further confirm chondrogenic phenotype on a protein level, Chm-1, Col II, and AGG protein levels were also measured using western blot. As shown in Fig. 1b, Chm-1 was specifically expressed in T-MSCs at a level higher than that of untransfected MSCs ($P < 0.01$). Meanwhile, Col II and AGG were significantly expressed in T-MSCs at levels higher than those of MSCs ($P < 0.01$). An immunofluorescence observation (Fig. 1c) indicated that T-MSCs expressed Chm-1 and Col II, whereas no staining could be observed in untransfected MSCs. These findings indicated that adenovirus-mediated transfection successfully generated Chm-1 gene-modified MSCs.

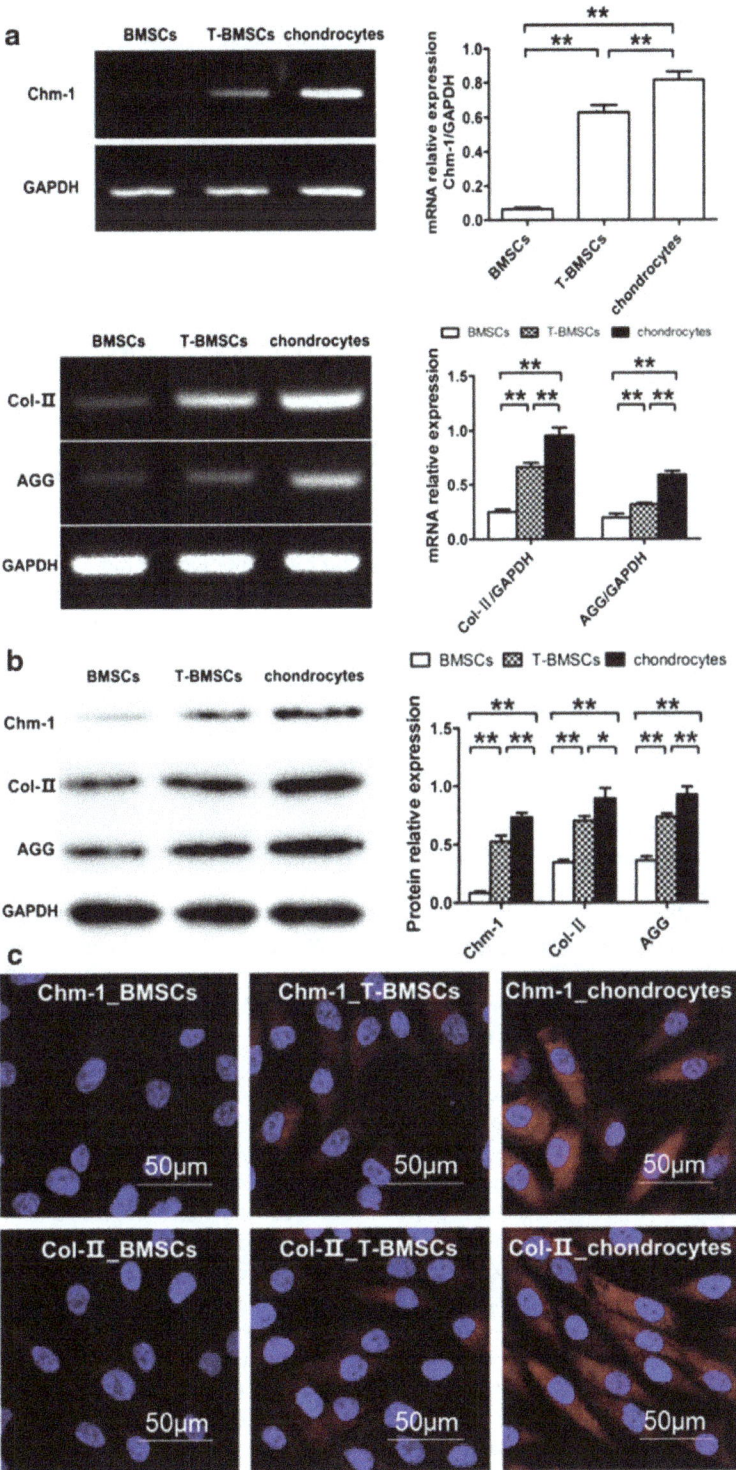

Fig. 1 (See legend on next page.)

(See figure on previous page.)
Fig. 1 a RT-PCR evaluation of *Chm-1*, *Col II*, and *AGG* gene expression in chondrocytes, MSCs, and T-MSCs cultured in vitro. *GAPDH* was used as an internal control. Intensity levels showing *Chm-1*, *Col II*, and *AGG* mRNA expression levels in T-MSCs and chondrocytes were significantly higher than those of MSCs (**$P < 0.01$, mean ± SD, $n = 4$). **b** Western blot assay for Chm-1, Col II, and AGG expression. Chm-1, Col II, and AGG expression were significantly upregulated in T-MSCs and chondrocytes. Analysis of band intensities indicated that Chm-1, Col II, and AGG expression levels higher in T-MSCs and chondrocytes than the levels in MSCs (**$P < 0.01$, mean ± SD, $n = 4$). **c** Immunofluorescence examination showed cell transduction efficiency 5 days after transduction. Chm-1 and Col II protein expression was evaluated by immunofluorescence examination. Data are represented as mean ± standard deviation from four independent sets of experiments. *$P \leq 0.05$, **$P \leq 0.01$, and ***$P \leq 0.001$ *BMSC* bone marrow mesenchymal stem cell, *Chm-1* Chondromodulin-1, *T-BMSC* Ad5-Chm-1 transfected BMSC

Construction of cell–scaffold complex

The coral scaffold is shown in Fig. 2a. Approximately 2×10^6 cells were seeded on each coral scaffold to form a cell–scaffold complex (Each scaffold loaded with i) chondrocytes, ii) 1:1 mixture of MSCs and chondrocytes, iii) MSCs, and iv) T-MSCs, respectively.). According to Fig. 2c, cell-seeding efficiency was around 75–78 %, and cells proliferated 1.32–1.45 times on coral scaffolds 5 days after seeding in each group. There were no significant differences among groups. Cell nuclei were stained with Hoechst 33342 and visualized by fluorescence microscope. The fluorescent micrographs (Fig. 2b) indicated that the internal structure of the coral scaffold was porous, and that cells were evenly distributed throughout the coral scaffolds. The results indicated that the scaffold was biocompatible and able to support the initial attachment and subsequent proliferation of MSCs in vitro.

In-vivo evaluation of the tissue-engineered cartilage

Figure 2d shows the gross appearance of specimens from different groups 1 and 2 months after implantation. Specimens from the chondrocyte-seeding (Fig. 2d,*i*) and chondrocyte–MSC-coseeding (Fig. 2d,*ii*) groups appeared light red. Specimens from the MSC-seeding group (Fig. 2d,*iii*) were dark red, and a thin layer of soft tissue and blood vessels could be observed clearly on the surfaces of the specimens. In contrast, specimens from the Chm-1-transfected MSCs (Fig. 2d,*iv*) could be separated easily from the adherent fibrous capsule and were light red.

H & E staining, TB staining, and MTS were performed to examine the tissue formation and vessel density in the scaffolds. H & E staining, TB staining, and MTS observation did not reveal obvious vascularization and bone formation in the chondrocyte-seeding (Fig. 3a) and chondrocyte–MSC-coseeding (Fig. 3b) groups. However, TB staining showed mature bone formation via endochondral ossification in scaffolds seeded only with MSCs (Fig. 3c, arrow), and MTS results (Fig. 3c) revealed active vascularization (arrow) in these specimens. By contrast, a large amount of mature cartilage formed in the pores of the coral scaffold, and bone formation and vascularization were not evident in the T-MSC specimens (Fig. 3d).

Expression of the chondrogenic-specific proteins Col II and Chm-1 was evaluated by immunohistology. As showed in Fig. 4, Col II (green fluorescence) and Chm-1 (green fluorescence) expression were evident in the chondrocyte-seeding (Fig. 4a) and chondrocyte–MSC-coseeding (Fig. 4b) groups but rarely in scaffolds seeded only with MSCs (Fig. 4c). The T-MSCs expressed abundant Col II (green fluorescence) and Chm-1 (green fluorescence), which was evenly distributed throughout the cells (Fig. 4d).

MTS-stained vessels were observed in the specimens after implantation for 1 and 2 months (Fig. 5). The blood vessel density of both groups continuously increased 1–2 months after implantation. Compared with the specimens seeded with MSCs (1 month 47.25 ± 3.59; 2 months 43.00 ± 4.24), the mean blood vessel densities in the chondrocyte–MSC-coseeding groups (1 month 4.50 ± 0.67; 2 months 12.25 ± 2.58) were significantly reduced at each time point ($P < 0.05$). The mean blood vessel density in the chondrocytes:MSCs = 1:1 group did not significantly differ from that in the chondrocyte-seeding group (1 month 4.25 ± 0.53; 2 months 8.00 ± 0.82; $P > 0.05$). MTS indicated that the mean blood vessel density in the T-MSC-seeding group (1 month 9.58 ± 1.85; 2 months 10.08 ± 1.64) was significantly lower than that in the MSC-seeding group (1 month 47.25 ± 3.59; 2 months 43.00 ± 4.24) at each time point ($P < 0.05$).

Discussion

MSCs could undergo chondrogenic differentiation in the presence of the appropriate growth factors and has been considered an attractive cell source for cartilage tissue engineering [8]. However, tissue-engineered cartilage constructed by chondrogenic induction of MSCs exhibits a hypertrophic phenotype and extensive calcification of the extracellular matrix after implantation [12, 13]. Mueller and Tuan [33] reported that the combination of TGF-β withdrawal, dexamethasone reduction, and thyroid hormone addition could induce hypertrophy of chondrogenic induced MSCs, accompanied by increased alkaline phosphatase activity, matrix mineralization, and changes in hypertrophy markers. They concluded that chondrogenically induced MSCs were functionally similar to growth plate chondrocytes, which underwent a differentiation program analogous to endochondral ossification. Vascular invasion is a histological marker of

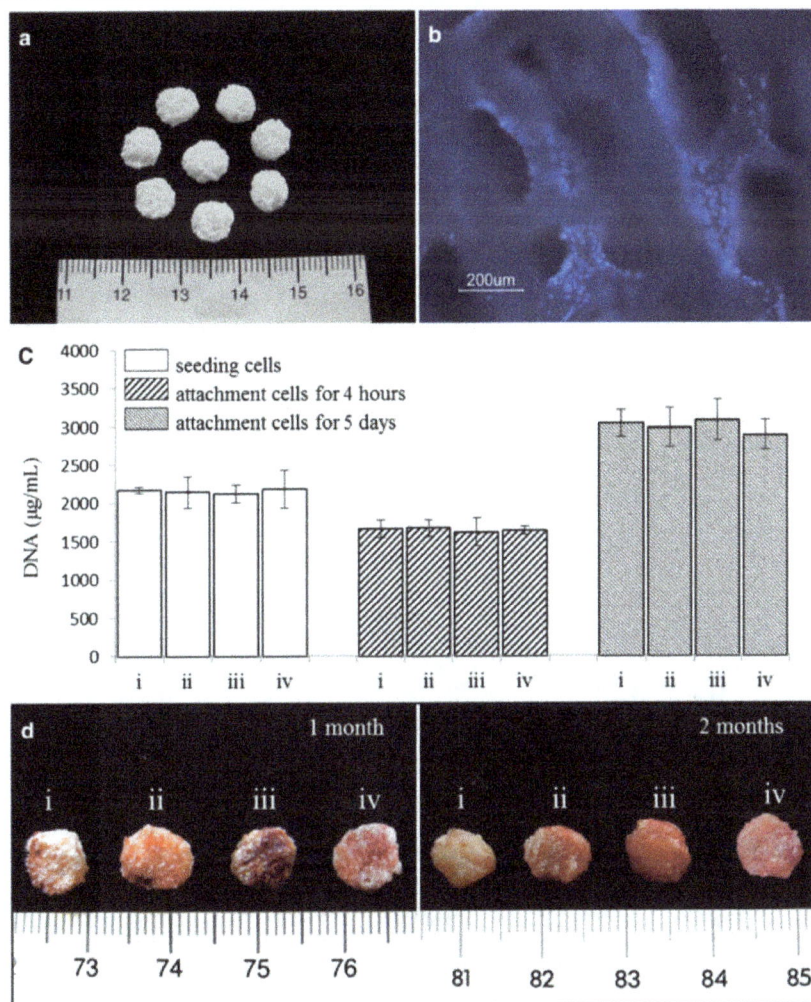

Fig. 2 a Natural coral scaffolds (8 mm in diameter and 2 mm in height). **b** T-MSCs and coral scaffold complex. Fluorescence microscope examination showed the attachment of T-MSCs on the coral scaffold. Nuclei were visualized by Hoechst 33342 staining. **c** Cell-seeding efficacy and proliferation on coral scaffolds ($n = 4$). Each scaffold was seeded with 2×10^6 cells which were *i* chondrocytes, *ii* 1:1 mixture of MSCs and chondrocytes, *iii* MSCs, and *iv* T-MSCs. The initial 2×10^6 cells before seeding from each group acted as control. There were no significant differences among groups ($P > 0.05$). **d** Representative macroscopic pictures of the cell–scaffold composites (*i* chondrocyte–coral composites, *ii* chondrocytes and MSCs coseeded into natural coral scaffolds in ratio of 1:1, *iii* MSC–coral composites, and *iv* T-MSC–coral composites) removed from animals after 1 month and 2 months (Color figure online)

endochondral ossification, and previous studies show that the degree of vascularization significantly differs between cartilage engineered with chondrocytes and cartilage engineered with MSCs [34], which indicate that the regulation of vascular formation may differ greatly between chondrocytes and MSCs. These drawbacks may significantly influence the treatment of cartilage defects with MSCs. Chm-1 is specifically expressed in the avascular area of some mesenchymal tissues, including certain ocular tissues, cardiac valve, and cartilage [23, 35–38]. Chm-1 has been demonstrated to inhibit angiogenesis, and many studies investigate the effects of Chm-1 and the

mechanisms by which it inhibits angiogenesis and disrupts the vasculature [39–41].

Microfracturing of the subchondral bone plate is a frequently employed approach in the clinic. The approach could guide MSCs to migrate into the defect to improve cartilage lesion repair [42]. Blanke et al. [25] reported that cartilage healing from microfracturing lacked the expression of Chm-1, and was associated with excessive matrix calcification and vascular ingrowth. And that additional transplantation of chondrocytes could significantly prevent matrix calcification and vascular ingrowth. In agreement with Blanke et al.'s study, we found that MSCs

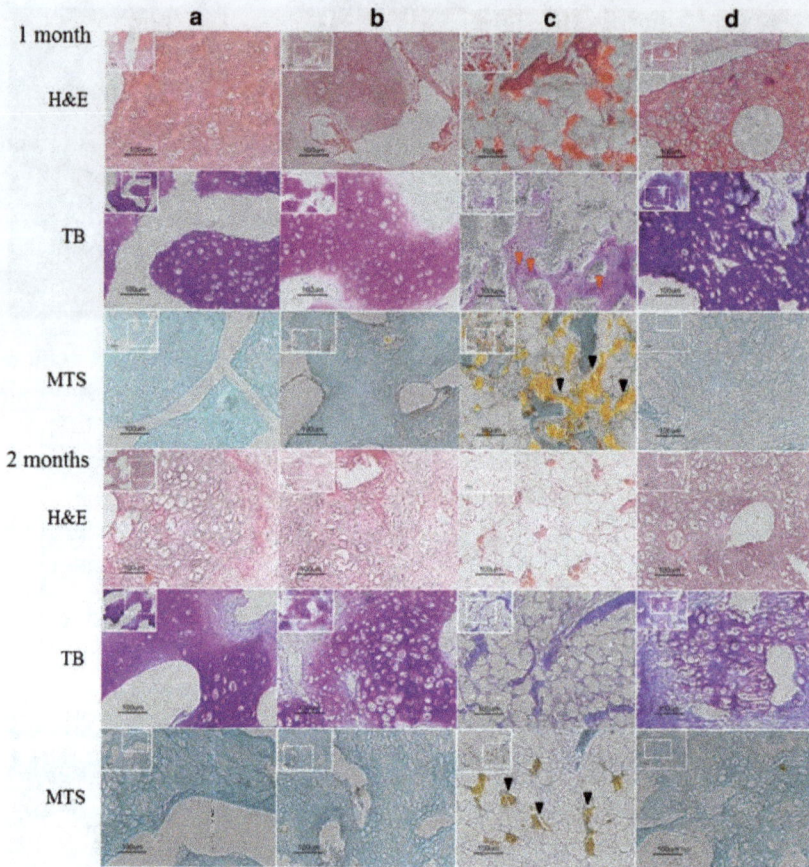

Fig. 3 Histologic analyses in chondrogenically differentiated cells and the vascularization in newly formed tissue in vivo. The specimens were transplanted into nude mice. Appearance of specimens 1 month and 2 months post transplantation for **a** chondrocyte–coral composites, **b** chondrocytes and MSCs coseeded into natural coral scaffolds in a ratio of 1:1, **c** MSC–coral composites, and **d** T-MSC–coral composites. The specimens were processed for histologic staining with H & E, TB, and MTS. *Red arrows* indicate endochondral ossification. *Black arrows* indicate vascular structures. *H&E* hematoxylin and eosin, *MTS* Masson's trichrome staining, *TB* toluidine blue (Color figure online)

expressed cartilage-specific genes including *Col II* and *AGG* (Fig. 1), however, unlike chondrocytes which expressed high levels of *Chm-1*, the expression of *Chm-1* was very low in MSCs (Fig. 1). This finding further demonstrated that native MSCs are not optimized cells for cartilage regeneration.

We then transfected MSCs with Ad5-Chm-1. Interestingly, transfected MSCs not only expressed a high level of Chm-1 but the expression of cartilage-specific genes was also significantly upregulated (Fig. 1). Previous studies have indicated that coculturing MSCs with chondrocytes leads to increased chondrogenic gene expression and ECM deposition in MSCs, and these phenotypic changes are considered to be the result of growth factors secreted by chondrocytes [43, 44]. According to the result of our study, Chm-1 is also a signaling molecule that regulates chondrogenic phenotype of MSCs.

Finally, we fabricated cartilage grafts with Chm-1-transfected MSCs. In this experiment, we used porous coral as a cell-seeding scaffold, which facilitates the vascularization and ossification of engineered tissue [45, 46]. The result showed that mature cartilage formed in the pores of the scaffold, and bone formation was not observed in the specimens 2 months after implantation (Figs. 2, 3, 4). The newly formed tissue also exhibited strong immunohistochemical staining for Chm-1. However, a large amount of bone and marrow tissue formed in the MSC-seeding group 2 months after implantation (Figs. 2, 3, 4). Importantly, the number of blood vessels in the engineered graft was similar to that in the chondrocyte-seeding group ($P > 0.05$) and significantly lower than that in the MSC-seeding group ($P < 0.05$) (Fig. 5). Klinger et al. [47] transfected osteochondral progenitor cells with Chm-1 and subsequently transplanted

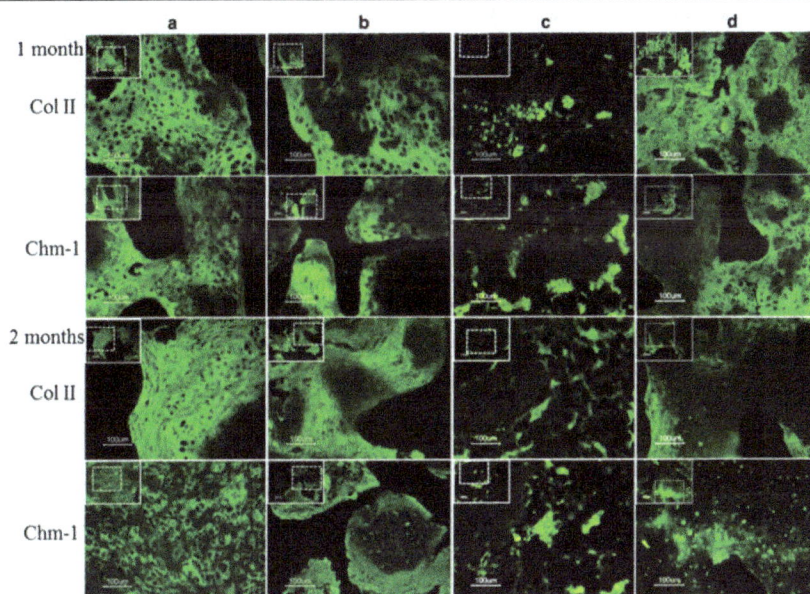

Fig. 4 Analysis of chondrogenic protein expression for neocartilage formation in the specimens. The specimens were transplanted into nude mice. Specimens 1 and 2 months post transplantation: **a** chondrocyte–coral composites; **b** chondrocytes and MSCs coseeded into natural coral scaffolds in a ratio of 1:1; **c** MSC–coral composites; and **d** T-MSC–coral composites. Specimens were processed to analyze the distribution of Col II (*green*) and Chm-1 (*green*) in cells by immunofluorescence (Color figure online)

cells into cartilage lesions of a joint and found that transfected osteochondral progenitor cells maintained chondrogenic phenotype and formed hyaline cartilage. The avascular and hypoxia environment of the joint made it difficult for them to obtain quantified data for the antivascular effect of Chm-1. Compared with Klinger et al., we ectopically implanted Chm-1-transfected MSCs with coral scaffold. Our results indicated that Chm-1 is critical in inhibiting vascularization and maintaining the chondrogenic phenotype of tissue-engineered cartilage from

MSCs, even in a microenvironment suitable for tissue vascularization and ossification.

Several genes have been considered as targets to facilitate cartilage formation by autologous cells, including TGF-β1, BMPs, IGF, and FGF-2 [43, 44]. The results of the current experiment indicated that *Chm-1* could not only upregulate chondrogenic phenotype of MSCs but could also prevent vascularization. Consequently, this gene efficiently maintained the chondrogenic phenotype of engineered cartilage. We conclude that Chm-1 gene-

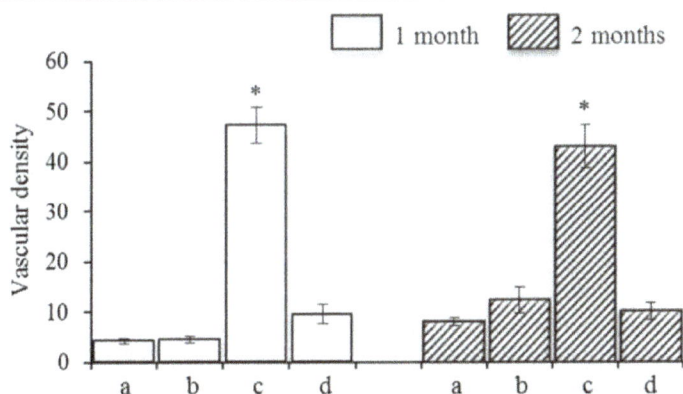

Fig. 5 Number of vessels in the specimens 1 month and 2 months post transplantation were statistically analyzed: **a** chondrocyte–coral composites; **b** chondrocytes and MSCs coseeded into natural coral scaffolds in a ratio of 1:1; **c** MSC–coral composites; and **d** T-MSC–coral composites. Vascular numbers of each specimen were counted per time point as the mean ± SD. Each bar represented four specimens (*$P < 0.05$)

modified MSCs may hold great potential in tissue-engineering applications for cartilage regeneration.

Conclusions

In summary, we transfected rabbit MSCs with Ad5-Chm-1 and seeded these cells into a natural coral tissue-engineering scaffold to investigate the effect of exogenous Chm-1 expression in MSCs. We report that Chm-1 inhibited vascularization and maintained chondrocyte phenotype in vivo in Ad5-Chm-1-transfected MSCs. These results demonstrated that Chm-1-modified MSCs may be an optimized cell source for cartilage tissue engineering.

Abbreviations

Ad5-Chm-1: adenovirus-containing Chm-1 gene; bp: base pairs; Chm-1: Chondromodulin-1; FMMU: Four Military Medical University; GFP: green fluorescent protein; H & E: hematoxylin and eosin; MOI: multiplicity of infection; MSC: mesenchymal stem cell; MTS: Masson's trichrome staining; RT-PCR: reverse transcription-PCR; TB: toluidine blue; TGF-β: transforming growth factor beta; T-MSC: Ad5-Chm-1-transfected mesenchymal stem cell.

Competing interests

The authors declare that they have no competing interests.

Authors' contributions

ZYC, JW, JZ, and WL contributed to the study design, study performance, data collection, and preparation of the manuscript. JHC and HML contributed to the study performance, data analysis and interpretation, and preparation and revision of the manuscript. FLC contributed to the study design, preparation, and revision of the manuscript. ZYC and JW contributed equally to this work and should be considered coauthors. All authors read and approved the final manuscript.

Acknowledgements

This work was supported by the National Natural Science Foundation of P.R. China (No. 31271026, 31300797), Shaanxi Province and Ministry of Education Natural Science Foundation (No. 2013JQ4022, 13JS107), Natural Science Basic Research Plan in Shaanxi Province of China (No. 2013JC2-03), and the District-serving Scientific Research Program funded by Shaanxi Provincial Education Department (No. 15JF032).

References

1. Yu H, Adesida AB, Jomha NM. Meniscus repair using mesenchymal stem cells—a comprehensive review. Stem Cell Res Ther. 2015;6:86.
2. Zhang L, Hu J, Athanasiou KA. The role of tissue engineering in articular cartilage repair and regeneration. Crit Rev Biomed Eng. 2009;37:1–57.
3. Gong Y, Su K, Lau TT, Zhou R, Wang DA. Microcavitary hydrogel-mediating phase transfer cell culture for cartilage tissue engineering. Tissue Eng Part A. 2010;16:3611–22.
4. Jeong CG, Hollister SJ. Mechanical and biochemical assessments of three-dimensional poly(1,8-octanediol-co-citrate) scaffold pore shape and permeability effects on in vitro chondrogenesis using primary chondrocytes. Tissue Eng Part A. 2010;16:3759–68.
5. Matricali GA, Dereymaeker GP, Luyten FP. Donor site morbidity after articular cartilage repair procedures: a review. Acta Orthop Belg. 2010;76:669–74.
6. Mackay AM, Beck SC, Murphy JM, Barry FP, Chichester CO, Pittenger MF. Chondrogenic differentiation of cultured human mesenchymal stem cells from marrow. Tissue Eng. 1998;4:415–28.
7. Yoo JU, Barthel TS, Nishimura K, Solchaga L, Caplan AI, Goldberg VM, et al. The chondrogenic potential of human bone-marrow-derived mesenchymal progenitor cells. J Bone Joint Surg Am. 1998;80:1745–57.
8. Frenz DA, Liu W, Williams JD, Hatcher V, Galinovic-Schwartz V, Flanders KC, et al. Induction of chondrogenesis: requirement for synergistic interaction of basic fibroblast growth factor and transforming growth factor-beta. Development. 1994;120:415–24.
9. Cao L, Yang F, Liu G, Yu D, Li H, Fan Q, et al. The promotion of cartilage defect repair using adenovirus mediated Sox9 gene transfer of rabbit bone marrow mesenchymal stem cells. Biomaterials. 2011;32:3910–20.
10. He CX, Zhang TY, Miao PH, Hu ZJ, Han M, Tabata Y, et al. TGF-beta1 gene-engineered mesenchymal stem cells induce rat cartilage regeneration using nonviral gene vector. Biotechnol Appl Biochem. 2012;59:163–69.
11. Katayama R, Wakitani S, Tsumaki N, Morita Y, Matsushita I, Gejo R, et al. Repair of articular cartilage defects in rabbits using CDMP1 gene-transfected autologous mesenchymal cells derived from bone marrow. Rheumatology (Oxford). 2004;43:980–85.
12. Pelttari K, Winter A, Steck E, Goetzke K, Hennig T, Ochs BG, et al. Premature induction of hypertrophy during in vitro chondrogenesis of human mesenchymal stem cells correlates with calcification and vascular invasion after ectopic transplantation in SCID mice. Arthritis Rheum. 2006;54:3254–66.
13. Mwale F, Girard-Lauriault PL, Wang HT, Lerouge S, Antoniou J, Wertheimer MR. Suppression of genes related to hypertrophy and osteogenesis in committed human mesenchymal stem cells cultured on novel nitrogen-rich plasma polymer coatings. Tissue Eng. 2006;12:2639–47.
14. Chen WH, Lai MT, Wu AT, Wu CC, Gelovani JG, Lin CT. In vitro stagespecific chondrogenesis of mesenchymal stem cells committed to chondrocytes. Arthritis Rheum. 2009;60:450–9.
15. Hendriks J, Riesle J, van Blitterswijk CA. Co-culture in cartilage tissue engineering. J Tissue Eng Regen Med. 2007;1:170–78.
16. Bian L, Zhai DY, Mauck RL, Burdick JA. Coculture of human mesenchymal stem cells and articular chondrocytes reduces hypertrophy and enhances functional properties of engineered cartilage. Tissue Eng Part A. 2011;17:1137–45.
17. Hwang NS, Varghese S, Puleo C, Zhang Z, Elisseeff J. Morphogenetic signals from chondrocytes promote chondrogenic and osteogenic differentiation of mesenchymal stem cells. J Cell Physiol. 2007;212:281–84.
18. Wu L, Leijten JC, Georgi N, Post JN, van Blitterswijk CA, Karperien M. Trophic effects of mesenchymal stem cells increase chondrocyte proliferation and matrix formation. Tissue Eng Part A. 2011;17:1425–36.
19. Kang N, Liu X, Guan Y, Wang J, Gong F, Yang X, et al. Effects of co-culturing BMSCs and auricular chondrocytes on the elastic modulus and hypertrophy of tissue engineered cartilage. Biomaterials. 2012;33:4535–44.
20. Meretoja VV, Dahlin RL, Kasper FK, Mikos AG. Enhanced chondrogenesis in co-cultures with articular chondrocytes and mesenchymal stem cells. Biomaterials. 2012;33:6362–69.
21. Hiraki Y, Tanaka H, Inoue H, Kondo J, Kamizono A, Suzuki F. Molecular cloning of a new class of cartilage-specific matrix, chondromodulin-I, which stimulates growth of cultured chondrocytes. Biochem Biophys Res Commun. 1991;175:971–77.
22. Hiraki Y, Inoue H, Iyama K, Kamizono A, Ochiai M, Shukunami C, et al. Identification of chondromodulin I as a novel endothelial cell growth inhibitor. Purification and its localization in the avascular zone of epiphyseal cartilage. J Biol Chem. 1997;272:32419–26.
23. Funaki H, Sawaguchi S, Yaoeda K, Koyama Y, Yaoita E, Funaki S, et al. Expression and localization of angiogenic inhibitory factor, chondromodulin-I, in adult rat eye. Invest Ophthalmol Vis Sci. 2001;42:1193–200.
24. Hiraki Y, Mitsui K, Endo N, Takahashi K, Hayami T, Inoue H, et al. Molecular cloning of human chondromodulin-I, a cartilage-derived growth modulating factor, and its expression in Chinese hamster ovary cells. Eur J Biochem. 1999;260:869–78.
25. Blanke M, Carl HD, Klinger P, Swoboda B, Hennig F, Gelse K. Transplanted chondrocytes inhibit endochondral ossification within cartilage repair tissue. Calcif Tissue Int. 2009;85:421–33.
26. Dong QS, Shang HT, Wu W, Chen FL, Zhang JR, Guo JP, et al. Prefabrication of axial vascularized tissue engineering coral bone by an arteriovenous loop: a better model. Mater Sci Eng C Mater Biol Appl. 2012;32:1536–41.
27. Cai L, Wang Q, Gu C, Wu J, Wang J, Kang N, et al. Vascular and micro-environmental influences on MSC-coral hydroxyapatite construct-based bone tissue engineering. Biomaterials. 2011;32:8497–505.
28. Geng W, Ma D, Yan X, Liu L, Cui J, Xie X, et al. Engineering tubular bone using mesenchymal stem cell sheets and coral particles. Biochem Biophys Res Commun. 2013;433:595–601.
29. Wu W, Cheng X, Zhao Y, Chen F, Feng X, Mao T. Tissue engineering of trachea-like cartilage grafts by using chondrocyte macroaggregate: experimental study in rabbits. Artif Organs. 2007;31:826–34.
30. Chen F, Chen S, Tao K, Feng X, Liu Y, Lei D, et al. Marrow-derived osteoblasts seeded into porous natural coral to prefabricate a vascularised bone graft in

the shape of a human mandibular ramus: experimental study in rabbits. Br J Oral Maxillofac Surg. 2004;42:532–37.

31. Peach MS, James R, Toti US, Deng M, Morozowich NL, Allcock HR, et al. Polyphosphazene functionalized polyester fiber matrices for tendon tissue engineering: in vitro evaluation with human mesenchymal stem cells. Biomed Mater. 2012;7:45016.

32. Facchiano F, Fernandez E, Mancarella S, Maira G, Miscusi M, D'Arcangelo D, et al. Promotion of regeneration of corticospinal tract axons in rats with recombinant vascular endothelial growth factor alone and combined with adenovirus coding for this factor. J Neurosurg. 2002;97:161–68.

33. Mueller MB, Tuan RS. Functional characterization of hypertrophy in chondrogenesis of human mesenchymal stem cells. Arthritis Rheum. 2008;58:1377–88.

34. Liu X, Sun H, Yan D, Zhang L, Lv X, Liu T, et al. In vivo ectopic chondrogenesis of BMSCs directed by mature chondrocytes. Biomaterials. 2010;31:9406–14.

35. Dietz UH, Ziegelmeier G, Bittner K, Bruckner P, Balling R. Spatio-temporal distribution of chondromodulin-I mRNA in the chicken embryo: expression during cartilage development and formation of the heart and eye. Dev Dyn. 1999;216:233–43.

36. Yoshioka M, Yuasa S, Matsumura K, Kimura K, Shiomi T, Kimura N, et al. Chondromodulin-I maintains cardiac valvular function by preventing angiogenesis. Nat Med. 2006;12:1151–59.

37. Shukunami C, Takimoto A, Miura S, Nishizaki Y, Hiraki Y. Chondromodulin-I and tenomodulin are differentially expressed in the avascular mesenchyme during mouse and chick development. Cell Tissue Res. 2008;332:111–22.

38. Shukunami C, Iyama K, Inoue H, Hiraki Y. Spatiotemporal pattern of the mouse chondromodulin-I gene expression and its regulatory role in vascular invasion into cartilage during endochondral bone formation. Int J Dev Biol. 1999;43:39–49.

39. Tsai AC, Pan SL, Sun HL, Wang CY, Peng CY, Wang SW, et al. CHM-1, a new vascular targeting agent, induces apoptosis of human umbilical vein endothelial cells via p53-mediated death receptor 5 up-regulation. J Biol Chem. 2010;285:5497–506.

40. Mera H, Kawashima H, Yoshizawa T, Ishibashi O, Ali MM, Hayami T, et al. Chondromodulin-1 directly suppresses growth of human cancer cells. BMC Cancer. 2009;9:166.

41. Shukunami C, Hiraki Y. Chondromodulin-I and tenomodulin: the negative control of angiogenesis in connective tissue. Curr Pharm Des. 2007;13:2101–12.

42. Henderson IJ, La Valette DP. Subchondral bone overgrowth in the presence of full-thickness cartilage defects in the knee. Knee. 2005;12:435–40.

43. Han F, Zhou F, Yang X, Zhao J, Zhao Y, Yuan X. A pilot study of conically graded chitosan-gelatin hydrogel/PLGA scaffold with dual-delivery of TGF-beta1 and BMP-2 for regeneration of cartilage-bone interface. J Biomed Mater Res B Appl Biomater. 2015;103:1344–53.

44. Orth P, Kaul G, Cucchiarini M, Zurakowski D, Menger MD, Kohn D, et al. Transplanted articular chondrocytes co-overexpressing IGF-I and FGF-2 stimulate cartilage repair in vivo. Knee Surg Sports Traumatol Arthrosc. 2011;19:2119–30.

45. Karageorgiou V, Kaplan D. Porosity of 3D biomaterial scaffolds and osteogenesis. Biomaterials. 2005;26:5474–91.

46. Hou R, Chen F, Yang Y, Cheng X, Gao Z, Yang HO, et al. Comparative study between coral-mesenchymal stem cells-rhBMP-2 composite and auto-bone-graft in rabbit critical-sized cranial defect model. J Biomed Mater Res A. 2007;80:85–93.

47. Klinger P, Surmann-Schmitt C, Brem M, Swoboda B, Distler JH, Carl HD, et al. Chondromodulin 1 stabilizes the chondrocyte phenotype and inhibits endochondral ossification of porcine cartilage repair tissue. Arthritis Rheum. 2011;63:2721–31.

8

Extracellular vesicles derived from mesenchymal stromal cells: a therapeutic option in respiratory diseases?

Soraia C. Abreu[1], Daniel J. Weiss[2] and Patricia R. M. Rocco[1*]

Abstract

Extracellular vesicles (EVs) are plasma membrane-bound fragments released from several cell types, including mesenchymal stromal cells (MSCs), constitutively or under stimulation. EVs derived from MSCs and other cell types transfer molecules (such as DNA, proteins/peptides, mRNA, microRNA, and lipids) and/or organelles with reparative and anti-inflammatory properties to recipient cells. The paracrine anti-inflammatory effects promoted by MSC-derived EVs have attracted significant interest in the regenerative medicine field, including for potential use in lung injuries. In the present review, we describe the characteristics, biological activities, and mechanisms of action of MSC-derived EVs. We also review the therapeutic potential of EVs as reported in relevant preclinical models of acute and chronic respiratory diseases, such as pneumonia, acute respiratory distress syndrome, asthma, and pulmonary arterial hypertension. Finally, we discuss possible approaches for potentiating the therapeutic effects of MSC-derived EVs so as to enable use of this therapy in clinical practice.

Background

In recent decades, the therapeutic potential and safety of mesenchymal stromal cells (MSCs) has been studied in the context of regeneration and immune modulation of injured tissues [1]. Many studies have demonstrated that, when systemically administered, MSCs are recruited to sites of inflammation through still-incompletely understood chemotactic mechanisms [2], stimulate endogenous repair of injured tissues [3], and modulate immune responses [4]. The beneficial effects of MSCs on tissue repair and regeneration are based on their paracrine activity, characterized by the capacity to secrete growth factors, cytokines, and chemokines, which orchestrate interactions within the microenvironment and influence tissue regeneration. These factors can inhibit apoptosis, stimulate proliferation, promote vascularization, and modulate the immune response [5]. Remarkably, conditioned medium collected from MSCs can convey many of these protective effects, suggesting that soluble factors rather than cell–cell contact are the major mechanism of MSC actions [6].

Notably, a growing body of literature suggests that many of these paracrine effects are mediated by extracellular vesicles (EVs) contained in the conditioned medium. EVs are small, spherical membrane fragments including exosomes, microvesicle particles, and apoptotic bodies in accordance with the recommendations of the International Society for Extracellular Vesicles (ISEV) [7]. The EVs are released by cells that are involved in cell-to-cell communication and are capable of altering the fate and phenotype of recipient cells [8]. The exosomes arise from intracellular endosomes, while the microvesicles originate directly from the plasma membrane. These particle types are secreted from a wide range of different cell types, including T and B lymphocytes, dendritic cells (DCs), mast cells, platelets, and MSCs derived from different tissues (bone marrow, placenta, as well as adipose and lung tissues), and can also be isolated in vivo from body fluids such as urine, serum, and bronchoalveolar lavage fluid (BALF) [9, 10]. Nevertheless, the classification of EVs differs depending on their origin, size, and contents (Table 1). Additionally, the number and nature of EVs may be affected by gender, age, circadian rhythms, fasting state, medication exposure, and physical activity [11]. However, whether these different classes of EVs represent distinct

* Correspondence: prmrocco@gmail.com
[1]Laboratory of Pulmonary Investigation, Carlos Chagas Filho Institute of Biophysics, Federal University of Rio de Janeiro, Av. Carlos Chagas Filho, 373, Ilha do Fundão, Rio de Janeiro, RJ 21941-902, Brazil
Full list of author information is available at the end of the article

Table 1 Characterization of extracellular vesicles

Extracellular vesicles	Origin	Size	Content	Markers
Exosomes	Multivesicular bodies	50–150 nm	Proteins, and lipids, DNA, mRNA and miRNA	CD63, CD81, CD9, heat-shock proteins, Alix, Tsg101, integrin, annexins and MHC classes I and II
Microvesicles	Plasma membrane	150–1000 nm	Proteins, and lipids, DNA, mRNA, miRNA and cell organelles.	Integrins, flotillins and tetraspanins
Apoptotic bodies	Membrane of dying cells	>1 μm	DNA, noncoding RNAs and cell organelles	Surface markers for macrophages

Alix ALG-2-interacting protein X, *MHC* major histocompatibility complex, *miRNA* microRNA, *Tsg101* tumor susceptibility gene 101

biological entities is not evident. Several parameters have been used to characterize the different classes of EVs, including size, ionic composition, sedimentation rate, flotation density on a sucrose gradient, lipid composition, protein cargo, and biogenesis pathway; however, most of these parameters are neither definitive nor exclusive to any specific class of EVs (Fig. 1) [7].

Exosomes range in size from 50 to 150 nm, have a homogeneous shape, and are defined as a subtype of EVs derived from specialized intracellular compartments, the multivesicular bodies (MVBs) [12]. Exosomes are constitutively released from cells, but their release is augmented significantly following activation by soluble agonists (cytokines, chemokines, and growth factors), as

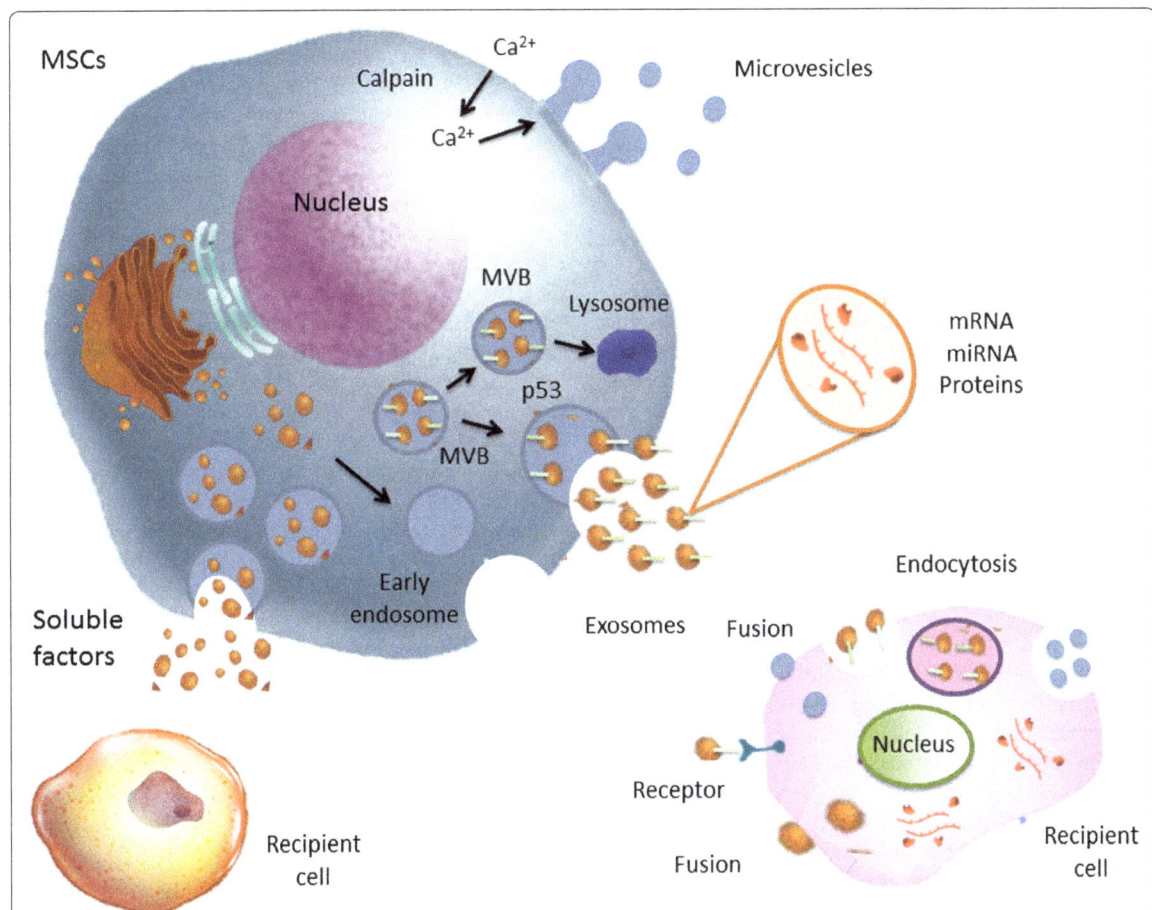

Fig. 1 Schematic representation of EVs biogenesis. Vesicles bud directly from the plasma membrane, whereas exosomes originate from ILVs that are generated by inward budding of the limiting membrane of a subgroup of late endosomes called multivesicular bodies (*MVBs*). MVBs can be directed towards the cell periphery and, after fusion with the plasma membrane, release their content into the extracellular space. *miRNA* microRNA, *MSC* mesenchymal stromal cell

well as physical, chemical (oxidative stress and hypoxia), and shear stresses [13]. In order to form an exosome, the limiting membrane of the MVBs buds inward, thus forming intraluminal vesicles (ILVs), which then fuse with the plasma membrane to release ILVs as exosomes. This process is mediated by p53-regulated exocytosis, which is dependent on cytoskeletal activation but independent of cell calcium influx [14]. In contrast, microvesicles range from 150 to 1000 nm in size and are more heterogeneous. They are released by budding of small cytoplasmic protrusions, a process dependent on calpain, cytoskeletal reorganization, and intracellular calcium concentration. Calcium ions are responsible for the asymmetric phospholipid distribution of the plasma membrane that yields microvesicle formation [14]. Finally, there is another type of EVs, larger than 1 μm: the apoptotic body, derived from dying cells. DNA, as a residue of the nucleus, is frequently present within these vesicles, as are noncoding RNAs and cell organelles [15].

The different EVs can be isolated from body fluids or in vitro cultured cells by specific standardized protocols, and characterized by differential ultracentrifugation, ultrafiltration, and immunoprecipitation with the use of antibody-loaded magnetic cell beads [16]. These procedures are critical because all types of vesicles, as well as membrane fragments, are normally present in the starting material and can contaminate specific EVs preparations. One major challenge in EVs research is therefore to standardize methods for isolation and analysis. Additionally, it is difficult to distinguish between exosomes and microvesicles because of their overlapping characteristics and the lack of discriminating markers [17]. Nevertheless, among the many subtypes of EVs, exosomes have emerged as physiologically relevant and powerful components of the MSC secretome [18].

The content of EVs consists of proteins, lipids, and nucleic acids; microvesicles and apoptotic bodies also have organellar contents. Since the effects of EVs usually depend on their cell of origin and may be influenced by physiological stress or pathological conditions, they could be used as biomarkers to diagnose, prognosticate, or predict diseases and their natural history [14]. Many reports have shown that the functions of EVs reflect, at least in part, those of their originating cells; differences between them occur because the EVs composition may be modified, suggesting that preferential packaging or exclusion of material occurs [19]. Information on the protein, lipid, and RNA expressions of EVs is collected in VESICLEPEDIA (http://www.microvesicles.org) [20], while the exosomes of different cell types and organisms are described in the ExoCarta database [21]. EVs play an important role in intercellular communication and are capable of modifying the activity of target cells through direct surface receptor interactions, receptor transfer between cells, protein delivery to target cells, or horizontal transfer of genetic information [22]. They are involved in cellular processes such as angiogenesis modulation, cell proliferation, and immune regulation [23]. EVs are therefore particularly attractive for their therapeutic potential, especially MSC-derived EVs, which appear to be an important tool to harness the clinical benefits of MSC therapy while using cell-free strategies based on the MSC secretome. These strategies may reduce the risks associated with engraftment of MSCs, such as possible immune reactions against MSCs and development of ectopic tissue. Since EVs carry a wide array of signals, several studies have been performed evaluating their implication in animal models of organ injury, including lung diseases. Nonetheless, comprehensive insight regarding the full scope of molecules packaged in MSC-derived EVs and their role in tissue regeneration has yet to be gained, and additional studies are needed to provide greater detail [9, 23].

Characteristics of MSC-derived EVs

MSC-derived EVs express surface molecules, such as CD29, CD73, CD44, and CD105, which are characteristic of their cells of origin. Among the MSC-derived EVs, the exosomes are those best characterized. Exosomes are known to conserve a set of proteins, including tetraspanins, involved in cell targeting (CD63, CD81, and CD9); heat-shock proteins Hsp60, Hsp70, and Hsp90 [24]; ALG-2-interacting protein X (Alix) and tumor susceptibility gene 101 (Tsg101), which are involved in their biogenesis from MVBs; integrins and annexins, which are important for transport and fusion [20]; and major histocompatibility complex classes I and II [25]. Microvesicles lack proteins of the endocytic pathway, but are rich in cholesterol and lipid raft-associated proteins, such as integrins and flotillins. Although tetraspanins are commonly used as unique markers for exosomes, they can be detected in microvesicles in some cases [26]. Several studies have been conducted evaluating the potential role of MSC-derived EVs in physiological and pathological conditions and their possible applications in the therapy of different diseases [12, 15]; however, few studies have evaluated the RNA and protein content of these vesicles.

MSC-derived EVs are enriched by distinct classes of RNAs that could be transferred to target cells and translated into proteins, resulting in an alteration of target cell behavior [27]. In particular, MSC-derived EVs contain transcripts involved in control of transcription (transcription factor CP2, clock homolog), cell proliferation (retinoblastoma-like 1, small ubiquitin-related modifier 1), and immune regulation (interleukin 1 receptor antagonist) [27]. Additionally, MSC-derived EVs contain noncoding RNA, microRNAs (miRNAs) that mediate posttranscriptional control of gene expression and, as such, modulate survival and metabolic activities of recipient cells [28]. These miRNAs can be present both in EVs and/or in their cells of origin [9]. The miRNAs detected in MSC-derived EVs are usually related to development, cell survival, and differentiation, while some MSC-derived EVs-enriched miRNAs are more closely associated with regulation of the immune system [9]. Comprehensive information on the complete RNA content of MSC-derived EVs is not currently available, however, and whether adult MSCs from different sources share similar RNA repertoires remains unknown. A recent study compared the RNA profile of exosomes released by adult MSCs from two different sources: adipose-derived MSCs (ASCs) and bone marrow-derived MSCs (BM-MSCs). Despite substantial similarity between the most represented RNAs in the ASC and BM-MSC exosomes, their relative proportions are different [29].

Proteome analysis may be equally important. Characterization of the content of BM-MSC-derived EVs identified several proteins, among which are mediators controlling self-renewal and differentiation. Interestingly, this analysis revealed a number of surface markers, such as platelet-derived growth factor receptor, epidermal growth factor receptor, and plasminogen activator, urokinase receptor; signaling molecules of the RAS-mitogen-activated protein kinase, Rho GTPase, and Cell division control protein 42 pathways; cell adhesion molecules; and additional MSC antigens [30], supporting a possible role for such vesicles in tissue repair. Treatment of cell-derived EVs with specific growth factors can change the phenotype and protein content of these vesicles; for example, ASCs treated with platelet-derived growth factor have been shown to produce EVs with enhanced angiogenic activity [31]. This wide distribution of biological activities gives MSC-derived EVs the potential to elicit diverse cellular responses and interact with many cell types.

Mechanisms of action and biological activities of EVs

EVs may interact with recipient cells by different mechanisms: interactions at the cell surface, internalization into endocytic compartments, and fusion with plasma membranes (Fig. 1) [32]. The efficiency of EVs uptake has been observed to correlate directly with intracellular and microenvironmental acidity [33]. Following ligand interaction, EVs may deliver their contents to the recipient cell that reprogrammed them. Recently, EVs from stem cells were demonstrated to shuttle a cysteine-selective transport channel (cystinosin) that restores function in mutant target cells [34]. EVs may also mediate the horizontal transfer of genetic information, such as subsets of mRNA and miRNA, from the cell of origin, thereby inducing alterations in the phenotype

and behavior of recipient cells by different pathways [35]. In this line, EVs produced by murine embryonic stem cells may reprogram hematopoietic progenitors by delivering not only proteins but also mRNA for several pluripotent transcription factors [36], whereas pretreatment of these EVs with RNase inhibited the observed biological effects, thus suggesting the contribution of EVs-derived mRNA [36]. Stem cells may therefore modulate their biological effects by delivering genetic information and altering the gene expression of target cells. Interestingly, the exchange of genetic information may be bidirectional: from injured cells to bone marrow-derived or resident stem cells; or from stem cells to injured cells. In this context, Dooner et al. [37] reported that bone marrow stem cells cocultured with injured lung cells expressed genes for lung-specific proteins, such as surfactant B and C, and Clara cell-specific proteins, which may be attributed to the transfer of lung-specific mRNAs to bone marrow cells via EVs released from the injured lung cells.

Additionally, EVs derived from injured and immune cells may induce stem cell recruitment and differentiation of resident stem cells present in several organs during adulthood, thus contributing to physiologic tissue repair [13]. Nevertheless, depending on their cells of origin, EVs can exert immunostimulatory or immunosuppressive effects [38]. Alveolar macrophages infected with mycobacteria release EVs containing pathogen-derived proinflammatory molecules and secrete Hsp70, which activates the nuclear factor-κB pathway by stimulating toll-like receptors (TLRs) [15], leading to the secretion of proinflammatory cytokines [14, 24]. On the other hand, EVs secreted by DCs are able to induce humoral responses against antigens processed by DCs before EVs purification, yielding strong protection against infection [39]. EVs may also modulate the function of target cells. For instance, EVs derived from lipopolysaccharide-activated monocytes induce apoptosis in target cells through transfer of caspase-1 [40]. Furthermore, proteomic analysis of damaged tissues usually reveals they are depleted of many rate-limiting ATP-generating enzymes, and are thus unable to utilize the restored oxygen supply to produce ATP. This depletion could be supplemented by the proteome of MSC-derived exosomes, which has a cargo rich in enzymatically active glycolytic enzymes and other ATP-generating enzymes, such as adenylate kinase and nucleoside-diphosphate kinase [41].

However, MSC-derived EVs have received more emphasis in the literature and have been most widely studied. In this line, EVs released from human MSCs have been shown to contain ribonucleoproteins involved in the intracellular trafficking of RNA and selected patterns of miRNA, suggesting dynamic regulation and compartmentalization of RNA involved in the development, regulation, regeneration, and cell differentiation, which contribute to recovery processes after injury of adult tissues (Fig. 2) [42]. Indeed, MSC-derived EVs exert an important inhibition in the differentiation and activation of T cells and their interferon-gamma (IFN-γ) release in vitro, as well as stimulating the secretion of anti-inflammatory cytokines (interleukin (IL)-10 and transforming growth factor beta (TGF-β)) and generation of regulatory T cells [43], suggesting that MSC-derived exosomes are relevant immunomodulatory therapeutic agents (Fig. 2). Additionally, treatment with MSC-derived EVs activates an M2 macrophage-like phenotype in the lung parenchyma, which is known to promote tissue repair and limit injury [44].

The immunomodulatory effects of BM-MSCs and derived EVs have been analyzed in vitro. BM-MSCs and their EVs exhibit similar inhibitory activity against B-cell proliferation, but EVs display less inhibitory activity on differentiation and antibody release of B cells compared with BM-MSCs. Moreover, BM-MSCs are more efficient than EVs at inhibiting T-cell proliferation. In one study, incubation of both T cells and B cells with EVs led to a decrease in granulocyte–macrophage colony-stimulating factor and IFN-γ and an increase in IL-10 and TGF-β compared with BM-MSCs [45].

Therapeutic potential of MSC-derived EVs in lung diseases

MSC-derived EVs have shown to be a promising therapy enabling tissue repair and wound healing. The effects of MSC-derived EVs can be potentiated under some conditions, such as exposure to hypoxia and coculture with animal or human serum obtained in pathologic conditions. These methods may induce the release and potentiate the effects of these EVs due to stimulation and the presence of cytokines and chemotactic and growth factors, which not only increase EVs release but also modify their content, leading to enhancement of beneficial effects.

EVs are also important vehicles for drug delivery because of their lipid bilayer and aqueous core, since they can carry both lipophilic and hydrophilic drugs [46]. Furthermore, EVs present several advantages for this purpose, such as: presence of protein and genetic materials, which enables active loading of biological material; high tolerability in the body due to the presence of inhibitors of complement and phagocytosis [30]; protection against degradative enzymes or chemicals; and ability to cross the plasma membrane to deliver their cargo to target cells [9, 47] and home to target tissues [9, 46]. Electroporation [48] and viral packaging strategies [49] have been used to load therapeutically

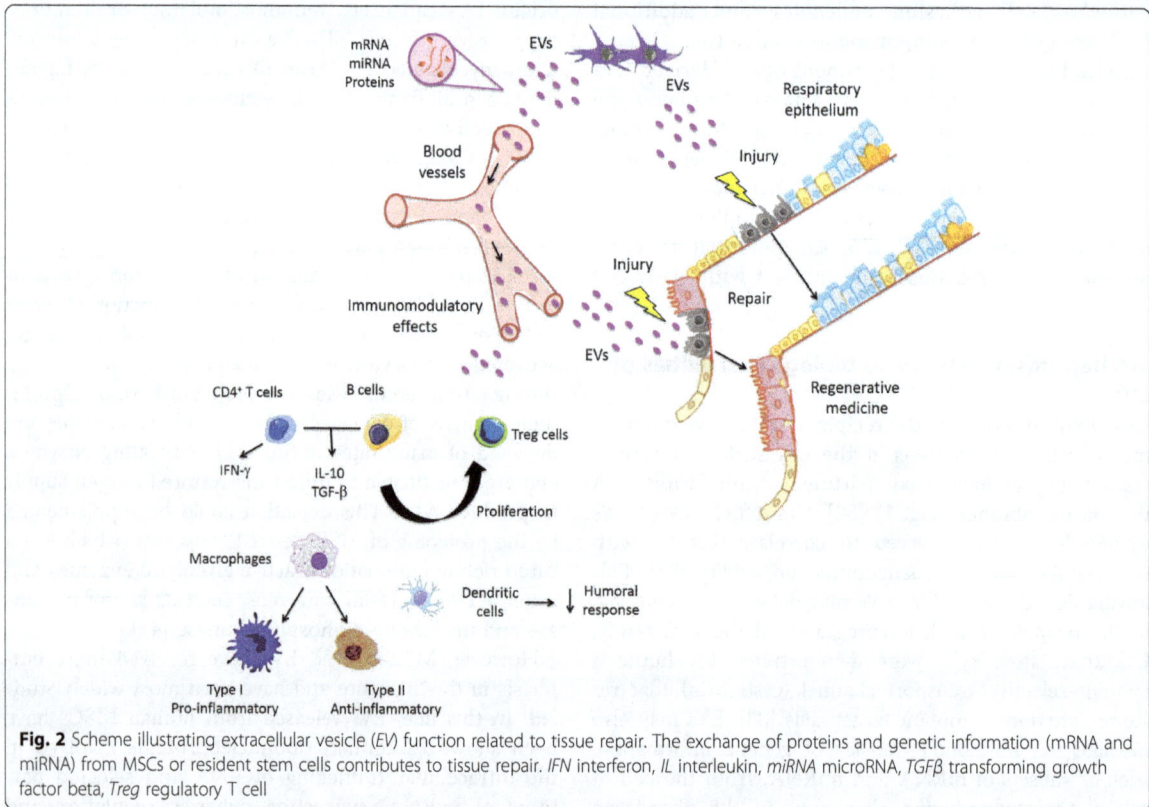

Fig. 2 Scheme illustrating extracellular vesicle (*EV*) function related to tissue repair. The exchange of proteins and genetic information (mRNA and miRNA) from MSCs or resident stem cells contributes to tissue repair. *IFN* interferon, *IL* interleukin, *miRNA* microRNA, *TGFβ* transforming growth factor beta, *Treg* regulatory T cell

active cargo molecules (e.g., small-molecule drugs or small interfering RNA (siRNA)) into EVs [48, 49].

While a predominant mechanism of MSCs in tissue repair through paracrine activity has already been suggested, some studies are being carried out to better understand the mechanisms associated with the beneficial effects of MSC-derived EVs in lung diseases, such as asthma, pulmonary arterial hypertension (PAH), acute respiratory distress syndrome (ARDS), and pneumonia (Table 2), and how they can be potentiated for translation to clinical practice.

Asthma

Asthma is a chronic inflammatory disease characterized by airway constriction and inflammation, which may lead to structural changes in the airways, often in response to allergens, infections, and air pollutants [50]. Even though several therapeutic strategies are currently available to reduce airway inflammation, no treatment has so far been able to hasten repair of the damaged lung [51]. In this line, some studies reported that MSCs reduced lung inflammation and remodeling in experimental allergic asthma [52–54].

EVs are released from several cells that are involved in allergies, including mast cells, DCs, T cells, and bronchial epithelial cells (BECs) in the lungs. For example, mast cell-derived EVs induce DC maturation, and DC-derived EVs can transport allergens and activate allergen-specific T-helper (Th) type 2 cells [55]. Among several potential mechanisms, BECs exposed to compressive stress – thus simulating the bronchoconstriction seen in asthma – produce EVs bearing tissue factor which may participate in promotion of subepithelial fibrosis and angiogenesis [56]. In short, available data indicate the potential contributions of BEC-derived EVs to the pathogenesis of asthma. Additionally, these findings may lead to the development of future treatments for asthma patients that target the inhibition of EVs secretion by these cells.

Several phenotypic and functional alterations have been observed in BALF EVs from asthmatics compared with healthy patients. These include higher expression of CD36, which has been implicated in bacterial recognition and may play a role in asthma exacerbations in response to bacterial infections [57], and that the EVs contain miRNAs, critical regulators of specific pathogenic events [58] which may act as biomarkers of lung diseases, such as the

Table 2 Effects of extracellular vesicles in lung diseases

Study	Model	Origin	Effects
Admyre et al., 2008 [55]	Allergic inflammation	Mast cell-derived EVs	DC maturation, allergen transportation, allergen-specific Th2 cell activation
Bakouboula et al., 2008 [63]	PAH	EVs released from stimulated or endothelial cells undergoing apoptosis	Increase in EVs release is directly related to PAH severity
Prado et al., 2008 [61]	Allergic inflammation	BALF-derived EVs from mice sensitized and challenged with ovalbumin	Inhibition of IgE response, Th2 cytokine production, and airway inflammation
Ionescu et al., 2012 [70]	Endotoxin-induced ARDS	CM from MSC	Increase in secretion of exosomes by MSCs and M2 macrophages, in part via IGF-1
Lee et al., 2012 [65]	Hypoxia-induced PAH	MSC-derived EVs	Reduced right ventricular systolic pressure and right ventricular hypertrophy
Torregrosa et al., 2012 [60]	Coculture of BECs with BALF EVs from asthmatic patients	EVs from BALF of asthmatic patients	Increased leukotriene and IL-8 release
Aliotta et al., 2013 [64]	Monocrotaline-induced PAH	Lung-derived and plasma-derived EVs from monocrotaline-induced PAH	Increased right ventricular mass and pulmonary vascular wall thickness
Zhu et al., 2014 [72]	Escherichia coli endotoxin-induced ARDS	EVs derived from hMSCs	Reduction in extravascular lung water, total protein levels in BALF, edema, neutrophil infiltration, associated with increased KGF expression
Cruz et al., 2015 [62]	Aspergillus hyphal extract-induced allergic inflammation	CM and EVs derived from hMSCs and mMSCs	More significant reduction of airway hyperresponsiveness, lung inflammation and CD4 T-cell Th2 and Th17 phenotype in both CM and EVs from hMSCs compared with mMSCs; inhibition of soluble mediators and EVs release reduced the beneficial effects of all treatments
Monsel et al., 2015 [78]	Escherichia coli pneumonia	MSC and MSC-derived EVs	Improved survival and reduced lung inflammation, protein permeability, and bacterial growth

ARDS acute respiratory distress syndrome, *BALF* bronchoalveolar lavage fluid, *BEC* bronchial epithelial cell, *CM* conditioned medium, *DC* dendritic cell, *EV* extracellular vesicle, *hMSC* human mesenchymal stem cell, *IGF-1* insulin-like growth factor-1, *IL* interleukin, *KGF* keratinocyte growth factor, *MSC* mesenchymal stem cell, *mMSC* mouse mesenchymal stem cell, *PAH* pulmonary arterial hypertension, *Th* T-helper

let-7 (let-7a–let-7e) and miRNA-200 (miR-200b and miR-141) families [59]. Further, incubation of BECs with BALF EVs from asthmatic patients resulted in increased leukotriene and IL-8 release [60].

Additionally, administration of BALF-derived EVs from mice sensitized and challenged with ovalbumin has been shown to inhibit IgE response, Th2 cytokine production, and airway inflammation in experimental asthma [61]. Similar behavior was observed with asthmatic serum-derived EVs, protecting against allergic airway inflammation and reducing BALF eosinophil counts, IgE levels, and Th2 response. EVs from different sources may therefore play a role in the development of asthma and allergy, either as a failure to induce effective tolerance or as enhancers of an already established response. In short, EVs may be therapeutic targets in anti-allergy treatment.

Recently, the therapeutic effects of EVs derived from human MSCs (hMSCs) and mouse MSCs (mMSCs) were investigated in experimental asthma. The authors observed that systemic administration of EVs from either hMSCs or mMSCs were each effective – in some cases, more effective than the administration of hMSCs or mMSCs themselves – in mitigating allergic airway hyperresponsiveness and lung inflammation, and altered the phenotype of antigen-specific CD4 T cells in a model of severe, acute, mixed Th2/Th17-mediated eosinophilic and neutrophilic airway allergic inflammation in immunocompetent mice. Additionally, blocking EVs release led to an absence of protective effects associated with both hMSCs and mMSCs [62].

Pulmonary arterial hypertension

PAH is a disease characterized by hyperplasia and hypertrophy of smooth muscle cells in small pulmonary arteries, associated with an increase in endothelial cell proliferation that leads to remodeling of pulmonary vessels and, consequently, an increase in mean pulmonary arterial pressure and right ventricular overload. Data obtained from patients with PAH show that the severity of PAH is related to an increase in circulating EVs released from stimulated or endothelial cells undergoing apoptosis, probably due to release of soluble vascular cellular adhesion molecule VCAM-1, and that proinflammatory markers, such as monocyte chemoattractant protein MCP-1 and highly specific C-reactive protein, were elevated in PAH patients. In addition, a further increase in endothelium-derived CD105 microparticles was observed in pulmonary arterial blood compared with venous blood in patients with PAH [63]. Inflammation plays an important role in the development of human PAH and there are several animal models of this condition, such as monocrotaline-induced and hypoxia-induced PAH in rodents.

Despite significant progress in elucidating PAH pathophysiology and treatment, few PAH therapies are available

and all have limited effectiveness. Many studies have therefore investigated the effects of MSC therapy in PAH, and demonstrated benefit. In a recent investigation, lung-derived and plasma-derived EVs generated from monocrotaline-induced PAH led to increased right ventricular mass and pulmonary vascular wall thickness, resulting in PAH-like changes in healthy mice This effect may be promoted directly by EVs on the pulmonary vasculature or by differentiation of bone marrow cells to endothelial progenitor cells that induce pulmonary vascular remodeling [64]. This suggests that EVs presented altered expressions of miRNAs involved in pulmonary vascular remodeling. Conversely, in hypoxia-induced PAH, MSC-derived EVs protected against elevation of right ventricular systolic pressure and development of right ventricular hypertrophy, whereas EVs-depleted medium and fibroblast-derived EVs had no effect. These beneficial effects of MSC-derived EVs can be related to suppression of hypoxic pulmonary macrophage influx and hypoxic activation of signal transducer and activator of transcription STAT3, combined with induction of proinflammatory and proproliferative mediators – including MCP-1 and hypoxia-inducible mitogenic factor HIMF – and increased pulmonary levels of the key miRNAs miR-17 and miR-204, the expressions of which are reduced in human pulmonary hypertension [65]. However, the animal models in which these effects were tested are not considered good representations of preclinical models of PAH. The beneficial effects observed with EVs treatment of PAH therefore require more in-depth investigation before they can be considered practice changing.

Acute respiratory distress syndrome

ARDS is a severe clinical condition characterized by alveolar-capillary damage, accumulation of protein-rich debris in the alveolar airspace, and progressive respiratory failure [66]. Although major improvements in treatment and supportive care of ARDS have been achieved, its mortality rate remains around 40 % [67].

Recently, some studies reported that MSCs can be a promising therapeutic approach for ARDS through paracrine effects [68–70]. Additionally, MSC-derived EVs have been shown to produce beneficial effects in experimental endotoxin-induced ARDS, reducing lung inflammation [71]. hMSC-derived EVs were therapeutically effective following *Escherichia coli* endotoxin-induced ARDS, thus reducing extravascular lung water, total protein levels in BALF, edema, and neutrophil infiltration. These beneficial effects were associated with an increase in keratinocyte growth factor (KGF) expression, as they were partially eliminated after delivery of EVs derived from KGF siRNA-pretreated MSCs [72]. Moreover, ischemic preconditioning can potentiate the protective effect of MSCs in endotoxin-induced ARDS through the secretion of exosomes since it confers strong protection against cell death and promotes their differentiation potential by activating multiple signaling pathways which open new avenues for therapeutic approaches [73].

Pneumonia

Bacterial pneumonia is among the main causes of respiratory failure in critically ill patients. Despite improvements in supportive care and appropriate antibiotic use, morbidity and mortality remain high [74]. Several studies have reported efficacy of MSCs in preclinical models of pneumonia due to their ability to secrete paracrine factors such as growth factors, anti-inflammatory cytokines, and antimicrobial peptides [75]. Outer-membrane vesicle release is a conserved phenomenon among pathogenic and nonpathogenic Gram-negative bacteria [76]. Nevertheless, little is known regarding Gram-positive EVs, especially their biogenesis and role in host–pathogen interactions. EVs from *Streptococcus pneumonia*, one of the leading causes of bacterial pneumonia worldwide, have only recently been characterized [77] and found to exhibit high immunogenicity due to the presence of the toxin pneumolysin.

Recently, in an in vivo model of *E. coli* pneumonia in mice, hMSC-derived EVs were as effective as their parent stem cells in improving survival and mitigating lung inflammation, protein permeability, and bacterial growth. The antimicrobial effect of hMSC-derived EVs was exerted in part through enhancement of monocyte phagocytosis of bacteria, which could be further increased by prestimulation of hMSCs with a TLR-3 agonist before EVs release. Uptake of hMSC-derived EVs through the CD44 receptor into injured human monocytes and alveolar epithelial cells was critical for their therapeutic effects. Another factor that should be stressed is that hMSC-derived EVs decreased tumor necrosis factor alpha secretion by lipopolysaccharide-primed human monocytes and restored intracellular ATP levels in injured human alveolar epithelial type II cells, suggesting immunomodulatory and metabolomic effects of EVs. Additionally, administration of a KGF neutralizing antibody abrogated the survival advantage mediated by hMSC-derived EVs, suggesting a possible mechanism for their therapeutic effect [78].

Conclusions

Several studies have reported that MSCs may repair damaged tissue by modifying target cell function through paracrine mechanisms without directly replacing injured cells. The role of EVs in this mechanism would be to exchange genetic material, which could explain the observed phenotypic and functional changes of MSCs [79]. This genetic material transfer may lead to the production of soluble factors, thus regulating cell proliferation, apoptosis, and/or inflammation and immune response.

EVs present many advantages over stem cells, such as homing ability to target tissue, preventing undesired accumulation in other organs, and absence of any innate toxicity or association with long-term maldifferentiated engrafted cells, tumor generation, or immune rejection after stem cell injection. However, the mechanisms associated with the beneficial effects induced by MSC-derived EVs require further investigation. In this line, the following points in particular warrant better evaluation: which signaling regulates the transfer of biologically active molecules within EVs, which surface receptors may yield selective specificity, and which stimuli are responsible for triggering EVs release. Understanding these EVs mechanisms may allow their use as diagnostic markers, for the delivery of drugs and genes, and as new therapeutic strategies. Although some studies have reported beneficial effects of MSC-derived EVs in asthma, ARDS, PAH, and pneumonia, many issues must be addressed before their use in clinical settings, including: the need for large-scale EVs production from MSCs; the need for criteria defining the potency of EVs, due to different preparations and MSC sources; the long-term effects of EVs; and the biodistribution of EVs in each respiratory disease.

Abbreviations

ARDS: Acute respiratory distress syndrome; ASC: Adipose-derived mesenchymal stromal cell; BALF: Bronchoalveolar lavage fluid; BEC: Bronchial epithelial cell; BM-MSC: Bone marrow-derived mesenchymal stromal cell; DC: Dendritic cell; EV: Extracellular vesicle; hMSC: Human mesenchymal stromal cell; Hsp: Heat-shock proteins; IFNγ: Interferon gamma; IL: Interleukin; ILV: Intraluminal vesicle; ISEV: International Society for Extracellular Vesicles; KGF: Keratinocyte growth factor; miRNA: MicroRNA; mMSC: Mouse mesenchymal stromal cell; MSC: Mesenchymal stromal cell; MVB: Multivesicular body; PAH: Pulmonary arterial hypertension; siRNA: Small interfering RNA; TGF-β: Transforming growth factor beta; Th: T-helper; TLR: Toll-like receptor.

Competing interests

The authors declare that they have no potential competing interests.

Authors' contributions

SCA, DJW, and PRMR are responsible for the concept of the review. SCA was responsible for writing the first draft of the manuscript. DJW and PRMR were responsible for assisting with study selection. DJW and PRMR were responsible for critical review of the manuscript. All authors read and approved the manuscript.

Acknowledgements

The authors would like to express their gratitude to Mrs Moira Elizabeth Schottler, Mr Filippe Vasconcellos, and Mrs Martha McMahan for their assistance in editing the article.
This study was supported by the Brazilian Council for Scientific and Technological Development (CNPq), the Rio de Janeiro State Research Foundation (FAPERJ), the Coordination for the Improvement of Higher Education Personnel (CAPES), and the Department of Science and Technology—Brazilian Ministry of Health (DECIT/MS).

Author details

[1]Laboratory of Pulmonary Investigation, Carlos Chagas Filho Institute of Biophysics, Federal University of Rio de Janeiro, Av. Carlos Chagas Filho, 373, Ilha do Fundão, Rio de Janeiro, RJ 21941-902, Brazil. [2]Department of

Medicine, Vermont Lung Center, College of Medicine, University of Vermont, 89 Beaumont Ave Given, Burlington, VT 05405, USA.

References

1. Lalu MM, McIntyre L, Pugliese C, Fergusson D, Winston BW, Marshall JC, Granton J, Stewart DJ. Safety of cell therapy with mesenchymal stromal cells (SafeCell): a systematic review and meta-analysis of clinical trials. PLoS One. 2012;7(10):e47559.

2. Hofmann NA, Ortner A, Jacamo RO, Reinisch A, Schallmoser K, Rohban R, Etchart N, Fruehwirth M, Beham-Schmid C, Andreeff M, Strunk D. Oxygen sensing mesenchymal progenitors promote neo-vasculogenesis in a humanized mouse model in vivo. PLoS One. 2012;7(9):e44468.

3. Bell GI, Meschino MT, Hughes-Large JM, Broughton HC, Xenocostas A, Hess DA. Combinatorial human progenitor cell transplantation optimizes islet regeneration through secretion of paracrine factors. Stem Cells Dev. 2012; 21(11):1863–76.

4. Zhao S, Wehner R, Bornhauser M, Wassmuth R, Bachmann M, Schmitz M. Immunomodulatory properties of mesenchymal stromal cells and their therapeutic consequences for immune-mediated disorders. Stem Cells Dev. 2010;19(5):607–14.

5. Waszak P, Alphonse R, Vadivel A, Ionescu L, Eaton F, Thebaud B. Preconditioning enhances the paracrine effect of mesenchymal stem cells in preventing oxygen-induced neonatal lung injury in rats. Stem Cells Dev. 2012;21(15):2789–97.

6. Goolaerts A, Pellan-Randrianarison N, Larghero J, Vanneaux V, Uzunhan Y, Gille T, Dard N, Planes C, Matthay MA, Clerici C. Conditioned media from mesenchymal stromal cells restore sodium transport and preserve epithelial permeability in an in vitro model of acute alveolar injury. Am J Physiol Lung Cell Mol Physiol. 2014;306(11):L975–85.

7. Lotvall J, Hill AF, Hochberg F, Buzas EI, Di Vizio D, Gardiner C, Gho YS, Kurochkin IV, Mathivanan S, Quesenberry P, Sahoo S, Tahara H, Wauben MH, Witwer KW, Thery C. Minimal experimental requirements for definition of extracellular vesicles and their functions: a position statement from the International Society for Extracellular Vesicles. J Extracell Vesicles. 2014;3:26913.

8. Camussi G, Deregibus MC, Bruno S, Cantaluppi V, Biancone L. Exosomes/microvesicles as a mechanism of cell-to-cell communication. Kidney Int. 2010;78(9):838–48.

9. Baglio SR, Pegtel DM, Baldini N. Mesenchymal stem cell secreted vesicles provide novel opportunities in (stem) cell-free therapy. Front Physiol. 2012;3:359.

10. Almqvist N, Lonnqvist A, Hultkrantz S, Rask C, Telemo E. Serum-derived exosomes from antigen-fed mice prevent allergic sensitization in a model of allergic asthma. Immunology. 2008;125(1):21–7.

11. Quesenberry PJ, Goldberg LR, Aliotta JM, Dooner MS, Pereira MG, Wen S, Camussi G. Cellular phenotype and extracellular vesicles: basic and clinical considerations. Stem Cells Dev. 2014;23(13):1429–36.

12. Chen J, Li C, Chen L. The role of microvesicles derived from mesenchymal stem cells in lung diseases. Biomed Res Int. 2015;2015:985814.

13. Bruno S, Camussi G. Role of mesenchymal stem cell-derived microvesicles in tissue repair. Pediatr Nephrol. 2013;28(12):2249–54.

14. Biancone L, Bruno S, Deregibus MC, Tetta C, Camussi G. Therapeutic potential of mesenchymal stem cell-derived microvesicles. Nephrol Dial Transplant. 2012;27(8):3037–42.

15. Fujita Y, Kosaka N, Araya J, Kuwano K, Ochiya T. Extracellular vesicles in lung microenvironment and pathogenesis. Trends Mol Med. 2015;21(9):533–42.

16. Fierabracci A, Del Fattore A, Luciano R, Muraca M, Teti A. Recent advances in mesenchymal stem cell immunomodulation: the role of microvesicles. Cell Transplant. 2015;24(2):133–49.

17. Gould SJ, Raposo G. As we wait: coping with an imperfect nomenclature for extracellular vesicles. J Extracell Vesicles. 2013;2: 10.3402/jev.v2i0.20389.

18. Kordelas L, Rebmann V, Ludwig AK, Radtke S, Ruesing J, Doeppner TR, Epple M, Horn PA, Beelen DW, Giebel B. MSC-derived exosomes: a novel tool to treat therapy-refractory graft-versus-host disease. Leukemia. 2014;28(4):970–3.

19. Katsuda T, Kosaka N, Takeshita F, Ochiya T. The therapeutic potential of mesenchymal stem cell-derived extracellular vesicles. Proteomics. 2013; 13(10–11):1637–53.

20. Chen J, Liu Z, Hong MM, Zhang H, Chen C, Xiao M, Wang J, Yao F, Ba M, Liu J, Guo ZK, Zhong J. Proangiogenic compositions of microvesicles derived from human umbilical cord mesenchymal stem cells. PLoS One. 2014;9(12): e115316.

21. Mathivanan S, Simpson RJ. ExoCarta: a compendium of exosomal proteins and RNA. Proteomics. 2009;9(21):4997–5000.

22. Tetta C, Bruno S, Fonsato V, Deregibus MC, Camussi G. The role of microvesicles in tissue repair. Organogenesis. 2011;7(2):105–15.

23. Akyurekli C, Le Y, Richardson RB, Fergusson D, Tay J, Allan DS. A systematic review of preclinical studies on the therapeutic potential of mesenchymal stromal cell-derived microvesicles. Stem Cell Rev. 2015;11(1):150–60.

24. Bruno S, Deregibus MC, Camussi G. The secretome of mesenchymal stromal cells: role of extracellular vesicles in immunomodulation. Immunol Lett. 2015;168(2):154–8.

25. Demayo F, Minoo P, Plopper CG, Schuger L, Shannon J, Torday JS. Mesenchymal-epithelial interactions in lung development and repair: are modeling and remodeling the same process? Am J Physiol Lung Cell Mol Physiol. 2002;283(3):L510–7.

26. Dale GL, Remenyi G, Friese P. Tetraspanin CD9 is required for microparticle release from coated-platelets. Platelets. 2009;20(6):361–6.

27. Tomasoni S, Longaretti L, Rota C, Morigi M, Conti S, Gotti E, Capelli C, Introna M, Remuzzi G, Benigni A. Transfer of growth factor receptor mRNA via exosomes unravels the regenerative effect of mesenchymal stem cells. Stem Cells Dev. 2013;22(5):772–80.

28. Eirin A, Riester SM, Zhu XY, Tang H, Evans JM, O'Brien D, van Wijnen AJ, Lerman LO. MicroRNA and mRNA cargo of extracellular vesicles from porcine adipose tissue-derived mesenchymal stem cells. Gene. 2014;551(1):55–64.

29. Baglio SR, Rooijers K, Koppers-Lalic D, Verweij FJ, Perez Lanzon M, Zini N, et al. Human bone marrow- and adipose-mesenchymal stem cells secrete exosomes enriched in distinctive miRNA and tRNA species. Stem Cell Res Ther. 2015;6:127.

30. Kim HS, Choi DY, Yun SJ, Choi SM, Kang JW, Jung JW, Hwang D, Kim KP, Kim DW. Proteomic analysis of microvesicles derived from human mesenchymal stem cells. J Proteome Res. 2012;11(2):839–49.

31. Lopatina T, Bruno S, Tetta C, Kalinina N, Porta M, Camussi G. Platelet-derived growth factor regulates the secretion of extracellular vesicles by adipose mesenchymal stem cells and enhances their angiogenic potential. Cell Commun Signal. 2014;12:26.

32. Phinney DG, Di Giuseppe M, Njah J, Sala E, Shiva S, St Croix CM, Stolz DB, Watkins SC, Di YP, Leikauf GD, Kolls J, Riches DW, Deiuliis G, Kaminski N, Boregowda SV, McKenna DH, Ortiz LA. Mesenchymal stem cells use extracellular vesicles to outsource mitophagy and shuttle microRNAs. Nat Commun. 2015;6:8472.

33. Parolini I, Federici C, Raggi C, Lugini L, Palleschi S, De Milito A, Coscia C, Iessi E, Logozzi M, Molinari A, Colone M, Tatti M, Sargiacomo M, Fais S. Microenvironmental pH is a key factor for exosome traffic in tumor cells. J Biol Chem. 2009;284(49):34211–22.

34. Iglesias DM, El-Kares R, Taranta A, Bellomo F, Emma F, Besouw M, Levtchenko E, Toelen J, van den Heuvel L, Chu L, Zhao J, Young YK, Eliopoulos N, Goodyer P. Stem cell microvesicles transfer cystinosin to human cystinotic cells and reduce cystine accumulation in vitro. PLoS One. 2012;7(8):e42840.

35. Camussi G, Deregibus MC, Cantaluppi V. Role of stem-cell-derived microvesicles in the paracrine action of stem cells. Biochem Soc Trans. 2013;41(1):283–7.

36. Ratajczak J, Miekus K, Kucia M, Zhang J, Reca R, Dvorak P, Ratajczak MZ. Embryonic stem cell-derived microvesicles reprogram hematopoietic progenitors: evidence for horizontal transfer of mRNA and protein delivery. Leukemia. 2006;20(5):847–56.

37. Dooner MS, Aliotta JM, Pimentel J, Dooner GJ, Abedi M, Colvin G, Liu Q, Weier HU, Johnson KW, Quesenberry PJ. Conversion potential of marrow cells into lung cells fluctuates with cytokine-induced cell cycle. Stem Cells Dev. 2008;17(2):207–19.

38. Bourdonnay E, Zaslona Z, Penke LR, Speth JM, Schneider DJ, Przybranowski S, Swanson JA, Mancuso P, Freeman CM, Curtis JL, Peters-Golden M. Transcellular delivery of vesicular SOCS proteins from macrophages to epithelial cells blunts inflammatory signaling. J Exp Med. 2015;212(5):729–42.

39. Qazi KR, Torregrosa Paredes P, Dahlberg B, Grunewald J, Eklund A, Gabrielsson S. Proinflammatory exosomes in bronchoalveolar lavage fluid of patients with sarcoidosis. Thorax. 2010;65(11):1016–24.

40. Sarkar A, Mitra S, Mehta S, Raices R, Wewers MD. Monocyte derived microvesicles deliver a cell death message via encapsulated caspase-1. PLoS One. 2009;4(9):e7140.

41. Lai RC, Yeo RW, Lim SK. Mesenchymal stem cell exosomes. Semin Cell Dev Biol. 2015;40:82–8.

42. Collino F, Deregibus MC, Bruno S, Sterpone L, Aghemo G, Viltono L, Tetta C, Camussi G. Microvesicles derived from adult human bone marrow and tissue specific mesenchymal stem cells shuttle selected pattern of miRNAs. PLoS One. 2010;5(7):e11803.

43. Mokarizadeh A, Delirezh N, Morshedi A, Mosayebi G, Farshid AA, Mardani K. Microvesicles derived from mesenchymal stem cells: potent organelles for induction of tolerogenic signaling. Immunol Lett. 2012;147(1–2):47–54.

44. Zhang B, Yin Y, Lai RC, Tan SS, Choo AB, Lim SK. Mesenchymal stem cells secrete immunologically active exosomes. Stem Cells Dev. 2014;23(11):1233–44.

45. Conforti A, Scarsella M, Starc N, Giorda E, Biagini S, Proia A, Carsetti R, Locatelli F, Bernardo ME. Microvescicles derived from mesenchymal stromal cells are not as effective as their cellular counterpart in the ability to modulate immune responses in vitro. Stem Cells Dev. 2014;23(21):2591–9.

46. Lai Y, Long Y, Lei Y, Deng X, He B, Sheng M, Li M, Gu Z. A novel micelle of coumarin derivative monoend-functionalized PEG for anti-tumor drug delivery: in vitro and in vivo study. J Drug Target. 2012;20(3):246–54.

47. Lai RC, Chen TS, Lim SK. Mesenchymal stem cell exosome: a novel stem cell-based therapy for cardiovascular disease. Regen Med. 2011;6(4):481–92.

48. Tian Y, Li S, Song J, Ji T, Zhu M, Anderson GJ, Wei J, Nie G. A doxorubicin delivery platform using engineered natural membrane vesicle exosomes for targeted tumor therapy. Biomaterials. 2014;35(7):2383–90.

49. Gyorgy B, Hung ME, Breakefield XO, Leonard JN. Therapeutic applications of extracellular vesicles: clinical promise and open questions. Annu Rev Pharmacol Toxicol. 2015;55:439–64.

50. Martinez FD, Vercelli D. Asthma. Lancet. 2013;382(9901):1360–72.

51. Papierniak ES, Lowenthal DT, Harman E. Novel therapies in asthma: leukotriene antagonists, biologic agents, and beyond. Am J Ther. 2013;20(1):79–103.

52. Abreu SC, Antunes MA, de Castro JC, de Oliveira MV, Bandeira E, Ornellas DS, et al. Bone marrow-derived mononuclear cells vs. mesenchymal stromal cells in experimental allergic asthma. Respir Physiol Neurobiol. 2013;187(2):190–8.

53. Goodwin M, Sueblinvong V, Eisenhauer P, Ziats NP, LeClair L, Poynter ME, Steele C, Rincon M, Weiss DJ. Bone marrow-derived mesenchymal stromal cells inhibit Th2-mediated allergic airways inflammation in mice. Stem Cells. 2011;29(7):1137–48.

54. Lathrop MJ, Brooks EM, Bonenfant NR, Sokocevic D, Borg ZD, Goodwin M, Loi R, Cruz F, Dunaway CW, Steele C, Weiss DJ. Mesenchymal stromal cells mediate Aspergillus hyphal extract-induced allergic airway inflammation by inhibition of the th17 signaling pathway. Stem Cells Transl Med. 2014;3(2):194–205.

55. Admyre C, Telemo E, Almqvist N, Lotvall J, Lahesmaa R, Scheynius A, Gabrielsson S. Exosomes—nanovesicles with possible roles in allergic inflammation. Allergy. 2008;63(4):404–8.

56. Park JA, Sharif AS, Tschumperlin DJ, Lau L, Limbrey R, Howarth P, Drazen JM. Tissue factor-bearing exosome secretion from human mechanically stimulated bronchial epithelial cells in vitro and in vivo. J Allergy Clin Immunol. 2012;130(6):1375–83.

57. Baranova INKR, Bocharov AV, Vishnyakova TG, Chen Z, Remaley AT, Csako G, et al. Role of human CD36 in bacterial recognition, phagocytosis, and pathogen-induced JNK-mediated signaling. J Immunol. 2008;181(10):7147–56.

58. Levanen B, Bhakta NR, Torregrosa Paredes P, Barbeau H, Hiltbrunner S, Pollack JL, Skold CM, Svartengren M, Grunewald J, Gabrielsson S, Eklund A, Larsson BM, Woodruff PG, Erle DJ, Wheelock AM. Altered microRNA profiles in bronchoalveolar lavage fluid exosomes in asthmatic patients. J Allergy Clin Immunol. 2013;131(3):894–903.

59. Pinkerton M, Chinchilli V, Banta E, Craig T, August A, Bascom R, Cantorna M, Harvill E, Ishmael FT. Differential expression of microRNAs in exhaled breath condensates of patients with asthma, patients with chronic obstructive pulmonary disease, and healthy adults. J Allergy Clin Immunol. 2013;132(1):217–9.

60. Torregrosa Paredes P, Esser J, Admyre C, Nord M, Rahman QK, Lukic A, Radmark O, Gronneberg R, Grunewald J, Eklund A, Scheynius A, Gabrielsson S. Bronchoalveolar lavage fluid exosomes contribute to cytokine and leukotriene production in allergic asthma. Allergy. 2012;67(7):911–9.

61. Prado NME, Segura E, Fernández-García H, Villalba M, Théry C, Rodríguez R, Batanero E. Exosomes from bronchoalveolar fluid of tolerized mice prevent allergic reaction. J Immunol. 2008;181(2):519–25.

62. Cruz FFBZ, Goodwin M, Sokocevic D, Wagner DE, Coffey A, Antunes M, Robinson KL, Mitsials SA, Kourembanas S, Thane K, Hoffman AM, McKenna DH, Rocco PRM, Weiss DJ. Systemic administration of human bone marrow-derived mesenchymal stromal cell extracellular vesicles ameliorates aspergillus hyphal extract-induced allergic airway inflammation in immunocompetent mice. Stem Cell Transl Med. 2015;4(11):1302–16.

63. Bakouboula B, Morel O, Faure A, Zobairi F, Jesel L, Trinh A, Zupan M, Canuet M, Grunebaum L, Brunette A, Desprez D, Chabot F, Weitzenblum E, Freyssinet JM, Chaouat A, Toti F. Procoagulant membrane microparticles correlate with the severity of pulmonary arterial hypertension. Am J Respir Crit Care Med. 2008;177(5):536–43.

64. Aliotta JM, Pereira M, Amaral A, Sorokina A, Igbinoba Z, Hasslinger A, El-Bizri R, Rounds SI, Quesenberry PJ, Klinger JR. Induction of pulmonary hypertensive changes by extracellular vesicles from monocrotaline-treated mice. Cardiovasc Res. 2013;100(3):354–62.

65. Lee C, Mitsialis SA, Aslam M, Vitali SH, Vergadi E, Konstantinou G, Sdrimas K, Fernandez-Gonzalez A, Kourembanas S. Exosomes mediate the cytoprotective action of mesenchymal stromal cells on hypoxia-induced pulmonary hypertension. Circulation. 2012;126(22):2601–11.

66. Matthay MA, Howard JP. Progress in modelling acute lung injury in a pre-clinical mouse model. Eur Respir J. 2012;39(5):1062–3.

67. Li L, Jin S, Zhang Y. Ischemic preconditioning potentiates the protective effect of mesenchymal stem cells on endotoxin-induced acute lung injury in mice through secretion of exosome. Int J Clin Exp Med. 2015;8(3):3825–32.

68. Gonzalez-Rey E, Anderson P, Gonzalez MA, Rico L, Buscher D, Delgado M. Human adult stem cells derived from adipose tissue protect against experimental colitis and sepsis. Gut. 2009;58(7):929–39.

69. Maron-Gutierrez T, Silva JD, Asensi KD, Bakker-Abreu I, Shan Y, Diaz BL, Goldenberg RC, Mei SH, Stewart DJ, Morales MM, Rocco PR, Dos Santos CC. Effects of mesenchymal stem cell therapy on the time course of pulmonary remodeling depend on the etiology of lung injury in mice. Crit Care Med. 2013;41(11):e319–33.

70. Ionescu L, Byrne RN, van Haaften T, Vadivel A, Alphonse RS, Rey-Parra GJ, Weissmann G, Hall A, Eaton F, Thebaud B. Stem cell conditioned medium improves acute lung injury in mice: in vivo evidence for stem cell paracrine action. Am J Physiol Lung Cell Mol Physiol. 2012;303(11):L967–77.

71. Lee JW, Fang X, Krasnodembskaya A, Howard JP, Matthay MA. Concise review: mesenchymal stem cells for acute lung injury: role of paracrine soluble factors. Stem Cells. 2011;29(6):913–9.

72. Zhu YG, Feng XM, Abbott J, Fang XH, Hao Q, Monsel A, et al. Human mesenchymal stem cell microvesicles for treatment of Escherichia coli endotoxin-induced acute lung injury in mice. Stem Cells. 2014;32(1):116–25.

73. Pasha Z, Wang Y, Sheikh R, Zhang D, Zhao T, Ashraf M. Preconditioning enhances cell survival and differentiation of stem cells during transplantation in infarcted myocardium. Cardiovasc Res. 2008;77(1):134–42.

74. Rubenfeld GD, Caldwell E, Peabody E, Weaver J, Martin DP, Neff M, et al. Incidence and outcomes of acute lung injury. N Engl J Med. 2005;353(16):1685–93.

75. Hackstein H, Lippitsch A, Krug P, Schevtschenko I, Kranz S, Hecker M, Dietert K, Gruber AD, Bein G, Brendel C, Baal N. Prospectively defined murine mesenchymal stem cells inhibit Klebsiella pneumoniae-induced acute lung injury and improve pneumonia survival. Respir Res. 2015;16:123.

76. MacDonald IA, Kuehn MJ. Offense and defense: microbial membrane vesicles play both ways. Res Microbiol. 2012;163(9–10):607–18.

77. Olaya-Abril A, Prados-Rosales R, McConnell MJ, Martin-Pena R, Gonzalez-Reyes JA, Jimenez-Munguia I, Gomez-Gascon L, Fernandez J, Luque-Garcia JL, Garcia-Lidon C, Estevez H, Pachon J, Obando I, Casadevall A, Pirofski LA, Rodriguez-Ortega MJ. Characterization of protective extracellular membrane-derived vesicles produced by Streptococcus pneumoniae. J Proteomics. 2014;106:46–60.

78. Monsel A, Zhu YG, Gennai S, Hao Q, Hu S, Rouby JJ, Rosenzwajg M, Matthay MA, Lee JW. Therapeutic effects of human mesenchymal stem cell-derived microvesicles in severe pneumonia in mice. Am J Respir Crit Care Med. 2015;192(3):324–36.

79. Quesenberry PJ, Aliotta JM. The paradoxical dynamism of marrow stem cells: considerations of stem cells, niches, and microvesicles. Stem Cell Rev. 2008;4(3):137–47.

vIL-10-overexpressing human MSCs modulate naïve and activated T lymphocytes following induction of collagenase-induced osteoarthritis

Eric Farrell[1,2*†], Niamh Fahy[2,3†], Aideen E Ryan[2,4,6], Cathal O Flatharta[2], Lisa O'Flynn[2,4,5], Thomas Ritter[2,4] and J Mary Murphy[2]

Abstract

Background: Recent efforts in osteoarthritis (OA) research have highlighted synovial inflammation and involvement of immune cells in disease onset and progression. We sought to establish the in-vivo immune response in collagenase-induced OA and investigate the ability of human mesenchymal stem cells (hMSCs) overexpressing viral interleukin 10 (vIL-10) to modulate immune populations and delay/prevent disease progression.

Methods: Eight-week-old male C57BL/6 mice were injected with 1 U type VII collagenase over two consecutive days. At day 7, 20,000 hMSCs overexpressing vIL-10 were injected into the affected knee. Control groups comprised of vehicle, 20,000 untransduced or adNull-transduced MSCs or virus alone. Six weeks later knees were harvested for histological analysis and popliteal and inguinal lymph nodes for flow cytometric analysis.

Results: At this time there was no significant difference in knee OA scores between any of the groups. A trend toward more damage in animals treated with hMSCs was observed. Interestingly there was a significant reduction in the amount of activated CD4 and CD8 T cells in the vIL-10-expressing hMSC group.

Conclusions: vIL-10-overexpressing hMSCs can induce long-term reduction in activated T cells in draining lymph nodes of mice with collagenase-induced OA. This could lead to reduced OA severity or disease progression over the long term.

Keywords: Collagenase-induced osteoarthritis, Mesenchymal stem cell, vIL-10, Cell therapy, Gene therapy, Xenogeneic

Background

Although osteoarthritis (OA) is typically characterised by loss or damage to articular cartilage, inflammation of the synovial membrane is a prevalent feature believed to contribute to both symptoms and disease progression. Thickening of the synovial membrane has been identified in patients with early-stage OA, and increased vascular density and cellular infiltration of the synovium is a prominent feature of disease pathogenesis [1]. Macrophages localised to the synovial lining are primary mediators of inflammation in the joint, responsible for the production of pro-inflammatory cytokines such as tumour necrosis factor alpha (TNF-α) and interleukin (IL)-1β, which can induce destructive processes in neighbouring cartilage [2, 3]. T lymphocytes are the most abundant infiltrating immune cells present in OA synovium, with both CD4[+] effector T cells and CD8[+] cytotoxic T cells observed in the sublining layer [4, 5]. Moreover, a role of CD4[+] T cells in the pathogenesis of OA has been identified, through the induction of MIP-1γ and osteoclastogenesis [6].

Mesenchymal stem cells (MSCs) have been considered a promising cell source for the repair of damaged cartilage in OA, due to their chondrogenic differential potential [7].

* Correspondence: e.farrell@erasmusmc.nl
†Equal contributors
[1]Department of Oral and Maxillofacial Surgery, Special Dental Care and Orthodontics, Erasmus MC, University Medical Centre, Room Ee1614, Erasmus MC, Wytemaweg 80, Rotterdam 3015CN, The Netherlands
[2]Regenerative Medicine Institute, National University of Ireland Galway, Galway, Ireland
Full list of author information is available at the end of the article

In addition to their multipotent nature, MSCs may enhance intrinsic tissue repair through the release of trophic factors which act to modulate inflammatory processes or recruit endogenous progenitor cells [8, 9]. MSCs have the ability to polarise pro-inflammatory macrophages towards an anti-inflammatory phenotype and suppress T-cell proliferation, and have been previously reported to exert anti-inflammatory effects on human osteoarthritic synovium in vitro [10–12]. Xenotransplantation of human MSCs (hMSCs) has enhanced meniscal regeneration and prevented OA progression in rats following hemimeniscectomy, with no significant difference observed between the reparative effects of xenogeneic or rat syngeneic MSCs [13]. Furthermore, allogeneic MSCs have been shown to prevent post-traumatic arthritis following intra-articular fracture in mice; however, an effect of hMSC delivery on synovial hyperplasia was not observed [14]. Although intra-articularly delivered adipose-derived stem cells have been reported to migrate to the synovium, reduce synovial lining thickness and decrease cartilage damage in a murine collagenase-induced OA model [15], the ability of bone marrow-derived hMSCs to alter the inflammatory environment in OA and subsequently delay disease progression requires further investigation.

One factor of particular interest to target OA-associated inflammatory processes is IL-10. IL-10 is a 34 kDa homodimeric cytokine produced by activated macrophages, T-helper type 2 cells and B cells, and is associated with diverse biological responses [12, 16–18]. IL-10 can suppress pro-inflammatory cytokine production; for example, TNF-α, IL-1β and IL-6 produced by activated macrophages, monocytes and T-helper type 1 cells respectively [17, 19]. Furthermore, IL-10 reduces monocyte expression of major histocompatibility complex (MHC) class II, resulting in decreased antigen presentation and subsequent inhibition of T-cell proliferation [20]. In addition to its immunosuppressive properties, IL-10 can stimulate B cells to increase MHC class II expression and immunoglobulin production, as well as enhance mast cell growth [21–23]. Human IL-10 exhibits 84 % amino acid sequence homology to an open reading frame product of Epstein–Barr virus, termed viral IL-10 (vIL-10) [24]. Although vIL-10 displays comparable immunosuppressive activity to human IL-10, it lacks particular stimulatory functions [22]. Furthermore, adenoviral-mediated gene transfer of vIL-10 or overexpression by retrovirally transduced MSCs has been reported to significantly decrease the frequency of arthritis, delay the onset and reduce the severity of arthritic symptoms in a murine collagen-induced rheumatoid arthritis model [25, 26].

In the present study, we have investigated the ability of xenogeneic hMSCs overexpressing vIL-10 to modulate the inflammatory environment and alter disease progression in a murine collagenase-induced OA model. Overexpression was induced using an adenoviral construct with transduced hMSCs referred to as AdIL-10 MSCs. This model is associated with joint ligament damage resulting in instability [27], as well as synovial activation compared with other instability-related OA models [28]. Intra-articular injection of 1 U collagenase was administered over two consecutive days to induce mild structural changes and OA progression in order to achieve a suitable model to evaluate the potential of an anti-inflammatory component to attenuate OA pathogenesis. Our findings highlight modulatory activity of hMSCs adenovirally transduced to overexpress vIL-10 on the levels of naive and activated CD4$^+$ and CD8$^+$ T cells in vivo during OA progression.

Methods

Isolation and culture of hMSCs

hMSCs were obtained from bone marrow aspirates taken from the iliac crest of healthy human donors. All procedures were performed with informed consent and approved by the Clinical Research Ethical Committee at University College Hospital, Galway, Ireland. The cells were isolated based on plastic adherence as described previously [29] and expanded in alpha-Minimum Essential Medium (α-MEM; Life Technologies, Zoetermeer, the Netherlands) containing 10 % pre-screened foetal bovine serum (FBS; Hyclone, South Logan, UT, USA), 1 % penicillin/streptomycin and 1 ng/ml recombinant human basic fibroblast growth factor (FGF2; PeproTech, London, UK). Upon reaching confluency, hMSCs were subcultured to passage 3 for adenoviral transduction. Three MSC donors were used for the in-vitro experiments and one of these was used for the in-vivo study with eight replicates per condition.

Adenoviral transduction of hMSCs

Adenoviral transduction of hMSCs was performed using the lanthanide-based method as described previously [30]. hMSCs were seeded at a density of 6×10^3 cells/cm^2 in a T175 flask 24 h prior to transduction. Calculated amounts of adenoviral vectors expressing vIL-10 (AdIL-10) or Null virus (AdNull) were added to serum-free α-MEM medium to generate a final multiplicity of infection of 100. LaCl$_3$ (Sigma-Aldrich, Wicklow, Ireland) was dissolved in deionised water to generate a 0.4 M stock solution and stored at 4 °C. A working solution of 0.04 mM LaCl$_3$ was generated following the addition of an appropriate volume of a 0.4 M stock to serum-free medium. An equal volume of 0.04 mM LaCl$_3$ solution was added to the virus solution, mixed gently and incubated at room temperature for 30 minutes. hMSCs were incubated with the LaCl$_3$/virus mixture for 3 h, following which cells were washed twice with serum-containing medium. Cells were harvested for subsequent experiments 48 h post transduction.

In-vitro assessment of immunomodulatory properties of AdIL-10 MSC CM

Macrophages were seeded at a density of 4×10^5 cells/ml in a 24-well plate, in RAW264.7 culture medium (Dulbecco's modified Eagle's medium (4500 ng/ml glucose), 2 mM glutamine, 10 % FBS and 1 % penicillin/streptomycin). Lipopolysaccharide (LPS; Sigma-Aldrich) was added to the cell culture medium at a final concentration of 0.5 ng/ml. Supernatant was harvested at 2, 4, 6, 12 and 24 h post LPS stimulation and centrifuged at $400 \times g$ for 5 minutes to pellet any debris. TNF-α levels in the supernatant were measured with a commercial human TNF-α DuoSet ELISA kit according to the manufacturer's instructions (R&D Systems, Minneapolis, Minnesota). To assess the effect of MSC conditioned medium (CM) ($n = 3$ donors) on TNF-α production by LPS-activated mouse macrophages, the vIL-10 concentration was quantified in the CM from AdIL-10-transduced MSCs by ELISA (purified rat anti-human vIL-10 (capture antibody); BD Biosciences; and biotinylated rat anti-human vIL-10 (detection antibody); BD Biosciences, Oxford England) diluted to ensure addition of 100 ng/ml vIL-10 per well. Equal volumes (compared with AdIL-10 MSC CM) of untransduced MSC CM and AdNull-transduced MSC CM were used as controls.

T-cell proliferation assays and mixed lymphocyte reaction cultures

Lymphocytes were obtained from the lymph nodes and spleens of C57BL/6 mice. Lymphocytes were washed with 0.1 % BSA/PBS and stained in pre-warmed (37 °C) 10 μM Vybrant CFDA SE (carboxy-fluorescein diacetate, succinmidyl ester (CFSE))/PBS staining solution (Invitrogen, Carlsbad, CA, USA) as per the manufacturer's instructions. Then 1×10^5 CFSE-stained T cells were stimulated with phorbol myristate acetate (5 ng/ml) and ionomycin (400 ng/ml) in T-cell medium (RPMI 1640 supplemented with 10 % FCS, 50 μM β-mercaptoethanol, 100 U/ml penicillin, 0.1 mg/ml streptomycin, 1 mM sodium pyruvate and 2 mM L-glutamine). Lymphocytes were co-cultured with human untransduced MSCs, AdNull MSCs and AdIL-10 MSCs ($n = 3$ donors) at MSC:T-cell ratios ranging from 1:400 to 1:10 in a humidified incubator for 4 days. Mouse T-cell proliferation (CFSE quantification) was measured by flow cytometry (FACS Canto; BD Biosciences).

For mouse mixed lymphocyte reactions, human untransduced MSCs, AdNull MSCs and AdIL-10 MSCs ($n = 3$ donors) were plated in 96-well round-bottom plates. CFSE-labelled untreated lymphocytes isolated from C57BL/6 mice were used as responders, at ratios of 1:5, 1:10 and 1:50 MSC:T cells. A total of 4×10^3, 2×10^4 or 4×10^4 stimulating cells were co-cultured with 2×10^5 CFSE-labelled responding lymphocytes for 5 days. T-cell proliferation was analysed on a FACS Canto (BD Biosciences).

OA model

All procedures were approved by the Animal Care and Research Ethics Committee of the National University of Ireland, Galway and conducted under licence issued by the Department of Health & Children, Ireland. Eight-week-old male C57BL/6 mice were supplied by Charles River Laboratories, Ballina, Ireland, UK. All animals were kept on a 12 h light/dark cycle with ad libitum access to water and standard laboratory chow. Eight animals were randomly assigned to each treatment group with the treatment leg (left or right) also chosen randomly. All animals were coded and scorers were blinded to the treatment.

To induce OA, 1 U of highly purified bacterial type VII collagenase (Sigma-Aldrich) in 6 μl of physiological saline (vehicle) was injected into the knee joint of mice twice over two consecutive days (totalling 2 U). Animals were anaesthetised using isoflurane and given a subcutaneous injection of buprenorphine (Temgesic, 0.01 mg/kg body weight) prior to collagenase injection. One week later animals were injected with one of five conditions in a volume of 6 μl: vehicle alone, 2×10^4 MSCs, AdIL-10-transduced MSCs, AdNull-transduced MSCs or the equivalent amount of virus used to transduce 2×10^4 cells suspended in the vehicle. The cell number was chosen based on ELISA-based analysis of the amount of vIL-10 produced by the hMSCs and corresponded to 100 ng/ml vIL-10 production. Seven weeks after the induction of OA, animals were euthanised and the joints were removed, fixed in 10 % formalin and subsequently decalcified in 10 % EDTA. Popliteal and inguinal lymph nodes were also harvested and analysed for T-cell markers by flow cytometry. A timeline of the model is provided in Additional file 1: Figure S1.

Histological scoring of sections

Paraffin sections (5 μm) were stained with safranin O and fast green. Sections were deparaffinised by immersion of slides twice in 100 % xylene for 5 minutes, and rehydrated to distilled water following immersion in 100 % ethanol for 2 minutes and for 1 minute in 95 % and 70 % ethanol. Sections were stained utilising 0.02 % fast green for 4 minutes, acetic acid for 3 seconds and 0.1 % safranin O for 6 minutes. Following dehydration, sections were mounted with DPX (all Sigma-Aldrich). To score the femurs and tibiae, six sections per animal, spaced approximately 100 μm apart, were scored by three independent, blinded observers. The scoring was based upon the semi-quantitative grading system designed by the OARSI working group [31] with damage ranging from 0 (no damage) to 6 (complete denudation of the cartilage). All scores for each knee compartment (lateral and medial tibia and femur) were added separately and averaged and were then averaged for the three observers. To score the amount of synovial inflammation, a grading of 0–3 was used, with 0 indicating normal synovium with one or two layers of cells and 3 being several cell layers thick. Grades 1 and 2 fell in between these two categories. Finally a total knee damage score was generated by summing all scores for the six compartments for each knee. Representative images of safranin O-stained slides are available in Additional file 2: Figure S2.

Thionine staining
Paraffin sections (5 μm) were de-waxed and redehydrated in xylene and decreasing concentrations of alcohol (100 %, 96 % and 70 % twice each for 5 minutes respectively). Samples were then stained in 0.04 % thionine in 0.01 M aqueous sodium acetate pH 4.5 for 5 minutes. They were then differentiated in 70 % ethanol for 10 seconds, rinsed in 96 % ethanol and then taken through 100 % ethanol and xylene. Slides were mounted with Entellan.

Multiplex ELISA
To assess monocyte chemoattractant protein-1 (MCP-1) levels in serum samples harvested at 7 weeks post induction of OA, a commercially available chemiluminescent array was utilised (16-plex mouse cytokine screen; Quansys Biosciences, Logan, Utah, US). Mice were randomised and serum from two animals per treatment group was pooled and stored at −80 °C until use. Briefly, an 8-point standard curve was prepared using antigen standard provided, and serum samples were diluted 1:2 in sample diluent. Then 50 μl of sample or standard was added to each well of a pre-antibody-coated 96-well plate. The plate was covered with a plate seal, and placed on a plate shaker at 500 RPM for 1 h at room temperature. Following sample incubation, the plate was washed three times with wash buffer and 50 μl per well of Detection Mix was added. The plate was covered with a new seal, and returned to the plate shaker for 1 h at 500 RPM at room temperature. Following six washes, 50 μl of 1× streptavidin–HRP was added per well, and the plate was incubated at room temperature for 15 minutes with shaking at 500 RPM. Following this incubation period, substrate A (hydrogen peroxide) and substrate B (signal enhancer) were mixed in a 1:1 ratio, and 50 μl was added per well. Imaging was performed immediately using a Fluorchem imager (Alpha Innotech, Dublin, Ireland) and the image was analysed using Q-View software, version 2.15 (Quansys Biosciences). All other cytokines in the array were below detectable levels.

Analysis of T-cell subsets from the draining lymph nodes
Popliteal and inguinal lymph nodes were removed from C57BL/6 mice (ipsilateral side). Single cell suspensions were obtained from each lymph node from each animal as described previously [32]. Mice were randomised and lymph nodes from two animals per treatment group were combined for T-cell analysis, with the exception of one AdIL-10 virus-treated animal which appears as a singlet due to the macroscopic observation of a large haematoma and swelling above the treated knee joint (thus creating a fifth data point in this group). Cell suspensions were stained according to Table 1 for cell surface expression of CD4, CD8 and CD25 to characterise individual T-cell subsets. CD4 identifies T-helper cells while CD8 identifies cytotoxic T cells. Cell surface expression of CD25 was used as a marker of T-cell activation [33]. For assessment of myeloid cells, popliteal lymph nodes from two animals and inguinal lymph nodes from four animals were pooled, because of low cell number. Cell suspensions were also stained for expression of CD11b (Alex Fluor® 647) and Ly6C (Peridinin Chlorophyll Protein Complex (PerCP)/Cy5.5) (Table 1; Biolegend, Dublin, Ireland) as markers of myeloid cells and inflammatory monocytes, respectively [34]. Samples were analysed on a FACS Canto cytometer (BD Biosciences).

Statistical analysis
In-vitro data were analysed by one-way or two-way ANOVA (Fig. 1) followed by Bonferroni's post-hoc test. Knee scores and FACS data were analysed by Kruskal–Wallis test followed by Dunn's multiple comparisons test.

Results
AdIL-10-transduced hMSCs are immunomodulatory and not immunogenic towards murine lymphocytes
To confirm the immunomodulatory potential of a xenogeneic source of MSCs, the ability of untransduced, AdNull-transduced and AdIL-10-transduced hMSCs to suppress the proliferation of stimulated C57BL/6 lymphocytes in vitro was determined. Untransduced, AdNull-transduced and AdIL-10-transduced hMSCs significantly reduced the proliferation of stimulated lymphocytes (at greater than five generations) compared with untreated cells (Fig. 1a; $p < 0.001$, $n = 3$). Furthermore, AdIL-10-transduced hMSCs showed a trend towards decreased proliferation compared with untransduced cells, but this effect was not significant. At three

Table 1 Flow cytometry antibodies used as markers of interest

Antibody	Flurophore	Antibody dilution factor	Biolegend catalogue number
Anti-mouse CD8a	PE/Cy7	1:100	100721
Anti-mouse CD4	APC	1:120	100411
Anti-mouse CD25	FITC	1:60	101907
Anti-mouse/human CD11b	Alexa Fluor® 647	1:60	101220
Anti-mouse Ly6C	PerCP/Cy5.5	1:50	128011

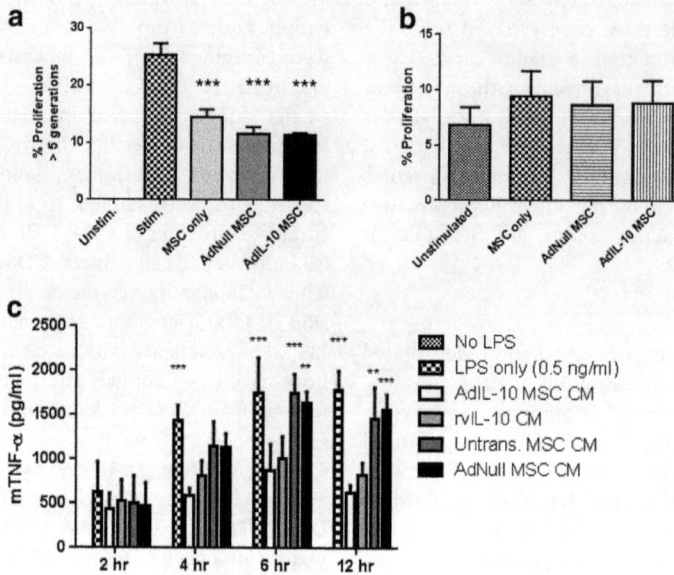

Fig. 1 AdIL-10-transduced MSCs are immunomodulatory and not immunogenic. **a** To confirm the immunomodulatory potential of hMSCs on activated murine lymphocytes, untransduced, AdNull-transduced and AdIL-10-transduced hMSCs were co-cultured with ionomycin and PMA-stimulated C57BL/6 lymphocytes for 4 days at a ratio of 1:10. Murine lymphocyte proliferation was measured by CFSE dilution and flow cytometric detection of CFSE fluorescent peaks. Untransduced, AdNull-transduced and AdIL-10-transduced MSCs reduced the proliferation of stimulated lymphocytes, confirming their immunosuppressive capacity. Values represent the mean ± SD of three biological replicates. Statistical analysis was performed using one-way ANOVA, followed by Bonferroni's multiple comparisons test. $***p < 0.001$, indicating a significant difference compared with non-co-cultured stimulated lymphocytes. **b** To assess potential immunostimulatory activity of hMSCs, untransduced, AdNull-transduced and AdIL-10-transduced hMSCs were co-cultured with unstimulated C57BL/6 lymphocytes for 5 days, at a ratio of 1:5. Despite their xenogeneic origin, hMSCs did not induce proliferation of unstimulated C57BL/ 6 lymphocytes beyond basal levels. **c** To confirm immunosuppressive activity of vIL-10 overexpressed by AdIL-10-transduced MSCs, LPS-stimulated murine macrophages (RAW264.7) were treated with conditioned medium (*CM*) from untransduced, AdNull-transduced or AdIL-10-transduced hMSCs (100 ng/ml vIL-10), or 100 ng/ml recombinant vIL-10 (rvIL-10) alone. CM harvested from AdIL-10-transduced MSCs as well as rvIL-10 significantly reduced TNF-α production by LPS-activated macrophages. Values represent the mean ± SD of three biological replicates. Statistical significance was determined by a one-way ANOVA with Bonferroni's multiple comparisons test. $*p < 0.05$ $**p < 0.005$, $***p < 0.001$. *Asterisks* indicate significantly different from AdIL-10 MSC CM. *LPS* lipopolysaccharide, *MSC* mesenchymal stem cell, *TNF-α* tumour necrosis factor alpha

or more generations, only AdIL-10-transduced MSCs showed a significant decrease in proliferation rate compared with stimulated but untreated splenocytes (data not shown). Given the xenogeneic source of the hMSCs we also tested the immunogenicity of these cells in culture with unstimulated splenocytes. No significant increase in splenocyte proliferation was observed with the addition of human cells whether these were MSCs, AdNull-transduced MSCs or AdIL-10-transduced MSCs (Fig. 1b; $p = 0.9854$, $n = 3$). This indicated that, at least in vitro, hMSCs were not immunogenic.

AdIL-10-transduced hMSC CM is anti-inflammatory

To confirm that vIL-10 present in CM harvested from transduced hMSCs was anti-inflammatory, the effect on TNF-α production by LPS-stimulated RAW264.7 macrophages was determined. Macrophages were stimulated with 0.5 ng/ml LPS alone, or co-incubated with LPS and AdIL-10, AdNull or untransduced MSC CM, or medium

containing an equivalent concentration of recombinant vIL-10 (100 ng/ml). At 4, 6 and 12 h post LPS stimulation there was a significant decrease in the amount of TNF-α production by AdIL-10 CM and recombinant vIL-10-treated macrophages compared with LPS only (Fig. 1c; 1.791 ± 0.21 ng/ml in LPS only vs 0.628 ± 0.08 ng/ml in AdIL-10-transduced MSC CM at 12 h, $p < 0.001$, $n = 3$). Treatment of the RAW264.7 cells with CM from untransduced or AdNull-transduced MSCs did not significantly reduce the levels of TNF-α production compared with samples treated with LPS only (1.791 ± 0.21 ng/ml in LPS only vs 1.456 ± 0.36 ng/ml in untransduced hMSC CM at 12 h).

vIL-10-overexpressing hMSCs induce long-lasting reduction in activated T cells

To determine the immunomodulatory effect of vIL-10-overexpressing hMSCs in vivo during OA development, popliteal and inguinal lymph nodes were harvested 6 weeks post hMSC injection and analysed by flow cytometry for CD4, CD8 and the activation marker CD25. Quantification of vIL-10 release from AdIL-10-transduced MSCs in culture prior to injection showed that AdIL-10-transduced cells produced 0.858 µg/ml of AdIL-10 vs 0 ng/ml in both the untransduced and AdNull-transduced MSCs. A significant effect of injection of AdIL-10-transduced MSCs on the presence and activation state of CD4$^+$ and CD8$^+$ T cells was observed.

In the popliteal lymph nodes we observed a significant reduction in the percentage of CD4$^+$ T cells in the AdIL-10 MSC group compared with the AdIL-10 only group (Fig. 2a; 43.9 ± 11.6 % vs 25 ± 2.8 %; $p = 0.0314$). In the inguinal lymph nodes both the AdIL-10 virus alone

and the AdIL-10-transduced MSCs had significantly lower percentages of activated CD4$^+$ T cells compared with vehicle (Fig. 2d; 13 ± 0.7 % and 12.2 ± 0.3 % vs 17.9 ± 0.6 % respectively; $p = 0.0021$) ($n = 4$ samples pooled from eight animals). There was also a significant reduction in the numbers of CD8$^+$ T cells in the popliteal (Fig. 3a; 32 ± 2.9 % in AdNull-transduced MSCs vs 19.7 ± 3.7 % in AdIL-10-transduced MSCs; $p = 0.0081$) and inguinal (Fig. 3c; 31.6 ± 0.7 % in vehicle vs 26.6 ± 1.3 % in AdIL-10-transduced MSCs; $p = 0.0238$) lymph nodes as well as a decrease in the number of activated CD8$^+$ T cells in the inguinal lymph nodes (Fig. 3d; 9.7 ± 1 % in hMSCs vs 4.7 ± 0.5 % in AdIL-10-transduced MSCs; $p = 0.0131$). Levels of T-cell-associated cytokines IFN-γ, IL-2 and

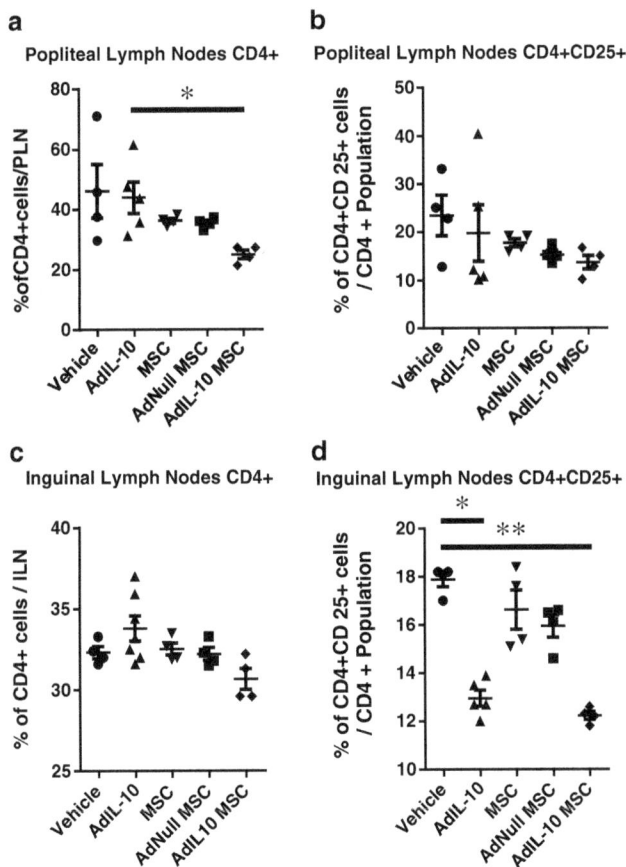

Fig. 2 AdIL-10-transduced MSCs reduce the number of CD4$^+$ T cells in the popliteal and inguinal lymph nodes 6 weeks after injection. CD4 and CD25 expression by lymphocytes isolated from the popliteal and inguinal lymph nodes, 6 weeks post treatment. **a, b** AdIL-10-transduced MSCs reduce CD4$^+$ T-cell levels in the popliteal lymph nodes compared with AdIL-10, with no significant difference in the number of activated (CD25$^+$) CD4 T cells between any experimental groups. **c, d** No significant difference was observed between any of the treatment groups in the amount of CD4$^+$ T cells present in the inguinal lymph nodes. However, AdIL-10 only and AdIL-10-transduced MSCs significantly decreased the amount of activated CD4$^+$ cells compared with vehicle-treated animals. Data points represent $n = 4$, pooled from eight animals. Data points in the AdIL-10 group represent $n = 5$, with three samples pooled from two animals and two single samples. Statistical significance was determined using a Kruskal–Wallis test, followed by Dunn's multiple comparisons test. *$p < 0.05$, **$p < 0.005$, ***$p < 0.001$. *Lines* indicate significant difference between the individual groups. *ILN* inguinal lymph node, *MSC* mesenchymal stem cell, *PLN* popliteal lymph node

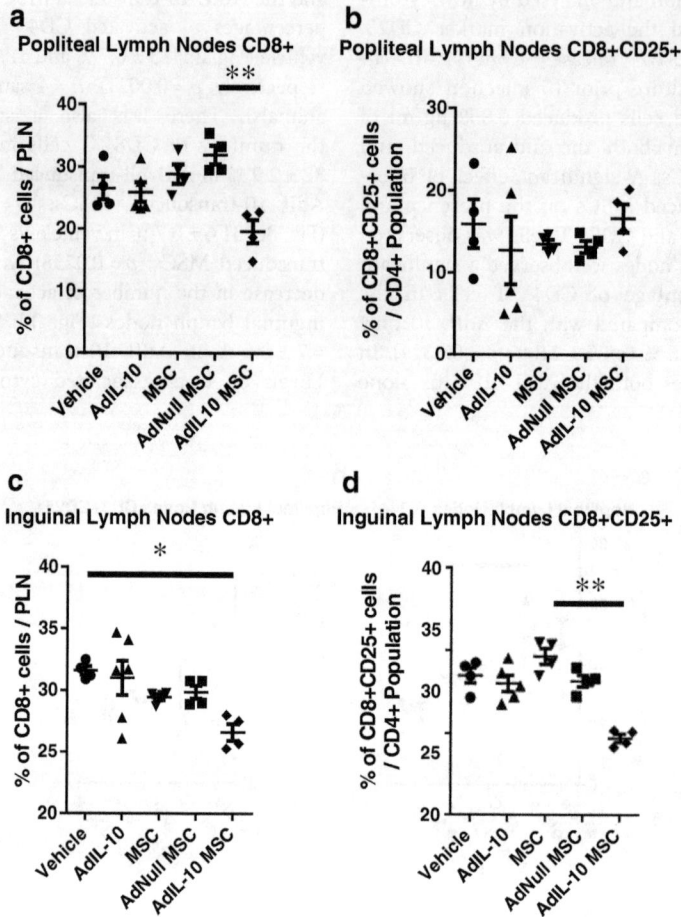

Fig. 3 AdIL-10-transduced MSCs reduce the number of CD8 T cells in the popliteal and inguinal lymph nodes 6 weeks after injection. CD8 and CD25 expression by lymphocytes isolated from the popliteal and inguinal lymph nodes, 6 weeks post treatment. **a, b** AdIL-10-transduced MSCs reduced the amount of CD8[+] T cells compared with AdNull-transduced MSCs in the popliteal lymph nodes, whereas no significant difference in the amount of activated CD8[+] T cells was observed. **c, d** AdIL-10-transduced MSCs significantly decreased the amount of CD8[+] cells in the inguinal lymph nodes and reduced the amount of activated CD8[+] T cells compared with MSCs alone. Data points represent $n = 4$, pooled from eight animals. Data points in the AdIL-10 group represent $n = 5$, with three samples pooled from two animals and two single samples. Statistical significance was determined using a Kruskal–Wallis test, followed by Dunn's multiple comparisons test. *$p < 0.05$, **$p < 0.005$, ***$p < 0.001$. *Lines* indicate significant difference between the two groups. *ILN* inguinal lymph node, *MSC* mesenchymal stem cell, *PLN* popliteal lymph node

IL-4 were undetectable in serum harvested from animals in all treatment groups at the experimental end point, as determined by multiplex ELISA (data not shown). Additionally, levels of CD11b[+] and CD11b[+] Ly6C hi cells in the popliteal lymph nodes (markers of pro-inflammatory monocytes) were similar in all pooled samples from each treatment group (Additional file 3: Figure S3A). However, a trend towards decreased levels of CD11b[+] and CD11b[+] Ly6C hi cells was observed in the inguinal lymph nodes following treatment with AdIL-10 virus alone. Levels of MCP-1 in serum harvested from animals at the experimental end point were similar between each treatment group (Additional file 3: Figure S3B).

Injection of hMSCs does not prevent OA development or progression

To assess the effect of AdIL-10-transduced hMSCs on OA progression, knees treated with collagenase at time 0 and hMSCs at day 7 were harvested, sectioned and stained with thionine 6 weeks after injection of the cells. At this time point there was no significant difference observed between the amount of damage in vehicle-treated knees compared with any other group in any compartment (Fig. 4a–d). All knees showed changes to both the cartilage and synovium, with more damage observed in the medial compartment compared with the lateral. However, significant inter-animal variability was seen in

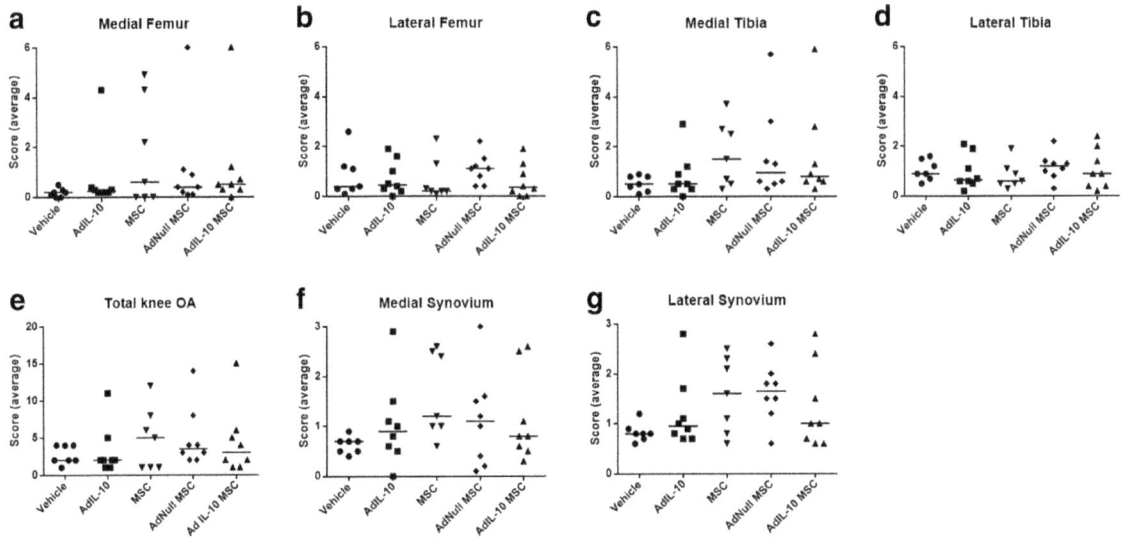

Fig. 4 Injection of hMSCs does not prevent OA development or progression. Following two consecutive injections of 1 U collagenase over 2 days (1 U per day), mice were injected with untransduced, AdNull-transduced or AdIL-10-transduced hMSCs (or AdIL-10 virus or vehicle as control) at 1 week post OA induction. After 6 weeks knees were harvested and analysed for cartilage damage and synovial hyperplasia. **a, b** Median damage scores in the medial and lateral femur (maximum score of 6 with three blinded observers, approximately six sections per knee). **c, d** Median damage scores in the medial and lateral tibia compartments (maximum score of 6 with three blinded observers, approximately six sections per knee). **e** Total damage within the four joint compartments (maximum score of 24). **f, g** Median scores of synovial hyperplasia in the medial and lateral synovial compartments (maximum score of 3 with three blinded observers, approximately six sections per knee). Mild OA was observed in all groups with no significant reduction in the treated groups. Overall there appeared to be more damage in the medial compartment. $n = 8$ animals per group. *MSC* mesenchymal stem cell, *OA* osteoarthritis

groups injected with AdIL-10, MSCs, AdNull MSCs and AdIL-10 MSCs for cartilage and synovium changes in the medial compartment (Fig. 4a, c) and synovial hyperplasia in the lateral compartment (Fig. 4f). Histologically, joints with high scores presented with significant osteophyte formation and loss of medial compartment cartilage (Fig. 5b). This appeared to occur more in the hMSC-treated knees, irrespective of whether the cells were transduced or not. The same pattern was observed in the synovial damage scores, with more inflammation observed in the samples that received injections of hMSCs compared with vehicle (Fig 4f, g). Although not significant, joints treated with AdIL-10-transduced MSCs showed less synovial changes compared with the MSC-treated and AdNull MSC-treated joints. Ultimately, groups with the least damage were those that were treated with vehicle or with virus only (Fig. 4e).

Discussion

Recently it has become clear that OA is a disease of the entire joint and not simply the cartilage [35]. Increasing evidence highlights the association between factors produced by inflamed synovium and the initiation and progression of the disease. Furthermore, inflammation of the synovial membrane with increased vascular density and cellular infiltration is a prominent feature of OA

pathogenesis [36]. The aim of this research was to investigate the potential of vIL-10-overexpressing hMSCs to mitigate this inflammation, via modulation of the immune response and to assess whether this modulation could delay or prevent the OA onset and progression.

It is clear from the literature that T lymphocytes play a role in the pathogenesis of OA [37] with cellular infiltration of the synovial tissue leading to induction of inflammatory factors [4, 5, 38] and osteoclastogenesis [6]. Hsieh et al. [39] have shown that CD8 knockout animals exhibit reduced OA, and an increase in the presence of CD3$^+$ cells has been demonstrated in the synovium of osteoarthritic animals [40]. Furthermore, many T-cell-associated factors, such as IL-17, have been shown to have negative effects upon joint cartilage [4, 41, 42]. While synovial inflammation is higher in rheumatoid arthritis compared with OA, osteoarthritic knees still have higher levels of inflammation compared with healthy controls [43]. Here we clearly demonstrate that the injection of vIL-10-overexpressing hMSCs leads to long-term reductions in the numbers of CD4$^+$ and CD8$^+$ T cells in the draining popliteal lymph nodes of the knees and activated CD4$^+$ and CD8$^+$ T cells in the inguinal lymph nodes. Also, injection of AdIL-10 virus alone led to similar outcomes in the levels of activated CD4$^+$ cells in the popliteal lymph nodes.

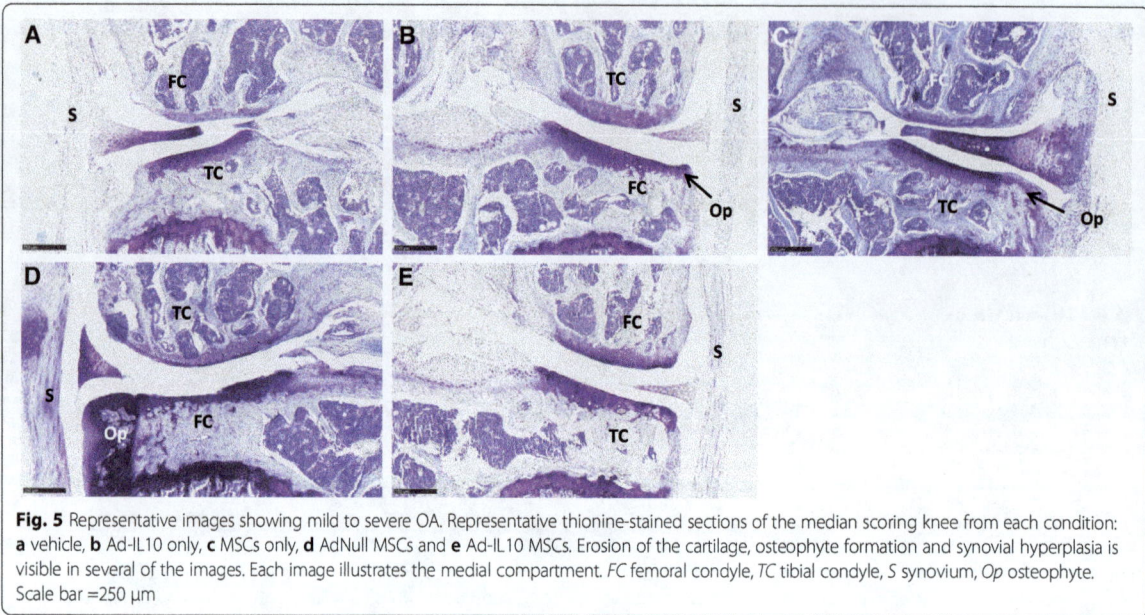

Fig. 5 Representative images showing mild to severe OA. Representative thionine-stained sections of the median scoring knee from each condition: **a** vehicle, **b** Ad-IL10 only, **c** MSCs only, **d** AdNull MSCs and **e** Ad-IL10 MSCs. Erosion of the cartilage, osteophyte formation and synovial hyperplasia is visible in several of the images. Each image illustrates the medial compartment. *FC* femoral condyle, *TC* tibial condyle, *S* synovium, *Op* osteophyte. Scale bar =250 μm

Furthermore, this was a long-lasting effect, with these reductions being observed 6 weeks after injection of hMSCs.

IL-10 is a pleiotropic cytokine that has been shown to have positive effects on the synthesis of collagen type II and aggrecan as well as reducing secretion of pro-inflammatory cytokines [44]. However, IL-10 has also been shown to induce proliferation and promote chondrogenic or hypertrophic differentiation of primary chondrocytes, highlighting a role for IL-10 in endochondral bone growth [45]. Elevated levels of IL-10 have been detected in cartilage, synovium and subchondral bone explants of patients with OA [46]. Despite the fact that IL-10 appears to play a protective role on cartilage in the joint, we did not observe a direct effect of vIL-10 overexpression on structural changes within the joint during development of OA. Further investigation is therefore required to evaluate the chondroprotective potential of vIL-10 and its ability to attenuate progression of OA. There is also the possibility that the presence of IL-10 in the osteoarthritic joint could lead to further cellular hypertrophy and a negative outcome since it is also associated with endochondral ossification [45]. However, no increased joint damage was seen in either vIL-10-treated group compared with the others. While it was not the focus of the study to understand subchondral bone changes, we cannot discount that there may have been differences in this region in vIL-10-treated animals. Furthermore, the effects of the inflammatory environment upon bone could be very different from those on cartilage. As such it is possible that this treatment, which had no clear effect on the joint cartilage, may have had subtle effects on the surrounding bone tissues.

MSCs are known to produce a host of paracrine factors that can reduce inflammation and induce tissue repair in a number of models [47]. However, whether they induce joint repair or simply reduce pain is not completely clear. It does appear that their mechanism of action is more related to paracrine effects and reduction of inflammation rather than by direct tissue generation/regeneration [48]. The exact efficacy of these MSCs in prevention or reversal is still unclear. While there have been some promising reports relating to the reduction of pain and even the repair of defects within the joint by MSCs, longer term and more comprehensive clinical trials are required in order to clearly identify the efficacy of MSC therapy. Schurgers et al. [49] previously reported differential effects of MSCs on T-cell proliferation in vitro vs in vivo using a collagen-induced arthritis (CIA) model. In contrast to these findings, we observed a decrease in T lymphocytes in vitro and in vivo in our OA model following treatment with hMSCs transduced to express vIL-10. However, even though we did not observe an immunogenic effect of untransduced or virally transduced hMSCs on mouse cells in vitro, we did observe that the majority of the severely damaged knees were in groups treated with the hMSCs or transduced hMSCs. Interestingly, fully genetically mismatched allogeneic MSCs have been shown to hold less immunosuppressive activity compared with a syngeneic or partially mismatched cell source, and have aggravated disease pathogenesis in a murine CIA model [50]. We hypothesise that our findings may be due to the xenogeneic nature of MSCs used in this study. It is possible that while there was a reduction in the levels of immune cells present and also in the TNF-α levels in vitro, vIL-10 overexpression was not sufficient to overcome these xenogeneic effects. Once again the effects on the bone density were not assessed and these effects should be taken into account for studies.

Our study has some limitations. Despite our observation of a long-term effect of vIL-10-overexpressing hMSCs on the T-cell number, we did not observe a large effect of this treatment on the severity and progression of OA. There are several possible reasons for this lack of effect on the severity and progression of OA. Firstly, the model used was designed to lead to mild OA. However, the degree of progression of the disease was slower/milder than expected at this time point based on pilot data (data not shown). Also there were large differences between animals in the levels of OA and joint damage. This led to a large amount of variation within and between groups. It is possible that a clearer picture might emerge with a slightly more severe OA model, larger animal numbers or a longer period between cell injection and joint harvest.

Conclusions

We present a long-term immunomodulatory effect of vIL-10-overexpressing hMSCs injected into the knees of osteoarthritic mice. While there was no clear reduction in disease onset or progression, we believe that this approach could be used in combination with reparative approaches to result in a long-term reduction in joint inflammation. As stated above, T lymphocytes are the most abundant infiltrating immune cells present in OA synovium and a role for $CD4^+$ T cells in the pathogenesis of OA has been elucidated (refs). Based on the reduction in activated T cells present in the draining lymph nodes of the knees, we believe that, in instances of OA with a high inflammatory component, treatment with a combination of syngeneic or perhaps allogeneic MSCs and vIL-10 could help to reduce symptoms and disease progression. We also feel that the results presented do somewhat call into question the use of xenogeneic cells for the investigation of MSC-mediated repair/protection of the joint following OA induction. This should be considered in the execution of similar experiments in the future.

Additional files

Additional file 1: Figure S1. Timeline illustrating the induction of OA in mice followed by injection of five different treatment conditions at day 7. Six weeks after treatment animals were euthanised and the joints harvested for histological scoring. Lymph nodes and blood were taken for flow cytometric and multiplex ELISA analyses respectively. (PDF 9 kb)

Additional file 2: Figure S2. Representative safranin O-stained sections of the median scoring knee from each condition as well as an untreated contralateral knee for comparison: **A** Vehicle, **B** Ad-IL10 only, **C** MSCs only, **D** AdNull MSCs, **E** Ad-IL10 MSCs and **F** untreated contralateral knee.

Erosion of the cartilage, osteophyte formation and synovial hyperplasia is visible in several of the images. Each image illustrates the medial compartment. *FC* femoral condyle, *TC* tibial condyle, *S* synovium, *Op* osteophyte. Scale bar =250 μm. (PDF 197 kb)

Additional file 3: Figure S3. AdIL-10-transduced MSCs do not reduce the levels of pro-inflammatory monocytes at 6 weeks post injection. **A** CD11b and Ly6C expression by myeloid cells isolated from the popliteal and inguinal lymph nodes at 6 weeks post treatment, as detected by flow cytometry. For the popliteal lymph nodes, data points represent $n = 4$, pooled from eight animals. Data points in the AdIL-10 group represent $n = 5$, with three samples pooled from two animals and two single samples. For inguinal lymph nodes, data points represent $n = 2$, pooled from eight animals. Data points in the AdIL-10 group represent $n = 3$, with one sample pooled from four animals, one sample pooled from three animals and one single sample. **B** Serum levels of MCP-1 at 6 weeks post treatment, as quantified utilising a chemiluminescent array. (PDF 47 kb)

Abbreviations

α-MEM: alpha-Minimum Essential Medium; CFSE: carboxy-fluorescein diacetate, succinmidyl ester; CIA: collagen-induced arthritis; CM: conditioned medium; FBS: foetal bovine serum; hMSC: human mesenchymal stem cell; IL: interleukin; LPS: lipopolysaccharide; MCP-1: monocyte chemoattractant protein-1; MHC: major histocompatibility complex class II; MSC: mesenchymal stem cell; OA: osteoarthritis; TNF-α: tumour necrosis factor alpha; vIL-10: viral interleukin-10.

Competing interests

The authors declare that they have no competing interests.

Authors' contributions

EF and NF acquired, analysed and interpreted data, drafted and edited the article. EF and JMM conceived this study and its design, interpreted the data, drafted and edited this article. AER, LOF and TR were responsible for the design and analysis of immune cell phenotypes ex vivo. COF was involved in design and carrying out of the animal model. All authors critically revised the article for important intellectual content and approved the final version to be submitted. EF, NF and JMM take responsibility for the integrity of the work as a whole, from inception to finished article.

Acknowledgements

The authors would like to acknowledge Yvonne Dooley and Swarna Raman for their assistance with histological preparations.

Role of the funding source

This work was supported by Science Foundation Ireland grant number 09/SRC/B1794, the European Union's 7th Framework Programme under grant agreement no. NMP3-SL-2010-245993 (GAMBA) and by a National University of Ireland travelling studentship. AR is supported by an Irish cancer research fellowship, CRF12RYA.

Author details

[1]Department of Oral and Maxillofacial Surgery, Special Dental Care and Orthodontics, Erasmus MC, University Medical Centre, Room Ee1614, Erasmus MC, Wytemaweg 80, Rotterdam 3015CN, The Netherlands. [2]Regenerative Medicine Institute, National University of Ireland Galway, Galway, Ireland. [3]Musculoskeletal Regeneration, AO Research Institute Davos (ARI), Davos, Switzerland. [4]College of Medicine, Nursing and Health Sciences, National University of Ireland Galway, Galway, Ireland. [5]Orbsen Therapeutics Ltd, Galway, Ireland. [6]Discipline of Pharmacology and Therapeutics, National University of Ireland Galway, Galway, Ireland.

References

1. Pessler F, et al. The synovitis of "non-inflammatory" orthopaedic arthropathies: a quantitative histological and immunohistochemical analysis. Ann Rheum Dis. 2008;67(8):1184–7.
2. Westacott CI, et al. Tumor necrosis factor alpha can contribute to focal loss of cartilage in osteoarthritis. Osteoarthr Cartil. 2000;8(3):213–21.

3.   Johnson K, et al. Interleukin-1 induces pro-mineralizing activity of cartilage tissue transglutaminase and factor XIIIa. Am J Pathol. 2001;159(1):149–63.

4.   Ishii H, et al. Characterization of infiltrating T cells and Th1/Th2-type cytokines in the synovium of patients with osteoarthritis. Osteoarthr Cartil. 2002;10(4):277–81.

5.   Pessler F, et al. A histomorphometric analysis of synovial biopsies from individuals with Gulf War Veterans' illness and joint pain compared to normal and osteoarthritis synovium. Clin Rheumatol. 2008;27(9):1127–34.

6.   Shen PC, et al. T helper cells promote disease progression of osteoarthritis by inducing macrophage inflammatory protein-1gamma. Osteoarthr Cartil. 2011;19(6):728–36.

7.   Yoo JU, et al. The chondrogenic potential of human bone-marrow-derived mesenchymal progenitor cells. J Bone Joint Surg Am. 1998;80(12):1745–57.

8.   Murphy JM, et al. Stem cell therapy in a caprine model of osteoarthritis. Arthritis Rheum. 2003;48(12):3464–74.

9.   Duffy MM, et al. Mesenchymal stem cell inhibition of T-helper 17 cell-differentiation is triggered by cell-cell contact and mediated by prostaglandin E2 via the EP4 receptor. Eur J Immunol. 2011;41(10):2840–51.

10. Yang SH, et al. Soluble mediators from mesenchymal stem cells suppress T cell proliferation by inducing IL-10. Exp Mol Med. 2009;41(5):315–24.

11. van Buul GM, et al. Mesenchymal stem cells secrete factors that inhibit inflammatory processes in short-term osteoarthritic synovium and cartilage explant culture. Osteoarthr Cartil. 2012;20(10):1186–96.

12. Nemeth K, et al. Bone marrow stromal cells attenuate sepsis via prostaglandin E-2-dependent reprogramming of host macrophages to increase their interleukin-10 production (vol 15, pg 42, 2009). Nat Med. 2009;15(4):462–2.

13. Horie M, et al. Intra-articular injection of human mesenchymal stem cells (MSCs) promote rat meniscal regeneration by being activated to express Indian hedgehog that enhances expression of type II collagen. Osteoarthr Cartil. 2012;20(10):1197–207.

14. Diekman BO, et al. Intra-articular delivery of purified mesenchymal stem cells from C57BL/6 or MRL/MpJ superhealer mice prevents posttraumatic arthritis. Cell Transplant. 2013;22(8):1395–408.

15. ter Huurne M, et al. Antiinflammatory and chondroprotective effects of intraarticular injection of adipose-derived stem cells in experimental osteoarthritis. Arthritis Rheum. 2012;64(11):3604–13.

16. Zdanov A, et al. Crystal structure of Epstein-Barr virus protein BCRF1, a homolog of cellular interleukin-10. J Mol Biol. 1997;268(2):460–7.

17. Fiorentino DF, Bond MW, Mosmann TR. Two types of mouse T helper cell. IV. Th2 clones secrete a factor that inhibits cytokine production by Th1 clones. J Exp Med. 1989;170(6):2081–95.

18. Evans JG, et al. Novel suppressive function of transitional 2 B cells in experimental arthritis. J Immunol. 2007;178(12):7868–78.

19. Hart PH, et al. Comparison of the suppressive effects of interleukin-10 and interleukin-4 on synovial fluid macrophages and blood monocytes from patients with inflammatory arthritis. Immunology. 1995;84(4):536–42.

20. Malefyt RD, et al. Interleukin-10 (IL-10) and viral-IL-10 strongly reduce antigen-specific human T-cell proliferation by diminishing the antigen-presenting capacity of monocytes via down-regulation of class-II major histocompatibility complex expression. J Exp Med. 1991;174(4):915–24.

21. Rousset F, et al. Interleukin 10 is a potent growth and differentiation factor for activated human B lymphocytes. Proc Natl Acad Sci U S A. 1992;89(5):1890–3.

22. Go NF, et al. Interleukin 10, a novel B cell stimulatory factor: unresponsiveness of X chromosome-linked immunodeficiency B cells. J Exp Med. 1990;172(6):1625–31.

23. Thompson-Snipes L, et al. Interleukin 10: a novel stimulatory factor for mast cells and their progenitors. J Exp Med. 1991;173(2):507–10.

24. Vieira P, et al. Isolation and expression of human cytokine synthesis inhibitory factor cDNA clones: homology to Epstein-Barr virus open reading frame BCRFI. Proc Natl Acad Sci U S A. 1991;88(4):1172–6.

25. Apparailly F, et al. Adenovirus-mediated transfer of viral IL-10 gene inhibits murine collagen-induced arthritis. J Immunol. 1998;160(11):5213–20.

26. Choi JJ, et al. Mesenchymal stem cells overexpressing interleukin-10 attenuate collagen-induced arthritis in mice. Clin Exp Immunol. 2008;153(2):269–76.

27. van Osch GJ, et al. Relation of ligament damage with site specific cartilage loss and osteophyte formation in collagenase induced osteoarthritis in mice. J Rheumatol. 1996;23(7):1227–32.

28. van Lent PL, et al. Active involvement of alarmins S100A8 and S100A9 in the regulation of synovial activation and joint destruction during mouse and human osteoarthritis. Arthritis Rheum. 2012;64(5):1466–76.

29. Coleman CM, et al. Growth differentiation factor-5 enhances in vitro mesenchymal stromal cell chondrogenesis and hypertrophy. Stem Cells Dev. 2013;22(13):1968–76.

30. Palmer GD, et al. A simple, lanthanide-based method to enhance the transduction efficiency of adenovirus vectors. Gene Therapy. 2008;15(5):357–63.

31. Pritzker KP, et al. Osteoarthritis cartilage histopathology: grading and staging. Osteoarthr Cartil. 2006;14(1):13–29.

32. Ryan AE, et al. Chondrogenic differentiation increases antidonor immune response to allogeneic mesenchymal stem cell transplantation. Mol Ther. 2014;22(3):655–67.

33. Caruso A, et al. Flow cytometric analysis of activation markers on stimulated T cells and their correlation with cell proliferation. Cytometry. 1997;27(1):71–6.

34. Ziegler-Heitbrock L, et al. Nomenclature of monocytes and dendritic cells in blood. Blood. 2010;116(16):e74–80.

35. Loeser RF, et al. Osteoarthritis: a disease of the joint as an organ. Arthritis Rheum. 2012;64(6):1697–707.

36. Fahy N, et al. Immune modulation to improve tissue engineering outcomes for cartilage repair in the osteoarthritic joint. Tissue Eng Part B Rev. 2015; 21(1):55–66. doi:10.1089/ten.TEB.2014.0098.

37. Sakkas LI, Platsoucas CD. The role of T cells in the pathogenesis of osteoarthritis. Arthritis Rheum. 2007;56(2):409–24.

38. Sakkas LI, et al. T cells and T-cell cytokine transcripts in the synovial membrane in patients with osteoarthritis. Clin Diagn Lab Immunol. 1998;5(4):430–7.

39. Hsieh JL, et al. CD8+ T cell-induced expression of tissue inhibitor of metalloproteinses-1 exacerbated osteoarthritis. Int J Mol Sci. 2013;14(10):19951–70.

40. Nakamura H, et al. T-cell mediated inflammatory pathway in osteoarthritis. Osteoarthr Cartil. 1999;7(4):401–2.

41. Honorati MC, et al. Interleukin-17, a regulator of angiogenic factor release by synovial fibroblasts. Osteoarthr Cartil. 2006;14(4):345–52.

42. Alsalameh S, et al. Cellular immune response toward human articular chondrocytes. T cell reactivities against chondrocyte and fibroblast membranes in destructive joint diseases. Arthritis Rheum. 1990;33(10):1477–86.

43. de Lange-Brokaar BJ, et al. Synovial inflammation, immune cells and their cytokines in osteoarthritis: a review. Osteoarthr Cartil. 2012;20(12):1484–99.

44. Wojdasiewicz P, Poniatowski LA, Szukiewicz D. The role of inflammatory and anti-inflammatory cytokines in the pathogenesis of osteoarthritis. Mediators Inflamm. 2014;2014:561459.

45. Jung YK, et al. Role of interleukin-10 in endochondral bone formation in mice: anabolic effect via the bone morphogenetic protein/Smad pathway. Arthritis Rheum. 2013;65(12):3153–64.

46. Hulejova H, et al. Increased level of cytokines and matrix metalloproteinases in osteoarthritic subchondral bone. Cytokine. 2007;38(3):151–6.

47. Song M, et al. The paracrine effects of mesenchymal stem cells stimulate the regeneration capacity of endogenous stem cells in the repair of a bladder-outlet-obstruction-induced overactive bladder. Stem Cells Dev. 2014;23(6):654–63.

48. van Buul GM, et al. Mesenchymal stem cells reduce pain but not degenerative changes in a mono-iodoacetate rat model of osteoarthritis. J Orthop Res. 2014;32(9):1167–74.

49. Schurgers E, et al. Discrepancy between the in vitro and in vivo effects of murine mesenchymal stem cells on T-cell proliferation and collagen-induced arthritis. Arthritis Res Ther. 2010;12(1):R31.

50. Sullivan C, et al. Genetic mismatch affects the immunosuppressive properties of mesenchymal stem cells in vitro and their ability to influence the course of collagen-induced arthritis. Arthritis Res Ther. 2012;14(4):R167.

10

How electromagnetic fields can influence adult stem cells: positive and negative impacts

Aleksandra Maziarz[1,2], Beata Kocan[1,2], Mariusz Bester[3], Sylwia Budzik[3], Marian Cholewa[3], Takahiro Ochiya[4] and Agnieszka Banas[1,2]*

Abstract

The electromagnetic field (EMF) has a great impact on our body. It has been successfully used in physiotherapy for the treatment of bone disorders and osteoarthritis, as well as for cartilage regeneration or pain reduction. Recently, EMFs have also been applied in in vitro experiments on cell/stem cell cultures. Stem cells reside in almost all tissues within the human body, where they exhibit various potential. These cells are of great importance because they control homeostasis, regeneration, and healing. Nevertheless, stem cells when become cancer stem cells, may influence the pathological condition. In this article we review the current knowledge on the effects of EMFs on human adult stem cell biology, such as proliferation, the cell cycle, or differentiation. We present the characteristics of the EMFs used in miscellaneous assays. Most research has so far been performed during osteogenic and chondrogenic differentiation of mesenchymal stem cells. It has been demonstrated that the effects of EMF stimulation depend on the intensity and frequency of the EMF and the time of exposure to it. However, other factors may affect these processes, such as growth factors, reactive oxygen species, and so forth. Exploration of this research area may enhance the development of EMF-based technologies used in medical applications and thereby improve stem cell-based therapy and tissue engineering.

Background

Many, if not all, tissues of the human body are thought to contain stem cells (called adult stem cells/adult tissue stem cells/progenitor cells) that are responsible for tissue regeneration and repair after injury. Adult stem cells are influenced by many biochemical and biophysical stimuli in their in vivo microenvironment, including fluid shear stress, hydrostatic pressure, substrate strains, trophic factors, the electromagnetic field (EMF), and so forth. Depending on the niche in which they reside, as well as the biochemical and biophysical stimuli, stem cells may differentiate or not into desired tissues [1–3]. These factors are of great importance because dysregulation of tissue regeneration and homeostasis may result in various pathological conditions, cancer being the most extensively described. Several studies have focused on

the circumstances that result in adult stem cells becoming cancer stem cells (tumor-initiating cells) that participate in carcinogenesis and metastasis. However, the nature of the interaction between adult and cancer stem cells and the mechanisms underlying the putative transition remain elusive. It is believed that during the initial stage of the pathological process, adult stem cells may be both "heroes" and "villains".

External environmental factors are commonly known to be simultaneously involved in pathological processes, making the maintenance of homeostasis a difficult challenge. Biophysical stimuli may cause downstream signaling towards pleiotropic processes in adult stem cells.

The EMF is pervasive throughout the environment and, owing to technological developments, seems to have great potential as a therapeutic tool. It has significant effects on cells, tissues, and many processes within organisms and plays an important role in biological processes involving adult stem cells, such as embryogenesis, regeneration, and wound healing [4], as well as in cell migration, DNA synthesis, and gene expression [5–7].

* Correspondence: agnieszkabanas@tlen.pl
[1]Laboratory of Stem Cells' Biology, Department of Immunology, Chair of Molecular Medicine, Faculty of Medicine, University of Rzeszow, ul. Kopisto 2a, 35-310 Rzeszow, Poland
[2]Centre for Innovative Research in Medical and Natural Sciences, Faculty of Medicine, University of Rzeszow, ul. Warzywna 1a, 35-310 Rzeszow, Poland
Full list of author information is available at the end of the article

However, the data regarding the influence of the EMF on adult stem cell biology are inconsistent.

Here, we review the current knowledge on the effects of EMFs on adult stem cells. Our goal is to present all available evidence for both the positive (stimulative and prodifferentiative) and negative (carcinogenic) impact of EMFs on stem cell biology.

Adult stem cells

Adult stem cells compose "a reservoir" of cells at various stages of development and possess the unique ability to self-renew and to differentiate into many types of specialized cells [8]. They play an important role in tissue regeneration and maintenance of homeostasis [1, 2, 9, 10]. Adult stem cells isolated and cultured ex vivo may differentiate under proper conditions and may give rise to multiple lineages in a controlled manner in vitro [9]. The cells can thus be used as an autologous source of cells for treatment of multiple modern-age diseases such as cardiovascular diseases [11], liver disease [12–16], and neurogenerative diseases [17]. What is more, the extracellular vesicles derived from adipose-derived mesenchymal stem cells (ASCs) [18–20] have been of particular interest due to their therapeutic activity.

On the other hand, adult stem cells under the influence of "improper stimuli" may contribute to carcinogenesis and pathological alterations, resulting in many chronic disorders. These stimuli may consist of biochemical and biophysical environmental factors which lead to imbalance in tissues and the stem cell niche. This initiates a cascade of degeneration, destruction, and anti-homeostatic processes, followed by diseases and finally death (Fig. 1).

The EMF as a therapeutic tool

EMF stimulation has been used successfully for the treatment of bone disorders for many years [5, 21–23]. It is clinically beneficial for bone fracture healing, treatment of osteoarthritis, and pain reduction [23]. The EMF stimulates osteogenesis, increases bone mineral density, decreases osteoporosis, and acts chondroprotectively [6, 23] (Table 1).

Endogenous electrical potentials and currents are generated in wounded tissues and they disappear when healing is complete. The EMF has a positive impact at different stages of healing (Fig. 2a). The processes affected by the EMF include cell migration and proliferation, expression of growth factors, nitric oxide signaling, cytokine modulation, and more. These effects have been observed using an EMF at low (30–300 kHz) and extremely low (3–30 Hz) frequencies.

Effects of the EMF on stem cells during early development

Imprinting of maternal and paternal genetic components occurs during early development and epigenetic mechanisms are involved in this phenomenon. Importantly, disruption of imprinting may lead to abortion or disease (e.g., malformation, cancer). Endogenous EMFs are present in developing and regenerating tissues and organs, either in the extracellular space or in the cell cytoplasm. Their strength ranges from a few to several hundred millivolts per millimeter [24]. The EMF, together with diffusible chemical gradients, leads to polarization and formation of spatial patterns in the developing embryo, creating the signals necessary for correct placement of the components

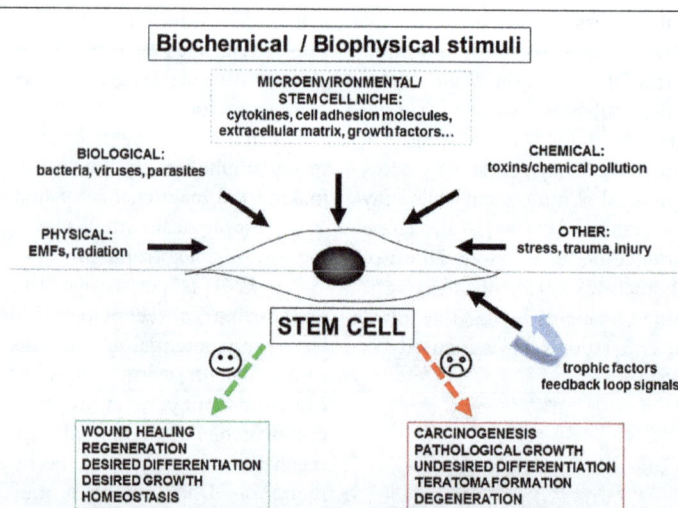

Fig. 1 Possible biochemical/biophysical stimuli affecting adult stem cells within the body that lead to physiological or pathological processes. The stimuli may lead towards positive, life-supporting processes (wound healing, regeneration, homeostasis) or negative, life-suppressing processes (carcinogenesis, degeneration). *EMF* electromagnetic field

Table 1 Effects of EMFs with different parameters on stem cell biology

Stem cell type	EMF characteristics	Exposure duration	Differentiation type	Stimulation effects	Reference
Sinusoidal EMF					
BM-MSCs	ELF-EMF Magnetic flux density: 1 mT Frequency: 50 or 100 Hz	Continuous for up to 8 days	Neurogenic	No effects on cell viability Increase in the expression of neuronal markers (NeuroD1, MAP2, NF-L) Stimulation of neural differentiation	Park et al. 2013 [17]
BM-MSCs	ELF-EMF Magnetic flux density: 1 mT Frequency: 50 Hz	Continuous for 12 days	Neurogenic	Inhibition of MSC growth Decrease of the neural stem cell marker expression (nestin) Increase of the neural cell marker expression (MAP2, NeuroD1, NF-L, and Tau)	Cho et al. 2012 [39]
BM-MSCs	ELF-EMF Magnetic flux density: 5 mT Frequency: 15 Hz	Three times a day (45 min every 8 h) for 21 days	Chondrogenic	More compact structure Varied effects on cartilage-specific marker expression (increase in COL. II, decrease in COL. X, or no impact on aggrecan, SOX9) Higher glycosaminoglycan/DNA content Improvement of chondrogenic differentiation in combination with growth factor treatment	Mayer-Wagner et al. 2011 [23]
BM-MSCs (derived from fetus)	ELF-EMF Magnetic flux density: 20 mT Frequency: 50 Hz	12 h/day for up to 23 days	Osteogenic	Decrease of MSC growth and metabolism No significant effect on MSC differentiation	Yan et al. 2010 [38]
ASCs	EMF Magnetic flux density: 1 mT Frequency: 30/45 Hz (positive differentiation conditions); 7.5 Hz (negative differentiation conditions)	8 h/day	Osteogenic	Alterations in ALP expression level Alterations in osteogenic differentiation level Alterations in the expression of osteogenic markers Enhancement of matrix mineralization	Kang et al. 2013 [6]
ESCs	Low-frequency EMF Magnetic flux density: 5 mT Frequency: 1, 10, and 50 Hz	30 min/day for 3, 5, or 7 days	–	Increase in cell proliferation rate, in a frequency-dependent manner (the highest rate in the 50 Hz group) Alterations in the cell cycle No effect on cell morphology and cell phenotype	Zhang et al. 2013 [35]
Combination of static and sinusoidal EMF					
CSCs	Static MF Magnetic flux density: 10 µT Sinusoidal ELF-EMF Magnetic flux density: 2.5 µT Frequency: 7 Hz (Ca^{2+} ICR)	Up to 5 days	Cardiogenic	Increase in metabolic activity Increase in proliferation rate Increase in the expression of cardiac markers (TnI, MHC, Nkx2.5) Decrease (SMA) or no change (VEGF, KDR) in the expression of vascular markersAlterations in the intracellular calcium distribution	Gaetani et al. 2009 [11]
CSCs/BM-MSCs	Static MF Magnetic flux density: 10 µT Sinusoidal ELF-EMF Frequency: 7 Hz (Ca^{2+} ICR)	For 5 days	Cardiogenic/osteogenic	Upregulation of cardiac markers (TnI, MHC) Downregulation of angiogenic markers (VEGF, KDR) Increase in the expression of osteogenic markers (ALP, OC, OPN) Alterations in plasma membrane morphology	Lisi et al. 2008 [43]

Table 1 Effects of EMFs with different parameters on stem cell biology (Continued)

				accompanied by a rearrangement in actin filaments	
Pulsed EMF BM-MSCs	Magnetic flux density: 1.1 mT Frequency: 5, 25, 50, 75, 100, and 150 Hz	30 min/day for 21 days	Osteogenic	Alterations in cell morphology, Increase in ALP expression and activity, Increase in the expression of osteogenic markers (COL I, OC), Stimulation of osteogenic differentiation, Enhancement of matrix mineralization	Luo et al. 2012 [7]
BM-MSCs	Magnetic flux density: 1.8–3 mT Frequency: 75 Hz	8 h/day for 14 days	Osteogenic	Acceleration of cell proliferation, Alterations in cell cycle, Increase in ALP expression level, Enhancement of the osteogenic differentiation	Esposito et al. 2012 [45]
BM-MSCs	Time of pulses: 300 µs (repetitive single quasi-rectangular pulses) Magnetic flux density: 0.13 mT Frequency: 7.5 Hz	2 h/day for 14 days	Osteogenic	Time-dependent alterations in cell proliferation rate, Stimulation of ALP activity at day 7, Enhancement of early osteogenic genes expression (Runx2/Cbfa1 and ALP) during the mid-stage of osteogenic differentiation	Tsai et al. 2009 [5]
BM-MSCs/osteoblast-like cells	Time of bursts: 5 ms Time of pulses: 5 µs Magnetic flux density: 0.1 mT Frequency:15 Hz	Continuous exposure	Osteogenic	Increase of matrix mineralization, No effect on ALP activity, Upregulation of several osteogenic marker genes (BMP-2, OC, OPG, IBSP, MMP-1, MMP-3), Stimulation of osteogenic differentiation	Jansen et al. 2010 [41]
BM-MSCs	Time of bursts: 5 ms Time of pulses: 1 µs Magnetic flux density: 0.1 mT Frequency:15 Hz	Continuous exposure	Osteogenic	Increase of cell viability rate, No effect on osteo-induction	Kaivosoja et al. 2015 [47]
BM-MSCs	Time of bursts: 4.5 ms Number of pulses: 20 Magnetic flux density: 1.8 mT (increase from 0 to 1.8 mT in 200 µs steps and then decrease to 0 mT in 25 µs steps during each pulse) Frequency: 15 Hz	8 h/day during culture period	Osteogenic, adipogenic, neurogenic	Enhancement of cell proliferation rate, Increase of cell densities, Alterations of cell cycle progression, No effect on the surface phenotype or multilineage differentiation potential	Sun et al. 2009 [21]
BM-MSCs	Time of bursts: 4.5 ms Number of pulses: 20 Magnetic flux density: 1.8 mT (increase from 0 to 1.8 mT in 200 µs steps and then decrease to 0 mT in 25 µs steps during each pulse) Frequency: 15 Hz	8 h/day during the culture period	Osteogenic	Increase in cell proliferation, Increase in ALP expression and activity	Sun et al. 2010 [33]
BM-MSCs/osteoblast-like cells	Time of bursts: 4.5 ms Number of pulses: 20	8 h/day	Osteogenic	Time-dependent alterations of osteogenic marker expression (BMP-2, Cbfa1, COL I, OC), Enhancement of matrix mineralization	Schwartz et al. 2009 [37]

Table 1 Effects of EMFs with different parameters on stem cell biology (Continued)

Cell type / EMF	Parameters	Exposure	Differentiation	Effects	Reference
	Magnetic flux density: 1.6 mT (increase from 0 to 1.6 mT in 200 μs steps and then decrease to 0 mT in 25 μs steps during each pulse) Frequency: 15 Hz			Surface-dependent decrease in cell number Increase in OPG expression level	
BM-MSCs/ASCs	Number of pulses: 10 Time of pulses: 13 ms Magnetic flux density: 1.5 mT Frequency: 75 Hz	Whole differentiation time (28 days)	Osteogenic	Increase in ALP activity Increase in OC expression Induction of ASC osteogenic differentiation Enhancement of matrix mineralization	Ongaro et al. 2014 [49]
BM-MSCs	Time of bursts: 4.5 ms Number of pulses: 20 Magnetic flux density: 1.6 mT (increase from 0 to 1.6 mT in 200 μs steps and then decrease to 0 mT in 25 μs steps during each pulse) Frequency: 15 Hz	8 h/day for 24 days	Osteogenic	Synergistic increase in ALP activity over that caused by BMP-2 Enhancement of the stimulatory effect of BMP-2 on OC	Schwartz et al. 2008 [40]
WJ-MSCs	Magnetic flux density: 1.8 or 3 mT Frequency: 75 Hz	8 h/day for up to 21 days	Chondrogenic	Increase in cell division Increase in cell densities Increase in COL II expression level Induction of early chondrogenic differentiation	Esposito et al. 2013 [36]
Sinusoidal PEMF ESCs	Magnetic flux density: 5 mT Frequency: 50 Hz	30 min/day for 14 days	—	Increase in proliferation rate	Bai et al. 2012 [32]
Low-frequency pulsed EMF (BEMER type) BM-MSCs/chondrocytes	Time of pulses: 30 ms Magnetic flux density: 35 μT (increase from 0 to 35 μT in 30 ms steps) Frequency: 30 Hz	Five times at 12-h intervals for 8 min	—	Impact on cell metabolism and cell matrix structure No increased expression of cancer-related genes	Walther et al. 2007 [48]
Pulsed EMF and single-pulse EMF ASCs	PEMF Time of bursts: 67.1 ms Number of pulses: 21 Time of pulses: 5.46 ms Magnetic flux density: 2 mT Frequency: 15 Hz SPEMF Time of bursts: 5 s Number of pulses: 30 Time of pulses: 5 ms Magnetic flux density: 1 T	PEMF: 8 h/day SPEMF: 3 min/day	Osteogenic/chondrogenic	No effects on cell viability Increase of the cartilaginous matrix deposition with both PEMF and SPEMF Enhancement of chondrogenic gene expression (SOX-9, COL II, and aggrecan) with both PEMF and SPEMF Enhancement of bone matrix gene expression (OC, COL I) only with PEMF	Chen et al. 2013 [42]

ALP alkaline phosphatase, ASC adipose tissue-derived mesenchymal stem cell, BM-MSC bone marrow-mesenchymal stem cell, BMP bone morphogenetic protein, COL collagen type, CSC cardiac stem cell, ELF extremely low frequency, EMF electromagnetic field, ESC epidermal stem cell, IBSP bone sialoprotein, ICR ion cyclotron resonance, MAP2 mitogen activated protein 2, MF magnetic field, MHC myosin heavy chain, MMP matrix metalloproteinase, ms millisecond, MSC mesenchymal stem cell, NeuroD1 neurogenic differentiation 1, NF-L low-molecular weight neurofilament, Nkx2.5 NK2 transcription factor related, locus 5, OC osteocalcin, OPG osteoprotegerin, OPN osteopontin, OSX osterix, PEMF pulsed electromagnetic field, Runx runt-related transcription factor, SMA smooth muscle actin, SOX9 sex-determining region Y box 9, SPEMF single-pulse electromagnetic field, Tau microtubule associated protein tau, TnI troponin I, VEGF vascular endothelial growth factor, WJ-MSC Wharton's jelly-mesenchymal stem cell

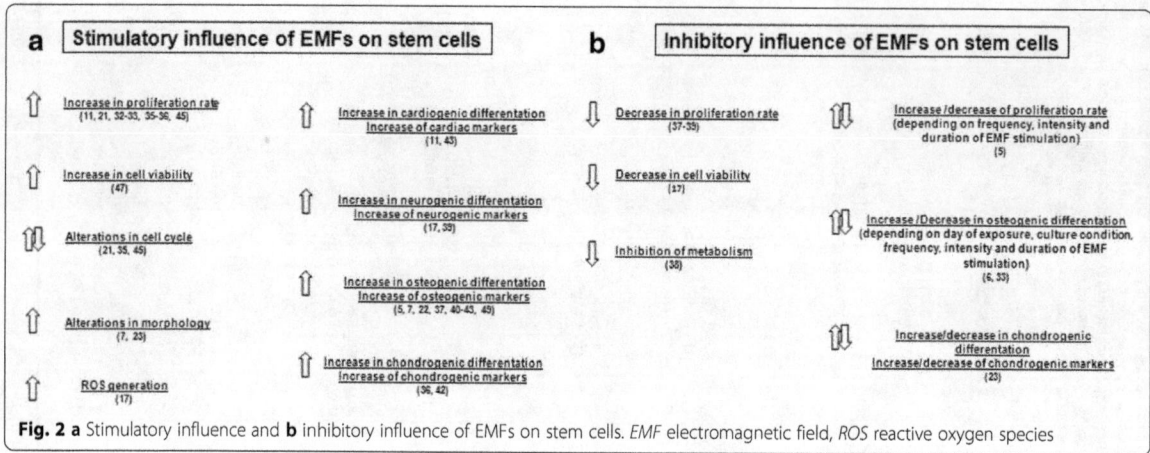

Fig. 2 a Stimulatory influence and **b** inhibitory influence of EMFs on stem cells. *EMF* electromagnetic field, *ROS* reactive oxygen species

within the developing organism. Importantly, exogenous EMFs applied in vitro have been shown to influence cell behavior. The success rate of assisted reproductive technologies has been observed to be rather low in comparison with natural methods. In addition, the incidence of congenital malformations (Wiedemann syndrome, Angelman syndrome) is also higher in newborns conceived using assisted reproductive technologies compared with those conceived naturally [25, 26]. One of the reasons for the success rate decrease and malformation increase may be the exposure of stem cells in early embryonic development to the EMF during incubation before embryo implantation. Exposure to the EMF may disturb the normal imprinting process. The fact that the vast majority of cloned embryos die during embryonic development, despite their normal chromosomal complementation, suggests that epigenetic reprogramming in reconstructed oocytes is incomplete [27].

A body of evidence indicates that EMF affects the gene expression and differentiation process through epigenetic mechanisms [28, 29]. Chromatin modifications are involved in mediating the effects of EMF stimulation [30].

Effects of the EMF on adult stem cells

Effects of the EMF on stem cell proliferation and the cell cycle

Scientific reports referring to the effects of the EMF on stem cell proliferation and the cell cycle have been inconsistent (Fig. 2a, b). Most research concerns human mesenchymal stem cells (MSCs). There have been numerous efforts to evaluate the effects of EMFs on different parameters; all of these are included and described precisely in Table 1. Consequently, we attempted to determine whether there is any general trend for selection of EMF characteristics and parameters in studies on human stem cell responses to EMF exposure (Fig. 3a, b). We gathered parameters of the EMF used in different

studies for a sinusoidal EMF (Fig. 4a) and for a pulsed electromagnetic field (PEMF) (Fig. 4b).

For instance, several studies have demonstrated that the EMF (sinusoidal as well as pulsed) increases the stem cell proliferation rate [11, 31–33] (Fig. 2a). Interestingly, when murine stromal stem cells were exposed to an EMF, different cellular responses were noticed depending on the gender [31]. Further studies concerning the significance of donor gender in human adult stem cell behavior after EMF stimulation would therefore be interesting.

An increase in cell proliferation was observed when the cell culture was exposed to an EMF during the active proliferation stage [34]. Zhang et al. [35] showed that a sinusoidal EMF at 50 Hz caused the largest increase of human epidermal stem cell proliferation after 7 days of exposure ($p < 0.05$) compared with other experimental groups and an untreated group. Sun et al. [21] revealed that proliferation of bone marrow mesenchymal stem cells (BM-MSCs) treated with a PEMF began earlier compared with untreated cells. The enhancement of cell proliferation resulted in 20–60 % higher cell densities during the exponential growth phase. What is more, PEMF treatment of Wharton's jelly mesenchymal stem cells triggered an increase in both cell division and cell density [36] (Table 1).

In contrast, Schwartz et al. [37] noted that PEMF treatment reduced the number of osteoblast-like cells cultured on a calcium phosphate surface by 40 %. It has also been reported that the EMF decreases the stem cell proliferation rate [38, 39] (Fig. 2b). However, we may suppose that the inhibition of MSC growth and metabolism is due to the higher EMF intensity value used by Yan et al. [38] in comparison with previous studies.

Tsai et al. [5] showed that PEMF stimulation did not alter proliferation of stem cells cultured in basal medium, while in osteogenic medium some differences occurred.

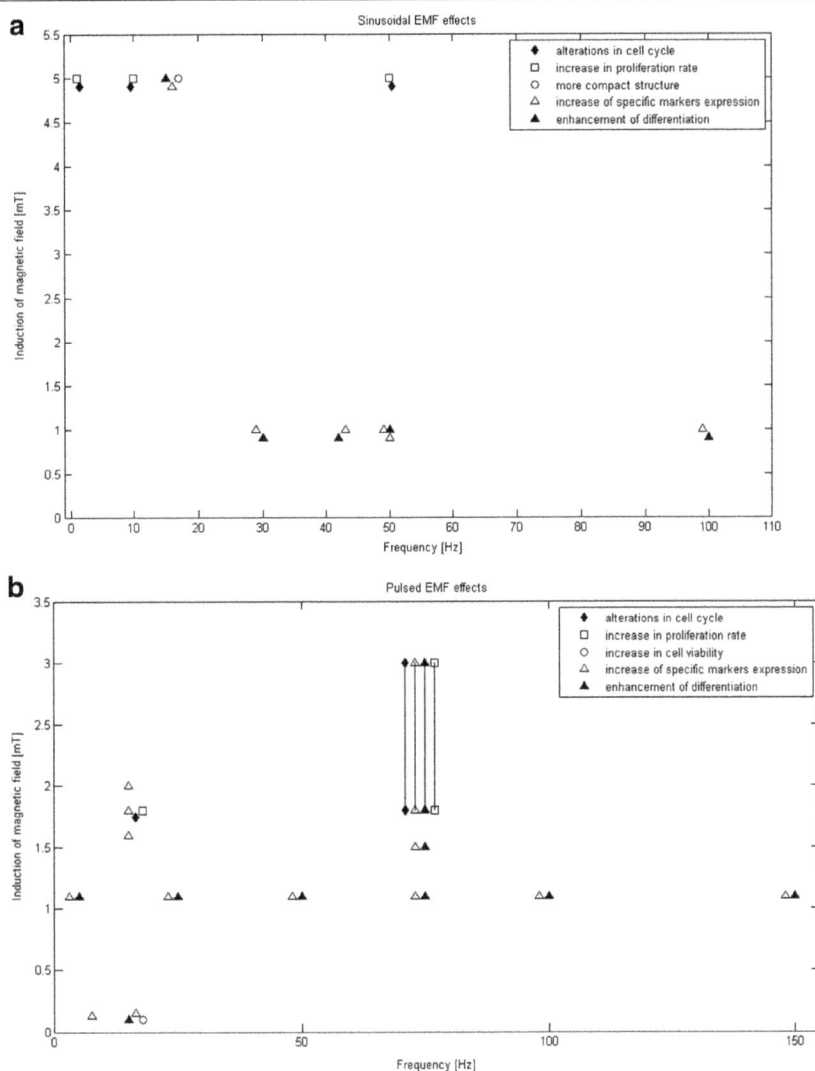

Fig. 3 a Selected sinusoidal EMF effects on stem cell biology that occur with established parameters of both frequency and induction of magnetic field. Effects include: alterations in cell cycle [35]; increase in cell proliferation rate [35]; more compact structure [23]; increase in specific markers' (neurogenic, osteogenic, chondrogenic) expression levels [6, 17, 23, 39]; and enhancement of differentiation (neurogenic, osteogenic, chondrogenic) [6, 17, 23]. **b** Selected pulsed EMF effects on stem cell biology that occur with established parameters of both frequency and induction of magnetic field. Effects include: alterations in cell cycle [21, 45]; increase in cell proliferation rate [21, 33, 36, 45]; increase in cell viability [47]; increase in specific markers' (osteogenic, chondrogenic) expression levels [5, 7, 33, 36, 37, 40–42, 45, 49]; and enhancement of differentiation (osteogenic, chondrogenic) [7, 36, 41, 45, 49]. *EMF* electromagnetic field

There was a significant increase in cell density in the untreated group compared with the PEMF-treated groups at day 7 (75 %; $p < 0.05$), whereas at day 10 the PEMF-treated groups showed an increase in proliferation (62 %; $p < 0.05$), in contrast to the control group (Table 1).

Because of its influence on proliferation, EMF stimulation also affects the cell cycle. Zhang et al. [35] showed an increase in the percentage of cells in the S phase, representing the DNA synthesis stage, and a decrease in the percentage of cells in the G1 phase ($p < 0.05$).

Moreover, these results were independent of the applied sinusoidal EMF frequency. Sun et al. [21] observed a 3–4 % ($p < 0.05$) increase in the proportion of cells in the G2/M phase during the first PEMF exposure and 4 h after the first PEMF stimulation. Then, 10 and 16 h after the first PEMF treatment, the percentage of cells in the G2/M phase and the S phase decreased by 8–12 % and 3–4 % ($p < 0.05$), respectively, whereas the proportion of cells in the G0/G1 phase, representing the newly divided cells, increased by 13–16 % ($p < 0.05$).

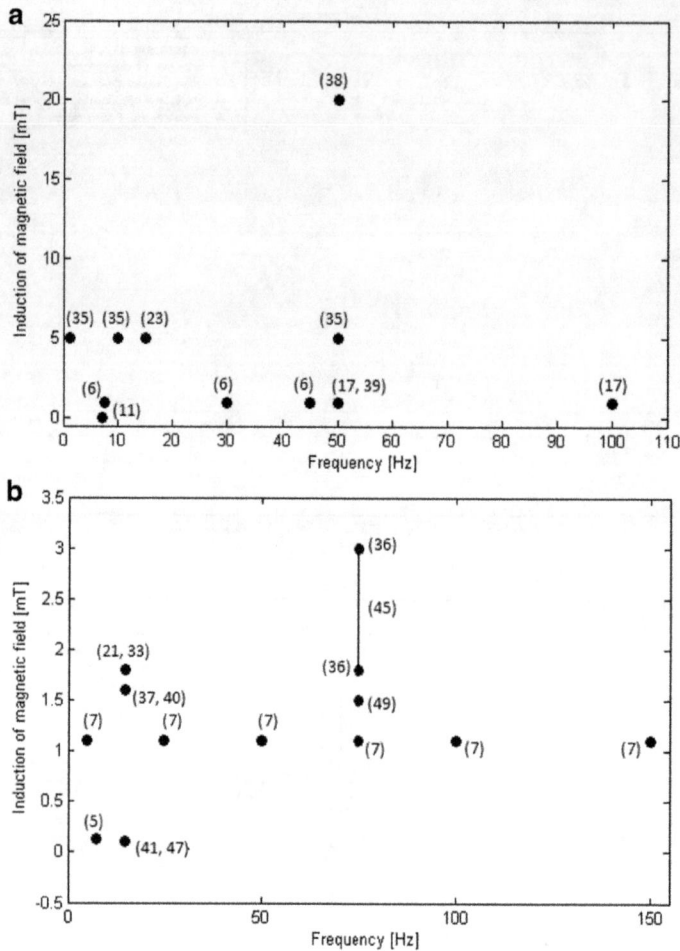

Fig. 4 Parameters of **a** sinusoidal EMF and **b** pulsed EMF mostly used in current studies together with references

Effects of the EMF on cell differentiation and marker expression

Numerous studies have been carried out on MSCs from different sources (Table 1). In most cases the differentiation was performed towards osteogenesis and chondrogenesis and the culture was grown in a medium containing differentiation factors. It has been reported that EMF stimulation affects the differentiation and the expression of specific markers (Table 1).

Many studies have shown the increase in osteogenic differentiation triggered by the EMF. Several studies have demonstrated an increase in alkaline phosphatase activity, an early marker of osteogenesis [5, 7, 33, 40]. Jansen et al. [41] observed higher expression levels of some osteogenic markers, such as bone morphogenetic protein BMP-2 (3.5-fold), transforming growth factor beta-1 (2.5-fold), matrix metalloproteinases MMP-1 (2.8-fold) and MMP-3 (2.1-fold), osteoprotegerin (1.7-fold), bone sialoprotein IBSP (twofold), and osteocalcin (OC; twofold). Interestingly, none of these markers was affected by a PEMF at the later stages of mineralization. Moreover, collagen type I (COL I) expression was steadily induced in the early stages of differentiation. In contrast, expression of receptor activator of NF-κB ligand (RANKL), which was insensitive to PEMF treatment in the early stages, was stimulated on day 14 ($p < 0.05$). Some investigations also showed higher expression of COL I and COL II, OC, runt-related transcription factor Runx2, and osterix in EMF-treated groups compared with control groups [5–7, 23, 33, 42, 43]. Moreover, studies performed by Creecy et al. [44] revealed that MSCs expressed both early (such as Runx2 and osterix) and late (osteopontin and OC) osteogenic genes as a function of level and duration of exposure to alternating electric current. The EMF stimulated matrix mineralization in comparison with untreated groups [6, 7, 33, 41].

The effect of the EMF depends on the external conditions of the cell culture. The EMF stimulated chondrogenic but not osteogenic differentiation when stem cells were cultured in a chondrogenic microenvironment. Some results suggest that the EMF affects the early stages of differentiation and reduces the time of differentiation [33, 36, 45].

Some studies have demonstrated alterations in neurogenic differentiation triggered by extremely low frequency (ELF)-EMF treatment. The expression of neural stem cell markers like nestin was thus decreased whereas neural cell markers such as mitogen-activated protein MAP2, neurogenic differentiation NeuroD1, low-molecular weight neurofilament NF-L, and microtubule-associated protein Tau were induced. Moreover, it was observed that the ELF-EMF accelerated the neural differentiation via reactive oxygen species (ROS)-induced epidermal growth factor receptor activation and, subsequently, the phosphorylation of Akt (known as protein kinase B) and cAMP response element-binding protein CREB. Based on these results, it has been suggested that EMF stimulation may induce neuronal differentiation without any chemicals or differentiation factors [17, 39]. Interestingly, Lee et al. [46] implied that ELF-EMF induces neural differentiation of BM-MSCs through activation of a ferritin-regulated mechanism.

The EMF has been reported to alter cardiac marker expression. Namely, troponin I, myosin heavy chain, connexin [43], and homeobox protein Nkx2.5 were upregulated ($p < 0.05$) by ELF-EMF treatment, tuned at the Ca^{2+} ion cyclotron energy resonance, compared with the untreated control. In contrast, vascular markers such as vascular endothelial growth factor and kinase domain receptor were downregulated or did not show any significant changes [11, 43].

However, we cannot clearly conclude how the EMF affects stem cell differentiation because the data concerning EMF stimulation of various markers' expression are inconsistent. Some studies have revealed that the EMF may cause both an increase and decrease in proliferation and differentiation, depending on the day of exposure, cell culture conditions, or characteristics of the EMF, such as frequency, intensity, and time of stimulation [5, 6, 39] (Fig. 2b, Table 1).

Other effects of the EMF on stem cells

EMF stimulation affects not only proliferation, the cell cycle, or differentiation of stem cells, but also other correlated processes. For instance, cells treated with ELF-EMF showed a tendency toward a more compact structure [23]. On the other hand, a PEMF changed the morphology of treated cells; stimulated cells were larger than control cells and became triangular and polygonal in shape, scales formed, and the cytoplasm contained abundant matrix and granular material compared with more immature untreated stem cells [7].

On the other hand, Hronik-Tupaj et al. [22] used alternating current electric fields for stimulation of BM-MSCs towards osteogenic differentiation. They observed upregulation of the stress markers heat shock proteins hsp27 and hsp70. Moreover, the increase in the hsp27 level was correlated with increased expression of lipofuscin, which is one of the aging or "wear-and-tear" pigments. These changes suggest a correlation between the expression of these markers and oxidative stress. They also observed higher levels of nicotinamide-adenine dinucleotide (NADH) and flavin-adenine dinucleotide and an increased redox ratio. Yan et al. [38] reported that ELF-EMF inhibits metabolism of treated MSCs.

Mechanism of the EMF influence on stem cells

The mechanism of the EMF (sinusoidal as well as pulsed) influence remains unclear. The EMF affects a number of biological processes whose functions are closely related to the properties of the cell membrane. The EMF may act on membrane potential through hyperpolarization or depolarization. An ELF-EMF [11, 23] and a PEMF [21, 33] may also modify the transmembrane ion channels. Reorientation of some molecules causes deformation of ion channels and alters the ion flow, especially of Ca^{2+}. Changes in intracellular Ca^{2+} levels affect the proliferation and differentiation of stem cells [6, 11]. The EMF may also influence signal transduction and intercellular communication [23].

Stem cells respond to the EMF differently depending on their state of differentiation. It is possible that the EMF (particularly PEMFs) modulates the activity of transcription factors and the level of cell cycle regulatory genes [33, 37, 40].

It is believed that one of the possible mechanisms involves the generation of ROS within the cell. Excessive concentration of ROS, such as superoxide anions (O_2^-) and hydrogen peroxide (H_2O_2), is considered to be cell destructive and results in inhibition of gene expression. In contrast, small amounts of ROS function as intracellular second messengers and activate signaling cascades involved in growth and differentiation of many cell types.

Some investigators imply that the ELF-EMF [17] and PEMF [37] act via a modification of signaling pathways, such as the extracellular signal regulated kinase pathway or phosphatidylinositol-4,5-bisphosphate 3-kinase/Akt signaling pathway. Park et al. [17] assumed that the ELF-EMF induced activation of NADH oxidase, which is involved in ROS production. The high level of ROS modifies signaling pathways by phosphorylation mechanisms.

Additionally, a weak EMF may accelerate electron transfer and thereby destabilize the hydrogen bonds of cellular macromolecules. This could explain the stimulation of transcription and protein expression, which has been observed after EMF exposure. However, the energy of a weak EMF is not sufficient to directly break a chemical bond in DNA. Therefore, it can be concluded that genotoxic effects are mediated by indirect mechanisms as microthermal processes, generation of ROS, or disturbance of DNA repair processes.

Conclusions

Adult stem cells are very important within our body because they are responsible for homeostasis, regeneration, aging, and so forth. Stem cells may respond differently to external stimulation such as the EMF/PEMF depending on cell type, cell density, differentiation stage, and type of medium, as well as the characteristics of the EMF. So far we have few data on the influence of the EMF on stem cell biology. More studies are therefore required because stem cells are responsible for multiple processes within the human body, both desired (e.g., wound healing, regeneration) and undesired (e.g., pathological growth, carcinogenesis).

The parameters of EMFs (frequency, magnetic flux density) and times of exposure used by different research groups are quite diverse with no clear rationale for why particular parameters are chosen. We demonstrated the parameters and the ranges of parameters used in different studies for a sinusoidal EMF (Fig. 4a) and a PEMF (Fig. 4b). The successful use of sinusoidal EMFs in differentiation studies has mainly involved an EMF with parameters of 1–5 mT, 10–50 Hz. The only study using a sinusoidal EMF [38] in which a higher intensity of EMF was used (20 mT) did not show any significant effect on osteogenic differentiation. Additionally, the authors observed a decrease in MSC growth and metabolism. Importantly, we have to remember that higher intensities of the EMF may result in microthermal processes as well as the generation of eddy currents; therefore, besides the EMF, we have to take into account additional stimulatory factors. Additionally, we suppose that stress/oxidative stress may be a very important factor.

On the other hand, the most commonly used range of PEMF was 0.1–3 mT, 15–75 Hz. For example, there were two studies on osteogenic differentiation using very similar parameters (0.1 mT, 15 Hz) but with different pulse times: 5 µs [41] and 1 µs [47]. This difference in pulse times resulted in different osteogenic induction outcomes: an increase in differentiation [41] or no effect [47]. Thus, we may conclude that many factors may influence intracellular processes, such as the time of pulses, time of exposure, type of stem cells, or experimental methodology. It is worth noting that a wide range of EMF parameters have been used, depending on

the desired effect. For instance, increases in cell proliferation were most evident at 5 mT, 50 Hz (for sinusoidal EMF), at 1.8 mT, 15 Hz (for PEMF), or at 1.8–3 mT, 75 Hz (for PEMF). In turn, the magnetic flux density used in most previous studies to enhance differentiation varied from 1 to 5 mT for sinusoidal EMF and from 0.1 to 3 mT for PEMF; the frequencies varied from 15 to 100 Hz for sinusoidal EMF and from 15 to 150 Hz for PEMF. This means that the aforementioned ranges of EMF parameters may be successfully used for stem cell-based therapies in which processes such as proliferation and differentiation are crucial. For example, the EMF has been shown to promote bone formation and therefore can be used in regenerative applications aimed at bone fracture healing [7]. Additionally, EMF stimulation of MSC chondrogenic potential during cartilage regeneration may result in beneficial effects [23]. What is more, EMF treatment can be used as an alternative tool for skin tissue engineering due to its positive impact on epidermal stem cell proliferation [32]. EMF modulation of stem cell differentiation into specific cell types promotes its application in cardiovascular disease [11] or neurodegenerative disorder [17] treatment.

Literature data concerning the influence of EMFs on stem cells with respect to carcinogenesis remain elusive. Defining the specific EMF range/characteristics inducing carcinogenesis would be very important. Walther et al. [48] did not observe any increase in cancer-related gene expression after low-frequency PEMF exposure. Radiofrequency EMFs have been suggested to trigger tumor promotion. However, the EMF mechanisms involved in induction of processes such as carcinogenesis and tumor formation are still under investigation and a lot of research needs to be done to explore this issue.

We hypothesize that some ranges of EMF parameters may promote regeneration but others may result in cancer formation, degeneration, and pathological alterations, depending on the stem cell type. These processes may be detected firstly at the epigenetic level, secondly at the genetic level, and finally at the proteomic and functional levels, leading towards either a positive or negative impact with respect to health and disease. To date, there are no data concerning this issue.

As a side comment, the number of cancer patients in our society is growing alarmingly. According to environmental health specialists, besides chemical pollution, this condition may be triggered by EMF exposure. Further studies are therefore required to explore this phenomenon at both in vitro and in vivo levels. We believe that EMF-based therapeutic applications may be used in the future for regenerative medicine approaches as well as in the "fight against cancer" or homeostasis restoration. More researchers, engineers, and medical doctors are required to improve the state of knowledge, working on stem cell biology, stem cell transplantation, biophysics, biochemistry, tissue engineering, engineering, regenerative medicine, oncology, and other areas to explore this phenomenon.

In conclusion, properly adjusted values of EMF frequencies, times of stimulation, as well as the microenvironmental niche may affect EMFs' impact on stem cell proliferation, differentiation, and migration to result in the desired therapeutic outcome. Additionally, this knowledge may help us to determine the best approach for using properly adjusted EMFs in future autologous stem cell-based therapy. Importantly, it is reasonable to check the impact of the EMF with respect to carcinogenesis.

Abbreviations
ASC: adipose-derived mesenchymal stem cell; BM-MSC: bone marrow-mesenchymal stem cell; COL: collagen; ELF: extremely low frequency; EMF: electromagnetic field; MSC: mesenchymal stem cell; OC: osteocalcin; PEMF: pulsed electromagnetic field; ROS: reactive oxygen species.

Competing interests
The authors declare that they have no competing interests.

Authors' contributions
AM drafted the manuscript and participated in the sequence alignment. BK was involved in drafting the manuscript and prepared Table 1 and Fig. 3a, b. MB prepared Fig. 4a, b and was involved in preparation of Table 1. SB prepared Fig. 4a, b. MC was involved in drafting the manuscript. TO participated in the manuscript design and draft, coordinated and revised it critically for important intellectual content, and gave final approval of the version to be published. AB provided the main idea, participated in its design and draft, prepared Figs. 1 and 2a, b, and coordinated and revised the manuscript critically for important intellectual content. All authors read and approved the final manuscript.

Acknowledgments
The study was performed within the project "Centre for Innovative Research in Medical and Natural Sciences" realized by University of Rzeszow, co-financed within the Regional Operational Program for the Podkarpackie Province for the years 2007–2013 (contract number UDA-RPPK.01.03.00-18-004/12-00).

Author details
[1]Laboratory of Stem Cells' Biology, Department of Immunology, Chair of Molecular Medicine, Faculty of Medicine, University of Rzeszow, ul. Kopisto 2a, 35-310 Rzeszow, Poland. [2]Centre for Innovative Research in Medical and Natural Sciences, Faculty of Medicine, University of Rzeszow, ul. Warzywna 1a, 35-310 Rzeszow, Poland. [3]Department of Biophysics, Faculty of Mathematics and Natural Sciences, University of Rzeszow, ul. Pigonia 1, 35-310 Rzeszow, Poland. [4]Division of Molecular and Cellular Medicine, National Cancer Center Research Institute, 5-1-1 Tsukiji, Chuo-ku 104-0045Tokyo, Japan.

References
1. Li L, Jiang J. Stem cell niches and endogenous electric fields in tissue repair. Front Med. 2011;5:40–4.
2. Pittenger MF, Mackay AM, Beck SC, Jaiswal RK, Douglas R, Mosca JD, et al. Multilineage potential of adult human mesenchymal stem cells. Science. 1999;284:143–7.
3. Govey PM, Loiselle AE, Donahue HJ. Biophysical regulation of stem cell differentiation. Curr Osteoporos Rep. 2013;11:83–91.
4. Hammerick KE, Longaker MT, Prinz FB. In vitro effects of direct current electric fields on adipose-derived stromal cells. Biochem Biophys Res Commun. 2010;397:12–7.
5. Tsai MT, Li WJ, Tuan RS, Chang WH. Modulation of osteogenesis in human mesenchymal stem cells by specific pulsed electromagnetic field stimulation. J Orthopaed Res. 2009;27:1169–74.
6. Kang KS, Hong JM, Kang JA, Rhie JW, Jeong YH, Cho DW. Regulation of osteogenic differentiation of human adipose-derived stem cells by controlling electromagnetic field conditions. Exp Mol Med. 2013;45:e6.
7. Luo F, Hou T, Zhang Z, Xie Z, Wu X, Xu J. Effects of pulsed electromagnetic field frequencies on the osteogenic differentiation of human mesenchymal stem cells. Orthopedics. 2012;35:e526–31.
8. Wagers AJ, Weissman IL. Plasticity of adult stem cells. Cell. 2004;116:639–48.
9. Doetsch F. A niche for adult neural stem cells. Curr Opin Genet Dev. 2003; 13:543–50.
10. Weiss ML, Troyer DL. Stem cells in the umbilical cord. Stem Cell Rev. 2006;2: 155–62.
11. Gaetani R, Ledda M, Barile L, Chimenti I, De Carlo F, Forte E, et al. Differentiation of human adult cardiac stem cells exposed to extremely low-frequency electromagnetic field. Cardiovasc Res. 2009;82:411–20.
12. Banas A, Teratani T, Yamamoto Y, Tokuhara M, Takeshita F, Quinn G, et al. Adipose tissue-derived mesenchymal stem cells as a source of humanhepatocytes. Hepatology. 2007;46:219–28.
13. Banas A, Teratani T, Yamamoto Y, Tokuhara M, Takeshita F, Osaki M, et al. IFATS collection: in vivo therapeutic potential of human adipose tissue mesenchymal stem cells after transplantation into mice with liver injury. Stem Cells. 2008;26(10):2705–12.
14. Banas A, Teratani T, Yamamoto Y, Tokuhara M, Takeshita F, Osaki M, et al. Rapid hepatic fate specification of adipose-derived stem cells and their therapeutic potential for liver failure. J Gastroenterol Hepatol. 2009;24(1):70–7.
15. Yamamoto Y, Banas A, Murata S, Ishikawa M, Lim CR, Teratani T, et al. A comparative analysis of the transcriptome and signal pathways in hepatic differentiation of human adipose mesenchymal stem cells. FEBS J. 2008; 275(6):1260–73.
16. Ochiya T, Yamamoto Y, Banas A. Commitment of stem cells into functional hepatocytes. Differentiation. 2010;79(2):65–73.
17. Park JE, Seo YK, Yoon HH, Kim CW, Park JK, Jeon S. Electromagnetic fields induce neural differentiation of human bone marrow derived mesenchymal stem cells via ROS mediated EGFR activation. Neurochem Int. 2013;62:418–24.
18. Katsuda T, Kosaka N, Takeshita F, Ochiya T. The therapeutic potential of mesenchymal stem cell-derived extracellular vesicles. Proteomics. 2013;13: 1637–53.
19. Katsuda T, Tsuchiya R, Kosaka N, Yoshioka Y, Takagaki K, Oki K, et al. Human adipose tissue-derived mesenchymal stem cells secrete functional neprilysin-bound exosomes. Sci Rep. 2013;3:1197.
20. Kurata H, Tamai R, Katsuda T, Ishikawa S, Ishii T, Ochiya T. Adipose tissue-derived mesenchymal stem cells in regenerative medicine treatment for liver cirrhosis—focused on efficacy and safety in preclinical and clinical studies. JSM Regen Med. 2014;2(1):1012.
21. Sun LY, Hsieh DK, Yu TC, Chiu HT, Lu SF, Luo GH, et al. Effect of pulsed electromagnetic field on the proliferation and differentiation potential of human bone marrow mesenchymal stem cells. Bioelectromagnetics. 2009; 30:251–60.
22. Hronik-Tupaj M, Rice WL, Cronin-Golomb M, Kaplan DL, Georgakoudi I. Osteoblastic differentiation and stress response of human mesenchymal stem cells exposed to alternating current electric fields. BioMed Eng. 2011;10:9.
23. Mayer-Wagner S, Passberger A, Sievers B, Aigner J, Summer B, Schiergens TS, et al. Effects of low frequency electromagnetic fields on the chondrogenic differentiation of human mesenchymal stem cells. Bioelectromagnetics. 2011;32:283–90.
24. Jaffe LF, Nuccitelli R. Electrical controls of development. Annu Rev Biophys Bioeng. 1977;6:445–76.
25. Maher ER, Afnan M, Barratt CL. Epigenetic risks related to assisted reproductive technologies. Hum Reprod. 2003;18:2508–11.
26. Jacobs S, Moley KH. Gametes and embryo epigenetic reprogramming after developmental outcome: implications for assisted reproductive technologies. Pediatr Res. 2005;58:437–46.
27. Rideout WM, Eggan K, Jaenisch R. Nuclear cloning and epigenetic reprogramming of the genome. Science. 2001;203:1003–8.
28. Ahuja YR, Vijayalakshmi V, Polasa K. Stem cell test: a practical tool in toxicogenomics. Toxicology. 2007;231:1–10.
29. Levin M. Bioelectromagnetics in morphogenesis. Bioelectromagnetics. 2003; 24:295–315.
30. Leone L, Podda MV, Grassi C. Impact of electromagnetic fields on stem cells: common mechanisms at the crossroad between adult neurogenesis and osteogenesis. Front Cell Neurosci. 2015;9:228.

31. Van Den Heuvel R, Leppens H, Nemethova G, Verschaeve L. Haemopietic cell proliferation in murine bone marrow cells exposed to extremely low frequency (ELF) electromagnetic fields. Toxicol In Vitro. 2001;15:351–5.

32. Bai WF, Zhang MS, Huang H, Zhu HX, Xu WC. Effects of 50 Hz electromagnetic fields on human epidermal stem cells cultured on collagen sponge scaffolds. Int J Radiat Biol. 2012;88:523–30.

33. Sun LY, Hsieh DK, Lin PC, Chiu HT, Chiou TW. Pulsed electromagnetic fields accelerate proliferation and osteogenic gene expression in human bone marrow mesenchymal stem cells during osteogenic differentiation. Bioelectromagnetics. 2010;31:209–19.

34. Diniz P, Shomura K, Soejima K, Ito G. Effects of pulsed electromagnetic field (PEMF) stimulation on bone tissue like formation are dependent on the maturation stages of the osteoblasts. Bioelectromagnetics. 2002;23:398–405.

35. Zhang M, Li X, Bai L, Uchida K, Bai W, Wu B, et al. Effects of low frequency electromagnetic field on proliferation of human epidermal stem cells: an in vitro study. Bioelectromagnetics. 2013;34:74–80.

36. Esposito M, Lucariello A, Costanzo C, Fiumarella A, Giannini A, Riccardi G, et al. Differentiation of human umbilical cord-derived mesenchymal stem cells, WJ-MSCs, into chondrogenic cells in the presence of pulsed electromagnetic fields. In Vivo. 2013;27:495–500.

37. Schwartz Z, Fisher M, Lohmann CH, Simon BJ, Boyan BD. Osteoprotegerin (OPG) production by cells in the osteoblast lineage is regulated by pulsed electromagnetic fields in cultures grown on calcium phosphate substrates. Ann Biomed Eng. 2009;37:437–44.

38. Yan J, Dong L, Zhang B, Qi N. Effects of extremely low-frequency magnetic field on growth and differentiation of human mesenchymal stem cells. Electromagn Biol Med. 2010;29:165–76.

39. Cho H, Seo YK, Yoon HH, Kim SC, Kim SM, Song KY, et al. Neural stimulation on human bone marrow-derived mesenchymal stem cells by extremely low frequency electromagnetic fields. Biotechnol Prog. 2012;28:1329–35.

40. Schwartz Z, Simon BJ, Duran MA, Barabino G, Chaudhri R, Boyan BD. Pulsed electromagnetic fields enhance BMP-2 dependent osteoblastic differentiation of human mesenchymal stem cells. J Orthop Res. 2008;26:1250–5.

41. Jansen JH, van der Jagt OP, Punt BJ, Verhaar JA, van Leeuwen JP, Weinans H, et al. Stimulation of osteogenic differentiation in human osteoprogenitor cells by pulsed electromagnetic fields: an in vitro study. BMC Musculoskelet Disord. 2010;11:188.

42. Chen CH, Lin YS, Fu YC, Wang CK, Wu SC, Wang GJ, et al. Electromagnetic fields enhance chondrogenesis of human adipose-derived stem cells in a chondrogenic microenvironment in vitro. J Appl Physiol. 2013;114:647–55.

43. Lisi A, Ledda M, de Carlo F, Pozzi D, Messina E, Gaetani R, et al. Ion cyclotron resonance as a tool in regenerative medicine. Electromagn Biol Med. 2008;27:127–33.

44. Creecy CM, O'Neill CF, Arulanandam BP, Sylvia VL, Navara CS, Bizios R. Mesenchymal stem cell osteodifferentiation in response to alternating electric current. Tissue Eng A. 2013;19:467–74.

45. Esposito M, Lucariello A, Riccio I, Riccio V, Esposito V, Riccardi G. Differentiation of human osteoprogenitor cells increases after treatment with pulsed electromagnetic fields. In Vivo. 2012;26:299–304.

46. Lee HN, Ko KN, Kim HJ, Rosebud Aikins A, Kim CW. Ferritin is associated with neural differentiation of bone marrow-derived mesenchymal stem cells under extremely low-frequency electromagnetic field. Cell Mol Biol (Noisy-le-Grand). 2015;61:55–9.

47. Kaivosoja E, Sariola V, Chen Y, Konttinen YT. The effect of pulsed electromagnetic fields and dehydroepiandrosterone on viability and osteo-induction of human mesenchymal stem cells. J Tissue Eng Regen Med. 2015;9:31–40.

48. Walther M, Mayer F, Kafka W, Schütze N. Effects of weak, low-frequency pulsed electromagnetic fields (BEMER type) on gene expression of human mesenchymal stem cells and chondrocytes: an in vitro study. Electromagn Biol Med. 2007;26:179–90.

49. Ongaro A, Pellati A, Bagheri L, Fortini C, Setti S, De Mattei M. Pulsed electromagnetic fields stimulate osteogenic differentiation in human bone marrow and adipose tissue derived mesenchymal stem cells. Bioelectromagnetics. 2014;32:426–36.

11

Hypoxic conditioned medium from mesenchymal stem cells promotes lymphangiogenesis by regulation of mitochondrial-related proteins

Chang Youn Lee[1†], Jin Young Kang[2†], Soyeon Lim[3], Onju Ham[4], Woochul Chang[5*] and Dae-Hyun Jang[6*]

Abstract

Background: Recently, cell-based therapeutic lymphangiogenesis has emerged and provided hope for lymphatic regeneration. Previous studies have demonstrated that secretomes of mesenchymal stem cells (MSCs) facilitate the regeneration of various damaged tissues. This study was conducted to evaluate the lymphangiogenic potential of hypoxic conditioned media (HCM) from MSCs.

Methods: To investigate the effects of MSC-secreted factors in starved human lymphatic endothelial cells (hLEC), hLECs were treated with endothelial basal medium (EBM)-2 (control), normoxic conditioned media (NCM), or HCM *in vitro* and *in vivo*.

Results: MSCs expressed lymphangiogenic factors including EGF, FGF2, HGF, IGF-1, and VEGF-A and -C. hLECs were treated with each medium. hLEC proliferation, migration, and tube formation were improved under HCM compared with NCM. Moreover, expression of mitochondrial-related factors, MFN1and 2, were improved in HCM-treated hLECs. Lymphedema mice injected with HCM showed markedly decreased lymphedema via increased lymphatic vessel formation when compared with EBM-2- or NCM-treated mice.

Conclusions: This study suggested that HCM from MSCs contain high levels of secreted lymphangiogenic factors and promote lymphangiogenesis by regulating mitochondrial-related factors. Thus, treatment with HCM may be a therapeutic strategy for lymphedema.

Keywords: Hypoxic conditioned media, Mesenchymal stem cells, Mitochondrial-related protein, Lymphangiogenesis, Lymphatic endothelial cells

Background

Lymphedema is a pathologic swelling (edema) that results from the accumulation of protein-rich fluid in the interstitial space because of congenital or acquired lymphatic system damage [1]. Secondary lymphedema is caused by disruption or obstruction of the normal lymphatic system in response to infection, trauma, or iatrogenic processes such as surgery (radical lymph node dissection) or radiation-related cancer therapy [2]. Secondary lymphedema is a lifelong condition that disturbs quality of life with signs and symptoms such as heaviness, discomfort, and impaired mobility of the limbs as well as a weakened immunity. To date, the only proposed treatment options have been pharmacotherapy, physiotherapy, and surgical treatments, such as lymph node transfer and lymphatic bypass. However, these methods require good compliance and lifelong care [3]. Recently, cell-based therapy for therapeutic lymphangiogenesis has emerged and made lymphatic regeneration possible [4]. Essential lymphatic growth factors, including epidermal growth factor (EGF), fibroblast growth

* Correspondence: wchang1975@pusan.ac.kr; dhjangmd@naver.com
†Equal contributors
[5]Department of Biology Education, College of Education, Pusan National University, Busan 46241, Republic of Korea
[6]Department of Rehabilitation Medicine, Incheon St. Mary's Hospital, College of Medicine, The Catholic University of Korea, Dongsu-ro 56, Bupyeong-gu, Incheon 21431, Republic of Korea
Full list of author information is available at the end of the article

factor (FGF), hepatocyte growth factor (HGF), insulin-like growth factor (IGF)-1, vascular endothelial growth factor (VEGF)-A, and VEGF-C, have been identified in several studies [5–10]. In addition, these growth factors have been shown to contribute to lymphangiogenesis in the damaged lymphatic areas [11].

Mesenchymal stem cells (MSCs) are multipotent cells that can be obtained from adult donors and are known to have a low risk of immune rejection [12]. Therefore, therapeutic applications using MSCs have been extensively studied and applied to various tissues in regenerative medicine owing to these beneficial effects. Moreover, MSC therapy has recently been introduced to the market. However, limitations such as poor survival, limited differentiation, and dedifferentiation of cells with passaging and donor site morbidity exist [13–16]. To avoid these limitations, researchers have turned their attention to new therapeutic approaches. MSCs are known to secrete various cytokines and growth factors, which show paracrine and autocrine activities for injured cells, particularly for those in hypoxic, apoptotic, or inflamed areas [17, 18]. The secreted factors have been demonstrated to have many beneficial therapeutic effects in various diseases, such as neurodegenerative diseases, cancers, and heart failure [19]. Several reports have shown that cytokines secreted from MSCs under normal growth conditions can promote lymphangiogenesis via VEGF-A and VEGF-C [20–22]. MSC-based therapy has been suggested to be the most promising stem cell therapy for lymphangiogenesis [4].

Some researchers showed the possibility of regulating stem cell paracrine actions via different culture methods [18]. As one of the modifications, MSCs exposed to hypoxia showed more protein secretion and greater paracrine effects. MSCs under hypoxia also showed increased proliferation and migration compared with those under normal growth conditions, and treatment with conditioned media from hypoxic MSCs exerted therapeutic effects on wound healing by enhancing the production of angiogenic paracrine factors including basic FGF, IGF, and VEGF [23]. Another study reported that conditioned media obtained from MSCs under hypoxia showed protective effects against cardiomyoblasts in hypoxia and angiogenic effects on endothelial cells [24].

Mitochondria are important organelles that maintain cellular homeostasis under stressful conditions, such as apoptotic stimuli, an increase in reactive oxygen species (ROS), and changes in intracellular calcium concentration [25–27]. To maintain their function, mitochondria continuously undergo fission and fusion. Mitofusins MFN1 and MFN2 in the mitochondrial outer membrane have been shown to cause mitochondrial membrane fusion by binding with OPA1 in the inner membrane, whereas dynamin-related protein 1 is mainly involved with mitochondrial fission based on the phosphorylation

status [28, 29]. Recent studies demonstrated that MFNs are necessary for angiogenic function in endothelial cells [30, 31]. In addition, VEGF-A is an important factor for angiogenesis, which can stimulate the activation of MFN-mediated signaling pathways [31]. We hypothesized that MFNs also play important roles in lymphangiogenesis.

We investigated the therapeutic ability of hypoxic conditioned media (HCM) from MSCs for lymphatic edema using in vitro and in vivo experimental systems. Higher expression of VEGF, a lymphangiogenic factor, was observed in hypoxic MSCs, and tube formation increased in human lymphatic endothelial cells (hLECs) treated with HCM. Overall, we confirmed that HCM have the ability to induce lymphangiogenesis in hLECs and electrocauterized mice.

Methods
Cell culture
Human bone marrow MSCs (hMSCs; catalogue number PT-2501) and adult human dermal lymphatic microvascular endothelial cells (hLECs) (HMVECs-dLyAd; catalogue number CC-2810) were purchased from Lonza (Basel, Switzerland). hMSCs were maintained at 37 °C in a humidified atmosphere containing 5 % CO_2. Culture media was composed of 10 % fetal bovine serum (Invitrogen, Waltham, MA, USA), Dulbecco's modified Eagle's medium—low glucose, 100 U/ml penicillin (Invitrogen), and 100 µg/ml streptomycin (Invitrogen). Media were replaced every 3 days. We used 7–10 passages of hMSCs for experiments. hLECs were cultured in Lonza EGM-2MV medium and replaced with fresh media every 2 days.

HCM from MSCs
hMSCs were incubated for 1 day after they were seeded in a 100 mm dish (1×10^6 cells/dish), washed twice with endothelial basal medium (EBM-2; Lonza), and then placed into a hypoxic chamber (Anaerobic Environment; ThermoForma, Waltham, MA, USA) containing 5 ml EBM-2 for 12 hours. The airtight humidified hypoxic chamber was maintained at 37 °C and continuously supplied with mixed gas (5 % CO_2, 10 % H_2, and 85 % N_2). The oxygen level in the chamber was ~0.5 %. Following incubation, the medium was collected and centrifuged at $1000 \times g$ for 10 minutes at 4 °C, after which the supernatant was transferred to a new tube. Similarly, normoxic conditioned medium (NCM) was derived from hMSC-cultured media under normoxic conditions for 12 hours with 5 ml EBM-2. Each medium was stored at −80 °C until used.

Lymphatic endothelial cells
Proliferation assay
Cell proliferation was measured using a Cell Counting Kit-8 (Dojindo, Kumamoto, Japan). hLECs were seeded

at 5×10^3 cells per well in 96-well culture plates and then cultured for 1 day. hLECs were then incubated with fresh EBM-2 for 12 hours. After the medium was removed, cells were washed twice with phosphate-buffered saline (PBS). Each group was subsequently treated with EBM-2, NCM, or HCM, respectively, after which the cells were incubated for 24 hours in a 37 °C humidified atmosphere incubator containing 5 % CO_2. Following incubation, Cell Counting Kit-8 was added to each well and samples were then incubated for 2 hours. Finally, the absorbance of water soluble formazan dye was measured at 450 nm using a microplate reader (Molecular Devices, Sunnyvale, CA, USA). All experiments were performed in triplicate.

Migration assay

hLECs (2×10^4 cells) were seeded into the upper chamber of a Transwell filter with 8 μm pores (Costar Corning, New York, NY, USA) coated with 10 μg/ml fibronectin. They were deprived of serum for 12 hours with EBM-2, after which EBM-2, NCM, or HCM were added to the lower chamber. Cells on the upper chambers were incubated at 37 °C for 9 hours under different stimuli. Following incubation, cells on the underside of the filter were stained with 0.25 % crystal violet. Nonmigrating cells on the upper side of the filter were removed with cotton swabs. The filter was then photographed using a digital microscope camera system (Olympus, Shinjuku, Japan), or stained cells were dissolved in 10 % acetic acid and transferred to a 96-well plate for colorimetric reading at 560 nm using a microplate reader (Molecular Devices).

Immunoblot analysis

hLECs were washed twice in ice-cold PBS, then lysed with lysis buffer (Cell Signaling, Danvers, MA, USA) containing the protease inhibitor cocktail PhosSTOP (Roche, Basel, Switzerland), and incubated at 4 °C for 30 minutes. Protein concentrations were determined using a BCA protein assay kit (Thermo Fisher Scientific, Inc., Waltham, MA, USA), after which they were separated in 10 % SDS–polyacrylamide gel and transferred to PVDF membrane (Millipore, Billerica, MA, USA). The membrane was then blocked using TBS-T (0.1 % Tween 20) containing 5 % (w/v) bovine serum albumin (BSA) for 1 hour at room temperature. The membranes were then washed with TBS-T and incubated with primary antibody overnight at 4 °C. The next morning, the membrane was washed three times with TBS-T for 5 minutes each and incubated with horseradish peroxidase-conjugated secondary antibody for 1 hour at room temperature. After extensive washing, bands were detected using enhanced chemiluminescence reagent (AbClon, Seoul, Republic of Korea) and band intensities were measured using a Photo-Image System (Molecular Dynamics, Sunnyvale, CA, USA). MFN1

(Novus, Littleton, CO, USA), MFN2 (Sigma, St. Louis, MO, USA), extracellular signal regulated kinase (ERK; Santa Cruz, CA, USA), p-ERK (Santa Cruz), lymphatic vessel endothelial hyaluronan receptor 1(LYVE-1; Novus), and beta-actin (Sigma) antibodies were used in the experiment.

Real-time PCR

Total RNA was isolated using TRIzol® Reagent (Life Technologies, Waltham, MA, USA). The total RNA quality and concentration were measured using Nano-Drop Lite (Thermo Fisher Scientific, Inc.). Single-stranded cDNA was synthesized from total RNA using a reverse transcription system (Promega, Fitchburg, WI, USA) according to the product guidelines. Amplification and detection of specific products were performed in a StepOnePlus Real-time PCR System (Life Technologies) using a FastStart Essential DNA Green Master (Roche). PCR conditions consisted of 95 °C for 10 minutes followed by 40 cycles of 95 °C for 10 seconds and 60 °C for 10 seconds. The threshold cycle (Ct) of each target gene was automatically defined and normalized to the control glyceraldehyde 3-phosphate dehydrogenase (GAPDH) (ΔCt value). The relative differences in the expression levels of each mRNA in VSMCs ($\Delta\Delta$Ct) were calculated and presented as fold induction ($2^{-\Delta\Delta Ct}$). Real-time PCR primers consisted of the following groups: EGF forward primer, 5′-GGT CTT GCT GTG GAC TGG AT-3′; EGF reverse primer, 5′-CTG CTA CAG CAA ATG GGT GA-3′; IGF-1 forward primer, 5′-TCA CCT TCA CCA GCT CTG C-3′; IGF-1 reverse primer, 5′-TGG TAG ATG GGG GCT GAT AC-3′; HGF forward primer, 5′-GCC TGA AAG ATA TCC CGA CA-3′; HGF reverse primer, 5′-GCC ATT CCC ACG ATA ACA AT-3′; FGF-2 forward primer, 5′-AGC GGC TGT ACT GCA AAA AC-3′; FGF reverse primer, 5′-CTT TCT GCC CAG GTC CTG TT-3′; VEGF-A forward primer, 5′-AGT CCA ACA TCA CCA TGC AG-3′; VEGF-A reverse primer, 5′-TTC CCT TTC CTC GAA CTG ATT T-3′; VEGF-C forward primer, 5′-CCT CAA CTC AAG GAC AGA AGA G-3′; VEGF-C reverse primer, 5′-CTG GCA GGG AAC GTC TAA TAA T-3′; GAPDH forward primer, 5′-ACA TCG CTC AGA CAC CAT G-3′; and GAPDH reverse primer, 5′-TGT AGT TGA GGT CAA TGA AGG G-3′.

Tube formation assay

A Matrigel-based tube formation assay was performed. Each well of a 96-well culture plate was coated with 50 μl of ECMatrix™ (Millipore) and then allowed to incubate for 1 hour at 37 °C. Next, hLECs were seeded onto the coated wells at a density of 1×10^4 cells/well, cultured with 500 μl of EBM-2, NCM, or HCM, and incubated at 37 °C under 5 % CO_2 for 12 hours. Following incubation, tube formation images were captured using a digital micro-scope camera system (Olympus).

Hind limb mouse model of lymphedema

The hind limb mouse model of lymphedema was obtained as described previously [11]. Eight-week-old BALB/c mice (Orient Bio Co., Seongnam, Korea) were assigned into groups of three. Normal mice were then anesthetized via subcutaneous injection of zoletil (4 mg/kg) with rompun (20 mg/kg). After a mouse was fully anesthetized, 0.5 % Evans blue solution was intradermally injected into the footpad of the hind limb to visualize the lymphatic vessels. A circumferential incision of the limb (thigh) was made to access lymph vessels that were subsequently electrocauterized (Bovie Medical Corporation, Clearwater, FL, USA). EBM-2, NCM, and HCM were then subcutaneously injected at the site of the damaged area on the second day after preparing the edema models and then every 3 days after that for a period of 24 days. The injection volume was 100 µl of 1-in-50 concentrated media. All animals were maintained under a 12-hour light–dark cycle condition and had free access to food and water. Experimental procedures were approved by the Committee for Care and Use of Laboratory Animals, Yonsei University College of Medicine, and performed in accordance with the Guidelines and Regulations for Animal Care.

Histological analysis

To measure lymphangiogenesis, mice were sacrificed 4 weeks after the surgery, after which a mouse's entire thigh region, including the site at which hMSC-conditioned medium was injected, was dissected. The hind limbs were excised from the sacrificed mouse, washed with PBS to remove the blood, and then fixed in 10 % formalin solution for 1 day at 4 °C. Samples were subsequently treated with decalcification solution (Calci-Clear Rapid; National Diagnostics, Atlanta, GA, USA) for 1 week. Tissue sections were sequentially mounted onto gelatin-coated glass slides to ensure that different stains could be used on successive tissue sections sliced through the injured area. After the sections were deparaffinized and rehydrated, samples were blocked using 2.5 % normal horse serum and incubated with an anti-LYVE-1 antibody. A biotinylated pan-specific universal secondary antibody and streptavidin/peroxidase complex reagent were used to treat the sections. Sections were subsequently stained with an antibody using a DAB substrate kit (Vector, Burlingame, CA, USA). Counterstaining was performed using 1 % methyl green, and dehydration was performed using 100 % n-butanol, ethanol, and xylene before mounting in VectaMount Mounting Medium (Vector, Burlingame, CA, USA). A coverslip was placed on top of each section, and the sections were

then observed using light microscopy. The images for hematoxylin and eosin (H & E) and LYVE-1 were obtained using virtual microscopy (BX51/dot Slide; Olympus). Images for LYVE-1 were detected using microscopy and transferred to a computer equipped with the MetaMorph software (version 4.6; Molecular Devices, Sunnyvale, CA, USA).

In vivo visualized lymphatic vessels

We observed the lymphatic vessels with fluorescence 3 weeks after surgery. The hind limb mouse model was anesthetized via a subcutaneous injection of zoletil (4 mg/kg) with rompun (20 mg/kg). After the mouse was fully anesthetized, 30 µl of fluorescein isothiocyanate (FITC)-dextran solution (1 mg/ml) was intradermally injected into the footpad of the hind limb to visualize the lymphatic vessels, after which the lymphatic vessels were observed under a fluorescence microscope (SteREO Discovery V12; Zeiss, Oberkochen, Germany).

Statistical analysis

Data are expressed as mean ± standard deviation (SD). Significance differences between groups were identified using Student's t test. Comparisons between more than two groups were made with one-way analysis of variance using Bonferroni's correction. $p < 0.05$ was considered statistically significant.

Results

HCM enhance lymphangiogenesis in hLECs

To investigate the ability of HCM to undergo lymphangiogenesis in vitro, we tested the proliferation, migration, and tube formation in hLECs. Under normal growth conditions, HCM did not cause any positive effects and was similar to NCM (data not shown). Therefore, hLECs were starved in serum-free EBM-2 culture for 12 hours to simulate in vivo conditions prior to NCM or HCM treatment, after which the effects of HCM on lymphangiogenesis were investigated. HCM markedly increased proliferation by up to twofold (Fig. 1a), as well as migration (Fig. 1b), compared with the control media and NCM. In addition, tube formation was promoted with HCM (Fig. 1c). We also examined LYVE-1 and ERK phosphorylation as proliferation markers, resulting in increased expression and phosphorylation under HCM treatment (Fig. 1d). The results showed that HCM have the ability to induce lymphangiogenesis in hLECs.

hMSCs induce the expression of pro-lymphangiogenesis factors under hypoxia

We next investigated several growth factors from hMSCs, which are known to induce lymphangiogenesis in hLECs. hMSCs exposed to 12 hours of hypoxia were lysed and used to estimate mRNA expression levels. The mRNA

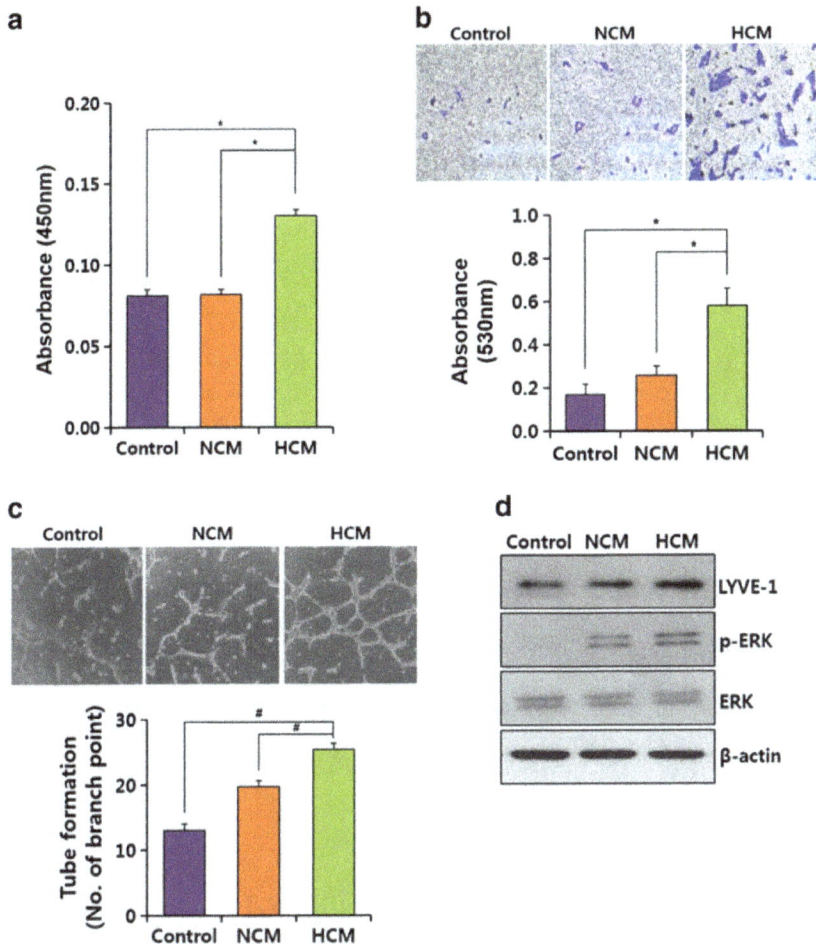

Fig. 1 hMSC-derived HCM stimulate hLECs in vitro. **a** hLEC proliferation was measured using the Cell Counting Kit-8 assay. **b** Representative images showed the results of migration assays of hLECs treated with NCM and HCM. **c** Representative photographs of tube formation in hLECs after treatment with NCM and HCM. **d** hLEC specific marker, LYVE-1, and phosphorylation of ERK were detected using western blot analysis. Values are represented as the mean of three measurements with the SD indicated by error bars. Control: EBM. *p <0.001, #p <0.05. *ERK* extracellular signal regulated kinase, *GAPDH* glyceraldehyde 3-phosphate dehydrogenase, *HCM* hypoxic conditioned media, *LYVE-1* lymphatic vessel endothelial hyaluronan receptor 1, *NCM* normoxic conditioned media

expression of six different pro-lymphangiogenesis factors increased, with VEGF-A and VEGF-C showing significant increases in gene expression (Fig. 2a). Therefore, we further examined whether the expression of VEGF-A and VEGF-C (important lymphangiogenic-stimulating factors) increased in response to HCM in hLECs. Both factors markedly increased under HCM compared with NCM, which is the medium normally used for hMSC cultivation (Fig. 2b, c).

HCM affect the expression of MFN1/2 in hLECs

VEGF is known to induce MFNs in vascular endothelial cells but not in hLECs. Therefore, we examined expression levels of mRNA and protein of MFNs in hLECs

under different conditions. HCM-treated cells showed the ability to induce both MFN1 and MFN2 expression compared with NCM-treated cells. NCM also increased MFN1 and MFN2 expression relative to the basal levels of these genes and proteins (Fig. 3a, b). We demonstrate the effect of lymphangiogenesis through experiments on an undisclosed MFN1/2 in hLECs. MFN1/2 small interfering RNAs (siRNAs) were used to demonstrate the relationship between MFN1/2 and lymphangiogenesis in the hLECs. Effects of MFN1/2 ablation were examined in hLECs using siRNA. Each siRNA was transfected into hLECs at a final concentration of 100 nM for 2 days. Ablation of both MFN1 and MFN2 led to an approximately 93 % and 80 % reduction in MFN1 and in MFN2

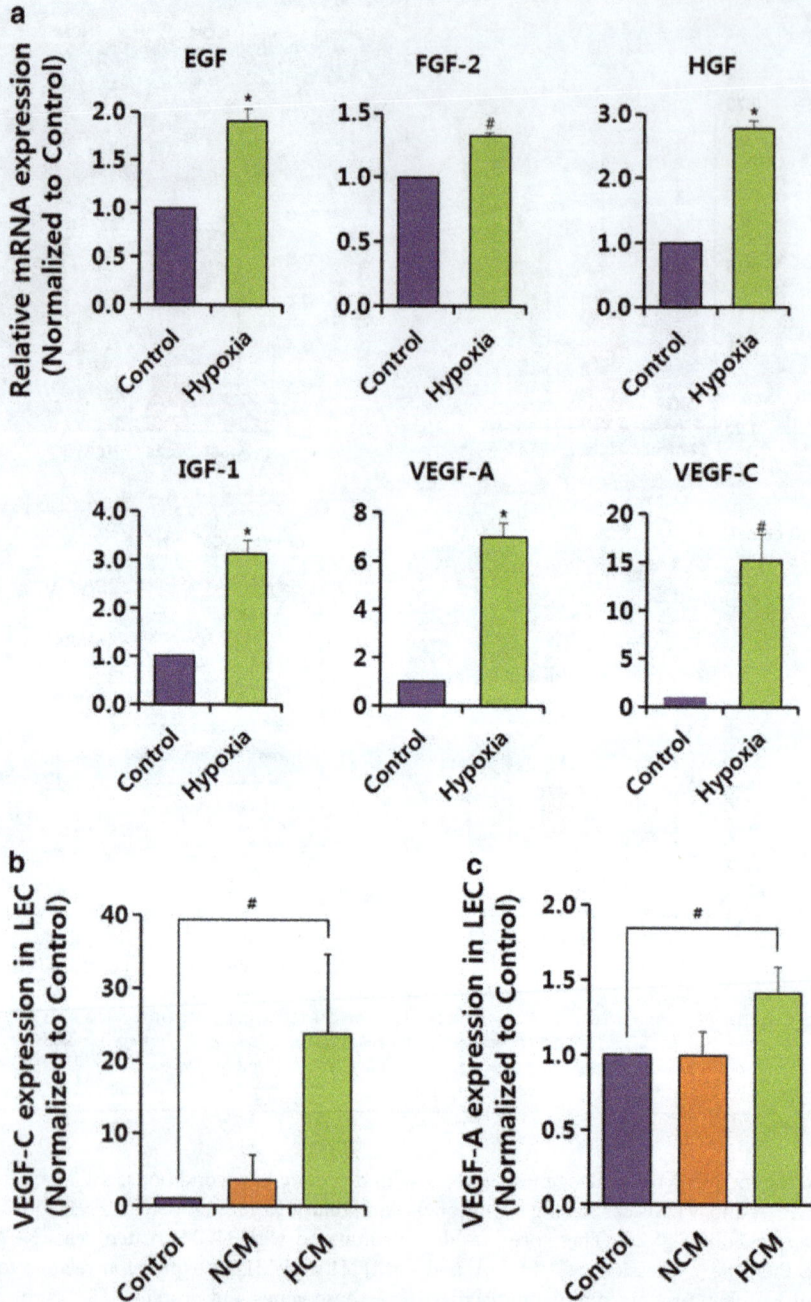

Fig. 2 Increased expression of lymphangiogenic factors of hMSCs in hypoxia. **a** Expression of lymphangiogenic factors (EGF, FGF, HGF, IGF-1, VEGF-A, and VEGF-C) of hMSCs in hypoxia. Expression of mRNAs was measured using quantitative RT-PCR. Gene expressions were normalized using GAPDH expression, respectively. **b, c** Expression of VEGF-C and VEGF-A was analyzed using quantitative RT-PCR in EBM, NCM, or HCM-treated hLECs. Control: EBM. Values are represented as the mean of three measurements with the SD indicated by error bars. *p <0.001, #p <0.05. *EGF* epidermal growth factor, *FGF* fibroblast growth factor, *HCM* hypoxic conditioned media, *HGF* hepatocyte growth factor, *IGF* insulin-like growth factor, *LEC* lymphatic endothelial cells, *NCM* normoxic conditioned media, *VEGF* vascular endothelial growth factor

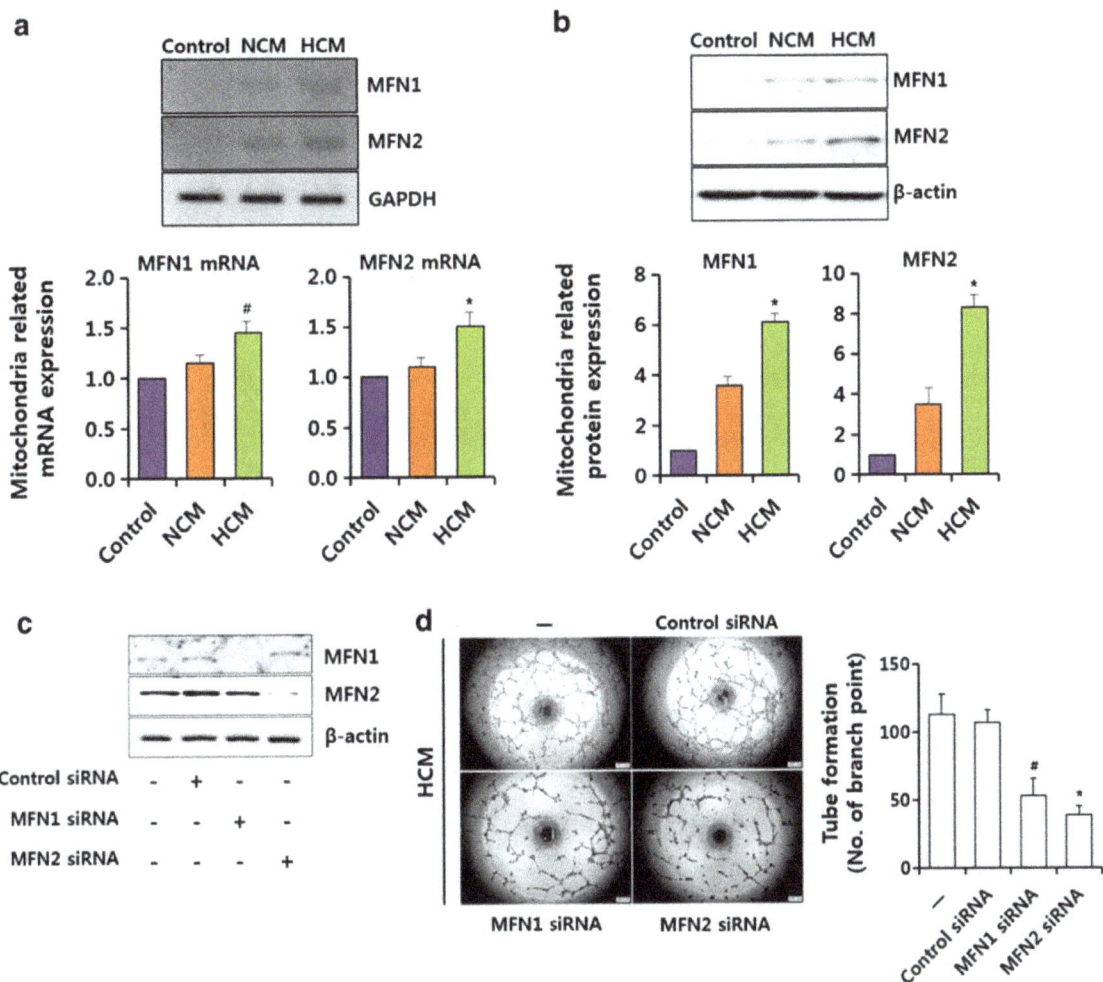

Fig. 3 Effects of HCM on expression of mitochondria-related genes in hLECs. **a** hLECs were starved in EBM for 12 hours and then subjected to hMSC-conditioned medium treatment for 12 hours. MFN1 and MFN2 mRNA expression were analyzed using RT-PCR. mRNA expression was normalized using GAPDH. **b** hLECs were starved in EBM for 12 hours and then subjected to hMSC-conditioned medium treatment for 12 hours. MFN1 and MFN2 expression was measured using western blot analysis. **c** Expression of MFN1/2 was detected in MFN1/2 siRNA-transfected hLECs and the expression of MFN1/2 by immunoblotting. **d** Representative photographs were tube formation of MFN1/2 siRNA-transfected hLECs after treatment of each media. Control: EBM. Values are represented as the mean of three measurements with the SD indicated by error bars. *p <0.001, #p <0.05. *GAPDH* glyceraldehyde 3-phosphate dehydrogenase, *HCM* hypoxic conditioned media, *MFN* mitofusin, *NCM* normoxic conditioned media, *siRNA* small interfering RNA

protein, respectively (Fig. 3c). Next, we detected the effects of MFN1/2 ablation on tube formation. As tube formation requires cellular migration on the Matrigel, we also investigated whether attenuation of MFN1/2 affects the ability for hLECs to migrate towards HCM. As a result, tube formation reduced in the MFN1 siRNA and MFN2 siRNA group compared with control group (Fig. 3d). We have identified the critical role of MFN1/2 in a positive effect on lymphangiogenesis by HCM from the results.

HCM promote in vivo lymphatic regeneration

To confirm the ability of HCM to induce lymphangiogenesis, a mouse model was generated using electrocauterization. The maximum swelling was examined 1–3 days after electrocauterization, and the thickness of the hind limb footpad was examined every 3 days for 27 days. As shown in Fig. 4a, footpad reduction was more effective in HCM-treated groups than in NCM-treated and control groups. H & E staining showed histological edema changes, with the control group exhibiting large

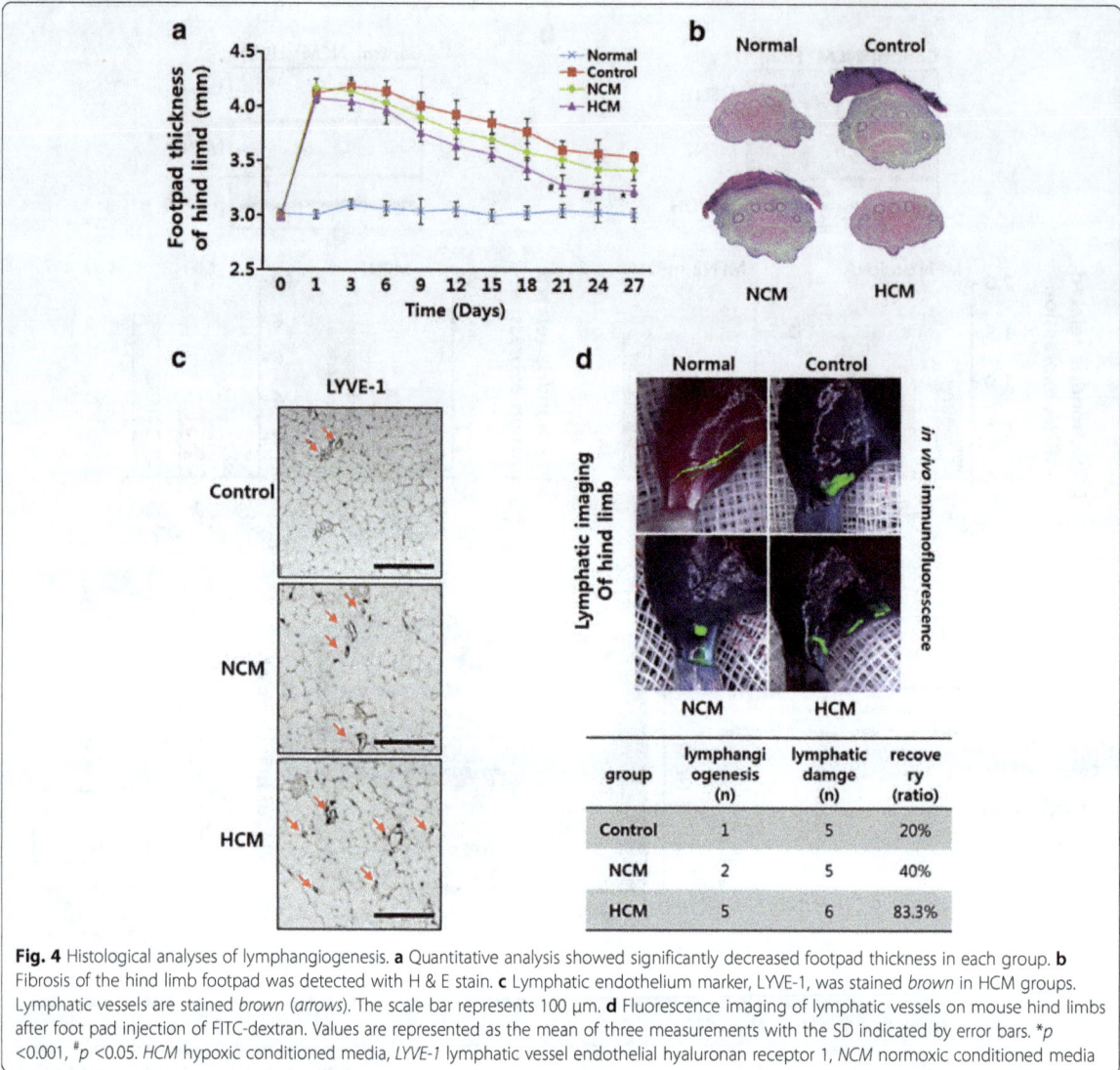

Fig. 4 Histological analyses of lymphangiogenesis. **a** Quantitative analysis showed significantly decreased footpad thickness in each group. **b** Fibrosis of the hind limb footpad was detected with H & E stain. **c** Lymphatic endothelium marker, LYVE-1, was stained *brown* in HCM groups. Lymphatic vessels are stained *brown* (*arrows*). The scale bar represents 100 μm. **d** Fluorescence imaging of lymphatic vessels on mouse hind limbs after foot pad injection of FITC-dextran. Values are represented as the mean of three measurements with the SD indicated by error bars. *p <0.001, #p <0.05. *HCM* hypoxic conditioned media, *LYVE-1* lymphatic vessel endothelial hyaluronan receptor 1, *NCM* normoxic conditioned media

cytosolic damage such as fibrosis. Although footpads from the NCM-treated groups showed less cytosolic damage, HCM-treated footpads showed marked recovery compared with normal footpads (Fig. 4b). Therefore, we further investigated whether these recovery effects were due to lymphatic regeneration. Each group of footpad samples was stained with LYVE-1, a marker of hLECs, and more LYVE-1-positive staining was observed in the HCM-treated group than in the control or NCM-treated groups (Fig. 4c). To confirm the regeneration effect of HCM in hLECs in vivo, we injected FITC-dextran to visualize the lymphatic vessels. NCM also showed an increased regeneration effect on lymphatic vessels of up to twofold when compared with that of the control, whereas

HCM showed a more significant lymphangiogenesis-promoting ability than the control (Fig. 4d).

Discussion

This study demonstrated for the first time that HCM promote lymphangiogenesis by MFN1 and MFN2 regulation in hLECs. Furthermore, the study demonstrated the potential impact that lymphangiogenesis may have on damaged lymphatic vessel repair in vivo.

MSCs appear to be the most feasible, potentially safe, and effective stem cells for therapeutic lymphangiogenesis via direct differentiation toward lymphatic lineage and cytokine secretion [4, 21, 32]. Because cytokines have short-lasting efficacies, secretory modification is needed. Cytokine secretion of MSCs can be induced by hypoxia, inflammatory stimuli, and three-dimensional culture configuration. Rehman et al. [33] demonstrated that paracrine factors released from stem cells under hypoxia demonstrated anti-apoptotic or pro-angiogenic effects. Several studies have reported that HCM contain cytokines and growth factors, such as EGF, FGF-2, HGF, IGF-1, VEGF-A, and VEGF-C [34, 35]. Several reports have also stated that these cytokines could augment lymphangiogenesis in animal models of lymphedema [5–10, 36]. Our results showed higher mRNA expressions of these cytokines in hMSC under hypoxia than under normoxic conditions (Fig 2; EGF, $p <0.001$; FGF-2, $p <0.05$; HGF, $p <0.001$; IGF-1, $p <0.001$; VEGF-A, $p <0.001$; VEGF-C, $p <0.05$).

We demonstrated that HCM has the ability to enhance hLEC proliferation and migration and to increase lymphatic vessel formation. Zhan et al. also showed that culture medium from hMSCs increased lymphatic vessel formation by inducing VEGF receptor-3 (VEGFR-3) in hLECs [37]. As shown in Fig. 2, increased mRNA expression of VEGF-A and VEGF-C by hypoxic hMSCs and HCM-treated hLECs was investigated. VEGFR-2 and VEGFR-3 are known to exist in hLECs and are activated by VEGF-A and VEGF-C, respectively [38, 39]. In addition, phosphorylation of ERK1/2 appears to be primarily caused by VEGF-A/VEGF receptor-2 upon hMSC stimulation in hLECs. Shimizu et al. [32] found that the expression of p-ERF was regulated by VEGF-C. We also showed that ERK phosphorylation and expression of LYVE-1 was increased by HCM in hLECs, which supports the increased VEGF-A expression in hLECs.

Mitochondrial function integrity is essential for organ homeostasis, particularly for rapidly dividing cells, such as hemopoietic precursor cells [40]. In this study, we ascertained the potential for HCM to promote lymphangiogenesis by regulating mitochondrial-related genes. Our results showed that mRNA expression and protein levels of MFN1 and MFN2 increased under HCM treatment in hLECs. Although the role of MFNs has been investigated in other cell types, such as endothelial cells, this is the first study, to the best of our knowledge, to investigate their roles in hLECs.

We focused on paracrine mediators of hMSCs cultured with HCM. Direct injection of HCM into a lymphedema mouse demonstrated lymphangiogenesis and edema improvement. As a precursor to future clinical trials, our study shows therapeutic potential for stimulation of lymphangiogenesis.

To use HCM, we should consider some additional details. According to a recent study [15] performed using hMSCs obtained from 20 people, there is a difference in

endothelial differentiation ability of the hMSCs. Lin et al. [16] reported differences in expression of VEGF and interleukin-8 among three donors. We believe that the donor cell variation should be considered to be applicable for patient therapeutics. Furthermore, the half-life of cytokines and growth factors, which are included in the conditioned media, has been reported to be very short in vivo [41, 42]. For this reason, conditioned media must be more frequently administered. Nevertheless, HCM are a promising treatment for damaged tissue regeneration.

Conclusions

We showed the effects of HCM from hMSCs on in vitro and in vivo lymphangiogenesis. Our results demonstrated an increase in hMSC lymphangiogenic cytokines, which can lead to proliferation, migration, and tube formations in hLECs, thereby showing hMSCs derived HCM potential therapeutic effects in the lymphedema mouse model. In addition, expression of MFN1 and MFN2 is suggested for their possible roles in lymphangiogenesis. Treatment with HCM may thus be a therapeutic strategy for lymphedema.

Abbreviations
Ct: Threshold cycle; EBM: Endothelial basal medium; EGF: Epidermal growth factor; ERK: Extracellular signal regulated kinase; FITC: Fluorescein isothiocyanate; FGF: Fibroblast growth factor; GAPDH: Glyceraldehyde 3-phosphate dehydrogenase; HCM: Hypoxic conditioned media; H & E: Hematoxylin and eosin; HGF: Hepatocyte growth factor; hLEC: Human lymphatic endothelial cell; hMSC: Human mesenchymal stem cell; IGF: Insulin-like growth factor; LYVE-1: lymphatic vessel endothelial hyaluronan receptor 1; MFN: Mitofusin; MSC: Mesenchymal stem cell; NCM: Normoxic conditioned media; PBS: Phosphate-buffered saline; ROS: Reactive oxygen species; SD: standard deviation; siRNA: Small interfering RNA; VEGF: Vascular endothelial growth factor; VEGFR: Vascular endothelial growth factor receptor.

Competing interests
The authors declare that they have no competing interests.

Authors' contributions
CYL and JYK carried out the experiments, designed the study, and drafted and revised the manuscript. SL and OH participated in the data analysis and drafted the manuscript. WC and D-HJ established the hypotheses, designed the study, and revised the manuscript. All authors agreed to be accountable for all aspects of the work in ensuring that questions related to the accuracy or integrity of any part of the work are appropriately investigated and resolved. All authors read and approved the manuscript.

Acknowledgements
This research was supported by the Catholic Medical Center Research Foundation made in the program year of 2013.

Author details
[1]Department of Integrated Omics for Biomedical Sciences, Graduate School, Yonsei University, Seoul 03722, Republic of Korea. [2]Department of Rehabilitation Medicine, National Traffic Injury Rehabilitation Hospital, College of Medicine, The Catholic University of Korea, Yangpyeong-gun 12564, Republic of Korea. [3]Institute for Bio-Medical Convergence, College of Medicine, Catholic Kwandong University, Gangneung-si 25601Gangwon-do, Republic of Korea. [4]Catholic Kwandong University International St. Mary's Hospital, Incheon Metropolitan City 22711, Republic of Korea. [5]Department of Biology Education, College of Education, Pusan National University, Busan

46241, Republic of Korea. [6]Department of Rehabilitation Medicine, Incheon St. Mary's Hospital, College of Medicine, The Catholic University of Korea, Dongsu-ro 56, Bupyeong-gu, Incheon 21431, Republic of Korea.

References

1. Oremus M, Walker K, Dayes I, Raina P. AHRQ technology assessments. Diagnosis and treatment of secondary lymphedema. Rockville: Agency for Healthcare Research and Quality (US); 2010.

2. Kerchner K, Fleischer A, Yosipovitch G. Lower extremity lymphedema update: pathophysiology, diagnosis, and treatment guidelines. J Am Acad Dermatol. 2008;59(2):324–31. doi:10.1016/j.jaad.2008.04.013.

3. Kim IG, Lee JY, Lee DS, Kwon JY, Hwang JH. Extracorporeal shock wave therapy combined with vascular endothelial growth factor-C hydrogel for lymphangiogenesis. J Vasc Res. 2013;50(2):124–33. doi:10.1159/000343699.

4. Qi S, Pan J. Cell-based therapy for therapeutic lymphangiogenesis. Stem Cells Dev. 2015;24(3):271–83. doi:10.1089/scd.2014.0390.

5. Marino D, Angehrn Y, Klein S, Riccardi S, Baenziger-Tobler N, Otto VI, et al. Activation of the epidermal growth factor receptor promotes lymphangiogenesis in the skin. J Dermatol Sci. 2013;71(3):184–94. doi:10.1016/j.jdermsci.2013.04.024.

6. Cao R, Bjorndahl MA, Gallego MI, Chen S, Religa P, Hansen AJ, et al. Hepatocyte growth factor is a lymphangiogenic factor with an indirect mechanism of action. Blood. 2006;107(9):3531–6. doi:10.1182/blood-2005-06-2538.

7. Chang LK, Garcia-Cardena G, Farnebo F, Fannon M, Chen EJ, Butterfield C, et al. Dose-dependent response of FGF-2 for lymphangiogenesis. Proc Natl Acad Sci U S A. 2004;101(32):11658–63. doi:10.1073/pnas.0404272101.

8. Yoon YS, Murayama T, Gravereaux E, Tkebuchava T, Silver M, Curry C, et al. VEGF-C gene therapy augments postnatal lymphangiogenesis and ameliorates secondary lymphedema. J Clin Invest. 2003;111(5):717–25. doi:10.1172/jci15830.

9. Karkkainen MJ, Haiko P, Sainio K, Partanen J, Taipale J, Petrova TV, et al. Vascular endothelial growth factor C is required for sprouting of the first lymphatic vessels from embryonic veins. Nat Immunol. 2004;5(1):74–80. doi:10.1038/ni1013.

10. Cursiefen C, Chen L, Borges LP, Jackson D, Cao J, Radziejewski C, et al. VEGF-A stimulates lymphangiogenesis and hemangiogenesis in inflammatory neovascularization via macrophage recruitment. J Clin Invest. 2004;113(7):1040–50. doi:10.1172/jci20465.

11. Hwang JH, Kim IG, Lee JY, Piao S, Lee DS, Lee TS, et al. Therapeutic lymphangiogenesis using stem cell and VEGF-C hydrogel. Biomaterials. 2011;32(19):4415–23. doi:10.1016/j.biomaterials.2011.02.051.

12. Ryan JM, Barry FP, Murphy JM, Mahon BP. Mesenchymal stem cells avoid allogeneic rejection. J Inflamm (Lond). 2005;2:8. doi:10.1186/1476-9255-2-8.

13. Zhang M, Methot D, Poppa V, Fujio Y, Walsh K, Murry CE. Cardiomyocyte grafting for cardiac repair: graft cell death and anti-death strategies. J Mol Cell Cardiol. 2001;33(5):907–21. doi:10.1006/jmcc.2001.1367.

14. Nygren JM, Jovinge S, Breitbach M, Sawen P, Roll W, Hescheler J, et al. Bone marrow-derived hematopoietic cells generate cardiomyocytes at a low frequency through cell fusion, but not transdifferentiation. Nat Med. 2004; 10(5):494–501. doi:10.1038/nm1040.

15. Portalska KJ, Groen N, Krenning G, Georgi N, Mentink A, Harmsen MC, et al. The effect of donor variation and senescence on endothelial differentiation of human mesenchymal stromal cells. Tissue Eng Part A. 2013;19(21–22): 2318–29. doi:10.1089/ten.TEA.2012.0646.

16. Lin X, Robinson M, Petrie T, Spandler V, Boyd WD, Sondergaard CS. Small intestinal submucosa-derived extracellular matrix bioscaffold significantly enhances angiogenic factor secretion from human mesenchymal stromal cells. Stem Cell Res Ther. 2015;6:164. doi:10.1186/s13287-015-0165-3.

17. Joyce N, Annett G, Wirthlin L, Olson S, Bauer G, Nolta JA. Mesenchymal stem cells for the treatment of neurodegenerative disease. Regen Med. 2010;5(6):933–46. doi:10.2217/Rme.10.72.

18. Madrigal M, Rao KS, Riordan NH. A review of therapeutic effects of mesenchymal stem cell secretions and induction of secretory modification by different culture methods. J Transl Med. 2014;12:260. doi:10.1186/s12967-014-0260-8.

19. Baraniak PR, McDevitt TC. Stem cell paracrine actions and tissue regeneration. Regen Med. 2010;5(1):121–43. doi:10.2217/Rme.09.74.

20. Maertens L, Erpicum C, Detry B, Blacher S, Lenoir B, Carnet O, et al. Bone marrow-derived mesenchymal stem cells drive lymphangiogenesis. Plos One. 2014;9(9):e106976. doi:10.1371/journal.pone.0106976.

21. Conrad C, Niess H, Huss R, Huber S, von Luettichau I, Nelson PJ, et al. Multipotent mesenchymal stem cells acquire a lymphendothelial phenotype and enhance lymphatic regeneration in vivo. Circulation. 2009; 119(2):281–9. doi:10.1161/CIRCULATIONAHA.108.793208.

22. Gil-Ortega M, Garidou L, Barreau C, Maumus M, Breasson L, Tavernier G, et al. Native adipose stromal cells egress from adipose tissue in vivo: evidence during lymph node activation. Stem Cells. 2013;31(7):1309–20. doi:10.1002/stem.1375.

23. Chen L, Xu Y, Zhao J, Zhang Z, Yang R, Xie J, et al. Conditioned medium from hypoxic bone marrow-derived mesenchymal stem cells enhances wound healing in mice. PLoS One. 2014;9(4):e96161. doi:10.1371/journal.pone.0096161.

24. Burlacu A, Grigorescu G, Rosca AM, Preda MB, Simionescu M. Factors secreted by mesenchymal stem cells and endothelial progenitor cells have complementary effects on angiogenesis in vitro. Stem Cells Dev. 2013;22(4): 643–53. doi:10.1089/scd.2012.0273.

25. Berridge MJ, Lipp P, Bootman MD. The versatility and universality of calcium signalling. Nat Rev Mol Cell Biol. 2000;1(1):11–21. doi:10.1038/35036035.

26. Green DR, Reed JC. Mitochondria and apoptosis. Science. 1998;281(5381): 1309–12. doi:10.1126/science.281.5381.1309.

27. Turrens JF. Mitochondrial formation of reactive oxygen species. J Physiol Lond. 2003;552(2):335–44. doi:10.1113/jphysiol.2003.049478.

28. Song ZY, Ghochani M, McCaffery JM, Frey TG, Chan DC. Mitofusins and OPA1 mediate sequential steps in mitochondrial membrane fusion. Mol Biol Cell. 2009;20(15):3525–32. doi:10.1091/mbc.E09-03-0252.

29. Han XJ, Lu YF, Li SA, Kaitsuka T, Sato Y, Tomizawa K, et al. CaM kinase I alpha-induced phosphorylation of Drp1 regulates mitochondrial morphology. J Cell Biol. 2008;182(3):573–85. doi:10.1083/jcb.200802164.

30. Zhou Q, Gensch C, Keller C, Schmitt H, Esser J, Moser M, et al. MnTBAP stimulates angiogenic functions in endothelial cells through mitofusin-1. Vasc Pharmacol. 2015;72:163–71. doi:10.1016/j.vph.2015.05.007.

31. Lugus JJ, Ngoh GA, Bachschmid MM, Walsh K. Mitofusins are required for angiogenic function and modulate different signaling pathways in cultured endothelial cells. J Mol Cell Cardiol. 2011;51(6):885–93. doi:10.1016/j.yjmcc.2011.07.023.

32. Shimizu Y, Shibata R, Shintani S, Ishii M, Murohara T. Therapeutic lymphangiogenesis with implantation of adipose-derived regenerative cells. J Am Heart Assoc. 2012;1(4):e000877. doi:10.1161/jaha.112.000877.

33. Rehman J, Traktuev D, Li J, Merfeld-Clauss S, Temm-Grove CJ, Bovenkerk JE, et al. Secretion of angiogenic and antiapoptotic factors by human adipose stromal cells. Circulation. 2004;109(10):1292–8. doi:10.1161/01.CIR.0000121425.42966.F1.

34. Crisostomo PR, Wang Y, Markel TA, Wang M, Lahm T, Meldrum DR. Human mesenchymal stem cells stimulated by TNF-alpha, LPS, or hypoxia produce growth factors by an NF kappa B- but not JNK-dependent mechanism. Am J Physiol Cell Physiol. 2008;294(3):C675–82. doi:10.1152/ajpcell.00437.2007.

35. Efimenko A, Starostina E, Kalinina N, Stolzing A. Angiogenic properties of aged adipose derived mesenchymal stem cells after hypoxic conditioning. J Transl Med. 2011;9:10. doi:10.1186/1479-5876-9-10.

36. Bjorndahl M, Cao R, Nissen LJ, Clasper S, Johnson LA, Xue Y, et al. Insulin-like growth factors 1 and 2 induce lymphangiogenesis in vivo. Proc Natl Acad Sci U S A. 2005;102(43):15593–8. doi:10.1073/pnas.0507865102.

37. Zhan J, Li YH, Yu J, Zhao YY, Cao WM, Ma J, Sun XX, Sun L, Qian H, Zhu W, Xu WR. Culture medium of bone marrow-derived human mesenchymal stem cells effects lymphatic endothelial cells and tumor lymph vessel formation. Oncol Lett. 2015;9:1221–1226. doi: 10.3892/Ol.2015.2868

38. Religa P, Cao RH, Bjorndahl M, Zhou ZJ, Zhu ZP, Cao YH. Presence of bone marrow-derived circulating progenitor endothelial cells in the newly formed lymphatic vessels. Blood. 2005;106(13):4184–90. doi: 10.1182/blood-2005-01-0226.

39. Dellinger MT, Brekken RA. Phosphorylation of Akt and ERK1/2 Is Required for VEGF-A/VEGFR2-Induced Proliferation and Migration of Lymphatic Endothelium. Plos One. 2011;6(12):e28947. doi: 10.1371/journal.pone.0028947.

40. Sack MN. Mitofusin function is dependent on the distinct tissue and organ specific roles of mitochondria. Mol Cell Cardiol. 2011;51(6):881–2. doi: 10.1016/j.yjmcc.2011.09.004.

41. Yde P, Mengel B, Jensen MH, Krishna S, Trusina A. Modeling the NF-kappaB mediated inflammatory response predicts cytokine waves in tissue. BMC Syst Biol. 2011;5:115. doi: 10.1186/1752-0509-5-115.

42. Khosravi A, Cutler CM, Kelly MH, Chang R, Royal RE, Sherry RM, et al. Determination of the elimination half-life of fibroblast growth factor-23. J Clin Endocrinol Metab. 2007;92(6):2374–7. doi:10.1210/jc.2006-2865.

12

Overexpression of the β2AR gene improves function and re-endothelialization capacity of EPCs after arterial injury in nude mice

Xiao Ke[1,2,†], Xiao-Rong Shu[1,2,†], Fang Wu[3], Qing-Song Hu[1,2], Bing-Qing Deng[1,2], Jing-Feng Wang[1,2] and Ru-Qiong Nie[1,2*]

Abstract

Background: Proliferation and migration of endothelial progenitor cells (EPCs) play important roles in restoring vascular injuries. β2 adrenergic receptors (β2ARs) are widely expressed in many tissues and have a beneficial impact on EPCs regulating neoangiogenesis. The aim of the present study was to determine the effect of overexpressing β2ARs in infused peripheral blood (PB)-derived EPCs on the re-endothelialization in injured vessels.

Methods: Induction of endothelial injury was performed in male nude mice that were subjected to wire-mediated injury to the carotid artery. Human PB-derived EPCs were transfected with an adenovirus serotype 5 vector expressing β2AR (Ad5/β2AR-EPCs) and were examined 48 h later. β2AR gene expression in EPCs was detected by real-time polymerase chain reaction and Western blot analysis. In vitro, the proliferation, migration, adhesion, and nitric oxide production of Ad5/β2AR-EPCs were measured. Meanwhile, phosphorylated Akt and endothelial nitric oxide synthase (eNOS), which are downstream of β2AR signaling, were also elevated. In an in vivo study, CM-DiI-labeled EPCs were injected intravenously into mice subjected to carotid injury. After 3 days, cells recruited to the injury sites were detected by fluorescent microscopy, and the re-endothelialization was assessed by Evans blue dye.

Results: In vitro, β2AR overexpression augmented EPC proliferation, migration, and nitric oxide production and enhanced EPC adhesion to endothelial cell monolayers. In vivo, when cell tracking was used, the number of recruited CM-DiI-labeled EPCs was significantly higher in the injured zone in mice transfused with Ad5/β2AR-EPCs compared with non-transfected EPCs. The degree of re-endothelialization was also higher in the mice transfused with Ad5/β2AR-EPCs compared with non-transfected EPCs. We also found that the phosphorylation of Akt and eNOS was increased in Ad5/β2AR-EPCs. Preincubation with β2AR inhibitor (ICI118,551), Akt inhibitor (ly294002), or eNOS inhibitor (L-NAME) significantly attenuated the enhanced in vitro function and in vivo re-endothelialization capacity of EPCs induced by β2AR overexpression.

Conclusions: The present study demonstrates that β2AR overexpression enhances EPC functions in vitro and enhances the vascular repair abilities of EPCs in vivo via the β2AR/Akt/eNOS pathway. Upregulation of β2AR gene expression through gene transfer may be a novel therapeutic target for endothelial repair.

Keywords: β2AR, Re-endothelialization, Proliferation, Migration, Endothelial progenitor cells, β2AR/Akt/eNOS

* Correspondence: nieruqiong@163.com
†Equal contributors
[1]Department of Cardiology, Sun Yat-sen Memorial Hospital of Sun Yat-sen University, No. 107, Yanjiangxi Road, Guangzhou, China
[2]Guangdong Province Key Laboratory of Arrhythmia and Electrophysiology, Guangzhou 510120, China
Full list of author information is available at the end of the article

Background

Coronary atherosclerotic heart diseases are the major cause of mortality worldwide in cardiovascular disease (CVD), yet current therapies only delay disease progression and improve lifestyle without addressing the fundamental problem of tissue loss. Several studies have shown that endothelial dysfunction is an early marker of atherosclerosis, and the activation of inflammatory processes and abnormalities in vascular homeostasis have been suggested to contribute to the development of atherosclerotic vascular disease [1]. Accumulating evidence indicates that the balance between endothelial injury and repair is a key component of atherosclerosis [2], and maintaining endothelial integrity is crucially important to preventing the initiation and development of atherosclerosis, coronary heart disease, and postangioplasty restenosis [3]. Thus, accelerated re-endothelialization might prevent the early stages of atherosclerosis and restenosis after angioplasty.

Endothelial progenitor cells (EPCs) mobilized from bone marrow into the peripheral blood (PB) have been shown to play an important role for vascular regeneration, endothelial repair, and replacement of dysfunctional endothelium by incorporating into the site of vessel injury, differentiating into endothelial cells (ECs), and releasing paracrine factors [4, 5]. Transplantation of EPCs is currently under intensive investigation in animal models and clinical research and have become a major focus of CVD treatment to accelerate re-endothelialization [6]. However, the beneficial effects of this practice have not been observed in patients with coronary artery disease, because the reparative capacity of EPCs appears to be limited by their poor survival environment [7]. Therefore, attempts to improve the function of transplanted EPCs with gene modifications may facilitate the repair of damaged endothelia and accelerate re-endothelialization.

Various molecules may be involved in the processes by which EPCs home to and restore damaged endothelium, such as nitric oxide (NO), stromal cell-derived factor 1a (SDF-1a), and vascular endothelial growth factor (VEGF). It is widely accepted that β-adrenoceptors exist on ECs and contribute to the regulation of vasomotor tone. Moreover, it is the β2 adrenergic receptor (β2AR), the most abundant βAR in the vasculature, that mediates the NO involved in relaxing vascular tone [8]. β2ARs, G protein-coupled receptors, are activated by adrenergic catecholamine to promote a series of intracellular signal transduction pathways that lead to multiple cell-specific responses [9]. It has been demonstrated that β2ARs are strongly expressed on EPCs and also mediate the homing and neovascularization capacity of EPCs to areas of ischemia [9]. Accordingly, the β2AR gene may be a valuable molecular target for gene therapies that use EPCs. However, β2ARs have not previously been shown to accelerate vascular repair via re-endothelialization mediated by

EPCs. Thus, we sought to determine whether β2AR gene transfer mediates the functional properties of EPCs during vascular injury.

In this study, we tested the therapeutic potential of β2AR gene transfer in EPCs by infusing transfected cells into nude mice after we induced an endothelial denudation injury. We also investigated the β2AR-mediated Akt/endothelial nitric oxide synthase/NO (Akt/eNOS/NO) signaling pathway that is related to both in vitro and in vivo biology of EPCs. These data demonstrate that the β2AR has an important role in EPC migration at the vascular injury site, and upregulating β2AR expression is a potential new therapeutic strategy that may improve the efficiency of EPC-induced re-endothelialization.

Methods

Ethics

The experimental research on humans in this study was performed in compliance with the Helsinki Declaration. All recruited patients consented to participate in this trial and to contribute their trial data for non-commercial purposes. The protocol of this trial was externally reviewed and approved by an anonymous independent ethical review committee to ensure that there were no serious ethical concerns. The animal procedures in this study complied with the Animal Care and Use Ethics Committees of Sun Yat-Sen University.

EPC culture and characterization

EPCs were isolated and cultured according to previously described methods [10, 11]. Briefly, PB mononuclear cells (MNCs) were isolated from healthy subjects (males from 25 to 35 years of age) by using Histopaque-1077 density gradient centrifugation at $400\,g$ for 30 min. The collected MNCs were washed three times with phosphate-buffered saline (PBS) (Jingmei Bio Tech Co. Ltd., Shenzhen, China). After the cells were purified, the MNCs were cultured on fibronectin-coated six-well plates in endothelial basal medium-2 (EBM-2) (CC-4176; Lonza, Basel, Switzerland) supplemented with EGM-2 Bullte Kit (Lonza) and 20 % fetal bovine serum (FBS) (Gibco, now part of Thermo Fisher Scientific, Waltham, MA, USA). After 4 days in culture, the non-adherent cells were abandoned. Adherent cells were cultured for 7 days and then were used for the following experiments.

EPCs were defined as cells that were dually positive when stained by using 1,1′-dioctadecyl-3,3,3′,3′-tetramethylindocarbocyanine (DiI)-acetylated low-density lipoprotein (ac-LDL) (20 μg/ml; Invitrogen, Carlsbad, CA, USA) and fluorescein isothiocyanate (FITC)-labeled BS-1 lectin (10 μg/ml; Sigma-Aldrich, St. Louis, MO, USA). Cultured EPCs were incubated with DiI-ac-LDL for 3 h at 37 °C; the cells then were washed in PBS, fixed in 4 % (vol/vol) paraformaldehyde (PFA) for 30 min, and incubated with FITC-labeled BS-1 lectin for 1 h. The cells were washed again and then incubated with 4′,6-dia-midino-2-phenylindole (DAPI), a nuclear counterstain. Double-positive cells were observed with a fluorescence microscope (×200 magnification; Olympus, Tokyo, Japan). Cells demonstrating double-positive fluorescence were identified as differentiating EPCs.

Flow cytometric analysis
The expression of endothelial marker proteins was examined in the cultured EPCs by using flow cytometric analysis with phycoerythrin (PE)-labeled monoclonal mouse anti-human antibodies recognizing CD31 (BD Pharmingen, San Diego, CA, USA), von Willebrand factor (vWF) (BD Pharmingen), kinase-insert domain receptor (KDR) (R&D Systems, Minneapolis, MN, USA), and CD14 (BD Pharmingen). To identify the cells that expressed these surface antigens, the EPCs were incubated for 40 min at 4 °C in a volume of 100 μl of solution containing an appropriate amount of PE-labeled antibody or corresponding IgG isotype control. At least 1×10^5 EPCs were acquired by using a flow cytometer (Beckman-Coulter, Fullerton, CA, USA).

Immunofluorescence
To characterize the expression of EC markers, EPCs were grown in fibronectin-coated six-well plates, and immunofluorescence analysis was performed by using rabbit polyclonal antibody against β2AR (Abcam, Cambridge, MA, USA) and mouse monoclonal antibody (mAb) against eNOS (Cell Signaling Technology, Boston, MA, USA). Briefly, the cells were washed in cold PBS three times and fixed in 4 % PFA for 30 min. Then the cells were washed again with PBS three times for 5 min each and incubated in 3 % bovine serum albumin (BSA) in PBS for 1 h. The cells were incubated with primary antibodies (anti-eNOS, anti-β2AR diluted 1:100 with 3 % BSA in PBS) at room temperature for 1 h. After the cells were washed three times for 5 min each in PBS on a shaker, the cells were exposed to goat anti-rabbit IgG (H + L) (catalog no. A-11011; Life Technologies, Carlsbad, CA, USA) and goat anti-mouse IgG (H + L) (catalog no. A-11011; Life Technologies) secondary antibodies for 1 h in the dark. The cells were washed again and incubated with DAPI to stain the EPC nuclei. Images were acquired by using a fluorescence microscope (×200 magnification; Olympus).

EPC gene transfer
An adenovirus sero-type 5 (Ad5) vector expressing the human β2AR gene (Ad5/β2AR) or enhanced green fluorescent protein (Ad5/EGFP) was used for gene delivery (purchased from GeneChem Company Ltd., Shanghai, China). To establish the appropriate virus concentration

for adenoviral gene transfer into EPCs, the effectiveness of different multiplicities of infection (MOIs) was evaluated in accordance with the instructions of the adenovirus manufacturer. Briefly, after the EPCs were cultured for 7 days, they were transduced with Ad5/β2AR and Ad5/EGFP in serum-free culture medium (MOI of approximately 500). The viruses were removed, and the cells were washed with PBS and incubated with EPC medium for another 48 h before subsequent experiments.

Real-time polymerase chain reaction and Western blot analysis
Total cellular RNA was isolated by using TRIzolreagent (Invitrogen). Double-stranded cDNA was synthesized by using an M-MLV Reverse Transcriptase cDNA Synthesis Kit (TaKaRa, Kusatsu, Shiga, Japan). Quantitative polymerase chain reaction (PCR) was carried out with Light Cycler 480 SYBR Green I Master Mix (Roche Diagnostics, Risch-Rotkreuz, Switzerland) in a Light Cycler 480 System. The cycling protocol for the PCR was as follows: 95 °C for 5 min, followed by 45 cycles of 95 °C for 10 s, 60 °C for 10 s, and 72 °C for 20 s. The primers used were as follows: β2AR: 5′-ATGGTGTGGATTGTGTCAGG-3′ (forward) and 5′-CAGGTCTCATTGGCATAGCA-3′ (reverse) and glyceraldehyde 3-phosphate dehydrogenase (GAPDH): 5′-GGTGGTCTCCTCTGACTTCAACA-3′ (forward) and 5′-GTTGCTGTAGCCAAATTCGTTGT-3′ (reverse).

EPCs were lysed with cell lysis buffer (Cell Signaling Technology) in accordance with the instructions of the manufacturer. Cell lysates were quantified by bicinchoninic acid (BCA) methods in accordance with the instructions of the manufacturer (Sangon Biotechnology, Shanghai, China). In total, 50 μg protein was subjected to SDS-PAGE and then transferred to polyvinylidene fluoride membranes. The following antibodies were used: rabbit anti-β2AR antibody (1:1000; Abcam, Cambridge, MA, USA), Phospho-Akt (Ser473) rabbit mAb (1:1000; Cell Signaling Technology), Phospho-eNOS (Ser1177) rabbit mAb (1:1000; Cell Signaling Technology), eNOS (49G3) rabbit mAb (1:1000; Cell Signaling Technology), Akt (C67E7) rabbit mAb (1:1000; Cell Signaling Technology), SDF-1 antibody (1:1000; Cell Signaling Technology), CXCR4(H-118) (1:1000; Santa Cruz Biotechnology, Inc., Dallas, TX, USA), and GAPDH (14C10) rabbit mAb (1:1000; Cell Signaling Technology). Proteins were visualized with horseradish peroxidase-conjugated anti-rabbit IgG (1:5000; Cell Signaling Technology). To detect the effect of stimulation of PB-derived EPCs with the selective β2AR agonist fenoterol (FENO) on the phosphorylation of Akt and eNOS, EPCs were pre-incubated with 10^{-8} M FENO (Sigma-Aldrich) for 6 h before proteins were harvested.

EPC proliferation and NO production

The effect of β2AR gene transfer into EPCs on cell proliferation was assessed by CCK8 (Dojindo Molecular Technologies, Kumamoto, Japan). EPCs were transduced with Ad5/β2AR or Ad5/EGFP, or they were not transduced (control). The EPCs were reseeded on 96-well plates. Briefly, the EPCs were seeded in 96-well plates (5×10^3 cells per well) in EBM-2 (Lonza, CC-4176) supplemented with 1 % FBS for 24 h. The medium then was replaced with 100 μl of fresh medium containing 10 μl of CCK8 solution, and the cells were incubated for another 2 h. The absorbance of each well then was determined at 450 nm by using an Infinite F200 Multimode plate reader. We used 0.3 μM ICI118,551 (Sigma-Aldrich), 10 μM LY294002 (Calbiochem, now part of EMD Millipore, Billerica, MA, USA), and 100 μM L-NAME (Calbiochem) to inhibit β2AR, Akt, and eNOS, respectively. The EPCs were preincubated with inhibitor for 30 min before FENO (10^{-8} M) stimulations.

NO secretion by EPCs was measured as the generation of nitrite. The cells were cultured with EBM-2 (growth factor-free) for 48 h after gene transfer. The supernatants were assayed to determine the level of NO by using a NO assay kit by the nitrate reductase method (Nanjing Jiancheng Institute of Biological Engineering, Nanjing, China).

In vitro EPC migration assays

EPC migration assays were performed by using a Transwell system (Corning Costar, Tewksbury, MA, USA) with 8-μm polycarbonate filter inserts in 24-well plates. Briefly, a total of 2×10^4 EPCs were suspended in 250 μl of EBM-2 medium supplemented with 1 % FBS, and the cells were incubated in the upper chamber for 30 min for each group. These groups included the control group, Ad5/β2AR-EPC group, and Ad5/EGFP-EPC group, which were not pretreated, and the Ad5/β2AR-EPC group and Ad5/EGFP-EPC group, which were pretreated with ICI118,551, LY294002, and L-NAME. The lower compartment of the modified Boyden chamber was placed in a 24-well culture plate in which each well was filled with 500 μl of EBM-2 supplemented with PBS or FENO (10^{-8} M). After the cells were cultured for 6 h, the cells that had migrated into the lower chamber were stained with DAPI. The transmigrated cells were randomly counted by an independent investigator who was blinded to treatments.

In vitro EPC adhesion assays

A monolayer of human umbilical vein endothelial cells (HUVECs) was prepared 48 h before the assay by plating 2×10^5 cells in each well of a four-well plate. The HUVECs were pretreated with or without 1 ng/ml tumor necrosis factor-α (TNF-α) (PeproTech, Rocky Hill, NJ, USA) for 12 h. Then 1×10^5 CM-DiI (Cell-Tracker™ CM-DiI, Invitrogen)-labeled EPCs was added to each well, and the cells were incubated for 3 h at 37 °C. The non-attached cells were gently removed with PBS, and the adherent EPCs were fixed in 4 % PFA and randomly counted by an independent investigator who was blinded to treatments.

Animal model and in vivo re-endothelialization assay

The carotid artery injuries and EPC transplantation were performed by using previously described methods [12, 13]. Male NRMInu/nu athymic nude mice (The Laboratory Animal Center of Sun Yat-sen University, Guangzhou, China) that were 6 to 8 weeks old were injected with human PB-derived EPCs. The animals were anaesthetized with ketamine (100 mg/kg intraperitoneally) and xylazine (5 mg/kg intraperitoneally). The surgeries were performed by using a stereoscopic microscope. The left carotid artery was exposed via a midline incision on the ventral side of the neck. The bifurcation of the carotid artery was located, and two ligatures were placed around the external carotid artery, which then was tied off with the distal ligature. An incision hole was made between the ligatures to introduce the denudation device. A curved flexible wire (0.35-mm diameter) was introduced into the common carotid artery and passed over the lining of artery three times to denude the endothelium. The wire then was removed, and the external carotid artery was tied off proximal to the incision hole with the proximal ligature.

EPCs (1×10^6 cells) that had been cultured for 7 days were resuspended in 100 μl of pre-warmed PBS (37 °C) and were transplanted 3 h after carotid artery injury via tail vein injection with a 27-G needle. The same volume of PBS was injected into placebo mice as a control. Three days after carotid artery injury, endothelial regeneration was evaluated by staining denuded areas with 50 μl of 5 % Evans blue dye via tail vein injection. To examine the homing of transplanted EPCs to the site of the injured carotid vessel, labeled EPCs (1×10^6) were incubated with CM-DiI (Cell Tracker™ CM-DiI; Invitrogen) in accordance with the instructions of the manufacturer. CM-DiI-labeled EPCs incorporated in the injured vessels were quantitatively analyzed under a fluorescence microscope (Olympus BX51).

Statistical analysis

All results are expressed as the mean ± standard error of the mean. Statistical significance was evaluated by means of Student's t test or analysis of variance. A P value of less than 0.01 was considered to denote statistical significance. All statistical analyses used SPSS statistical software (SPSS version 13.0; IBM Corporation, Armonk, NY, USA).

Results

Characterization of EPCs and endogenous expression of β2AR on EPCs

Recently, EPCs were classified into two distinct types: early and late EPCs. Early EPCs appear after 5 to 7 days, and late EPCs appear after 14 to 21 days [14]. A beneficial effect on vascular repair after injury has been shown for early EPCs [15]. After 7 days of culture on fibronectin-coated plates, PB-derived EPCs had a spindle-shaped morphology. Cellar immunostaining showed that most of the adherent cells have double-positive staining for the uptake of DiI-ac-LDL and for the binding of FITC-lectin, indicating that these cells possess the functional properties of ECs (Fig. 1a). In addition, flow cytometric analysis of EC antigens revealed that 40.82 ± 3.98 % of the adherent cells were positive for CD31, 50.69 ± 4.76 % for vWF, 96.56 ± 8.76 % for VEGF (VEGFR2/KDR), and 55.47 ± 4.75 % for the monocytic maker CD14 (Fig. 1b). All of these characteristics indicate that the cultured adherent cells were appropriately identified as EPCs, as previously described [12, 16]. We performed immunocytochemistry for colocalization of β2ARs and the EPC marker eNOS. Fluorescence microscopy revealed that β2ARs were found to localize to the cell membrane of EPCs, and these results were confirmed by using a positive marker, eNOS, which is also expressed in EPCs (Fig. 1c). In addition, β2AR expression on the EPCs did not increase when the cells were stimulated with the selective β2AR agonist FENO for 6 or 12 h, as shown in the Western blot analyses. These results are consistent with the findings of Galasso et al. (Fig. 1d) [9].

Overexpression of β2AR in transfected EPCs

Adenovirus effectively mediated the transfection of the β2AR gene into EPCs. To upregulate the β2AR gene in EPCs, we transduced EPCs by using Ad5/β2AR gene transfection. At 24 h after transfection, EGFP expression was detected by fluorescent microscopy, and the transfection efficiency was approximately 80 % (data not shown). At 48 h after transfection, our results showed that β2AR mRNA was significantly increased in the Ad5/β2AR-EPC group compared with the Ad5/EGFP-EPC group or in controls by using real-time PCR (9.73 ± 3.56 versus 0.89 ± 0.65 or 1.00 ± 0.47; $P < 0.01$). These results were further confirmed by using Western blot analysis (0.40 ± 0.03 versus 0.18 ± 0.03 or 0.2 ± 0.01; $P < 0.01$) (Fig. 2a, b).

β2AR gene transfer in EPCs upregulates cell proliferation, migration and adhesion; NO production; SDF-1/CXCR4 expression in vitro

We used CCK8 assays to examine whether overexpressing β2AR affected proliferation in EPCs that were stimulated with FENO. The rate of proliferation was significantly higher in the Ad5/β2AR-transduced EPCs than in the Ad5/EGFP-transduced EPCs or the non-transduced EPCs (Fig. 3a). We also detected NO production by analyzing NO levels in the conditioned media at 48 h after β2AR gene transfer. As shown in Fig. 3b, the level of NO production was higher in the Ad5/β2AR-transduced EPCs than in the Ad5/EGFP-transduced EPCs or the non-transduced EPCs.

In the migration assays, there were no differences in basal migration capacity among Ad5/β2AR-transduced EPCs, Ad5/EGFP-transduced EPCs, and non-transduced EPCs. However, the rate of EPC migration was significantly higher in Ad/β2AR-transduced EPCs that were induced by FENO stimulation than in the Ad5/EGFP transduced EPCs and non-transduced EPCs that were stimulated by FENO (28.2 ± 4.2 versus 20.2 ± 5.5 or 20.6 ± 4.6; $P < 0.01$) (Fig. 3c-e).

TNF-α can enhance the expression of adhesion molecules in ECs [16]. To determine the function of β2AR in EPC adhesion, we investigated its role in the adhesion of EPCs cultured on mature HUVEC monolayers. Similar to the results of the migration assay, without TNF-α stimulation, there was no difference in the adhesiveness of EPCs cultured on HUVECs among the Ad5/β2AR-transduced, Ad5/EGFP-transduced, and non-transduced EPC groups. However, when the HUVECs were activated with TNF-α, the adhesive ability of the Ad5/β2AR-transduced EPCs was greater than the adhesive ability of the Ad5/EGFP-transduced EPCs and the non-transduced EPCs (61.2 ± 7.2 versus 27.4 ± 4.1 or 28.6 ± 4.9; $P < 0.01$) (Fig. 3d-f). Therefore, these data suggested that β2AR mediated the biological function of EPCs in vitro and that β2AR overexpression would increase the regulatory capacity of EPCs.

The chemokine stromal-derived factor (SDF-1) and its unique receptor CXCR4 (SDF-1 and CXCR4 axis) play an important role invasculogenesis, neovascularization, and re-endothelialization [17, 18]. SDF-1 and CXCR4 are also involved in the biological functions of EPCs, including migration, adhesion, mobilization, and homing. Interestingly, in our study, Western blot analysis revealed that SDF-1 and CXCR4 levels were significantly higher at both time points in the cells that were stimulated with FENO than in the control EPCs that were not stimulated with FENO, and this difference was stronger in the Ad5/β2AR-transduced EPC group (Fig. 3g). These results suggest that β2AR stimulation promotes the expression of members of the SDF-1/CXCR4 axis in EPCs.

β2AR gene transfer increases the re-endothelialization capacity of EPCs in vivo

The carotid endothelium injury in mice was confirmed by Evans blue staining (Fig. 4a). To assess the capability

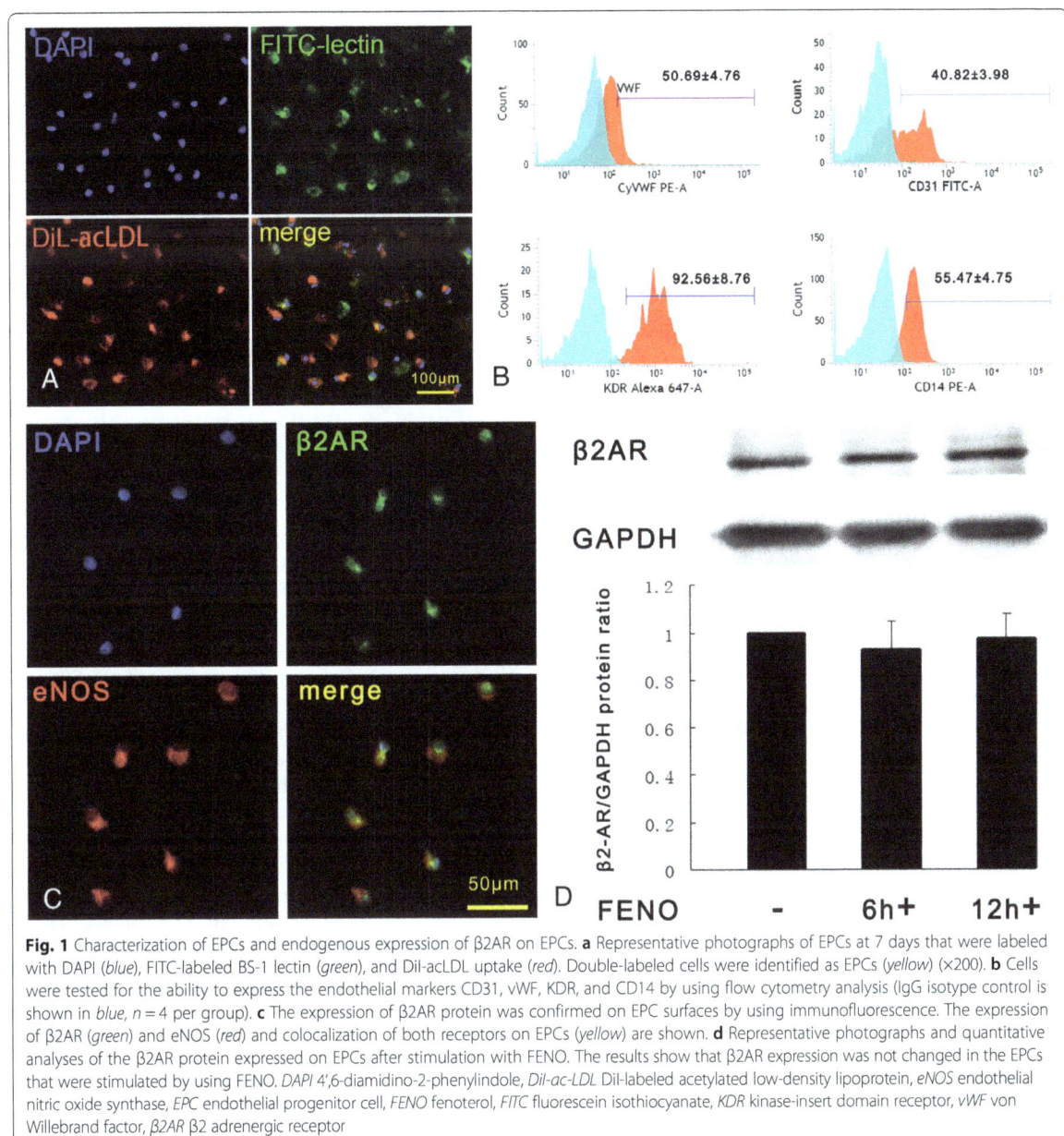

Fig. 1 Characterization of EPCs and endogenous expression of β2AR on EPCs. **a** Representative photographs of EPCs at 7 days that were labeled with DAPI (*blue*), FITC-labeled BS-1 lectin (*green*), and Dil-acLDL uptake (*red*). Double-labeled cells were identified as EPCs (*yellow*) (×200). **b** Cells were tested for the ability to express the endothelial markers CD31, vWF, KDR, and CD14 by using flow cytometry analysis (IgG isotype control is shown in *blue*, $n = 4$ per group). **c** The expression of β2AR protein was confirmed on EPC surfaces by using immunofluorescence. The expression of β2AR (*green*) and eNOS (*red*) and colocalization of both receptors on EPCs (*yellow*) are shown. **d** Representative photographs and quantitative analyses of the β2AR protein expressed on EPCs after stimulation with FENO. The results show that β2AR expression was not changed in the EPCs that were stimulated by using FENO. *DAPI* 4′,6-diamidino-2-phenylindole, *Dil-ac-LDL* Dil-labeled acetylated low-density lipoprotein, *eNOS* endothelial nitric oxide synthase, *EPC* endothelial progenitor cell, *FENO* fenoterol, *FITC* fluorescein isothiocyanate, *KDR* kinase-insert domain receptor, *vWF* von Willebrand factor, *β2AR* β2 adrenergic receptor

of transplanted EPCs on endothelium recovery, we injected PB-derived EPCs into a nude mice model in which we caused wire-denuded carotid arteries. PBS, Ad5/β2AR-transduced EPCs, or Ad5/EGFP-transduced EPCs were injected into nude mice through the tail vein. Notably, compared with PBS, treatment with non-transduced EPCs or with Ad5/EGPF-transduced EPCs substantially increased re-endothelialization of denuded carotid arteries in nude mice; however, injecting the Ad5/β2AR-transduced EPCs resulted in a larger re-endothelialization area in the injured carotid arteries

$(82.1 \pm 4.2$ % versus 44.4 ± 6.5 % or 46.8 ± 4.3 % versus 26.6 ± 7.5 %, $n = 5$, $P < 0.01$; Fig. 4b). To determine whether labeled EPCs were able to home and incorporate into the sites of vascular endothelium injury, each mice was injected with 1×10^6 CM-DiI-labeled EPCs after carotid artery injury. DiI-labeled EPCs were identified as red fluorescent cells. Data showed that there were more homing EPCs in the Ad5/β2AR-transduced EPC group than in the Ad5/EGFP-transduced EPC group that incorporated into the FITC-lectin-positive endothelial layer (Fig. 4c, d).

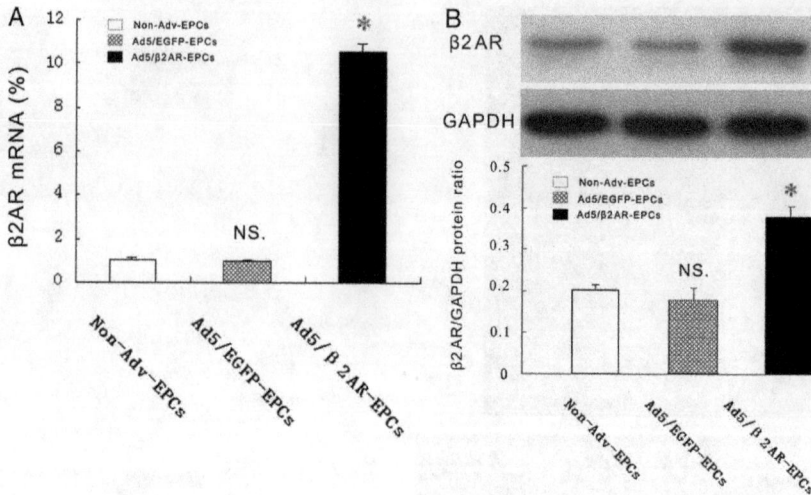

Fig. 2 Overexpression of β2AR in transfected EPCs. **a** β2AR mRNA levels were measured in non-Adv EPCs, Ad5/EGFP EPCs, and Ad5/β2AR EPCs by using real-time polymerase chain reaction and normalized against β-actin ($n = 5$ per group). *$P < 0.01$ versus non-Adv EPCs or Ad5/EGFP EPCs. *NS* not significant versus non-Adv-EPCs. **b** β2AR protein levels were significantly higher in the Ad5/β2AR EPC group than in the non-Adv EPC or Ad5/EGFP-EPC group. *$P < 0.01$ versus non-Adv EPCs or Ad5/EGFP EPCs. *NS* not significant versus non-Adv-EPCs ($n = 5$ per group). *Ad5* adenovirus serotype 5, *EGFP* enhanced green fluorescent protein, *EPC* endothelial progenitor cell, *β2AR* β2 adrenergic receptor

The β2AR/Akt/eNOS pathway improves EPC function in vitro and the endothelial repair capacity of EPCs in vivo

The activations of Akt and eNOS are both important to determining the number and function of active EPCs. The activation of βARs in cardiac progenitor cells (CPCs) was previously shown to promote the proliferation and survival of CPCs and to be associated with the increased phosphorylation of Akt and eNOS. Therefore, we examined whether the β2AR/Akt/eNOS signaling pathway plays an important role in the proliferation, NO production, migration, and adhesion of EPCs. We used inhibitors, including ICI118,551 (a selective inhibitor of β2AR), LY294002 (an Akt inhibitor), and L-NAME (an eNOS inhibitor), to inhibit the activation of the β2AR/Akt/eNOS signaling pathway.

Following stimulation with FENO (10^{-8} M) for 6 h, the phosphorylation levels of Akt and eNOS were enhanced in the Ad5/β2AR-transduced EPCs compared with those in the Ad5/EGFP-transduced EPCs and in the non-transduced EPCs. Furthermore, the increases in Akt and eNOS phosphorylation that were observed in the Ad5/β2AR-transduced EPCs were inhibited by pre-incubation for 1 h with 0.3 μM of ICI118,551, 10 μM LY294002 or 100 μM L-NAME (Fig. 5a, b). Interestingly, we also found that β2AR overexpression-mediated effects on proliferation, NO production, migration, and adhesion were all significantly inhibited by pretreatment with ICI118,151, LY294002, and L-NAME (Fig. 5c-f). These results indicate that the β2AR/Akt/eNOS signaling pathway is at least partially responsible for the function of EPCs in vitro. We also investigated whether the β2AR/Akt/eNOS signaling pathway is associated with the re-endothelialization capacity of EPCs in vivo. In agreement with the in vitro results, pretreatment with inhibitors of β2AR, Akt, and eNOS attenuated the enhanced re-endothelialization in the mice that were transplanted with Ad5/β2AR-transduced EPCs (Fig. 5g).

Discussion

The present study demonstrates that β2AR expression on EPCs is involved in the restoration of damaged endothelium. Ad5/β2AR gene transfer treatment enhanced β2AR expression in EPCs and increased the capacity of EPCs to migrate, adhere, proliferate, and secrete NO in vitro and the re-endothelialization capacity of EPCs in vivo. Moreover, the increase in re-endothelialization capacity of EPCs is closely correlated with upregulation of the β2AR/Akt/eNOS pathway. Our present study is the first to demonstrate that β2ARs play a crucial role in regulating the vascular endothelial repair function of EPCs. At the molecular level, these effects were found to be at least partially associated with β2AR/Akt/eNOS pathway activation.

Endothelial dysfunction is known to play pivotal roles in degenerative vascular disease. Restoration of endothelial integrity is an important technique that can be used to cure vascular disease. Since Asahara et al. [19] first cultivated EPCs, numerous experiments have shown that EPCs mediated endothelial restoration in many processes, suggesting that they may also be useful for treating vascular injury. The term EPCs has been applied interchangeably

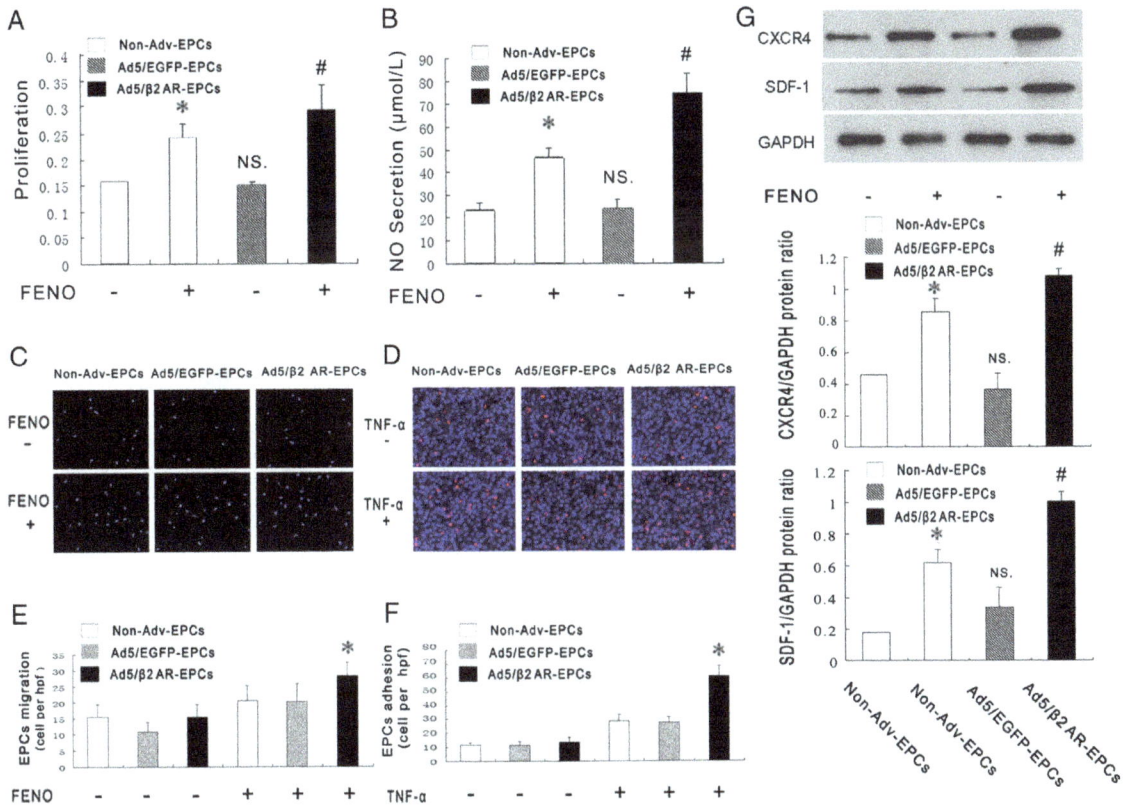

Fig. 3 β2AR gene transfer in EPCs upregulates cell proliferation, migration, and adhesion; NO production; and SDF-1/CXCR4 expression in vitro.
a Proliferation was measured in cultured EPCs by using CCK8 assays after 24 h of stimulation with FENO (10^{-8} M). The rate of proliferation in the Ad5/β2AR EPC group was significantly higher than the rate in the non-Adv-EPC, non-Adv- EPC + FENO, or Ad5/EGFP-EPC groups (0.29 ± 0.05, 0.16 ± 0.00, 0.24 ± 0.03, and 0.15 ± 0.01, respectively; *$P < 0.01$ versus non-Adv EPCs or Ad5/EGFP-EPCs, #$P < 0.01$ versus non-Adv-EPCs, non-Adv-EPCs + FENO, or Ad5/EGFP-EPCs; $n = 5$ per group). **b** NO secretion was assayed in EPCs by using the nitrate reductase method after β2AR gene transfer. The production of NO was significantly higher in the Ad5/β2AR-EPCs than in the non-Adv-EPCs, non-AdvEPCs + FENO, or Ad5/EGFP-EPCs (74.42 ± 8.68, 24.01 ± 2.68, 46.77 ± 4.08, and 24.24 ± 3.71, respectively; *$P < 0.01$ versus non-Adv-EPCs or Ad5/EGFP-EPCs, and #$P < 0.01$ versus non-Adv-EPCs, non-Adv-EPCs + FENO, or Ad5/EGFP-EPCs; $n = 5$ per group). **c** and **e** Quantitative analyses (**c**) and representative photographs (**e**) showing the migratory activity observed in EPCs (*$P < 0.01$ versus non-Adv-EPCs + FENO or Ad5/EGFP EPCs + FENO; $n = 5$ per group). **d** and **f** Quantitative analyses (**d**) and representative photographs (**f**) of CM-Dil-labeled EPCs that adhered to human umbilical vein endothelial cells with or without stimulation with TNF-α (1 μg/ml; *$P < 0.01$ versus non-Adv-EPCs + TNF-α or Ad5/EGFP-EPCs + TNF-α; $n = 5$ per group). **g** Overexpression of β2AR increased the expression of SDF-1 and CXCR4 in EPCs. Representative photograph and quantification analysis for SDF-1 and CXCR4 expression in EPCs (*$P < 0.01$ versus non-Adv EPCs without FENO activation, #$P < 0.01$ versus non-Adv-EPCs with FENO activation, and NS versus non-Adv-EPCs without FENO activation, $n = 5$ per group). Ad5 adenovirus serotype 5, Dil-ac-LDL Dil-labeled acetylated low-density lipoprotein, EGFP enhanced green fluorescent protein, EPC endothelial progenitor cell, FENO fenoterol, NO nitric oxide, NS not significant, SDF-1a stromal cell-derived factor 1a, TNF-α tumor necrosis factor-α, β2AR β2 adrenergic receptor

to a variety of cell populations by different investigators, suggesting that EPCs are not a single type of cell population. The definition of EPCs remains controversial. Many studies have attempted to identify cell surface markers that are unique to EPCs and to distinguish them from mature ECs and from myeloid-monocytic cells; however, these attempts have met with little success. Broadly speaking, two approaches to identify EPCs have predominated: (1) identification of cells bearing surface markers that indicate both cellular naïveté and endothelial origin and (2) inference of the presence of

endothelial precursors with a given cell population by the identification of cells bearing mature endothelial characteristics after a period of culture under angiogenic conditions [20]. In our experiment, according to the method of Hur et al. [14], isolated MNCs were resuspended by using the EGM-2 BulletKit system. After 5 to 7 days, attached cells were elongated and had a spindle shape. The recognition of cell surface antigen markers by flow cytometric analysis has confirmed the cells as early EPCs, consistent with the results of other studies [12, 21].

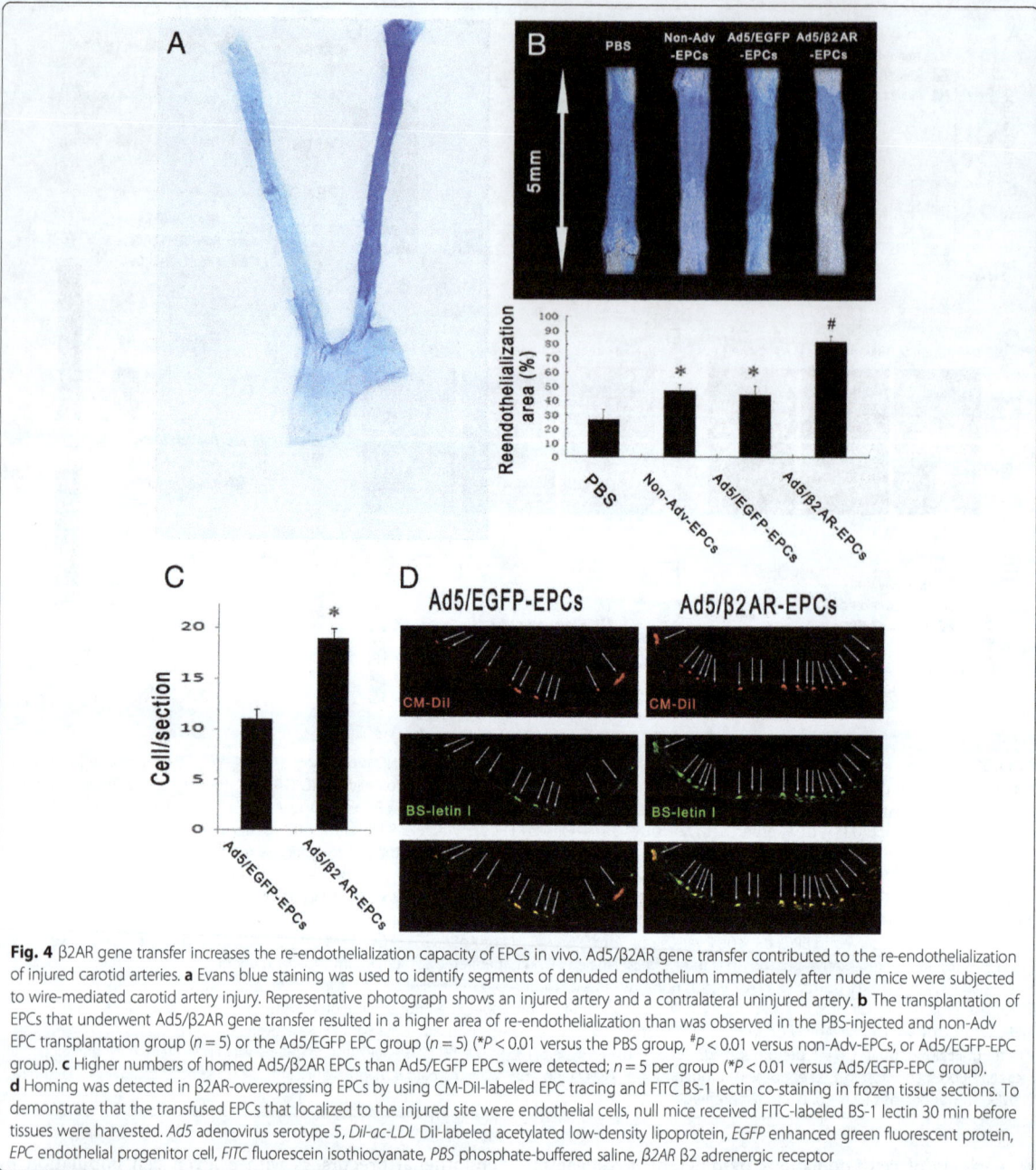

Fig. 4 β2AR gene transfer increases the re-endothelialization capacity of EPCs in vivo. Ad5/β2AR gene transfer contributed to the re-endothelialization of injured carotid arteries. **a** Evans blue staining was used to identify segments of denuded endothelium immediately after nude mice were subjected to wire-mediated carotid artery injury. Representative photograph shows an injured artery and a contralateral uninjured artery. **b** The transplantation of EPCs that underwent Ad5/β2AR gene transfer resulted in a higher area of re-endothelialization than was observed in the PBS-injected and non-Adv EPC transplantation group ($n = 5$) or the Ad5/EGFP EPC group ($n = 5$) (*$P < 0.01$ versus the PBS group, #$P < 0.01$ versus non-Adv-EPCs, or Ad5/EGFP-EPC group). **c** Higher numbers of homed Ad5/β2AR EPCs than Ad5/EGFP EPCs were detected; $n = 5$ per group (*$P < 0.01$ versus Ad5/EGFP-EPC group). **d** Homing was detected in β2AR-overexpressing EPCs by using CM-DiI-labeled EPC tracing and FITC BS-1 lectin co-staining in frozen tissue sections. To demonstrate that the transfused EPCs that localized to the injured site were endothelial cells, null mice received FITC-labeled BS-1 lectin 30 min before tissues were harvested. *Ad5* adenovirus serotype 5, *DiI-ac-LDL* DiI-labeled acetylated low-density lipoprotein, *EGFP* enhanced green fluorescent protein, *EPC* endothelial progenitor cell, *FITC* fluorescein isothiocyanate, *PBS* phosphate-buffered saline, *β2AR* β2 adrenergic receptor

Maintenance of the normal number and function of EPCs in the systemic circulation is now known to be an important novel endogenous vascular repair factor. In animal and clinical research, transplanted EPCs can home to impaired arteries, promoting re-endothelialization and reducing neointima formation. Various molecules may be involved in the process of EPC homing. EPCs and ECs have previously been shown to express both β1ARs and β2ARs [8, 9]. β2ARs are G protein-coupled receptors

that can induce cell proliferation and promote cell survival in many tissues, and stimulation of endothelial β2ARs has been shown to activate eNOS and release of NO in human umbilical vein endothelium [22]. Previous studies have shown that EPCs and ECs harvested from β2AR-knockout (β2AR-KO) mice are impaired in their abilities to migrate and stimulate network formation on Matrigel in vitro [8, 9]. Moreover, when β2AR-KO ECs and EPCs were injected

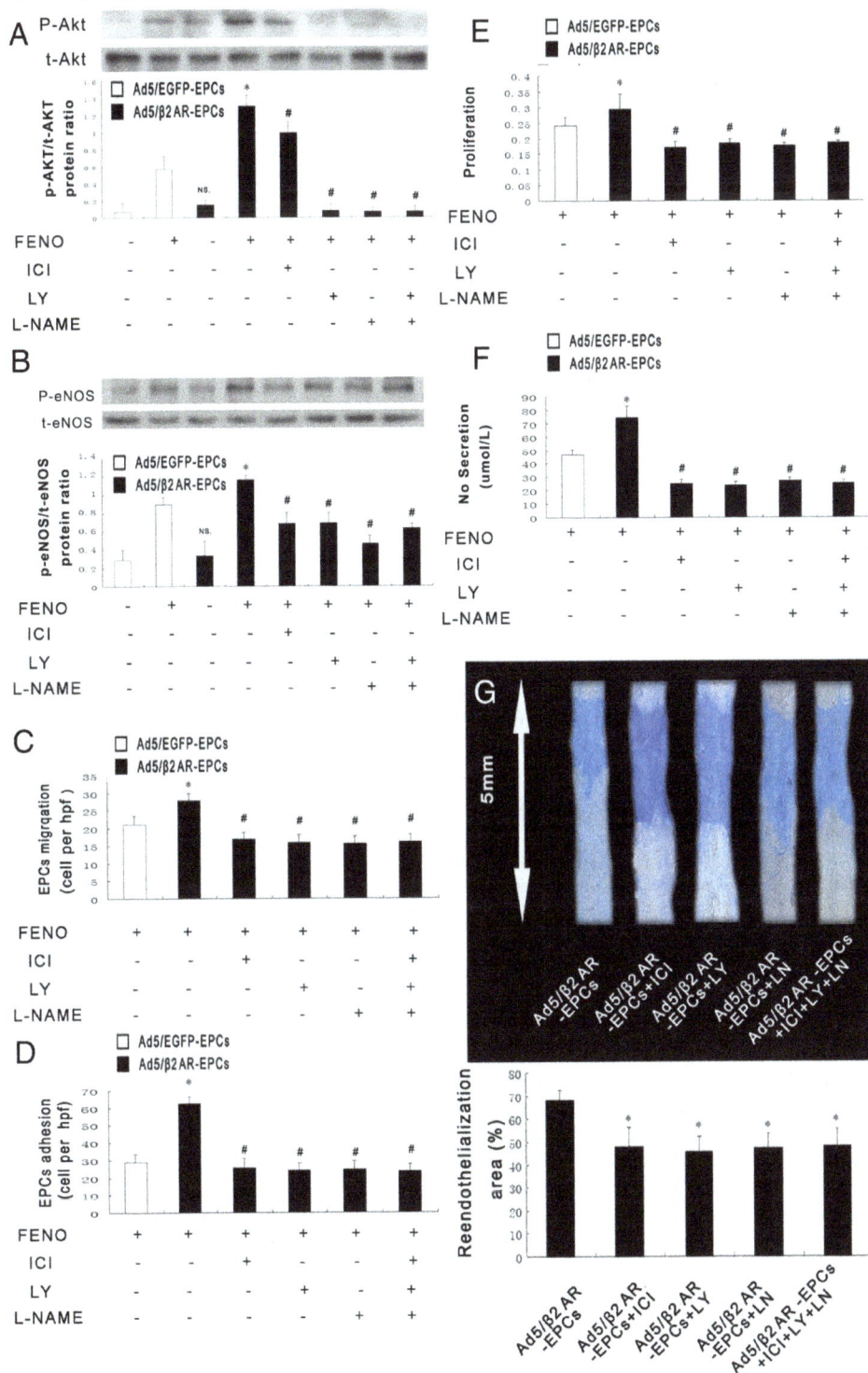

Fig. 5 (See legend on next page.)

(See figure on previous page.)
Fig. 5 The β2AR/Akt/eNOS pathway improves EPC function in vitro and the endothelial repair capacity of EPCs in vivo. **a** Representative photograph and quantification analysis of Akt phosphorylation in EPCs (*$P < 0.01$ versus Ad5/EGFP-EPCs with FENO activation, #$P < 0.05$ versus Ad5/β2AR-EPCs with FENO activation, and NS versus Ad5/EGFP-EPCs; $n = 5$ per group). **b** Representative photograph and quantification analysis of eNOS phosphorylation in EPCs (*$P < 0.01$ versus Ad5/EGFP-EPCs with FENO activation, #$P < 0.01$ versus Ad5/β2AR-EPCs with FENO activation, and NS versus Ad5/EGFP-EPCs; $n = 5$ per group). **c** Quantification analysis of migration (*$P < 0.01$ versus Ad5/EGFP-EPCs with FENO activation and #$P < 0.01$ versus Ad5/β2AR-EPCs with FENO activation; $n = 5$ per group). **d** Quantification analysis of adhesion (*$P < 0.05$ versus Ad5/EGFP-EPCs with FENO activation and #$P < 0.05$ versus Ad5/β2AR-EPCs with FENO activation; $n = 5$ per group). **e** Quantification analysis of proliferation (*$P < 0.05$ versus Ad5/EGFP-EPCs with FENO activation and #$P < 0.05$ versus Ad5/β2AR-EPCs with FENO activation; $n = 5$ per group). **f** Quantification analysis of NO production (*$P < 0.05$ versus Ad5/EGFP-EPCs with FENO activation and #$P < 0.05$ versus Ad5/β2AR-EPCs with FENO activation; $n = 5$ per group). **g** The re-endothelialized area at day 3 after carotid injury in nude mice with transplantation of Ad5/β2AR-EPCs that were pre-treated with β2AR or Akt or eNOS inhibitors (*$P < 0.01$ versus Ad5/β2AR-EPCs; $n = 5$ per group). *Ad5* adenovirus serotype 5, *EGFP* enhanced green fluorescent protein, *eNOS* endothelial nitric oxide synthase, *EPC* endothelial progenitor cell, *FENO* fenoterol, *NO* nitric oxide, *NS* not significant, *β2AR* β2 adrenergic receptor

into ischemic hind limbs, no significant amelioration of neovascularization was noted [8, 9]. However, whether β2ARs mediate the capacity of EPCs to re-endothelialize damaged arteries following an injury has never previously been tested. Given the results described in prior reports, we hypothesized that β2AR overexpression in human EPCs would result in the functional enhancement of EPCs in vitro and that β2AR gene upregulation would contribute to the restoration of endothelial injury in vivo.

To address these assumptions, we first investigated the effect of β2AR overexpression on the proliferation, migration, adhesion, and NO secretion of EPCs. We found that when EPCs were stimulated with FENO, the rates of proliferation and migration were significantly higher in EPCs that overexpressed β2ARs, indicating that FENO-induced functions were enhanced by β2AR overexpression in EPCs. Furthermore, our results showed that TNF-α-induced adhesion and NO secretion were enhanced by β2AR overexpression in EPCs. Given the close association between β2AR and EPC function, deeper insight into the contribution of β2AR expression in EPCs to accelerated re-endothelialization may be of clinical importance for the treatment of CVDs. We investigated the effect of infusing cultured PB-derived EPCs in a previously described nude mouse model of carotid artery injury. As shown in Fig. 4b, β2AR overexpression in the infused EPCs resulted in a significantly larger area of re-endothelialization, suggesting that β2AR signaling might be an important molecular mechanism that contributes to enhanced re-endothelialization by EPCs. Notably, the SDF-1α/CXCR4 interaction plays an important role in the regulation of a variety of cellular functions, including cell migration, proliferation, survival, and angiogenesis. Previous studies have demonstrated that the SDF-1a/CXCR4 axis is crucial for the therapeutic integrity of EPCs and their ability to home to injured vessels. For example, investigators showed that SDF-1a was involved in ischemia-mediated mobilization and homing in EPCs after vascular injury [18]. It has also been reported that CXCR4 expression restored the functions of EPCs in hypertensive patients [23]. Our data showed that stimulation with β2AR promoted EPC function and significantly increased SDF-1a and CXCR4 expression in cultured EPCs, especially in EPCs that overexpressed β2AR. These data collectively suggest that β2AR is an important potential therapeutic target and that β2AR-mediated EPC functions are dependent, at least in part, on the production of SDF-1a/CXCR4 and may otherwise be equal to the SDF/CXCR4 axis in the biology of EPCs. Therefore, our results indicate that transplantation of EPCs that have enhanced β2AR expression could be used as a novel therapy for treating vascular endothelial injuries.

In regard to how β2AR gene transfer modulates re-endothelialization, several lines of evidence have shown that activation of β2ARs promoted the proliferation and survival of cardiac progenitor cells in association with increased eNOS phosphorylation [24, 25]. Akt, a multifunctional regulator of cell survival, is a downstream effector of phosphatidylinositol-3 kinase (PI3K). It is widely accepted that activating Akt phosphorylation stimulates eNOS phosphorylation at Ser1177 and increases endothelial NO production, which is involved in the mobilization of stem and progenitor cells [26]. Our current data demonstrate that this signal paradigm also exists in EPCs because β2AR stimulation increased the phosphorylation of Akt and eNOS, which is consistent with increased EPC proliferation, migration, and adhesion function in vitro and with accelerated re-endothelialization in vivo. These cellular effects are blocked by β2AR/Akt/eNOS inhibition. Furthermore, we upregulated the expression of β2AR via Ad5/β2AR gene transfer and found that the increased re-endothelialization capacity that was induced in the cultured EPCs that overexpressed β2AR was blocked by the β2AR inhibitor ICI111, 181; the Akt inhibitor LY294002; and the eNOS inhibitor L-NAME. These data strongly support the hypothesis that β2AR plays a crucial role in the regulation of vascular repairs by circulating EPCs. We believe that this study is also the first to provide data indicating that mechanisms involving β2AR signaling underlie the functions of EPCs in vitro and their capacity to re-endothelialize injuries in vivo.

The present study has some limitations. First, although we used a β2AR inhibitor to explore the important functions of EPCs in the vitro and in vivo experiments, we did not focus on exploring the functional impairments that were observed in the β2AR-KO EPCs during homing to impaired arteries. Second, β2AR gene transfer results in accelerated re-endothelialization after vascular injury, but the adenovirus-mediated β2AR gene overexpression should be verified to determine its clinical safety. Finally, we used a relatively simple model in nude mice to demonstrate that β2AR gene transfer resulted in EPCs with enhanced repair capacities. Whether the β2AR/Akt/eNOS signaling pathway contributes more or less to EPCs to mediate endothelial repair relative to the contributions of other signaling pathways needs to be further investigated.

Conclusions

Our study is the first to provide direct evidence that β2AR expression in EPCs is an important molecular target for therapeutic studies and that β2AR activation may be responsible for accelerated EPCs homing during injury. More importantly, our study demonstrates that overexpressing the β2AR gene in EPCs results in rapid re-endothelialization and largely improved post-injury vascular repairs and that these processes are mediated by the β2AR/Akt/eNOS signaling pathway. Stem cell therapies have shown great potential as strategies aimed at increasing the efficiency of vascular regeneration; therefore, β2AR gene may become a novel therapeutic molecular target in clinical studies aimed at improving cardiovascular care. These scientific questions deserve further investigation.

Abbreviations

Ad5: adenovirus serotype 5; BSA: bovine serum albumin; CVD: cardiovascular disease; DAPI: 4′,6-diamidino-2-phenylindole; DiI-ac-LDL: DiI-labeled acetylated low-density lipoprotein; EBM-2: endothelial basal medium-2; EC: endothelial cell; EGFP: enhanced green fluorescent protein; eNOS: endothelial nitric oxide synthase; EPC: endothelial progenitor cell; FBS: fetal bovine serum; FENO: fenoterol; FITC: fluorescein isothiocyanate; GAPDH: glyceraldehyde 3-phosphate dehydrogenase; HUVEC: human umbilical vein endothelial cell; ICI: ICI118,551 (a selective inhibitor of β2AR); KDR: kinase-insert domain receptor; KO: knockout; LY: LY294002 (an Akt inhibitor); mAb: monoclonal antibody; MNC: mononuclear cell; MOI: multiple of infection; NO: nitric oxide; PB: peripheral blood; PBS: phosphate-buffered saline; PCR: polymerase chain reaction; PE: phycoerythrin; PFA: paraformaldehyde; SDF-1a: stromal cell-derived factor 1a; TNF-α: tumor necrosis factor-α; VEGF: vascular endothelial growth factor; vWF: von Willebrand factor; β2AR: β2 adrenergic receptor.

Competing interests

The authors declare that they have no competing interests.

Authors' contributions

XK and R-QN designed and conducted experiments, analyzed the data, and compiled the main manuscript. X-RS and FW conducted immunohistochemical experiments and evaluated enzyme activities. Q-SH and B-QD conducted animal experiments. J-FW conducted physiological measurements and prepared samples. All authors read and approved the final manuscript.

Acknowledgements
This study was supported by a grant from the National Natural Science Foundation of China (NO.81370309).

Author details
[1]Department of Cardiology, Sun Yat-sen Memorial Hospital of Sun Yat-sen University, No. 107, Yanjiangxi Road, Guangzhou, China. [2]Guangdong Province Key Laboratory of Arrhythmia and Electrophysiology, Guangzhou 510120, China. [3]Department of Geriatric, The First Affiliated Hospital of Sun Yat-sen University, Guangzhou 510080, China.

References

1. Sen T, Aksu T. Endothelial progenitor cell and adhesion molecules determine the quality of the coronary collateral circulation/endothelial progenitor cells (CD34 + KDR+) and monocytes may provide the development of good coronary collaterals despite the vascular risk factors and extensive atherosclerosis. Anadolu Kardiyol Derg. 2012;12:447–8.
2. Mannarino E, Pirro M. Endothelial injury and repair: a novel theory for atherosclerosis. Angiology. 2008;59:69S–72.
3. Inoue T, Croce K, Morooka T, Sakuma M, Node K, Simon DI. Vascular inflammation and repair: implications for re-endothelialization, restenosis, and stent thrombosis. JACC Cardiovasc Interv. 2011;4:1057–66.
4. Yin Y, Liu H, Wang F, Li L, Deng M, Huang L, et al. Transplantation of cryopreserved human umbilical cord blood-derived endothelial progenitor cells induces recovery of carotid artery injury in nude rats. Stem Cell Res Ther. 2015;6:37.
5. Zhu S, Malhotra A, Zhang L, Deng S, Zhang T, Freedman NJ, et al. Human umbilical cord blood endothelial progenitor cells decrease vein graft neointimal hyperplasia in SCID mice. Atherosclerosis. 2010;212:63–9.
6. Li X, Chen C, Wei L, Li Q, Niu X, Xu Y, et al. Exosomes derived from endothelial progenitor cells attenuate vascular repair and accelerate reendothelialization by enhancing endothelial function. Cytotherapy. 2016;18:253–62.
7. Alexandru N, Popov D, Dragan E, Andrei E, Georgescu A. Circulating endothelial progenitor cell and platelet microparticle impact on platelet activation in hypertension associated with hypercholesterolemia. PLoS One. 2013;8:e52058.
8. Iaccarino G, Ciccarelli M, Sorriento D, Galasso G, Campanile A, Santulli G, et al. Ischemic neoangiogenesis enhanced by beta2-adrenergic receptor overexpression: a novel role for the endothelial adrenergic system. Circ Res. 2005;97:1182–9.
9. Galasso G, De Rosa R, Ciccarelli M, Sorriento D, Del GC, Strisciuglio T, et al. β2-adrenergic receptor stimulation improves endothelial progenitor cell-mediated ischemic neoangiogenesis. Circ Res. 2013;112:1026–34.
10. Hill JM, Zalos G, Halcox JP, Schenke WH, Waclawiw MA, Quyyumi AA, et al. Circulating endothelial progenitor cells, vascular function, and cardiovascular risk. N Engl J Med. 2003;348:593–600.
11. Hristov M, Erl W, Weber PC. Endothelial progenitor cells: mobilization, differentiation, and homing. Arterioscler Thromb Vasc Biol. 2003;23:1185–9.
12. Chen L, Wu F, Xia WH, Zhang YY, Xu SY, Cheng F, et al. CXCR4 gene transfer contributes to in vivo reendothelialization capacity of endothelial progenitor cells. Cardiovasc Res. 2010;88:462–70.
13. Zhang XY, Su C, Cao Z, Xu SY, Xia WH, Xie WL, et al. CXCR7 upregulation is required for early endothelial progenitor cell-mediated endothelial repair in patients with hypertension. Hypertension. 2014;63:383–9.
14. Hur J, Yoon CH, Kim HS, Choi JH, Kang HJ, Hwang KK, et al. Characterization of two types of endothelial progenitor cells and their different contributions to neovasculogenesis. Arterioscler Thromb Vasc Biol. 2004;24:288–93.
15. Giannotti G, Doerries C, Mocharla PS, Mueller MF, Bahlmann FH, Horváth T, et al. Impaired endothelial repair capacity of early endothelial progenitor cells in prehypertension: relation to endothelial dysfunction. Hypertens. 2010;55:1389–97.
16. Nishiwaki Y, Yoshida M, Iwaguro H, Masuda H, Nitta N, Asahara T, et al. Endothelial E-selectin potentiates neovascularization via endothelial progenitor cell-dependent and -independent mechanisms. Arterioscler Thromb Vasc Biol. 2007;27:512–8.
17. Walter DH, Haendeler J, Reinhold J, Rochwalsky U, Seeger F, Honold J, et al. Impaired CXCR4 signaling contributes to the reduced neovascularization capacity of endothelial progenitor cells from patients with coronary artery disease. Circ Res. 2005;97:1142–51.

18. Yin Y, Zhao X, Fang Y, Yu S, Zhao J, Song M, et al. SDF-1alpha involved in mobilization and recruitment of endothelial progenitor cells after arterial injury in mice. Cardiovasc Pathol. 2010;19:218–27.

19. Asahara T, Murohara T, Sullivan A, Silver M, van der Zee R, Li T, et al. Isolation of putative progenitor endothelial cells for angiogenesis. Science. 1997;275:964–7.

20. Padfield GJ, Newby DE, Mills NL. Understanding the role of endothelial progenitor cells in percutaneous coronary intervention. J Am Coll Cardiol. 2010;55:1553–65.

21. Xia WH, Yang Z, Xu SY, Chen L, Zhang XY, Li J, et al. Age-related decline in reendothelialization capacity of human endothelial progenitor cells is restored by shear stress. Hypertension. 2012;59:1225–31.

22. Xu B, Li J, Gao L, Ferro A. Nitric oxide-dependent vasodilatation of rabbit femoral artery by beta(2)-adrenergic stimulation or cyclic AMP elevation in vivo. Br J Pharmacol. 2000;129:969–74.

23. Liu X, Zhang GX, Zhang XY, Xia WH, Yang Z, Su C, et al. Lacidipine improves endothelial repair capacity of endothelial progenitor cells from patients with essential hypertension. Int J Cardiol. 2013;168:3317–26.

24. Duda DG, Fukumura D, Jain RK. Role of eNOS in neovascularization: NO for endothelial progenitor cells. Trends Mol Med. 2004;10:143–5.

25. Khan M, Mohsin S, Avitabile D, Siddiqi S, Nguyen J, Wallach K, et al. β-adrenergic regulation of cardiac progenitor cell death versus survival and proliferation. Circ Res. 2013;112:476–86.

26. Aicher A, Heeschen C, Mildner-Rihm C, Urbich C, Ihling C, Technau-Ihling K, et al. Essential role of endothelial nitric oxide synthase for mobilization of stem and progenitor cells. Nat Med. 2003;9:1370–6.

13

LL-37 stimulates the functions of adipose-derived stromal/stem cells via early growth response 1 and the MAPK pathway

Yoolhee Yang[1], Hyunju Choi[1], Mira Seon[1], Daeho Cho[2*†] and Sa Ik Bang[1,3*†]

Abstract

Background: LL-37 is a naturally occurring antimicrobial peptide found in the wound bed and assists wound repair. No published study has characterized the role of LL-37 in the function(s) of human mesenchymal stem cells (MSCs). This study investigated the functions of adipose-derived stromal/stem cells (ASCs) activated by LL-37 by performing both in vitro assays with cultured cells and in vivo assays with C57BL/6 mice with hair loss.

Methods: Human ASCs were isolated from healthy donors with written informed consent. To examine the effects of LL-37 on ASC function, cell proliferation and migration were measured by a cell counting kit (CCK-8) and a Transwell migration assay. Early growth response 1 (EGR1) mRNA expression was determined by microarray and real-time PCR analyses. The protein levels of EGR1 and regenerative factors were analyzed by specific enzyme-linked immunosorbent assays and western blotting.

Results: LL-37 treatment enhanced the proliferation and migration of human ASCs expressing formyl peptide receptor like-1. Microarray and real-time PCR data showed that EGR1 expression was rapidly and significantly increased by LL-37 treatment. LL-37 treatment also enhanced the production of EGR1. Moreover, small interfering RNA-mediated knockdown of EGR1 inhibited LL-37-enhanced ASC proliferation and migration. Activation of mitogen-activated protein kinases (MAPKs) was essential not only for LL-37-enhanced ASC proliferation and migration but also EGR1 expression; treatment with a specific inhibitor of extracellular signal-regulated kinase, p38, or c-Jun N-terminal kinase blocked the stimulatory effect of LL-37. EGR1 has a strong paracrine capability and can influence angiogenic factors in ASCs; therefore, we evaluated the secretion levels of vascular endothelial growth factor, thymosin beta-4, monocyte chemoattractant protein-1, and stromal cell-derived factor-1. LL-37 treatment increased the secretion of these regenerative factors. Moreover, treatment with the conditioned medium of ASCs pre-activated with LL-37 strongly promoted hair growth in vivo.

Conclusions: These findings show that LL-37 increases EGR1 expression and MAPK activation, and that preconditioning of ASCs with LL-37 has a strong potential to promote hair growth in vivo. This study correlates LL-37 with MSC functions (specifically those of ASCs), including cell expansion, cell migration, and paracrine actions, which may be useful in terms of implantation for tissue regeneration.

Keywords: LL-37, Mesenchymal stem cells, Adipose-derived stromal/stem cells, Early growth response 1, Cell migration, Proliferation, Paracrine actions, MAPK pathway, Hair growth, Regeneration

* Correspondence: cdhkor@sookmyung.ac.kr; si55.bang@samsung.com
†Equal contributors
2Department of Life Science, Sookmyung Women's University, Seoul, Korea
1Department of Plastic Surgery, Samsung Medical Center, Sungkyunkwan University School of Medicine, Seoul, Korea
Full list of author information is available at the end of the article

Background

Cell therapy using adult multipotent stromal cells or mesenchymal stem cells (MSCs) is clinically used to repair and regenerate various damaged tissues [1, 2]. Adipose-derived stromal/stem cells (ASCs) are multipotent mesenchymal cells isolated from adipose tissue and are accessible, abundant, and self-replenishing [3]. These cells are also capable of differentiating into multiple mesenchymal lineages, including osteoblasts, adipocytes, chondrocytes, and other types of cells. The differentiation, proliferation, and direct migration of MSCs into local damaged tissue undergoing regeneration are regarded as the primary mechanisms underlying the actions of MSCs [3, 4]. Besides these actions, the strong paracrine effects of various growth factors and cytokines secreted by MSCs is a key mechanism underlying MSC-mediated tissue regeneration and repair [5, 6]. Therefore, it is important to understand the molecular mechanism controlling the machinery that underlies these strong paracrine effects in MSCs.

Early growth response 1 (EGR1), a member of the immediate-early gene family, encodes a zinc finger transcription factor and is rapidly induced by mitogens and growth factors [7]. Once induced, EGR1 plays a pivotal role in the expression and production of various growth factors, cytokines, and other bioactive molecules, resulting in the stimulation of cellular functions for tissue repair and regeneration. Interestingly, EGR1 might be an activation marker of human MSCs, which highly express the *EGR1* gene, and have multiple roles involving angiogenesis and mitogenesis [8, 9]. Strong induction of EGR1 is mediated by the mitogen-activated protein kinase (MAPK) pathway, a crucial signaling pathway associated with cell migration and proliferation [7, 10].

LL-37 is a naturally occurring 37-amino acid sequence synthesized from the C-terminus of human cationic antimicrobial protein 18 (hCAP-18) [11] and is widely found in various body fluids and cell types including epithelial cells and immune cells [12, 13]. Secretion of LL-37/hCAP-18 is significantly elevated at the wound bed, where this peptide demonstrates proliferative, angiogenic, and immunomodulatory activities through the MAPK pathway [14]. Besides participating in innate host defense [11, 15], this peptide also has wound-healing effects [16, 17] and is a potent chemoattractant for various cell types including immune cells through activation of formyl peptide receptor like-1 (FPRL1), its main receptor [18]. A recent study by Krasnodembskaya et al. showed that human MSCs possess direct antimicrobial activity, which is mediated in part by secretion of the human cathelicidin hCAP-18/LL-37 [19].

Many studies reported ASC-mediated tissue regeneration in various damaged tissues [1, 20] and LL-37 is an important mediator of the repair and regeneration of wounds,

bones, islets, and other damaged tissues [16, 21, 22]. However, the precise effect of LL-37 on adjacent human ASCs has not been identified. In the present study, we hypothesized that LL-37 enhances their therapeutic potential by activating ASCs via EGR1 and MAPK signaling. Our findings indicate that LL-37 may be used as a preconditioning agent before ASC transplantation for tissue regeneration.

Methods

Cell culture

Subcutaneous adipose tissue was obtained during elective surgeries with written informed consent, as approved by the Samsung Medical Center Institutional Review Board. All donors were < 40 years old and did not have diabetes or acute inflammation. The mean body mass index of the donors was 25.2 ± 3.64. Human ASCs were isolated according to a previous protocol [23] and cultured in low-glucose Dulbecco's modified Eagle's medium supplemented with 10 % fetal bovine serum, 100 U/mL penicillin, and 100 µg/mL streptomycin at 37 °C in a humidified atmosphere containing 5 % CO_2. ASCs were characterized by the presence of the cell surface markers CD73, CD90, and CD105 and the absence of CD11b, CD34, CD45, and HLA-DR [24].

Cell viability and proliferation assays

Cells were treated with human LL-37 (Phoenix Pharmaceuticals, USA) for 48 hr under serum deprivation conditions. Cell viability was determined by Trypan blue staining. Cell proliferation was measured with the cell counting kit (CCK)-8 according to the manufacturer's protocol (Dojindo, Japan). ASCs (5×10^3 cells/well) were treated with 2.5–20 µg/mL LL-37 for 24 and 48 hr prior to adding CCK-8 solution. Absorbance at 450 nm was determined with a multi-plate reader (Molecular Devices, CA, USA).

Migration assay

A cell migration assay was performed using Transwell plates (8 µm pore size; Costar, Corning, NY, USA) according to a previous study [25]. Briefly, ASCs were suspended in serum-free medium and 100 µL of the cell suspension (7×10^5 cells/mL) was added to each upper well. LL-37 at the indicated concentrations (5, 10, and 20 µg/mL) was placed in the lower wells of a 24-well tissue culture plate. After incubation for 6 hr at 37 °C, cells that had migrated were stained with 0.15 % crystal violet and counted in five random microscopy fields using the Scanscope scanning system (Aperio Scanscope, CA, USA).

Small interfering RNA transfection

Cells were transfected with EGR1-targeting small interfering RNA (siRNA) (Santa Cruz Biotechnology, USA) or negative control siRNA (Bioneer, Daejeon, Korea) using Lipofectamine RNAi (Invitrogen, CA, USA). Briefly, when cells reached 60–70 % confluency, siRNA (final concentration, 100 nM) was combined with Lipofectamine RNAi and allowed to complex for 20 min. The transfection mixture was then applied to ASCs and incubated for 6 hr at 37 °C. Subsequently, cells were maintained in complete medium for 36 hr before being subjected to the cell migration assay and enzyme-linked immunosorbent assays (ELISAs).

Western blot analysis

Total protein from cell lysates was separated by SDS-PAGE and transferred to a nitrocellulose membrane, which was incubated with the corresponding primary antibodies (anti-EGR1 (Santa Cruz Biotechnology) and anti-GAPDH (Cell Signaling, USA)) overnight at 4 °C. Washed membranes were incubated for 1 hr with a horseradish peroxidase-conjugated anti-mouse secondary antibody. Bands were visualized using an enhanced chemiluminescence detection system (Amersham Biosciences, Piscataway, NJ, USA). The band intensities were quantified using TotalLab software (UK).

Fluorescence-activated cell sorting

FPRL1 expression in ASCs from four donors was evaluated by surface staining. Cells were washed with fluorescence-activated cell sorting (FACS) buffer, stained with a mouse anti-FPRL1 antibody (R&D Systems, USA), and then stained with anti-mouse IgG-FITC (Sigma, USA). Labeled cells were measured with a FACSCalibur instrument (Becton Dickinson Biosciences, CA, USA) and analyzed using the Win MDI program (Win MDI version 2.8).

Immunostaining

Immunofluorescence analysis was performed as previously described [23]. Briefly, ASCs (5×10^3 cells/well) were seeded on four-well Lab-Tek II chamber slides (Nalge Nunc International, IL, USA), treated with LL-37 for 48 hr, fixed, and permeabilized for 20 min with 0.1 % Triton X-100 prepared in phosphate-buffered saline. After washing, cells were blocked and incubated with an anti-proliferating cell nuclear antigen (PCNA) antibody (Abcam, Cambridgeshire, UK) for 1 hr. Next, cells were incubated with goat anti-mouse IgG-Alexa Fluor 488 (Invitrogen) for 30 min at 37 °C. A nucleic acid dye (DAPI; 0.5 μg/mL) was added to stain the nuclei. PCNA immunofluorescence was detected with an LSM700 confocal microscope system (Carl Zeiss, NY, USA; 400× objective).

ELISA

Cells were plated in six-well plates (300,000 cells/well) and treated with 20 μg/mL LL-37 in serum-free medium. After 48 hr, the culture medium was collected and ELISAs were performed for thymosin beta-4 (TB4; Immune Diagnostics, Canada), vascular endothelial growth factor (VEGF; Invitrogen), monocyte chemoattractant protein-1 (MCP-1; eBioscience, USA), stromal cell-derived factor-1 (SDF-1; RayBiotech, USA), and EGR1 (EIAab, China), in accordance with the manufacturers' recommendations. For detection of EGR1, the cells were lysed by the addition of 200 μL of cell lysis buffer after LL-37 treatment for 6 hr.

Real-time PCR

Total RNA was extracted using TRIzol reagent (Invitrogen). After removing possible DNA contamination by DNAse treatment of the extracted RNA, cDNA was synthesized using 2 μg of total RNA and SuperScript II reverse transcriptase according to the manufacturer's instructions. The primers used are provided in Additional file 1: Table S1. For real-time PCR, quantitative PCR was performed using the 7900 Real-Time PCR System (Applied Biosystems, Foster City, CA, USA) and the Power SYBR Green qPCR Master Mix Kit (Life Technologies, CA, USA). The cycling profile for real-time PCR (50 cycles) was as follows: 95 °C for 10 min, 95 °C for 15 sec, and 60 °C for 60 sec. The comparative threshold cycle (Ct) method (i.e., $2^{-\Delta\Delta Ct}$) was used to calculate fold amplification.

Microarray

The microarray data are accessible in the Gene Expression Omnibus (GEO) database under accession number GSE76392. For the microarray, human ASCs were treated with 20 μg/mL LL-37 for 1 hr. Each total RNA sample (200 ng) was labeled and amplified using the Low Input Quick Amp Labeling Kit (Agilent Technologies, CA, USA). Cy3-labeled aRNAs were resuspended in 100 μL of hybridization solution (Agilent Technologies). Labeled aRNAs were placed on the Agilent SurePrint G3 Human GE 4x44K array (Agilent Technologies) and covered with a Gasket 8-plex slide (Agilent Technologies). Slides were hybridized for 17 hr at 65 °C. The hybridized slides were sequentially washed in 2× SSC containing 0.1 % SDS for 2 min, 1× SSC for 3 min, and 0.2× SSC for 2 min at room temperature. Finally, the slides were centrifuged at 3000 rpm for 20 sec to dry. Arrays were analyzed using an Agilent scanner with associated software. Gene expression levels were calculated with Feature Extraction v10.7.3.1 (Agilent Technologies). The relative signal intensities of each gene were generated using the Robust Multi-Array Average algorithm. Data were processed based on the quantile normalization method using GeneSpring GX 13.0 (Agilent Technologies). This normalization method aims to make the distribution of intensities for each array in a set of arrays the same. The normalized and log-transformed intensity values were analyzed using Gene-Spring GX 13.0. Fold-change filters included the requirement that the genes be present in at least 200 % of controls for upregulated genes and less than 50 % of controls for downregulated genes. Hierarchical cluster analysis was conducted using the Cluster 3.0 program, the Euclidean distance, and average linkage algorithm.

Animals and in vivo hair growth test

Five-week-old male C57BL/6 mice were purchased from SLC Inc. (Haruno, Japan) and allowed to adapt to their new environment for 2 weeks. The fur on the backs of 7-week-old mice ($n = 14$) was shaved with hair clippers and removed with hair removal cream. Conditioned medium (CM) was topically applied daily for up to 18 days. Pigment darkening and the hair growth rate were monitored every 3 days for 2 weeks. Hair growth was evaluated by three independent dermatological scientists. The hair growth score was determined using the method described by Vegesna et al. [26, 27]. The animal experiments were approved by the Institutional Animal Care and Use Committee of Samsung Medical Center (approval number: SMC-IACUC-2013-0103-003) and all experiments followed regulatory standards.

Plasmid construction and selection of stably transfected cell lines

hEGR1 was subcloned into the *Hind* III/*Xho* I sites of pcDNA3.1(+). After digestion of the pcDNA3.1(+)-hEGR1 plasmid with the restriction enzymes, the resulting *hEGR1* gene along with the *BamH*I and *Sfi*I sites was inserted into the multiple cloning site of the expression plasmid pLenti6/V5-D-TOPO (Invitrogen), which contains a cytomegalovirus promoter upstream of the inserted gene. The resulting plasmid was named pLenti6/V5-hEGR1.

The lentiviral expression system based on four plasmids was obtained from Invitrogen. Briefly, 2.5 µg of pLP1, 2.5 µg of pLP2, 2.5 µg of pLP/VSVG, and 2.5 µg of the lentiviral vector pLenti6/V5-hCAMP were co-transfected with Fugene6 (Roche) into HEK293T cells cultured in 10 cm plates. The medium was changed every 24 hr. After 48 hr, supernatants were pooled, filtered through a 0.45-µm filter, and centrifuged at 6000 rpm at 4 °C for 16 hr. The pellet was resuspended in 1 mL of complete medium and stored at −80 °C. The resulting recombinant lentivirus was added to ASCs cultured in 10 cm plates. At 48 hr after seeding, the medium was replaced with complete medium containing 10 µg/mL blasticidin as a selective agent. At 18 days after transfection, selected cell colonies were split into 60 mm petri dishes. Isolated cell colonies were cultured and expanded.

Statistical analysis

Statistical significance was estimated using the Student's *t*-test. Mean differences were considered to be significant when $P < 0.05$.

Results

LL-37 increases the proliferation and migration of human ASCs in a dose-dependent manner

LL-37 reportedly activates functions such as proliferation and migration through FPRL1, a G-protein-coupled receptor, in many cell types [18]; therefore, several donor pools of ASCs were examined for expression of FPRL1. Flow cytometric data confirmed that FPRL1 was expressed on ASCs in each donor pool (Fig. 1a).

To investigate the role of LL-37 in ASC proliferation, we examined the effect of LL-37 on the proliferative ability of ASCs using the CCK-8 assay. LL-37 treatment markedly stimulated ASC proliferation in a dose-dependent manner without causing cytotoxicity (Fig. 1b and Additional file 2: Figure S1a). Furthermore, we determined the number of proliferating cells by performing immunofluorescence staining of PCNA, a nuclear protein associated with cell proliferation. The proportion of ASCs with positively stained nuclei (green fluorescence) was markedly enhanced by LL-37 treatment (Fig. 1c).

To investigate the effects of LL-37 on ASC migration, a Transwell migration assay was performed. LL-37 treatment rapidly increased human ASC migration within 6 hr; this effect was dose-dependent, with maximal stimulation at 20 µg/mL among the concentrations tested (Fig. 1d). ASCs were pretreated with pertussis toxin (Ptx), a Gαi inhibitor, or a neutralizing anti-LL-37 antibody (αLL-37) before activation with LL-37. Ptx or αLL-37 treatment prior to LL-37 stimulation significantly inhibited ASC migration and proliferation (Fig. 1e and f). Taken together, these data suggest that LL-37 induces ASC proliferation and migration through the Gαi-coupled receptor FPRL1.

EGR1 is critical for LL-37-enhanced ASC migration and proliferation

We investigated the target genes of LL-37 that induce ASC proliferation and migration using the Agilent human 4x44K array, a human signaling pathway finder. Microarray analysis showed that LL-37 treatment significantly increased EGR1 expression by 18.47-fold (Table 1 and Additional file 3: Table S2). LL-37 treatment also increased the levels of several genes (including *EGR2*, early growth response 2; *KLF10*, Krupple-like factor 10; *FOS*; and *CTGF*, connective tissue growth factor) linked to various functions, such as the cell cycle, cell migration, cell proliferation, and transcription. To confirm the effect of LL-37 treatment on EGR1 expression in ASCs, cells were treated with 10 or 20 µg/mL LL-37 for

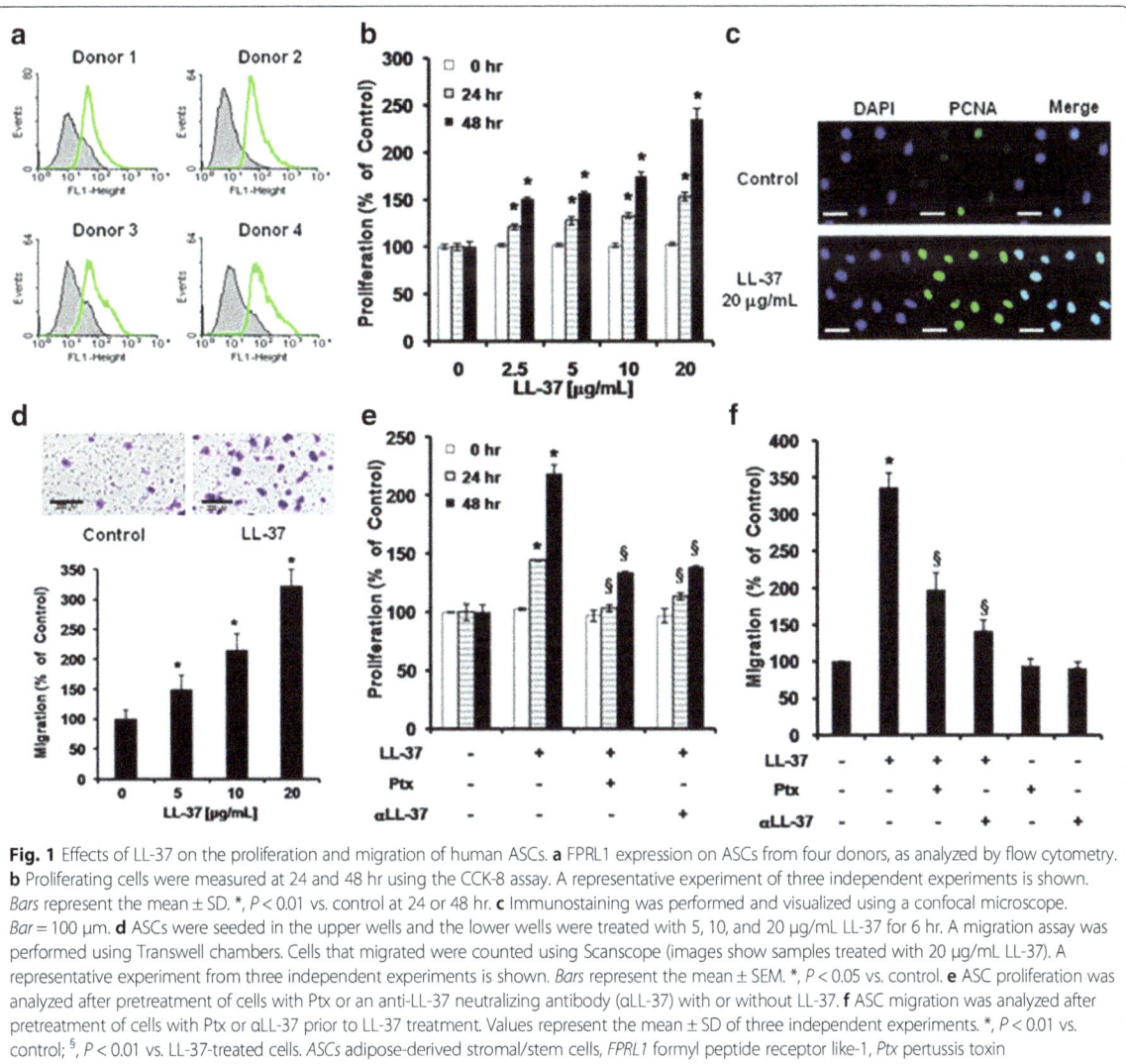

Fig. 1 Effects of LL-37 on the proliferation and migration of human ASCs. **a** FPRL1 expression on ASCs from four donors, as analyzed by flow cytometry. **b** Proliferating cells were measured at 24 and 48 hr using the CCK-8 assay. A representative experiment of three independent experiments is shown. *Bars* represent the mean ± SD. *, *P* < 0.01 vs. control at 24 or 48 hr. **c** Immunostaining was performed and visualized using a confocal microscope. *Bar* = 100 μm. **d** ASCs were seeded in the upper wells and the lower wells were treated with 5, 10, and 20 μg/mL LL-37 for 6 hr. A migration assay was performed using Transwell chambers. Cells that migrated were counted using Scanscope (images show samples treated with 20 μg/mL LL-37). A representative experiment from three independent experiments is shown. *, *P* < 0.01 vs. control. **e** ASC proliferation was analyzed after pretreatment of cells with Ptx or an anti-LL-37 neutralizing antibody (αLL-37) with or without LL-37. **f** ASC migration was analyzed after pretreatment of cells with Ptx or αLL-37 prior to LL-37 treatment. Values represent the mean ± SD of three independent experiments. *, *P* < 0.01 vs. control; §, *P* < 0.01 vs. LL-37-treated cells. *ASCs* adipose-derived stromal/stem cells, *FPRL1* formyl peptide receptor like-1, *Ptx* pertussis toxin

different amounts of time and then real-time PCR analysis was performed. LL-37 treatment significantly increased the EGR1 mRNA level, which peaked at 1 hr and returned to the basal level at about 6 hr (Fig. 2a). Because LL-37 treatment considerably increased the mRNA level of EGR1, we attempted to ascertain whether it also increased the protein level of EGR1 via western blot analysis. LL-37 treatment rapidly increased the protein level of EGR1 in ASCs in a time-dependent manner (Fig. 2b). In addition to western blot analysis, ELISA analysis of cell lysates demonstrated that EGR1 production was significantly increased by LL-37 treatment (Additional file 2: Figure S1b)

Next, to study the involvement of EGR1 in LL-37-enhanced ASC proliferation and migration, ASCs were transfected with EGR1-targeting siRNA. We confirmed

EGR1 gene silencing at the protein level (Fig. 2c). LL-37-enhanced proliferation and migration of ASCs were markedly attenuated by EGR1-targeting siRNA transfection (Fig. 2d and e). These data suggest that EGR1 is important for LL-37-enhanced migration and proliferation of ASCs.

Involvement of the MAPK pathway in LL-37-induced ASC proliferation and migration

The MAPK pathway plays an important role in the migration and proliferation of MSCs and cancer cells [10, 28]. To elucidate the signaling mechanism involved in LL-37-induced ASC proliferation and migration in detail, we examined the effect of LL-37 on the MAPK pathway in human ASCs. Treatment with LL-37 significantly increased the levels of phosphorylated

Table 1 List of genes up-regulated by LL-37

Gene name	Accession No	Fold increase	Classification
Early growth response 1 (*EGR1*)	NM_001964	18.47	Transcription, Immune response
Early growth response 2 (*EGR2*)	NM_000399	6.13	Transcription
High mobility group AT-hook 2 (*HMGA2*)	NM_003483	4.72	Transcription
Response gene to complement 32 (*RGC32*)	NM_014059	4.33	Cell cycle
Solute carrier family 20 (phosphate transporter) (*SLC20A1*)	NM_005415	4.19	Signal transduction
Dual specificity phosphatase 1 (*DUSP1*)	NM_004417	3.94	Cell cycle
Dual specificity phosphatase 6 (*DUSP6*)	NM_001946	3.74	Cell cycle, Signal transduction
Connective tissue growth factor (*CTGF*)	NM_001901	3.71	Cell migration, Cell growth, Cell adhesion, Signal transduction
Kruppel-like factor 10 (*KLF10*)	NM_005655	3.31	Transcription, Signal transduction, Cell proliferation
Dachsous 1 (Drosophila) (*DCHS1*)	NM_003737	3.05	Cell adhesion
Tumor necrosis factor receptor superfamily, member 12A (*TNFRSF12A*)	NM_016639	2.98	Apoptosis, Cell adhesion
Pleckstrin homology-like domain, family A, member 2 (*PHLDA2*)	NM_003311	2.94	Apoptosis
v-fos FBJ murine osteosarcoma viral oncogene homolog (*FOS*)	NM_005252	2.80	Transcription, Inflammatory response, Immune response
Tribbles homolog 2 (Drosophila) (*TRIB2*)	NM_021643	2.56	Cell adhesion
Tubulin, beta 2C (*TUBB2C*)	NM_006088	2.52	Transport, Apoptosis
Ras association (RalGDS/AF-6) and pleckstrin homology domains 1 (*RAPH1*)	NM_213589	2.46	Signal transduction
Myeloid cell leukemia sequence 1 (BCL2-related) (*MCL1*)	NM_021960	2.34	Apoptosis
SH2B adaptor protein 3 (*SH2B3*)	NM_005475	2.26	Cell cycle, Signal transduction
Cysteine-rich, angiogenic inducer, 61 (*CYR61*)	NM_001554	2.19	Cell adhesion, Cell proliferation
Sphingosine kinase 1 (*SPHK1*)	NM_021972	2.10	Cell migration, Apoptosis, Cell cycle, Cell proliferation
Ras association (RalGDS/AF-6) and pleckstrin homology domains 1 (*RAPH1*)	NM_213589	2.01	Signal transduction

The fold increase indicates the increase in expression in comparison to control cells, as determined by a human cDNA microarray

extracellular signal-regulated kinase (ERK) 1/2, p38, and c-Jun N-terminal kinase (JNK), but did not change the total levels of these proteins (Fig. 3a). To further determine the involvement of kinase phosphorylation in the increased migration, proliferation, and EGR1 production of ASCs, cells were treated with LL-37 in the presence and absence of a specific inhibitor of ERK (PD98059), p38 (SB203580), or JNK (SP600125). All these MAPK inhibitors (PD98059, SB203580, and SP600125) significantly reduced LL-37-induced EGR1 production (Fig. 3b). In addition, each of the inhibitors attenuated the LL-37-induced increase in ASC migration and proliferation (Fig. 3c and d). Taken together, these data suggest that: 1) LL-37 induces multiple signaling pathways, including those linked with ERK, p38, and JNK phosphorylation; and 2) these kinases are involved in the regulation of human ASC migration and proliferation. These results suggest that LL-37 stimulates ASC function through a MAPK-dependent mechanism.

LL-37 treatment enhances growth factor production in human ASCs, and treatment with the CM of ASCs pre-activated with LL-37 stimulates hair regeneration in vivo

It was recently reported that EGR1 has a strong paracrine capability and can influence angiogenic factors in ASCs [8, 29]; therefore, we first examined whether LL-37 regulates the expression of various regenerative growth factors and bioactive molecules in ASCs. LL-37 treatment considerably upregulated the mRNA expression of VEGF, TB4, MCP-1, and SDF-1 (Fig. 4a–d). We next confirmed the direct secretion of these growth factors by ASCs using ELISAs. The secretion levels of VEGF, TB4, MCP-1, and SDF-1 were significantly increased by LL-37 treatment (Fig. 4e–h).

To further evaluate the paracrine effects of LL-37 in vivo, the hair growth of C57BL/6 mice with hair loss was tested in vivo. We compared hair growth after topically applying CM of ASCs pre-activated with or without LL-37 (20 μg/mL). After 2 weeks, the hair growth

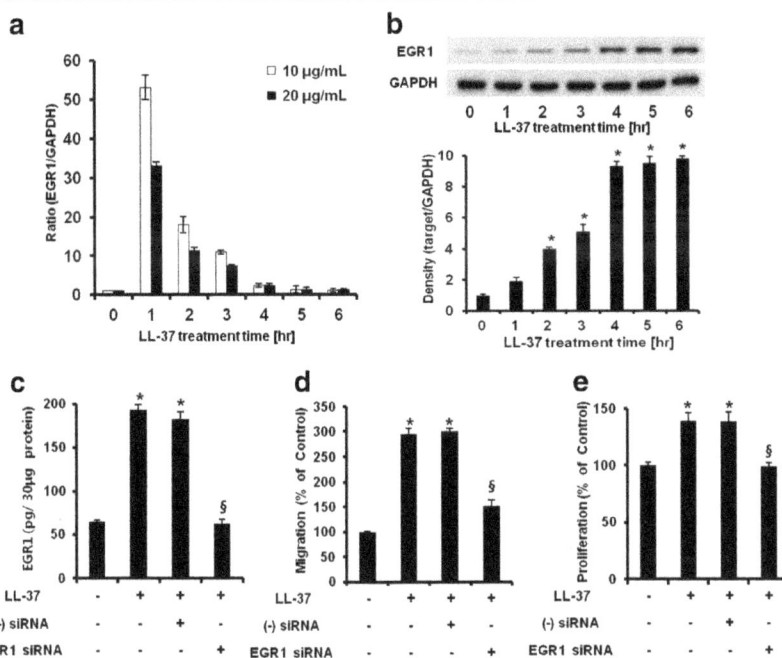

Fig. 2 EGR1 is critical for human LL-37-enhanced ASC migration and proliferation. **a** EGR1 mRNA expression was determined by real-time PCR. EGR1 mRNA was detected after treatment with 10 and 20 μg/mL LL-37 for 0–6 hr. **b** ASCs were incubated with 20 μg/mL LL-37 for 0–6 hr. Cell lysates were collected for western blot analysis. The EGR1 protein level was increased by LL-37 treatment in a time-dependent manner. **c** ASCs were treated with LL-37 and transfected with siRNA targeting EGR1 or a negative control sequence. After stabilization, cells were collected and lysed by the addition of 200 μL of cell lysis buffer. EGR1-targeting siRNA transfection was quantified by performing an EGR1-specific ELISA of cell lysates. **d** Cells that migrated after EGR1-targeting siRNA transfection were imaged by microscopy and counted using Scanscope. Results are expressed as the mean ± SD of three independent experiments. *, $P < 0.01$ vs. control; §, $P < 0.01$ vs. LL-37-treated cells transfected with negative control siRNA. **e** The effects on human ASC proliferation were similar to those on migration. A representative experiment from three independent experiments is shown. *Bars* represent the mean ± SD. *, $P < 0.01$ vs. control; §, $P < 0.01$ vs. LL-37-treated cells transfected with negative control siRNA. *EGR1* early growth response 1, *ASC* adipose-derived stromal/stem cells, *siRNA* small interfering RNA, *ELISA* enzyme-linked immunosorbent assay

score was higher in the group treated with CM of ASCs than in the negative control group treated with medium alone. Moreover, the hair growth score was higher in the group treated with CM of ASCs pre-activated with LL-37 than in the group treated with CM of non-activated ASCs (Fig. 4i and j), with dark pigmentation and hair regeneration observed on the initially pink hairless skin. Treatment with minoxidil (MNX), which is widely used to treat hair loss, was also effective for hair regeneration. Specifically, 95–100 % of hair was regenerated to full length in the group treated with CM of ASCs pre-activated with LL-37 in comparison to the MNX-treated group, whereas the level of regeneration was lower in the group treated with CM of non-activated ASCs (30–35 %; Fig. 4i). Additionally, to elucidate the biological function of EGR1 in vitro and in vivo, EGR1 was lentivirally introduced into human ASCs. Cell lines stably expressing the control vector and EGR1 were designated pLenti6/V5-D-TOPO and pLenti6/V5-hEGR1, respectively.

As expected, the secretion levels of VEGF, TB4, MCP-1, and SDF-1 were significantly higher in the medium of cells overexpressing EGR1 than in the medium of wild type (WT) cells or those expressing the control vector (Additional file 2: Figure S2a–2d). To confirm the paracrine effects of stable EGR1-overexpressing cells in vivo, we performed an additional investigation of hair regeneration. Mice were treated with CM of WT ASCs, ASCs expressing the control vector, or EGR1-overexpressing ASCs. Compared with CM of WT ASCs and ASCs expressing the control vector, treatment with CM of EGR1-overexpressing ASCs considerably increased hair regeneration (Additional file 2: Figure S2e and 2f).

Taken together, LL-37 might mediate paracrine actions by stimulating the secretion of growth factors such as VEGF, TB4, SDF-1, and MCP-1 by ASCs. In addition, CM from ASCs pre-activated with LL-37 and EGR1-overexpressing cells strongly promotes hair growth in vivo.

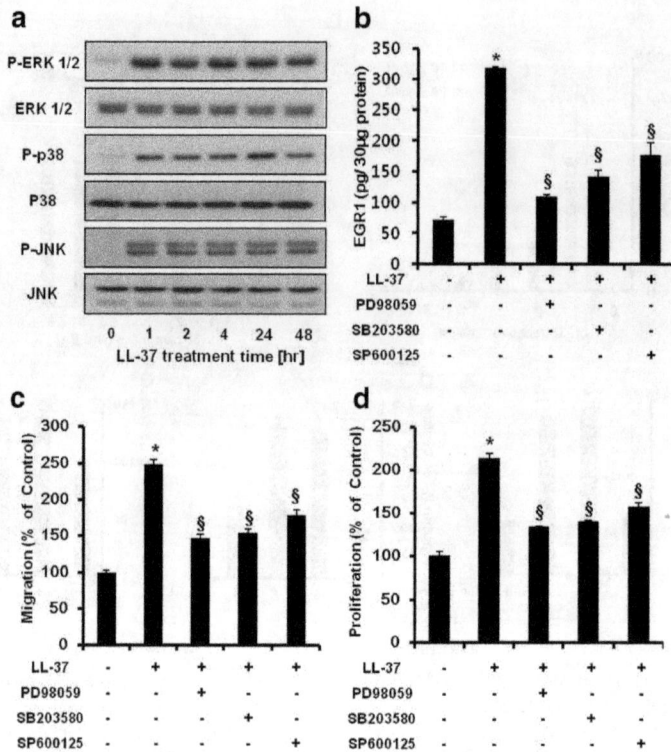

Fig. 3 Involvement of the MAPK pathway in LL-37-induced ASC proliferation and migration. **a** Human ASCs were treated with 20 μg/mL LL-37 for 1, 2, 4, 24, and 48 hr. After cell lysis, the levels of phosphorylated ERK1/2, p38, and JNK were determined by western blot analysis. The levels of total ERK1/2, p38, and JNK were used to confirm equal loading of the cell lysates. LL-37 treatment significantly increased the phosphorylation of ERK1/2, p38, and JNK. **b** ASCs were pretreated with or without a specific inhibitor of ERK1/2 (PD98059), p38 (SB203580), or JNK (SP600125) for 1 hr and then treated with 20 μg/mL LL-37. EGR1 protein levels in cell lysates were analyzed by an EGR1-specific ELISA. **c** LL-37-enhanced migration decreased in human ASCs treated with PD98059, SB203580, or SP600125. **d** LL-37-enhanced ASC proliferation was inhibited by treatment with specific inhibitors of ERK1/2, p38, or JNK (similar to ASC migration). A representative experiment from three independent experiments is shown. *Bars* represent the mean ± SD. *, $P < 0.01$ vs. control; §, $P < 0.01$ vs. LL-37-treated ASCs. *MAPK* mitogen-activated protein kinase, *ASCs* adipose-derived stromal/stem cells, *EGR1* early growth response-1

Discussion

Many recent studies report clinical trials of ASCs (or CM from ASCs) for the treatment of vascular injury and for the promotion of bone, skin, and hair regeneration [1, 2, 30]. However, the regenerative effects of MSCs are limited because they vary according to the donor, the donor's age, the sampling site, and the culture techniques and conditions. Therefore, it is essential to develop an activator of MSCs. LL-37 is a naturally occurring antimicrobial peptide found in the wound bed, where it assists wound repair and regeneration [15–17]; it is important to elucidate the relationship between LL-37 and adjacent ASCs for tissue regeneration and repair. Therefore, we investigated the functions of ASCs activated by LL-37 exposure.

We demonstrated that human ASCs expressed FPRL1 and that LL-37 treatment significantly enhanced the proliferation and migration of these cells. Next, we searched for the factor(s) that underlies LL-37-enhanced migration

and proliferation of ASCs. Initially, we thought that TB4 and MCP-1 were candidates for such factors because LL-37 treatment considerably enhanced their production (Fig. 4f and h), which facilitated ASC migration (unpublished data). However, these are not the main factors, although this is only one possible mechanism underlying LL-37-enhanced migration of ASCs. This is because LL-37 treatment rapidly promoted the migration of human ASCs within 6 hr (Fig. 1d), whereas TB4 and MCP-1 production was noticeably changed much later, after about 24 hr (Fig. 4f and h). Our microarray and real-time PCR data showed that EGR1 mRNA expression rapidly responded to LL-37 stimulation within 1 hr (Fig. 2a and Table 1). Moreover, western blot and ELISA analyses revealed that LL-37 treatment also significantly increased EGR1 protein levels by about 5–10-fold at 6 hr (Fig. 2b and Additional file 2: Figure S1b), which regulated ASC migration (Fig. 2d). Some previous studies reported that EGR1 mRNA was rapidly induced within 1–2 hr and that

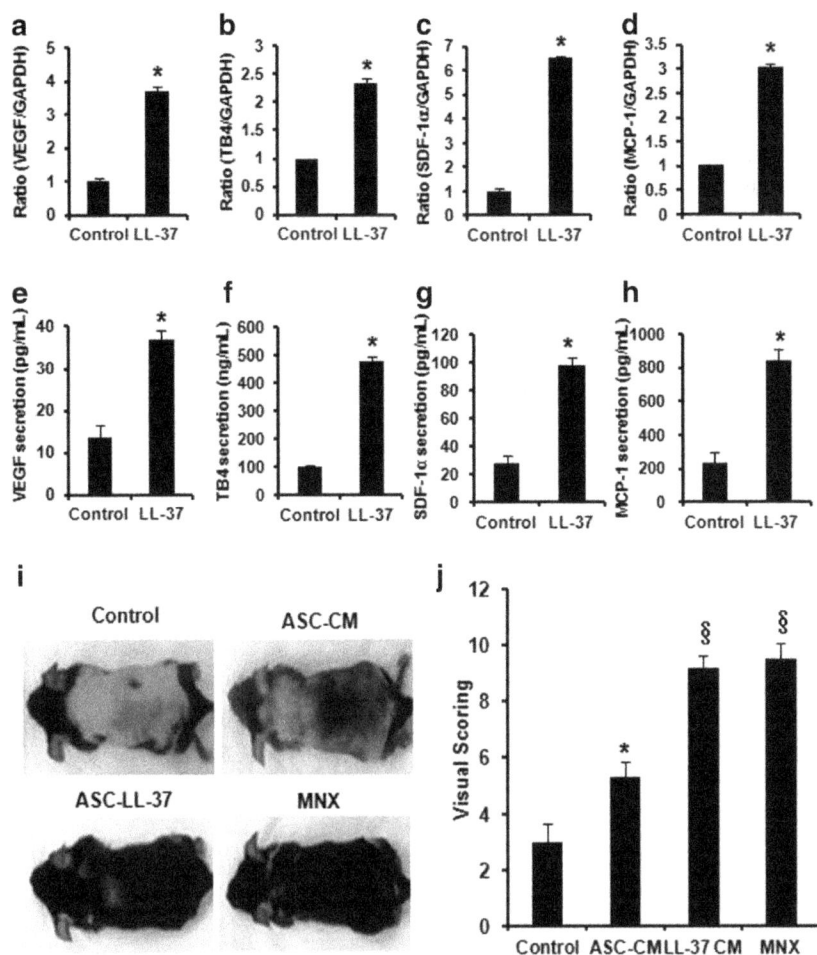

Fig. 4 Preconditioning with LL-37 stimulates hair regeneration in vivo and production of regenerative factors in vitro. **a** VEGF, (**b**) TB4, (**c**) SDF-1α, and (**d**) MCP-1 mRNA expression was determined by real-time PCR. **e** VEGF, (**f**) TB4, (**g**) SDF-1α, and (**h**) MCP-1 proteins were detected using the supernatants of confluent cell cultures. ASCs were treated with 20 μg/mL LL-37 for 48 hr, and then the protein levels of VEGF, TB4, SDF-1α, and MCP-1 were analyzed using specific ELISAs. A representative experiment from three independent experiments is shown. *Bars* represent the mean ± SD. *, $P < 0.05$ vs. control. **i** Hair was removed from the backs of C57BL/6 mice and the hair growth rate was monitored for 3 weeks. CM of human ASCs pretreated with or without LL-37 (20 μg/mL) was topically applied daily for up to 18 days to mice with hair loss. Gross views observed by photographs. **j** Hair growth was scored as described in the Materials and methods section. *, $P < 0.05$ for control vs. group treated with ASC CM. §, $P < 0.05$ for group treated with ASC CM vs. group treated with CM of ASCs pre-activated with LL-37. *VEGF* vascular endothelial growth factor, *TB4* thymosin beta-4, *SDF-1α* stromal cell-derived factor-1α, *MCP-1* monocyte chemoattractant protein-1, *ASC* adipose-derived stromal/stem cell, *ELISA* enzyme-linked immunosorbent assay, *CM* conditioned medium

EGR1 protein was also relatively rapidly induced at the same time [29, 31]. Interestingly, our data showed that EGR1 protein induction was a little sluggish, although EGR1 mRNA expression was immediately increased within 1 hr. However, as in other reports, EGR1 protein was slowly induced at 5–6 hr [32–34]. It is possible that the optimal time of EGR1 induction differs according to the type of stimulus (e.g., cytokine, growth factor, and oxidative stress) or the cell type. A recent study by Min et al. suggested that EGR1 controls hematopoietic stem cell migration and expansion [35]. Similarly, our data showed

that EGR1-targeting siRNA transfection inhibited LL-37-enhanced proliferation and migration of ASCs (Fig. 2c–e). Thus, these results suggest that EGR1 is important for the regulation of LL-37-enhanced proliferation and migration of human ASCs.

In this study, LL-37 promoted the expansion and migration of human ASCs at all concentrations, but was most effective at 20 μg/mL (Fig. 1b and d). Interestingly, EGR1 expression was highest in ASCs treated with a lower concentration (10 μg/mL) of LL-37 (Fig. 2a), suggesting that, in addition to EGR1, other factors and mechanisms mediate the effects of LL-37 on the proliferation and migration of human ASCs. For example, LL-37 treatment also considerably increased interleukin (IL)-8 expression and secretion in human ASCs (Additional file 2: Figure S3a and S3b). We previously demonstrated that IL-8 induced by TB4 plays a key role in ASC proliferation [23]. Moreover, IL-8 stimulated ASC migration, and IL-8 knockdown using IL-8-specific siRNA inhibited cell migration (unpublished data). IL-8 promotes angiogenesis and migration of bone marrow-derived MSCs [36], and may also affect multiple functions of human ASCs, including their proliferation and migration. Further investigations are necessary to clarify the other mechanisms underlying LL-37-induced proliferation and migration of human ASCs and the factors involved. Taken together, these data suggest that EGR1 is an important, but not the only, factor for LL-37-mediated ASC migration and proliferation.

It was recently reported that EGR1 has a strong paracrine capability and can influence angiogenic factors in ASCs [8, 29]; therefore, we investigated the enhanced secretion of TB4, VEGF, MCP-1, and SDF-1 in response to LL-37 stimulation (Fig. 4a–h). These growth factors and cytokines are associated with the regeneration and repair of various tissues, including skin, bone, and skeletal muscle [37–40]. Among them, TB4 directly stimulates hair growth and proliferation of hair follicle dermal papilla cells. Furthermore, TB4 accelerates hair growth via activation, migration, and differentiation of hair follicle stem cells [38, 41]. Recently, VEGF was shown to stimulate hair growth by facilitating the supply of nutrients to the hair follicle, increasing the follicular diameter [42, 43]. Based on these reports, TB4 and VEGF are regarded as the main factors in the hair growth promotion effect of the CM of LL-37-activated ASCs. Interestingly, one of these regenerative factors, MCP-1, is a pro-inflammatory cytokine. However, it also promotes healing in diabetic wounds, dental pulp, and skeletal muscle by increasing cell migration, angiogenesis, or macrophage infiltration, leading to a higher regenerative potential [39, 44, 45]. Therefore, MCP-1 seems to be a key cytokine mediating regenerative effects as well as a pro-inflammatory cytokine, and its action might depend on the microenvironment in the human body. LL-37-enhanced regenerative factors also influence their microenvironment through paracrine actions [46]. In support of this, CM from ASCs pre-activated with LL-37 strongly promoted hair growth as efficiently as MNX (95–100 % of hair), which is a

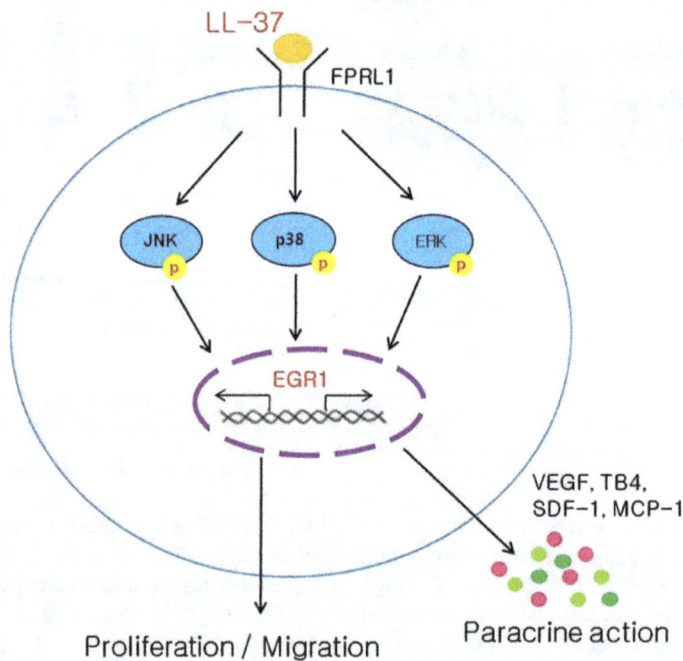

Fig. 5 Scheme representing the functions of ASCs after LL-37 treatment. LL-37 enhances ASC proliferation and migration via the EGR1 and MAPK pathways. Furthermore, LL-37 stimulates the secretion of growth factors such as VEGF, TB4, SDF-1, and MCP-1 in human ASCs. CM from ASCs preconditioned with LL-37 strongly promotes hair growth in vivo. LL-37 can be modulated to activate human ASCs and may provide a therapeutic approach for tissue regeneration by enforcing the functions of ASCs (e.g., expansion, migration, and paracrine actions). *ASCs* adipose-derived stromal/stem cells, *EGR1* early growth response 1, *MAPK* mitogen-activated protein kinase, *VEGF* vascular endothelial growth factor, *TB4* thymosin beta-4, *SDF-1α* stromal cell-derived factor-1α, *MCP-1* monocyte chemoattractant protein-1, *CM* conditioned medium

commercially available treatment for hair loss (Fig. 4i and j). Besides these paracrine actions, we previously demonstrated that the SDF-1/CXCR4 axis is essential for human dermal fibroblast (HDF) migration, resulting in wound-healing effects [25]. This suggests that CXCR4-expressing HDFs can migrate along the concentration gradient of SDF-1 secreted by human ASCs in response to LL-37 at wound sites, leading to wound repair and regeneration. Liu et al. and Neuhaus et al. reported that SDF-1 also enhances EGR1 expression in endothelial cells and angiogenesis in ischemic regions [47, 48]. Based on these reports, LL-37-induced SDF-1 might also enhance EGR1 expression in human ASCs. Therefore, it is reasonable to conclude that LL-37 can gradually accelerate other mechanisms via autocrine loops and paracrine actions on adjacent cells such as HDFs. Taken together, LL-37 may activate human ASCs via autocrine and paracrine actions.

Conclusions

In summary, these results demonstrate that LL-37 enhances ASC proliferation and migration via the EGR1 and MAPK pathways (Fig. 5). Furthermore, LL-37 stimulates the secretion of growth factors such as VEGF, TB4, SDF-1, and MCP-1. CM from ASCs preconditioned with LL-37 strongly promotes hair growth in vivo. Therefore, LL-37 can be modulated to activate human ASCs and may provide a therapeutic approach to tissue regeneration by enforcing the functions of ASCs (e.g., expansion, migration, and paracrine actions).

Additional files

Additional file 1: Table S1. List of primers used for real-time PCR. (PDF 106 kb)

Additional file 2: Figure S1. Effects of LL-37 on the viability and EGR1 production of human ASCs. **Figure S2.** EGR1-overexpressing ASCs secrete higher levels of the regenerative factors VEGF, TB4, SDF-1α, and MCP-1, and the CM of these cells stimulates hair regeneration in vivo. **Figure S3.** LL-37 increases the expression and production of IL-8 in human ASCs. (PDF 267 kb)

Additional file 3: Table S2. List of upregulated genes in ASCs exposed to LL-37 by microarray analysis. (XLS 45 kb)

Abbreviations

ASC: adipose-derived stromal/stem cell; CCK-8: cell counting kit-8; CM: conditioned medium; CTGF: connective tissue growth factor; EGR1: early growth response 1; EGR2: early growth response 2; ELISA: enzyme-linked immunosorbent assay; ERK: extracellular signal-regulated kinase; FACS: fluorescence-activated cell sorting; FPRL1: formyl peptide receptor like-1; hCAP-18: human cationic antimicrobial protein 18; HDF: human dermal fibroblast; IL: interleukin; JNK: c-Jun N-terminal kinase; KLF10: Krupple-like factor 10; MAPK: mitogen-activated protein kinase; MCP-1: monocyte chemoattractant protein-1; MNX: minoxidil; MSC: mesenchymal stem cell; PCNA: proliferating cell nuclear antigen; Ptx: pertussis toxin; SDF-1: stromal cell-derived factor-1; siRNA: small interfering RNA; TB4: thymosin beta-4; VEGF: vascular endothelial growth factor; WT: wild type; αLL-37: anti-LL-37 antibody.

Competing interests
The authors declare that they have no competing interests.

Authors' contributions
YY contributed to conception and design of the study, data analysis and interpretation, collection and assembly of data, and drafting and revision of the manuscript. HC carried out the molecular work involving western blotting, siRNA transfection, and ELISAs as well as the animal work and contributed to drafting of the manuscript. MS performed the cell culture, characterization of ASCs, the migration assay, and the proliferation assay with statistical analysis, and participated in drafting of the manuscript. DC contributed to analysis and interpretation of the data, financial support, and helped to revise the manuscript. SB conceived and designed the study, provided financial support, and gave final approval of the manuscript. All authors read and approved the final manuscript.

Acknowledgments
We thank Jaehee Kim for support with the revision of the article and analysis of the data. This work was supported by a Samsung Biomedical Research Institute grant (SMO 1131631) and the Korea Drug Development Fund (KDDF), which is funded by the Ministry of Science, ICT and Future Planning, the Ministry of Trade, Industry & Energy, and the Ministry of Health & Welfare of the Republic of Korea (KDDF-201404-04).

Author details
[1]Department of Plastic Surgery, Samsung Medical Center, Sungkyunkwan University School of Medicine, Seoul, Korea. [2]Department of Life Science, Sookmyung Women's University, Seoul, Korea. [3]Bio-Med Translational Research Center, Samsung Medical Center, Seoul, Korea.

References
1. Bruder SP, Fink DJ, Caplan AI. Mesenchymal stem cells in bone development, bone repair, and skeletal regeneration therapy. J Cell Biochem. 1994;56(3):283–94.
2. Khosrotehrani K. Mesenchymal stem cell therapy in skin: why and what for? Exp Dermatol. 2013;22(5):307–10.
3. Zuk PA, Zhu M, Ashjian P, De Ugarte DA, Huang JI, Mizuno H, et al. Human adipose tissue is a source of multipotent stem cells. Mol Biol Cell. 2002;13(12):4279–95.
4. Maijenburg MW, van der Schoot CE, Voermans C. Mesenchymal stromal cell migration: possibilities to improve cellular therapy. Stem Cells Dev. 2012;21(1):19–29.
5. Lee JW, Fang X, Krasnodembskaya A, Howard JP, Matthay MA. Concise review: mesenchymal stem cells for acute lung injury: role of paracrine soluble factors. Stem Cells. 2011;29(6):913–9.
6. Liang X, Ding Y, Zhang Y, Tse HF, Lian Q. Paracrine mechanisms of mesenchymal stem cell-based therapy: current status and perspectives. Cell Transplant. 2014;23(9):1045–59.
7. Tarcic G, Avraham R, Pines G, Amit I, Shay T, Lu Y, et al. EGR1 and the ERK-ERF axis drive mammary cell migration in response to EGF. FASEB J. 2012;26(4):1582–92.
8. Tamama K, Barbeau DJ. Early growth response genes signaling supports strong paracrine capability of mesenchymal stem cells. Stem Cells Int. 2012;2012:428403.
9. Caplan AI. Why are MSCs therapeutic? New data: new insight. J Pathol. 2009;217(2):318–24.
10. Ben-Chetrit N, Tarcic G, Yarden Y. ERK-ERF-EGR1, a novel switch underlying acquisition of a motile phenotype. Cell Adh Migr. 2013;7(1):33–7.
11. Murakami M, Ohtake T, Dorschner RA, Schittek B, Garbe C, Gallo RL. Cathelicidin anti-microbial peptide expression in sweat, an innate defense system for the skin. J Invest Dermatol. 2002;119(5):1090–5.
12. Frohm Nilsson M, Sandstedt B, Sorensen O, Weber G, Borregaard N, Stahle-Backdahl M. The human cationic antimicrobial protein (hCAP18), a peptide antibiotic, is widely expressed in human squamous epithelia and colocalizes with interleukin-6. Infect Immun. 1999;67(5):2561–6.
13. Larrick JW, Hirata M, Balint RF, Lee J, Zhong J, Wright SC. Human CAP18: a novel antimicrobial lipopolysaccharide-binding protein. Infect Immun. 1995;63(4):1291–7.

14. Niyonsaba F, Ushio H, Nagaoka I, Okumura K, Ogawa H. The human beta-defensins (-1, -2, -3, -4) and cathelicidin LL-37 induce IL-18 secretion through p38 and ERK MAPK activation in primary human keratinocytes. J Immunol. 2005;175(3):1776–84.

15. Kai-Larsen Y, Agerberth B. The role of the multifunctional peptide LL-37 in host defense. Front Biosci. 2008;13:3760–7.

16. Carretero M, Escamez MJ, Garcia M, Duarte B, Holguin A, Retamosa L, et al. In vitro and in vivo wound healing-promoting activities of human cathelicidin LL-37. J Invest Dermatol. 2008;128(1):223–36.

17. Heilborn JD, Nilsson MF, Kratz G, Weber G, Sorensen O, Borregaard N, et al. The cathelicidin anti-microbial peptide LL-37 is involved in re-epithelialization of human skin wounds and is lacking in chronic ulcer epithelium. J Invest Dermatol. 2003;120(3):379–89.

18. De Y, Chen Q, Schmidt AP, Anderson GM, Wang JM, Wooters J, et al. LL-37, the neutrophil granule- and epithelial cell-derived cathelicidin, utilizes formyl peptide receptor-like 1 (FPRL1) as a receptor to chemoattract human peripheral blood neutrophils, monocytes, and T cells. J Exp Med. 2000; 192(7):1069–74.

19. Krasnodembskaya A, Song Y, Fang X, Gupta N, Serikov V, Lee JW, et al. Antibacterial effect of human mesenchymal stem cells is mediated in part from secretion of the antimicrobial peptide LL-37. Stem Cells. 2010;28(12):2229–38.

20. Cashman TJ, Gouon-Evans V, Costa KD. Mesenchymal stem cells for cardiac therapy: practical challenges and potential mechanisms. Stem Cell Rev. 2013;9(3):254–65.

21. Pound LD, Patrick C, Eberhard CE, Mottawea W, Wang GS, Abujamel T, et al. Cathelicidin antimicrobial peptide: a novel regulator of islet function, islet regeneration, and selected gut bacteria. Diabetes. 2015;64(12):4135–47.

22. Kittaka M, Shiba H, Kajiya M, Fujita T, Iwata T, Rathvisal K, et al. The antimicrobial peptide LL37 promotes bone regeneration in a rat calvarial bone defect. Peptides. 2013;46:136–42.

23. Jeon BJ, Yang Y, Kyung Shim S, Yang HM, Cho D, Ik BS. Thymosin beta-4 promotes mesenchymal stem cell proliferation via an interleukin-8-dependent mechanism. Exp Cell Res. 2013;319(17):2526–34.

24. Dominici M, Le Blanc K, Mueller I, Slaper-Cortenbach I, Marini F, Krause D, et al. Minimal criteria for defining multipotent mesenchymal stromal cells. The International Society for Cellular Therapy position statement. Cytotherapy. 2006;8(4):315–7.

25. Yang Y, Shim SK, Kim HA, Seon M, Yang E, Cho D, et al. CXC chemokine receptor 4 is essential for Lipo-PGE1-enhanced migration of human dermal fibroblasts. Exp Dermatol. 2012;21(1):75–7.

26. Vegesna V, O'Kelly J, Uskokovic M, Said J, Lemp N, Saitoh T, et al. Vitamin D3 analogs stimulate hair growth in nude mice. Endocrinology. 2002;143(11):4389–96.

27. Jung MK, Ha S, Huh SY, Park SB, Kim S, Yang Y, et al. Hair-growth stimulation by conditioned medium from vitamin D3-activated preadipocytes in C57BL/6 mice. Life Sci. 2015;128:39–46.

28. Yang Y, Cheon S, Jung MK, Song SB, Kim D, Kim HJ, et al. Interleukin-18 enhances breast cancer cell migration via down-regulation of claudin-12 and induction of the p38 MAPK pathway. Biochem Biophys Res Commun. 2015;459(3):379–86.

29. Kerpedjieva SS, Kim DS, Barbeau DJ, Tamama K. EGFR ligands drive multipotential stromal cells to produce multiple growth factors and cytokines via early growth response-1. Stem Cells Dev. 2012;21(13):2541–51.

30. Park BS, Kim WS, Choi JS, Kim HK, Won JH, Ohkubo F, et al. Hair growth stimulated by conditioned medium of adipose-derived stem cells is enhanced by hypoxia: evidence of increased growth factor secretion. Biomed Res. 2010;31(1):27–34.

31. Cheng JC, Chang HM, Leung PC. Egr-1 mediates epidermal growth factor-induced downregulation of E-cadherin expression via Slug in human ovarian cancer cells. Oncogene. 2013;32(8):1041–9.

32. Kwapiszewska G, Chwalek K, Marsh LM, Wygrecka M, Wilhelm J, Best J, et al. BDNF/TrkB signaling augments smooth muscle cell proliferation in pulmonary hypertension. Am J Pathol. 2012;181(6):2018–29.

33. Zhang X, Liu Y. Suppression of HGF receptor gene expression by oxidative stress is mediated through the interplay between Sp1 and Egr-1. Am J Physiol Renal Physiol. 2003;284(6):F1216–25.

34. Spohn D, Rossler OG, Philipp SE, Raubuch M, Kitajima S, Griesemer D, et al. Thapsigargin induces expression of activating transcription factor 3 in human keratinocytes involving Ca2+ ions and c-Jun N-terminal protein kinase. Mol Pharmacol. 2010;78(5):865–76.

35. Min IM, Pietramaggiori G, Kim FS, Passegue E, Stevenson KE, Wagers AJ. The transcription factor EGR1 controls both the proliferation and localization of hematopoietic stem cells. Cell Stem Cell. 2008;2(4):380–91.

36. Wang L, Li Y, Chen X, Chen J, Gautam SC, Xu Y, et al. MCP-1, MIP-1, IL-8 and ischemic cerebral tissue enhance human bone marrow stromal cell migration in interface culture. Hematology. 2002;7(2):113–7.

37. Liu YS, Ou ME, Liu H, Gu M, Lv LW, Fan C, et al. The effect of simvastatin on chemotactic capability of SDF-1alpha and the promotion of bone regeneration. Biomaterials. 2014;35(15):4489–98.

38. Philp D, Kleinman HK. Animal studies with thymosin beta, a multifunctional tissue repair and regeneration peptide. Ann N Y Acad Sci. 2010;1194:81–6.

39. Wood S, Jayaraman V, Huelsmann EJ, Bonish B, Burgad D, Sivaramakrishnan G, et al. Pro-inflammatory chemokine CCL2 (MCP-1) promotes healing in diabetic wounds by restoring the macrophage response. PLoS One. 2014;9(3):e91574.

40. Niyaz M, Gurpinar OA, Oktar GL, Gunaydin S, Onur MA, Ozsin KK, et al. Effects of VEGF and MSCs on vascular regeneration in a trauma model in rats. Wound Repair Regen. 2015;23(2):262–7.

41. Philp D, St-Surin S, Cha HJ, Moon HS, Kleinman HK, Elkin M. Thymosin beta 4 induces hair growth via stem cell migration and differentiation. Ann N Y Acad Sci. 2007;1112:95–103.

42. Bassino E, Gasparri F, Giannini V, Munaron L. Paracrine crosstalk between human hair follicle dermal papilla cells and microvascular endothelial cells. Exp Dermatol. 2015;24(5):388–90.

43. Gnann LA, Castro RF, Azzalis LA, Feder D, Perazzo FF, Pereira EC, et al. Hematological and hepatic effects of vascular epidermal growth factor (VEGF) used to stimulate hair growth in an animal model. BMC Dermatol. 2013;13:15.

44. Hayashi Y, Murakami M, Kawamura R, Ishizaka R, Fukuta O, Nakashima M. CXCL14 and MCP1 are potent trophic factors associated with cell migration and angiogenesis leading to higher regenerative potential of dental pulp side population cells. Stem Cell Res Ther. 2015;6:111.

45. Zhang J, Xiao Z, Qu C, Cui W, Wang X, Du J. CD8 T cells are involved in skeletal muscle regeneration through facilitating MCP-1 secretion and Gr1(high) macrophage infiltration. J Immunol. 2014;193(10):5149–60.

46. Sassoli C, Pini A, Chellini F, Mazzanti B, Nistri S, Nosi D, et al. Bone marrow mesenchymal stromal cells stimulate skeletal myoblast proliferation through the paracrine release of VEGF. PLoS One. 2012;7(7):e37512.

47. Neuhaus T, Stier S, Totzke G, Gruenewald E, Fronhoffs S, Sachinidis A, et al. Stromal cell-derived factor 1alpha (SDF-1alpha) induces gene-expression of early growth response-1 (Egr-1) and VEGF in human arterial endothelial cells and enhances VEGF induced cell proliferation. Cell Prolif. 2003;36(2):75–86.

48. Liu H, Liu S, Li Y, Wang X, Xue W, Ge G, et al. The Role of SDF-1-CXCR4/CXCR7 Axis in the Therapeutic Effects of Hypoxia-Preconditioned Mesenchymal Stem Cells for Renal Ischemia/Reperfusion Injury. PLoS ONE. 2012;7(4):e34608.

14

Role of VEGF-A in angiogenesis promoted by umbilical cord-derived mesenchymal stromal/stem cells: in vitro study

Irina Arutyunyan[1,2], Timur Fatkhudinov[1,3,4*], Evgeniya Kananykhina[1,2], Natalia Usman[1,3], Andrey Elchaninov[1,2], Andrey Makarov[1,3], Galina Bolshakova[1], Dmitry Goldshtein[5] and Gennady Sukhikh[1]

Abstract

Background: Mesenchymal stromal/stem cells derived from human umbilical cord (UC-MSCs) uniquely combine properties of embryonic and postnatal MSCs and may be the most acceptable, safe, and effective source for allogeneic cell therapy e.g. for therapeutic angiogenesis. In this report we describe pro-angiogenic properties of UC-MSCs as manifested *in vitro*.

Methods: UC-MSCs were isolated from human Wharton's jelly by enzymatic digestion. Presence of soluble forms of VEGF-A in UC-MSC-conditioned media was measured by ELISA. Effects of the conditioned media on human umbilical vein-derived endothelial EA.hy926 cells proliferation were measured by MTT-assay; changes in cell motility and directed migration were assessed by scratch wound healing and transwell chamber migration assays. Angiogenesis was modeled *in vitro* as tube formation on basement membrane matrix. Progressive differentiation of MSCs to endothelioid progeny was assessed by CD31 immunostaining.

Results: Although no detectable quantities of soluble VEGF-A were produced by UC-MSCs, the culture medium, conditioned by the UC-MSCs, effectively stimulated proliferation, motility, and directed migration of EA.hy926 cells. In 2D culture, UC-MSCs were able to acquire CD31[+] endothelial cell-like phenotype when stimulated by EA.hy926-conditioned media supplemented with VEGF-A165. UC-MSCs were capable of forming unstable 2D tubular networks either by themselves or in combinations with EA.hy926 cells. Active spontaneous sprouting from cell clusters, resulting from disassembling of such networks, was observed only in the mixed cultures, not in pure UC-MSC cultures. In 3D mode of sprouting experimentation, structural support of newly formed capillary-like structures was provided by UC-MSCs that acquired the CD31[+] phenotype in the absence of exogenous VEGF-A.

Conclusion: These data suggest that a VEGF-A-independent paracrine mechanism and at least partially VEGF-A-independent differentiation mechanism are involved in the pro-angiogenic activity of UC-MSCs.

Keywords: Mesenchymal stromal cells, Multipotent, Wharton jelly, Endothelial cells, In vitro techniques, Vascular endothelial growth factor-A, Angiogenesis inducing agents, CD31 antigen, Cell migration assays, Extracellular matrix

* Correspondence: tfat@yandex.ru
[1]Research Center for Obstetrics, Gynecology and Perinatology of Ministry of Healthcare of the Russian Federation, 4 Oparina Street, Moscow 117997, Russia
[3]Pirogov Russian National Research Medical University, Ministry of Healthcare of the Russian Federation, 1 Ostrovitianov Street, Moscow 117997, Russia
Full list of author information is available at the end of the article

Background

The concept of therapeutic angiogenesis stems from understanding the importance of adequate microvascular supply for growth and regeneration of affected tissues; it refers to actions performed to facilitate revascularization of ischemic tissues. As long as the direct delivery of exogenous cytokines and growth factors is ineffective, primarily because of their rapid elimination in vivo [1], expert opinions agree that the most promising approach for therapeutic angiogenesis is represented by stem cell therapy using multipotent mesenchymal stromal/stem cells (MSCs) because it comprises simultaneous activation of multiple mechanisms (paracrine, replacement, trophic, immunomodulatory) to provide support on different stages of formation and maturation of blood vessels [2–4]. Most of the research in this field is performed on bone marrow-derived MSCs (which represent a 'gold standard') or adipose tissue-derived MSCs; both lineages have certain angiogenic potential and implement it in a similar manner [2, 5]. The field of therapeutic angiogenesis is currently expanded by using MSCs from other sources, importantly from the umbilical cord and placenta. Perinatal stem cells share characteristics with both embryonic and adult stem cells because they may exhibit pluripotency as well as multipotent tissue maintenance, thus representing a bridge between embryonic and adult stem cells [6]. Umbilical cord-derived MSCs (UC-MSCs) have distinct biological properties: they are highly proliferative and enriched in transcriptionally active genes related to liver or cardiovascular system development and function. Besides, UC-MSCs exhibit superior grades of plasticity and immunomodulatory activity, lack tumorigenicity, and are considered the best resource for allogeneic transplantation [7–10].

Although therapeutic efficacy of UC-MSC transplantation to ischemic tissue is demonstrated in vivo [11], the understanding of how these cells implement their proangiogenic potential is far from complete, and there is a certain controversy among reports on this subject. The inner space of umbilical cord, net of large vessels, is occupied by special connective tissue, the Wharton jelly, which is very loose and rich in gel-like ground substance. Complete absence of microcirculatory vessels from Wharton jelly may indicate some anti-angiogenic properties of this microenvironment. The assumption is partially supported by recent in vitro studies. For instance, Kuchroo et al. [12] show that UC-MSCs do not produce detectable amounts of soluble vascular endothelial growth factor (VEGF)-A protein (strictly speaking, they probably do because the corresponding gene is actively transcribed, but apparently this secreted VEGF-A is saturated by soluble VEGF-A-specific receptors that are also secreted by UC-MSC and act as a buffer); at the same time, the authors observe certain stimulating influences of UC-MSCs on umbilical vein-derived endothelial cells in vitro and suggest the existence of some alternative, VEGF-independent mechanism for this stimulation. A comparative study by Amable et al. [13] describe a shifted balance of pro-angiogenic and anti-angiogenic factors in UC-MSC secretome, as compared with MSCs from other conventional sources. Ways in which UC-MSCs interact with endothelial cells and their own responses to inducers of endothelial differentiation represent an open issue. In the current study, the pro-angiogenic activity of human UC-MSCs is challenged in vitro by modeling of angiogenesis using 2D and 3D artificial matrices. Several specifically addressed problems include the importance of VEGF-A, modes of UC-MSC cooperation with umbilical vein-derived endothelial EA.hy926 cells, and UC-MSC ability to acquire $CD31^+$ phenotypes under various stimulations.

Methods

Cell cultures

The study involving human material was approved by the Ethics Committee at the Research Center for Obstetrics, Gynecology and Perinatology. Written informed consent was obtained from all participants prior to the study.

MSCs were isolated from human umbilical cords ($n = 5$). The material was rinsed in phosphate-buffered saline (PBS) with 1 mg/ml cefazolin (Sintez, Kurgan, Russia) and cut into 3–4 cm pieces. After removal of blood vessels and amnion, the Wharton jelly was chopped into smaller fragments with scissors. The fragments were incubated with 200 U/ml collagenase type I (PanEco, Moscow, Russia) and 40 U/ml dispase (Invitrogen, Waltham, MA, USA) for 60 minutes at 37 °C. After the addition of fetal bovine serum (FBS; GE Healthcare, Pittsburg, PA, USA), the digested mixture was centrifuged at $1000 \times g$ for 10 minutes at room temperature. Finally, the digested pieces were washed with serum-free Dulbecco's modified Eagle's medium (DMEM; PanEco) and cultured in growth medium (DMEM/F12 supplemented with 10 % FBS and 1 % penicillin–streptomycin (PanEco)) in a humidified incubator at 37 °C under a 5 % CO_2 atmosphere.

UC-MSCs were characterized according to the minimal criteria to define human MSCs as proposed by the Mesenchymal and Tissue Stem Cell Committee of the International Society for Cellular Therapy [14]. For immunophenotype analysis, cells were labeled for 30 minutes at room temperature using the BD Stemflow™ hMSC Analysis Kit (BD Biosciences, Pharmingen, San Diego, CA, USA). After being fixed with 4 % paraformaldehyde (SERVA Electrophoresis, Heidelberg, Germany), the cells were analyzed on a FACScalibur using CellQuest software (BD Biosciences). The StemPro® Adipogenesis Differentiation Kit, the StemPro® Osteogenesis

Differentiation Kit, and the StemPro® Chondrogenesis Differentiation Kit (Gibco, Life Technologies, Carlsbad, CA, USA) were used to demonstrate the differentiation capacity of UC-MSCs in accordance with the manufacturer's instructions.

Human endothelial EA.hy926 cells were derived from the American Type Culture Collection (Manassas, VA, USA). Established in 1983 by fusing primary human umbilical vein endothelial cells (HUVEC) with a thioguanine-resistant clone of the human lung adenocarcinoma cell line A549/8, EA.hy926 cells represent a widely-used endothelial cell line expressing endothelin-1, Weibel-Palade bodies, prostacyclin, factor VIII-related antigen, and endothelial adhesion molecules ICAM-1 and VCAM-1 [15]. This line was chosen for its highly specific functions that are characteristic of the human vascular endothelium combined with advantages of immortality, stability through passage number, and high reproducibility of the properties [16, 17].

Immunofluorescence

Cells were fixed with 4 % paraformaldehyde (SERVA Electrophoresis) for 10 minutes at room temperature. After two washes with PBS, the cells were blocked for 5 minutes with Protein Block (Abcam, Cambridge, MA, USA) at room temperature and then incubated overnight at 4 °C with antibodies against CD31 (ab24590; Abcam). After washing with PBS, the cells were incubated with fluorescein isothiocyanate (FITC)-conjugated antimouse IgG (ab6810; Abcam) for 1 hour in the dark. Cell nuclei were stained with 4′,6-diamidino-2-phenylindole (DAPI; Sigma-Aldrich, St. Louis, MO, USA). The cells were observed under the Leica DM 4000 B fluorescent microscope (Leica Microsystems, Heidelberg, Germany).

Preparation of conditioned media

At 100 % confluence, the cells (UC-MSCs or EA.hy926) were washed with serum-free DMEM, and the media were replaced with fresh growth media. After 24, 48, or 72 hours, the media were collected and centrifuged at $2800 \times g$ for 5 minutes, filtered through a 0.22 μm filter (GE Osmonics Labstore, Minnetonka, MN, USA), and were then stored at −70 °C until VEGF-A quantification. The media conditioned by UC-MSCs or EA.hy926 cells for 72 hours were used in subsequent experiments.

VEGF-A quantification

Media conditioned by EA.hy926 cells or UC-MSCs were collected after 24, 48, or 72 hours. VEGF-A-121 and VEGF-A-165 were quantified using a commercial enzyme-linked immunosorbent assay kit (#8784; Vector-Best, Novosibirsk, Russia) in accordance with the instructions of the manufacturer. Data analysis was performed using the online application (http://elisaanalysis.com/app).

Endothelial cell proliferation assay

EA.hy926 cells were seeded in a 96-well plate (3×10^3 cells in 200 μl of growth media per well). After 1, 2, or 3 days the media were replaced with UC-MSC-conditioned media, UC-MSC-conditioned media supplemented with 200 ng/ml anti-VEGF antibody (ab9570; Abcam), or fresh growth media (control wells). At day 4 the cell proliferation was measured by 3-(4,5-dimethylthiazol-2-yl)-2,5-diphenyltetrazolium bromide (MTT) assay. MTT (Sigma-Aldrich) stock solution was added to each well (to a final MTT concentration of 1.5 mg/ml). The plate was returned to a cell culture incubator for 2 hours. When the purple precipitate was clearly visible under the microscope, 100 μl of dimethyl sulfoxide (DMSO; Sigma-Aldrich) were added. After 15 minutes, the absorbance in each well was measured at 570 nm in a Multiskan GO microplate spectrophotometer (Thermo Fisher Scientific, Waltham, MA, USA). The reference wavelength was 650 nm.

Endothelial cell transwell migration assay

The migration of EA.hy926 cells to UC-MSC-released chemoattractants was measured by transwell chamber migration assay. UC-MSCs were seeded in a 24-well plate (10^5 cells in 600 μl of growth media per well). One-half of the wells with the cells contained anti-VEGF antibody (ab9570; Abcam) in 200 ng/ml final concentration. The same volume of growth media without cells was added to control wells. After 24 hours, inserts with a polycarbonate membrane (pore size of 8 μm) (#35224; SPL Life Sciences, Pochun, South Korea) were installed in the plate. EA.hy926 cells were seeded in the upper chambers (10^5 cells in 250 μl of growth media). After 24, 48, or 72 hours, nonmigrating cells in the upper chamber were attentively removed with cotton swabs, and cells on the lower surface of the membrane were fixed with 4 % paraformaldehyde (SERVA Electrophoresis) and stained with DAPI (Sigma-Aldrich). The total numbers of migrated cells were then counted in eight randomly selected fields for each insert (magnification × 100) using LAS AF v.3.1.0 build 8587 (Leica Microsystems).

Endothelial cell scratch healing assay

EA.hy926 cells were seeded in a 96-well plate (3×10^4 cells in 100 μl of growth media per well). After 24 hours, each confluent cell monolayer was scratched with a Wound-Maker™ tool (Essen Bioscience, Ann Arbor, MI, USA), which created 96 homogeneous scratch wounds without cell irritation. After washing with PBS, 100 μl of fresh growth media, UC-MSC-conditioned media, or UC-MSC-conditioned media supplemented with anti-VEGF antibody (ab9570; Abcam) in 200 ng/ml final concentration were added to the wells. A 36-hour time-lapse movie was created by IncuCyte ZOOM® Live-Cell Imaging Platform

(Essen BioScience). Wound confluence was measured using an automated Cell Migration software module (Essen BioScience).

In vitro tube formation assay

In this experiment we used BD Matrigel™ Basement Membrane Matrix Phenol Red Free (#356237; BD Biosciences). This matrix is highly enriched in laminin-1, collagen IV, heparan sulfate, proteoglycan, entactin/nidogen, and various growth factors. Although it does not contain all of the signature components of an endothelial basement membrane, Matrigel promotes tube formation in vitro for all endothelial cells tested to date [18].

Prior to the experiment, UC-MSCs were labeled with PKH26 (yellow–orange fluorescent dye) and EA.hy926 cells were labeled with PKH67 (green fluorescent dye) (Sigma-Aldrich) according to the manufacturer's instructions. A total of 150 µl of chilled Matrigel was added to a 48-well plate and incubated at 37 °C for 30 minutes. UC-MSCs, or EA.hy926 cells, or 1:1 mixed UC-MSC-EA.hy926 cells (total 35×10^3 cells per well) were suspended in 500 µl of growth media and were added to the solidified Matrigel. Additionally, EA.hy926 cells suspended in 500 µl of UC-MSC-conditioned media or UC-MSC-conditioned media supplemented with 200 ng/ml of anti-VEGF antibody (ab9570; Abcam) were used. After incubation on Matrigel at 37 °C in a 5 % CO_2 chamber, morphological changes were observed under an Axiovert 40 CFL inverted microscope (Carl Zeiss, Jena, Germany). Six representative fields for each well were photographed. Images were analyzed using AxioVs40 4.8.2.0 (Zeiss, Oberkochen, Germany) to determine the length of the tubes and the number of branch points (magnification × 50).

Endothelial differentiation of UC-MSCs in monolayer

After UC-MSCs formed the confluent monolayer, the growth media were replaced with the induction media. Three endothelial induction media were used for UC-MSC culture: EA.hy926-conditioned media mixed 1:1 with growth media; EA.hy926-conditioned media mixed 1:1 with growth media supplemented with VEGF-A-165 (#583702; BioLegend, San Diego, CA, USA), 50 ng/ml; and growth media supplemented with VEGF-A-165. Contents of FCS in the control and differentiation media were reduced to 5 % to avoid excessive cell growth. The media were replaced twice a week. At day 21, UC-MSCs were fixed with 4 % paraformaldehyde (SERVA Electrophoresis) and stained with CD31 antibodies as already described.

Endothelial differentiation of UC-MSCs in Matrigel

Whole mount immunostaining of the 3D structure, formed by the cells in Matrigel, proved impossible due to nonspecific absorption of the antibodies by the matrix.

For this reason, the analysis was performed on cryosections of secondary sprouting networks. Structures formed in Matrigel were embedded in Tissue-Tek® OCT Compound (Sakura Finetek, Torrance, CA, USA) and cut into 5–7 µm sections using a cryostat. The sections were stained with CD31 antibodies as already described.

Statistical analysis

Data are expressed as mean ± standard deviation (SD). Student's t test was used for pairwise comparisons between groups of normally distributed values, whereas the Mann–Whitney test was applied for distributions differing from normal. Multiple comparisons were done by either one-way analysis of variance (ANOVA) or ANOVA on ranks (for the cases of unconfirmed normality); $p < 0.05$ was considered statistically significant.

Results

Characterization of cell cultures

The cells isolated from Wharton jelly of the human umbilical cord were plastic adherent with a spindle-shaped, fibroblast-like morphology (Fig. 1b). Flow cytometry analysis showed that these cells were positive for the MSC markers CD105, CD73, and CD90 and were negative for CD11b, CD19, CD45, CD34, and HLA-DR (Fig. 1a).

UC-MSCs demonstrated multipotent differentiation potential. Sudan III staining of neutral lipid vacuoles showed that UC-MSCs could differentiate into adipocytes. Alizarin Red S staining of calcium compound crystals showed that UC-MSCs could differentiate into osteoblasts. Positive staining of mucopolysaccharides by Alcian blue indicated that UC-MSCs could differentiate into chondrocytes (Fig. 1b).

EA.hy926 cells formed a monolayer of closely apposed small polygonal cells. EA.hy926 cells were found to be positive for CD31 as endothelial marker (Fig. 1c).

VEGF-A-121 and VEGF-A-165 quantification in the conditioned media

Within 3 days, the concentration of soluble forms of VEGF-A (VEGF-A-121 and VEGF-A-165) in EA.hy926-conditioned media gradually increased from 57.3 ± 7.7 pg/ml to 229.0 ± 24.9 pg/ml, whereas in UC-MSC-conditioned media the contents of VEGF-A did not change, remaining at the level of the growth medium (Fig. 1d).

The influence of UC-MSC-conditioned media on endothelial cell proliferation

Results of the MTT assay indicated that EA.hy926 cells incubated in UC-MSC-conditioned media had a significant increase in cell viability after a 24-hour incubation when compared with cells incubated with growth media alone ($p < 0.05$). Consequently, the difference between the

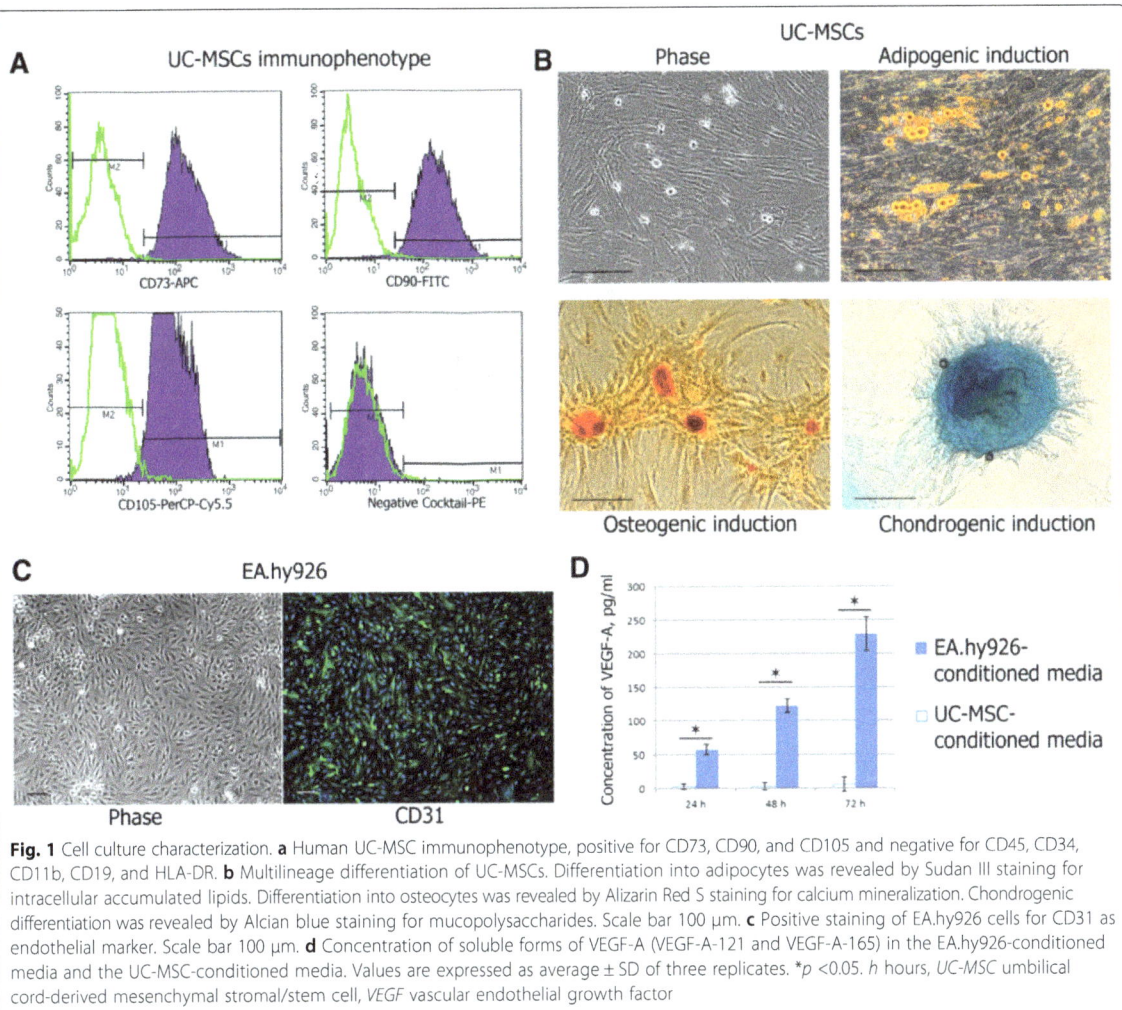

Fig. 1 Cell culture characterization. **a** Human UC-MSC immunophenotype, positive for CD73, CD90, and CD105 and negative for CD45, CD34, CD11b, CD19, and HLA-DR. **b** Multilineage differentiation of UC-MSCs. Differentiation into adipocytes was revealed by Sudan III staining for intracellular accumulated lipids. Differentiation into osteocytes was revealed by Alizarin Red S staining for calcium mineralization. Chondrogenic differentiation was revealed by Alcian blue staining for mucopolysaccharides. Scale bar 100 μm. **c** Positive staining of EA.hy926 cells for CD31 as endothelial marker. Scale bar 100 μm. **d** Concentration of soluble forms of VEGF-A (VEGF-A-121 and VEGF-A-165) in the EA.hy926-conditioned media and the UC-MSC-conditioned media. Values are expressed as average ± SD of three replicates. *$p < 0.05$. *h* hours, *UC-MSC* umbilical cord-derived mesenchymal stromal/stem cell, *VEGF* vascular endothelial growth factor

two media increased, and 72 hours of incubation resulted in almost 1.6-fold excess absorption in wells with UC-MSC-conditioned media compared with growth media. The addition of the VEGF-neutralizing antibody to the UC-MSC-conditioned media did not significantly attenuate the EA.hy926 cell proliferation as compared with the UC-MSC-conditioned media treatment ($p < 0.05$) (Fig. 2a).

The influence of UC-MSC-conditioned media on endothelial cell migration

Endothelial cell migration was evaluated using a transwell chamber migration assay. EA.hy926 cells migrated from the upper chamber to the lower surface of membrane through 8 μm pores when the lower chamber contained only the growth culture medium. When the lower chamber was seeded with UC-MSCs, the efficacy of EA.hy926 cell-directed migration increased significantly

at all time points ($p < 0.05$). The addition of the VEGF-neutralizing antibody to the lower chambers seeded with UC-MSCs did not significantly attenuate the EA.hy926 cell migration as compared with UC-MSC-conditioned media treatment ($p < 0.05$) (Fig. 2b).

Additionally we used a scratch wound healing assay of tissue-culture cell monolayers to measure the influence of UC-MSC-conditioned media on endothelial cell EA.hy926 migration. UC-MSC-conditioned media stimulated the motility of endothelial cells during the wound recovery; the difference became significant at 8 hours after scratching. For example, at 18 hours after scratching, wound confluence in the control wells was 52.68 ± 5.80 % while in the wells with UC-MSC-conditioned media this index reached 89.74 ± 5.63 %, and in the wells with UC-MSC-conditioned media supplemented with the VEGF-neutralizing antibody it reached 90.04 ± 5.26 %. The dynamics of wound confluence, illustrated

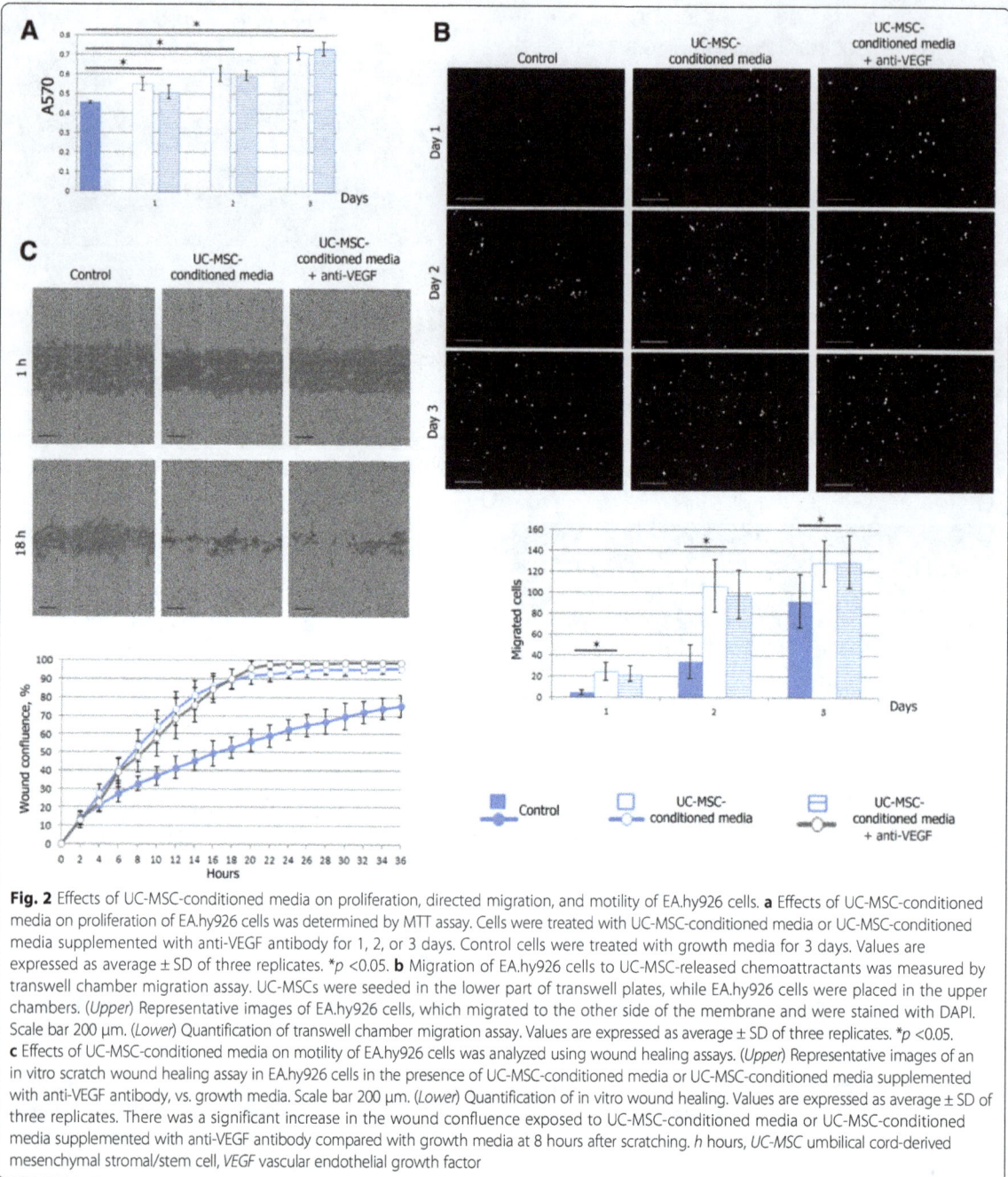

Fig. 2 Effects of UC-MSC-conditioned media on proliferation, directed migration, and motility of EA.hy926 cells. **a** Effects of UC-MSC-conditioned media on proliferation of EA.hy926 cells was determined by MTT assay. Cells were treated with UC-MSC-conditioned media or UC-MSC-conditioned media supplemented with anti-VEGF antibody for 1, 2, or 3 days. Control cells were treated with growth media for 3 days. Values are expressed as average ± SD of three replicates. *p <0.05. **b** Migration of EA.hy926 cells to UC-MSC-released chemoattractants was measured by transwell chamber migration assay. UC-MSCs were seeded in the lower part of transwell plates, while EA.hy926 cells were placed in the upper chambers. (*Upper*) Representative images of EA.hy926 cells, which migrated to the other side of the membrane and were stained with DAPI. Scale bar 200 μm. (*Lower*) Quantification of transwell chamber migration assay. Values are expressed as average ± SD of three replicates. *p <0.05. **c** Effects of UC-MSC-conditioned media on motility of EA.hy926 cells was analyzed using wound healing assays. (*Upper*) Representative images of an in vitro scratch wound healing assay in EA.hy926 cells in the presence of UC-MSC-conditioned media or UC-MSC-conditioned media supplemented with anti-VEGF antibody, vs. growth media. Scale bar 200 μm. (*Lower*) Quantification of in vitro wound healing. Values are expressed as average ± SD of three replicates. There was a significant increase in the wound confluence exposed to UC-MSC-conditioned media or UC-MSC-conditioned media supplemented with anti-VEGF antibody compared with growth media at 8 hours after scratching. *h* hours, *UC-MSC* umbilical cord-derived mesenchymal stromal/stem cell, *VEGF* vascular endothelial growth factor

with representative images from the time-lapse recording, is given in Fig. 2c.

Tube formation assay

The angiogenic capability of various cell types was assessed using an in vitro capillary-like structure (tube) formation assay on basement membrane matrix. As shown in Fig. 3a, both UC-MSCs and endothelial EA.hy926 cells were able

to form the networks on Matrigel, but the parameters of the networks were different. UC-MSCs and UC-MSCs mixed 1:1 with EA.hy926 cells already began sprouting 1 hour after cell seeding. These networks were unstable and began to disintegrate 3 hours later, just as EA.hy926 cells started sprouting. Additionally there was a difference between the network structures: EA.hy926 cells formed fine meshes with a greater number of branch points and shorter

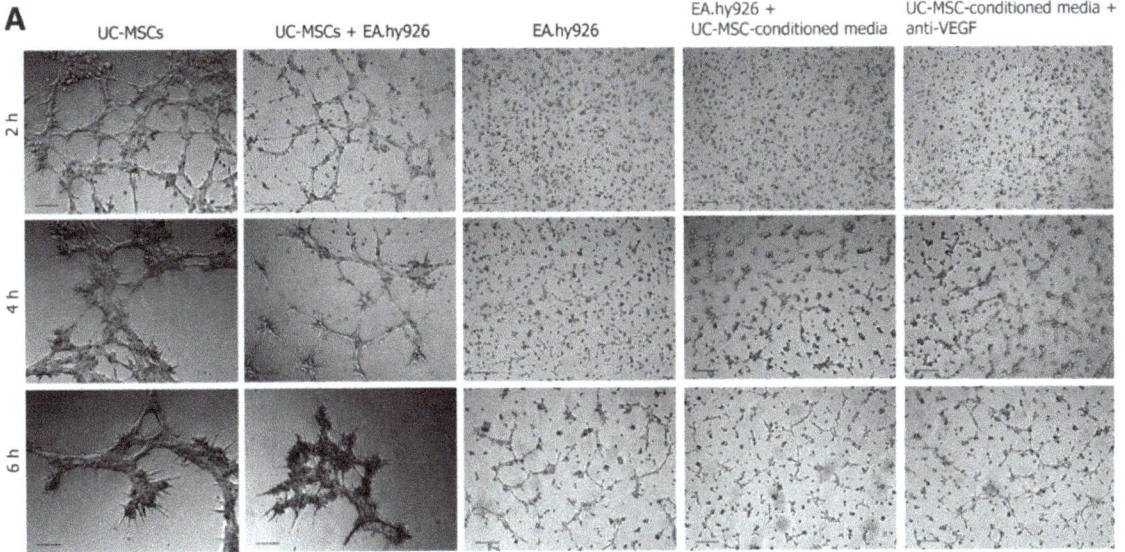

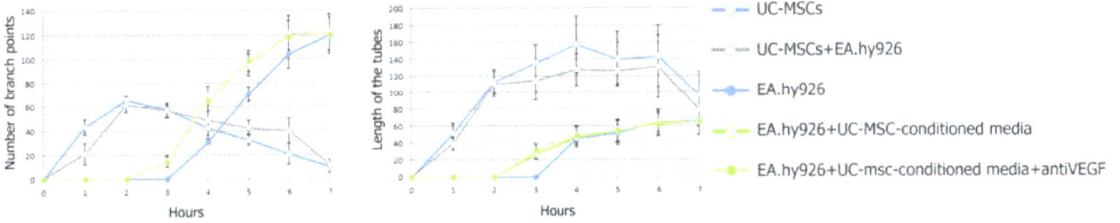

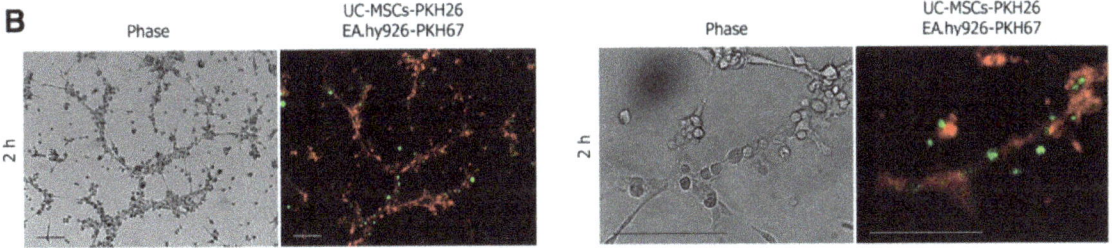

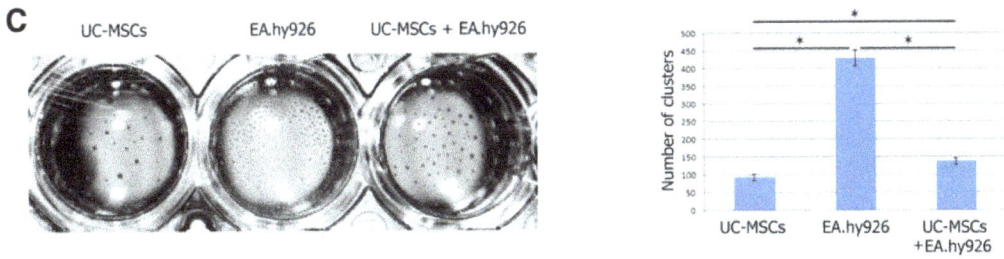

Fig. 3 (See legend on next page.)

tubes, while the UC-MSCs and UC-MSC-EA.hy926 cell mix formed coarse meshes with fewer branch points and longer tubes. The addition of UC-MSC-conditioned media (regardless of the presence of the VEGF-neutralizing antibody) to the wells with EA.hy926 cells did not significantly alter the parameters of the networks, but contributed to their formation 1 hour earlier (Fig. 3a).

Surprisingly, we found that PKH26-labeled UC-MSCs became the basis of a mixed culture network, while PKH67-labeled EA.hy926 cells were only associated with it (Fig. 3b).

In all groups, the networks were unstable and disintegrated into tight clusters for 24 hours. These clusters were not stationary structures. In a few days, they were capable of limited movement and fusion. The movement of cell clusters stopped in about 5–7 days, and the number of clusters was different between the groups: 430.0 ± 21.2 per well for EA.hy926 cells, 137.1 ± 9.2 per well for UC-MSC-EA.hy926 cell mix, and 93.0 ± 9.2 for UC-MSCs (Fig. 3c).

Further, we observed that the clusters formed by the EA.hy926 cells and UC-MSC-EA.hy926 cell mix, but not by UC-MSCs alone, became centers of secondary sprouting. Sprouting cells had a typical elongated shape. Gradually, the isolated sprouting centers joined into a single, very stable (follow-up of more than 30 days) 3D network with a plurality of branch points, often dichotomic (Fig. 4a). Moreover, in this mixed culture network only PKH26-labeled UC-MSCs formed sprouts while PKH67-labeled EA.hy926 cells stayed in the centers of the clusters (Fig. 4b).

Endothelial differentiation of UC-MSCs in monolayer
UC-MSCs cultured in complete fresh growth medium exhibited both the shape and the wave-like arrangement of the MSCs in the monolayer; the cell growth was restricted by mutual contact inhibition, and neither of them differentiated to the CD31+ phenotype.

When cultured in EA.hy926-conditioned medium, the cells retained the same characteristics but grew to higher densities. When cultured in EA.hy926-conditioned medium supplemented with VEGF-A-165, they formed distinct tubular structures (three to five per

35 mm dish), assembled from several dozens of narrow stretched cells with elongated nuclei, positively stained with CD31 antibody.

Finally, using VEGF-A-165 as a single growth supplement (except for the serum), added to complete fresh medium, led to a mosaic loss of contact inhibition. The cells started to grow in multiple layers, but formed no tubular structures and stayed CD31− (Fig. 5a).

Endothelial differentiation of UC-MSCs in Matrigel
Immunostaining of cryosections showed that, upon coculturing with EA.hy926 in Matrigel, the UC-MSCs started to express CD31 spontaneously, without additional VEGF-A-165 supplement (Fig. 5b).

Discussion
Weak secretion of VEGF-A by UC-MSCs derived from Wharton jelly may be related to the unusual structure of loose connective tissue forming Wharton jelly, and, in particular, the lack of blood capillaries in it. Although early events of hematopoiesis and capillary formation in this tissue are described in detail, by 7–9 weeks of development the hematopoiesis in Wharton jelly ceases, and the capillaries undergo regression [19]. It is plausible that these changes, as well as subsequent maintenance of the anti-angiogenic environment, are accompanied, or mediated, by low concentrations of soluble VEGF-A in the intercellular spaces. According to some authors, the Wharton jelly-derived UC-MSCs are able to secrete soluble forms of VEGF-A [20]; however, the majority of the reports (including this one) mention the almost complete absence of VEGF-A protein from the UC-MSC-conditioned culture medium as a specific feature reflecting VEGF-A deficiency of UC-MSC secretome. Typical levels of VEGF-A secretion reported for UC-MSCs are 10^2 less than for adipose tissue-derived MSCs and 10^3 less than for bone marrow-derived MSCs, despite detectable levels of transcription of the corresponding gene [12, 13]. Nevertheless, UC-MSCs can effectively accelerate migration and promote tube formation from endothelial cells in vitro—the effect is mediated by the UC-MSC-conditioned media; that is, by in vitro modeling of in vivo paracrine mechanisms [12].

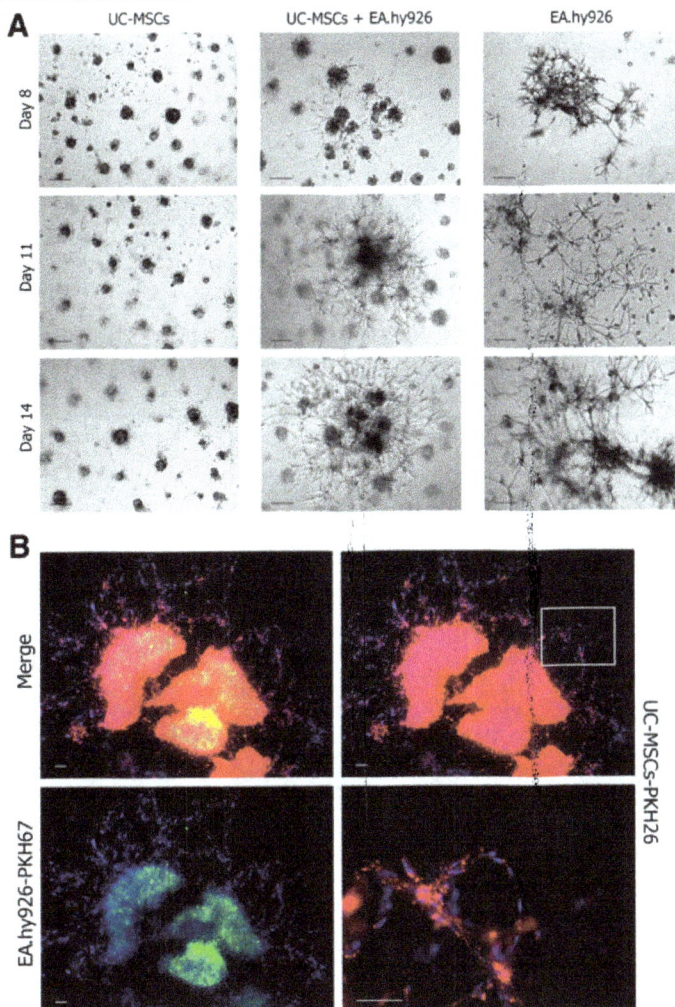

Fig. 4 Secondary sprouting in Matrigel. **a** Clusters formed in Matrigel by the EA.hy926 cells and UC-MSC-EA.hy926 cell mix, but not by UC-MSCs alone, became centers of secondary sprouting. Gradually, the isolated sprouting centers joined into a single, very stable 3D network. Scale bar 100 μm. **b** In a mixed culture network only PKH26-labeled UC-MSCs (*red*) formed sprouts while PKH67-labeled EA.hy926 cells (*green*) stayed in the centers of the clusters. Scale bar 100 μm. *UC-MSC* umbilical cord-derived mesenchymal stromal/stem cell (Color figure online)

UC-MSCs implement their pro-angiogenic potential via some VEGF-A-independent mechanism. Why should this matter? The problem is that intermediate results of clinical trials using VEGF-A-121 or VEGF-A-165 (as active exogenous proteins or in the form of genetic constructs) have been qualified as rather contradictory: the effects sometimes deviate from those expected [1, 21]. This may be explained by duality that originates from the level of VEGF-A binding to its receptor, vascular endothelial growth factor receptor (VEGFR). By acting via VEGFR2, VEGF-A increases survival and proliferation of endothelial cells, as well as recruitment of other progenitors to the site of injury, thus supporting formation and maturation of new blood vessels; in contrast, the VEGFR1-mediated action of VEGF-A is anti-angiogenic [1]. In some circumstances, excessive influx of VEGF-A (either exogenous for cells in vitro, or endogenous for ischemia models in vivo) may dysregulate intrinsic VEGFR balance of target cells and switch them to VEGFR1 expression or may upregulate soluble VEGFR1 expression that can operate as a negative feedback system, thereby undermining the entire positive effect of the treatment [22–24]. Moreover, the qualitative and quantitative balance of VEGFR1 and VEGFR2 proteins may vary in human populations, further complicating the proper choice of treatment for some cases [23]. Considering this, any VEGF-A-based pro-angiogenic therapy may prove to be more reliable and lead to more

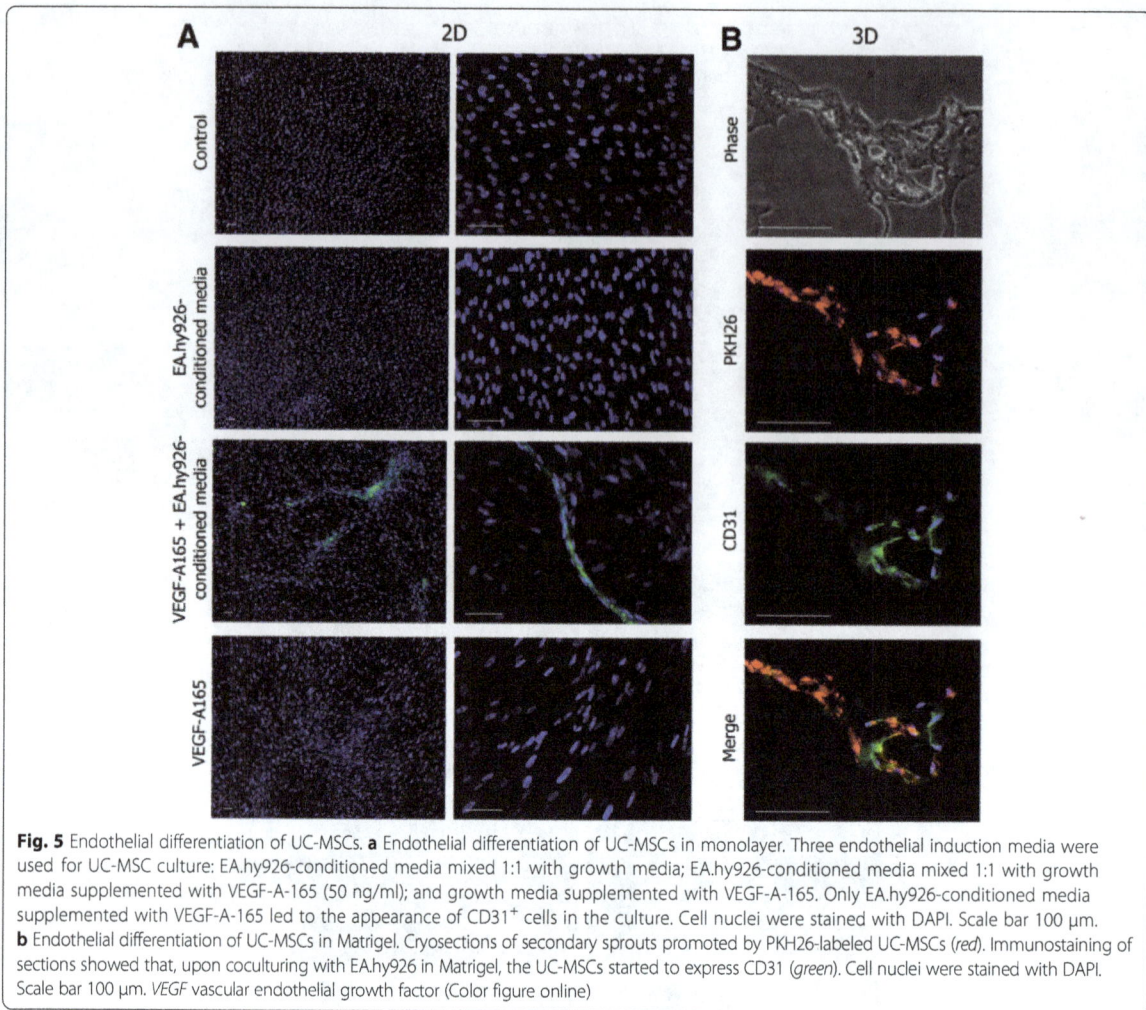

Fig. 5 Endothelial differentiation of UC-MSCs. **a** Endothelial differentiation of UC-MSCs in monolayer. Three endothelial induction media were used for UC-MSC culture: EA.hy926-conditioned media mixed 1:1 with growth media; EA.hy926-conditioned media mixed 1:1 with growth media supplemented with VEGF-A-165 (50 ng/ml); and growth media supplemented with VEGF-A-165. Only EA.hy926-conditioned media supplemented with VEGF-A-165 led to the appearance of CD31+ cells in the culture. Cell nuclei were stained with DAPI. Scale bar 100 μm. **b** Endothelial differentiation of UC-MSCs in Matrigel. Cryosections of secondary sprouts promoted by PKH26-labeled UC-MSCs (*red*). Immunostaining of sections showed that, upon coculturing with EA.hy926 in Matrigel, the UC-MSCs started to express CD31 (*green*). Cell nuclei were stained with DAPI. Scale bar 100 μm. *VEGF* vascular endothelial growth factor (Color figure online)

predictable results when supported by some additional VEGF-A-independent line of intervention (e.g., using UC-MSCs).

Published evidence for the stimulating influence of MSCs on endothelial cell proliferation is rather controversial because of the variety of sources and methods of obtaining the cells [2]. For example, it is shown that bone marrow-derived MSCs (including cells cultured under hypoxic conditions) have no effect on EA.hy926 cell growth [25]. In our experiments, the UC-MSC-conditioned media stimulated proliferation of EA.hy926 cells; this is consistent with results reported by Choi et al. [26] for a different endothelial line (HUVEC). The absence of VEGF-A from UC-MSC-conditioned media suggests that the endothelial cells respond to a different sort of inducer (possibly VEGF-B, the positive effect of which on EA.hy926 cell proliferation was confirmed in a

recent study [27], but it is still questionable whether VEGF-B is produced by MSCs).

The transwell systems are widely used to assess chemotaxis, which plays an important role during the early stages of angiogenesis. In our setting, UC-MSCs secreted chemoattractants for EA.hy926 cells. Similar results have been reported previously for other cell lines – HUVEC, human microvascular endothelial cell line HMEC1, and mouse neural crest-derived cell line N2a, the effect of UC-MSCs being more pronounced compared with bone marrow-derived MSCs [26, 28, 29].

Similarly, the UC-MSC-conditioned medium increased the mobility of EA.hy926 endothelial cells in the monolayer scratch experiments; Bronckaers et al. [2] demonstrate the same effect for bone marrow-derived MSCs. The "scratch assay" is a common way to analyze proliferation as combined with directed migration of the cells in vitro [30]; exactly these processes play a central role in angiogenesis [31].

UC-MSCs thus secrete factors that may attract endothelial and progenitor cells, while stimulating their mobility; what kind of factors in particular could be partly deduced from the literature. Interleukin (IL)-8 is shown to induce cytoskeleton rearrangement and directed migration of EA.hy926 cells by activation of p38 mitogen-activated protein kinase signaling [32]. The rate of migration of endothelial cells in vitro is also shown to depend on hepatocyte growth factor (HGF) and monocyte chemoattractant protein-1 (MCP-1) levels in UC-MSC-conditioned media [28]. These findings are consistent with other research showing that secretion of IL-8, HGF, and MCP-1 by UC-MSCs is significantly more intensive than by MSCs derived from bone marrow or adipose tissue [13, 20].

Published data on the possibilities of endothelial differentiation of MSCs themselves are rather contradictory. There exist several protocols that differ in the composition of inducers (most widely used is VEGF-A-165 at 50 ng/ml), duration of the process of differentiation (takes from 2 to 28 days), and selection of molecular markers for the control immunostaining (CD31, von Willebrand factor (vWF), vascular endothelial cadherin (VE-cadherin), and VEGFR2 are the most common); accordingly, the final products of these protocols vary greatly [33]. At the same time, designating the differentiated MSCs as fully mature and functional endothelial cells is considered inaccurate; it is therefore more correct to define these cells as endothelial-like cells [2].

In our experiments, the UC-MSCs were capable of differentiation to the CD31$^+$ phenotype under the influence of differentiation medium containing VEGF-A-165 as an essential, although insufficient, inducer. In contrast, Choi et al. [26] observe no expression of endothelial markers by UC-MSCs after treatment with complex differentiation media containing epidermal growth factor (EGF), VEGF, basic fibroblast growth factor, insulin-like growth factor-1, hydrocortisone, and some other potential inducers. In yet another study, UC-MSCs treated with media containing VEGF, EGF, and hydrocortisone started to express endothelial markers (vWF, VE-cadherin, and VEGFR2) uniformly, without any changes in cell organization or cell morphology [34]. Possibilities of endothelial differentiation of MSCs in vivo are even more questionable [2]. One of the reasons for this is the low level of VEGF-A in ischemic tissues: it is about 10^3 times lower than in standard endothelial differentiation media (50 ng/ml) [35], and roughly corresponds to the VEGF-A level in the EA.hy926-conditioned media (Fig. 1d).

The networks formed in coculture of UC-MSCs with EA.hy926 cells on Matrigel were similar to the networks formed by pure UC-MSCs (judging by their assembly time, length of the tubes, and branching point number). The core of the mixed networks was composed of the PKH26-labeled MSCs, while the PKH67-labeled EA.hy926 cells were associated with the outer surface of this core (Fig 3a). Such an arrangement of cell types in mixed networks differs from that reported previously: other authors attribute only a minor role to MSCs [26, 36, 37]. The inconsistency probably relates to a different proportion of cell types taken for the network priming.

All tubular networks observed in our setting were unstable. Independently of whether they were formed by UC-MSCs combined with EA.hy926 cells or either of the cell types on their own, the networks underwent spontaneous disassembling in the course of 24 hours, producing tight cell clusters; this is consistent with previously published data [26, 36, 38]. These clusters, resulting from 2D network disassembly, subsequently turned into sprouting centers producing a single stable 3D network. Similar results are reported by Portalska et al. [38], who observed in vitro assembly of blood vessel-like structures from bone marrow-derived MSCs predifferentiated to an endothelial-like phenotype: the cells started to form a network with a 20-hour delay (as compared with the native undifferentiated MSCs), and this network remained stable for at least 7 days. In pure cultures of endothelial cells, as well in two of the five mixed cultures, the sprouting occurred invariably, but all five pure UC-MSC cultures showed no signs of the sprouting. This confirms the idea that individual MSC cultures, equally complying with the standards, may show morphological and functional variation [39].

Overall, the results indicate that the ability of UC-MSCs to participate in sprouting, manifested in cocultures with EA.hy926 cells on Matrigel, is a consequence of their differentiation to an endothelial-like phenotype (especially given that the signals from the local environment, either through cell–cell contact, soluble factors, or cell–matrix interactions, profoundly influence MSC endothelial differentiation [37]). We also assume that the low reproducibility of sprouting indexes between individual cultures of UC-MSCs is caused mainly by unequal susceptibility of these cultures to specific endothelial differentiation stimuli. Notably, acquisition of the CD31$^+$ phenotype by UC-MSCs in long-term coculture with EA.hy926 cells on Matrigel occurred in the absence of exogenous VEGF-A. It is therefore possible that the role of VEGF-A in endothelial differentiation of MSCs is not so significant as believed previously; this inference is substantiated by the fact that MSCs do not express membrane-anchored VEGF receptors, although VEGF-A can signal through platelet-derived growth factor receptors [40].

Conclusions

Many of the studies investigating the paracrine factors secreted from MSCs derived from various sources have reported the presence of VEGF-A and have implicated its importance in angiogenesis. In this study, we confirmed that MSCs derived from Wharton jelly of the human umbilical cord produced no detectable quantities of soluble VEGF-A (VEGF-A-121 and VEGF-A-165); despite this, culture medium conditioned by UC-MSCs effectively stimulated the proliferation, motility, and directed migration of endothelial EA.hy926 cells. These data suggest that a VEGF-A-independent paracrine mechanism is involved in the pro-angiogenic activity of UC-MSCs.

In our experiments, the UC-MSCs were capable of differentiation to an endothelial cell-like CD31$^+$ phenotype under the influence of differentiation medium containing VEGF-A-165 as an essential, although insufficient, inducer. However, we found that the ability of UC-MSCs to participate in secondary sprouting, as manifested in long-term cocultures with EA.hy926 cells on Matrigel, is a consequence of their differentiation to an endothelial-like CD31$^+$ phenotype in the absence of exogenous VEGF-A. We can assume that signals from the local environment, either through UC-MSC–EA.hy926 cell contact or UC-MSC–basement membrane matrix interactions, profoundly influenced UC-MSC differentiation. Endothelial differentiation as one of the proposed mechanisms of action for UC-MSC transplantation can thus also be partially VEGF-A independent.

The conclusions of this study have practical applications in the field of pro-angiogenic therapy: VEGF-A-based therapy supported by an additional VEGF-A-independent line of intervention (e.g., using UC-MSC transplantation) may have higher efficacy.

Abbreviations

ANOVA: Analysis of variance; DAPI: 4′,6-Diamidino-2-phenylindole; DMEM: Dulbecco's modified Eagle's medium; DMSO: Dimethyl sulfoxide; EGF: Epidermal growth factor; FBS: Fetal bovine serum; FITC: Fluorescein isothiocyanate; HGF: Hepatocyte growth factor; HUVEC: Human umbilical vein endothelial cells; IL: Interleukin; MCP-1: Monocyte chemoattractant protein-1; MSC: Mesenchymal stromal/stem cell; MTT: 3-(4,5-Dimethylthiazol-2-yl)-2,5-diphenyltetrazolium bromide; PBS: Phosphate-buffered saline; SD: Standard deviation; UC-MSC: Umbilical cord-derived mesenchymal stromal/stem cell; VE-cadherin: Vascular endothelial cadherin; VEGF: Vascular endothelial growth factor; VEGFR: Vascular endothelial growth factor receptor; vWF: Von Willebrand factor.

Competing interests

The authors declare that they have no competing interests.

Authors' contributions

IA and TF designed the study, coordinated the research, wrote most of the text, and composed the figures. IA, EK, AE, NU, and AM performed the experiments, and collected and analyzed the data. AE performed the statistical analysis. NU, GB, and DG participated in the manuscript writing. GS conceived of the study. All authors have been involved in drafting and consequent critical revision of all versions of the manuscript and the figures for

important intellectual content and way of presentation. All authors read and approved the manuscript.

Acknowledgements

The authors thank Irina Teveleva for careful proofreading of the text. This work was supported by Ministry of Education and Science of the Russian Federation (unique identification code: RFMEFI61314X0008).

Author details

[1]Research Center for Obstetrics, Gynecology and Perinatology of Ministry of Healthcare of the Russian Federation, 4 Oparina Street, Moscow 117997, Russia. [2]Scientific Research Institute of Human Morphology, 3 Tsurupa Street, Moscow 117418, Russia. [3]Pirogov Russian National Research Medical University, Ministry of Healthcare of the Russian Federation, 1 Ostrovitianov Street, Moscow 117997, Russia. [4]Laboratory of Regenerative Medicine, Research Center for Obstetrics, Gynecology and Perinatology, 4 Oparin Street, Moscow 117997, Russia. [5]Research Center of Medical Genetics, 1 Moskvorechie Street, Moscow 115478, Russia.

References

1. Ylä-Herttuala S, Rissanen TT, Vajanto I, Hartikainen J. Vascular endothelial growth factors: biology and current status of clinical applications in cardiovascular medicine. J Am Coll Cardiol. 2007;49(10):1015–26.
2. Bronckaers A, Hilkens P, Martens W, Gervois P, Ratajczak J, Struys T, et al. Mesenchymal stem/stromal cells as a pharmacological and therapeutic approach to accelerate angiogenesis. Pharmacol Ther. 2014;143(2):181–96.
3. Liew A, O'Brien T. Therapeutic potential for mesenchymal stem cell transplantation in critical limb ischemia. Stem Cell Res Ther. 2012;3(4):28.
4. Yan J, Tie G, Xu TY, Cecchini K, Messina LM. Mesenchymal stem cells as a treatment for peripheral arterial disease: current status and potential impact of type II diabetes on their therapeutic efficacy. Stem Cell Rev. 2013;9(3):360–72.
5. Lin RZ, Moreno-Luna R, Zhou B, Pu WT, Melero-Martin JM. Equal modulation of endothelial cell function by four distinct tissue-specific mesenchymal stem cells. Angiogenesis. 2012;15(3):443–55.
6. Taghizadeh RR, Cetrulo KJ, Cetrulo CL. Wharton's Jelly stem cells: future clinical applications. Placenta. 2011;32 Suppl 4:S311–5.
7. De Kock J, Najar M, Bolleyn J, Al Battah F, Rodrigues RM, Buyl K, et al. Mesoderm-derived stem cells: the link between the transcriptome and their differentiation potential. Stem Cells Dev. 2012;21(18):3309–23.
8. Gauthaman K, Fong CY, Suganya CA, et al. Extra-embryonic human Wharton's jelly stem cells do not induce tumorigenesis, unlike human embryonic stem cells. Reprod Biomed Online. 2012;24(2):235–46.
9. Li X, Bai J, Ji X, Li R, Xuan Y, Wang Y. Comprehensive characterization of four different populations of human mesenchymal stem cells as regards their immune properties, proliferation and differentiation. Int J Mol Med. 2014;34(3):695–704.
10. Lv F, Lu M, Cheung KM, Leung VY, Zhou G. Intrinsic properties of mesemchymal stem cells from human bone marrow, umbilical cord and umbilical cord blood comparing the different sources of MSC. Curr Stem Cell Res Ther. 2012;7(6):389–99.
11. Santos Nascimento D, Mosqueira D, Sousa LM, et al. Human umbilical cord tissue-derived mesenchymal stromal cells attenuate remodeling after myocardial infarction by proangiogenic, antiapoptotic, and endogenous cell-activation mechanisms. Stem Cell Res Ther. 2014;5(1):5.
12. Kuchroo P, Dave V, Vijayan A, Viswanathan C, Ghosh D. Paracrine factors secreted by umbilical cord-derived mesenchymal stem cells induce angiogenesis in vitro by a VEGF-independent pathway. Stem Cells Dev. 2015;24(4):437–50.
13. Amable PR, Teixeira MV, Carias RB, Granjeiro JM, Borojevic R. Protein synthesis and secretion in human mesenchymal cells derived from bone marrow, adipose tissue and Wharton's jelly. Stem Cell Res Ther. 2014;5(2):53.
14. Dominici M, Le Blanc K, Mueller I, Slaper-Cortenbach I, Marini F, Krause D, et al. Minimal criteria for defining multipotent mesenchymal stromal cells. The International Society for Cellular Therapy position statement. Cytotherapy. 2006;8(4):315.
15. Edgell CJ, McDonald CC, Graham JB. Permanent cell line expressing human factor VIII-related antigen established by hybridization. Proc Natl Acad Sci U S A. 1983;80(12):3734–7.

16. Aranda E, Owen GI. A semi-quantitative assay to screen for angiogenic compounds and compounds with angiogenic potential using the EA.hy926 endothelial cell line. Biol Res. 2009;42(3):377–89.

17. Bouïs D, Hospers GA, Meijer C, Molema G, Mulder NH. Endothelium in vitro: a review of human vascular endothelial cell lines for blood vessel-related research. Angiogenesis. 2001;4(2):91–102.

18. Arnaoutova I, George J, Kleinman HK, Benton G. The endothelial cell tube formation assay on basement membrane turns 20: state of the science and the art. Angiogenesis. 2009;12(3):267–74.

19. Robaczyński J. Development of capillaries in Wharton's jelly of the human umbilical cord in the course of ontogenesis. Folia Morphol (Warsz). 1967; 26(4):371–82.

20. Edwards SS, Zavala G, Prieto CP, Elliott M, Martínez S, Egaña JT, et al. Functional analysis reveals angiogenic potential of human mesenchymal stem cells from Wharton's jelly in dermal regeneration. Angiogenesis. 2014;17(4):851–66.

21. Shimamura M, Nakagami H, Taniyama Y, Morishita R. Gene therapy for peripheral arterial disease. Expert Opin Biol Ther. 2014;14(8):1175–84.

22. Imoukhuede PI, Popel AS. Quantification and cell-to-cell variation of vascular endothelial growth factor receptors. Exp Cell Res. 2011;317(7):955–65.

23. Imoukhuede PI, Dokun AO, Annex BH, et al. Endothelial cell-by-cell profiling reveals temporal dynamics of VEGFR1 and VEGFR2 membrane-localization following murine hindlimb ischemia. Am J Physiol Heart Circ Physiol. 2013; 304(8):H1085–93.

24. Saito T, Takeda N, Amiya E, Nakao T, Abe H, Semba H, et al. VEGF-A induces its negative regulator, soluble form of VEGFR-1, by modulating its alternative splicing. FEBS Lett. 2013;587(14):2179–85.

25. Burlacu A, Grigorescu G, Rosca AM, Preda MB, Simionescu M. Factors secreted by mesenchymal stem cells and endothelial progenitor cells have complementary effects on angiogenesis in vitro. Stem Cells Dev. 2013;22(4):643–53.

26. Choi M, Lee HS, Naidansaren P, Kim HK, O E, Cha JH, et al. Proangiogenic features of Wharton's jelly-derived mesenchymal stromal/stem cells and their ability to form functional vessels. Int J Biochem Cell Biol. 2013;45(3):560–70.

27. Zhang GH, Qin R, Zhang SH, Zhu H. Effects of vascular endothelial growth factor B on proliferation and migration in EA.Hy926 cells. Mol Biol Rep. 2014; 41(2):779–85.

28. Shen C, Lie P, Miao T, Yu M, Lu Q, Feng T, et al. Conditioned medium from umbilical cord mesenchymal stem cells induces migration and angiogenesis. Mol Med Rep. 2015;12(1):20–30.

29. Hsieh JY, Wang HW, Chang SJ, Liao KH, Lee IH, Lin WS, et al. Mesenchymal stem cells from human umbilical cord express preferentially secreted factors related to neuroprotection, neurogenesis, and angiogenesis. PLoS One. 2013;8(8):e72604.

30. Rodriguez LG, Wu X, Guan JL. Wound-healing assay. Methods Mol Biol. 2005;294:23–9.

31. Isner JM, Vale P, Symes J, Losordo DW, Asahara T. Angiogenesis and cardiovascular disease. Dialog Cardiovasc Med. 2001;6(3):145–72.

32. Lai Y, Liu XH, Zeng Y, Zhang Y, Shen Y, Liu Y. Interleukin-8 induces the endothelial cell migration through the Rac 1/RhoA-p38MAPK pathway. Eur Rev Med Pharmacol Sci. 2012;16(5):630–8.

33. Vater C, Kasten P, Stiehler M. Culture media for the differentiation of mesenchymal stromal cells. Acta Biomater. 2011;7(2):463–77.

34. Chen MY, Lie PC, Li ZL, Wei X. Endothelial differentiation of Wharton's jelly-derived mesenchymal stem cells in comparison with bone marrow-derived mesenchymal stem cells. Exp Hematol. 2009;37(5):629–40.

35. Jiang Q, Ding S, Wu J, Liu X, Wu Z. Norepinephrine stimulates mobilization of endothelial progenitor cells after limb ischemia. PLoS One. 2014;9(7):e101774.

36. Blocki A, Wang Y, Koch M, Peh P, Beyer S, Law P, et al. Not all MSCs can act as pericytes: functional in vitro assays to distinguish pericytes from other mesenchymal stem cells in angiogenesis. Stem Cells Dev. 2013;22(17):2347–55.

37. Lozito TP, Kuo CK, Taboas JM, Tuan RS. Human mesenchymal stem cells express vascular cell phenotypes upon interaction with endothelial cell matrix. J Cell Biochem. 2009;107(4):714–22.

38. Portalska K, Leferink A, Groen N, Fernandes H, Moroni L, van Blitterswijk C, et al. Endothelial differentiation of mesenchymal stromal cells. PLoS One. 2012;7(10):e46842.

39. Pacini S, Petrini I. Are MSCs angiogenic cells? New insights on human nestin-positive bone marrow-derived multipotent cells Front. Cell Dev Biol. 2014;2:20.

40. Ball SG, Shuttleworth CA, Kielty CM. Vascular endothelial growth factor can signal through platelet-derived growth factor receptors. J Cell Biol. 2007; 177(3):489–500.

15

The trans-well coculture of human synovial mesenchymal stem cells with chondrocytes leads to self-organization, chondrogenic differentiation, and secretion of TGFβ

Eva Johanna Kubosch[1], Emanuel Heidt[1], Anke Bernstein[1], Katharina Böttiger[1] and Hagen Schmal[2,3]*

Abstract

Background: Synovial mesenchymal stem cells (SMSC) possess a high chondrogenic differentiation potential, which possibly supports natural and surgically induced healing of cartilage lesions. We hypothesized enhanced chondrogenesis of SMSC caused by the vicinity of chondrocytes (CHDR).

Methods: Human SMSC and CHDR interactions were investigated in an in-vitro trans-well monolayer coculture over a time period of up to 21 days. Protein expression was analyzed using histology, immunostaining, or enzyme-linked immunosorbent assay. Additionally, mRNA expression was assessed by quantitative PCR.

Results: After 7 days, phase-contrast microscopy revealed cell aggregation of SMSC in coculture with CHDR. Afterwards, cells formed spheres and lost adherence. However, this phenomenon was not observed when culturing SMSC alone. Fluorescence labeling showed concurrent collagen type II expression. Addition of transforming growth factor beta (TGFβ) to the cocultures induced SMSC aggregation in less time and with higher intensity. Additionally, alcian blue staining demonstrated enhanced glycosaminoglycan expression around SMSC aggregates after 1 and 2 weeks. Although TGFβ mRNA was expressed in all SMSC, the protein was measured with constantly increasing levels over 21 days only in supernatants of the cocultures. Considering the enhanced mRNA levels following supplementation with TGFβ, a positive feedback mechanism can be supposed. In line with the development of a chondrogenic phenotype, aggrecan mRNA expression increased after 7 and 14 days in the cocultures with and without TGFβ. Coculture conditions also amplified collagen type II mRNA expression after 2 weeks without and already after 1 week with TGFβ. There was no difference in collagen type I and type X expression between SMSC alone and the coculture with CHDR. Expression of both collagens increased following addition of TGFβ. mRNA data correlated with the intensity of immunofluorescence staining.

Conclusions: Paracrine effects of CHDR induce a chondrogenic phenotype in SMSC possibly mimicking joint homeostasis. Coculture approaches may lead to a better understanding of cellular interactions with potential implications for cartilage repair procedures.

Keywords: Synovial mesenchymal stem cell, Coculture, Differentiation, Chondrocytes, Synovium, Chondrogenesis

* Correspondence: hagen.schmal@freenet.de
[2]Department of Orthopaedics and Traumatology, Odense University Hospital, Sdr. Boulevard 29, 5000 Odense C, Denmark
[3]Department of Clinical Research, University of Southern Denmark, Odense, Denmark
Full list of author information is available at the end of the article

Background

Cartilage lesions have a limited capacity for repair and cause osteoarthritis (OA), so the search for treatment alternatives is ongoing. Many current approaches focus on the use of mesenchymal stem cells, which may play a significant role in both natural and surgically supported cartilage repair. Previously it has been described that natural cartilage repair can occur, especially in osteochondral defects; however, "the repair was mediated wholly by the proliferation and differentiation of mesenchymal cells of the marrow" [1]. More recently, a possible role for synovial mesenchymal stem cells (SMSC) was highlighted [2]. Injection of bone marrow-derived mesenchymal stem cells (BMSC) in osteoarthritic knees resulted in a long-term improvement of clinical outcome parameters [3], and SMSC were successfully used for arthroscopically assisted cartilage repair resulting in improved MRI features, histology, and clinical outcome [4]. Different studies suggested that SMSC have the best chondrogenic potential compared with mesenchymal stem cells derived from other tissue sources [5]. Adherence of SMSC to cartilage was mediated by hyaluronan [6], a possible mechanism for how these cells may be enriched in cartilage lesions. When SMSC migrate or are surgically placed at the site of a cartilage defect, they are in the direct vicinity of chondrocytes (CHDR) in their natural habitat. For this specific situation, coculture models are a powerful instrument to define and clarify cell–cell interactions. Until now the emphasis of SMSC/CHDR cocultures was to show effects in acute [7] or chronic [8] inflammation. Hereby, it could be demonstrated that SMSC were able to secret typical cartilage markers such as aggrecan and decisively influence the course of inflammation. Furthermore, pellet cocultures of mesenchymal stem cells, usually bone marrow derived, and CHDR resulted in formation of hyaline structured cartilage showing partially an even higher quality than the CHDR control group [9]. This phenomenon was independent of certain culture conditions and cell sources [10]. The disadvantage of this experimental approach is the missing possibility to differentiate between paracrine effects and cell–cell interactions. The hypothesis of the study was that CHDR are able to induce a chondrogenic differentiation of synovial stem cells. Since we presumed paracrine signals originating from CHDR causing this phenomenon, a coculture model was chosen where the cells were separated by a filter. This model does not allow direct cell contact and mimics the biological situation of cells collecting in a lesion with only marginal contact to the original cartilage layer. For evaluation, the markers of chondrogenesis aggrecan and collagen type II, the marker of dedifferentiation collagen type I, and the hypertrophy marker collagen type X were analyzed on RNA and protein levels. Emphasis of the

study was the histological observation of cell organization, the time frame, and the influencing cytokines.

Methods

Isolation, culture, and importance of SMSC in inflammation have been described previously [7]. The cells' preparation protocols were approved by the Ethics Committee of the University of Freiburg as part of the "Tissue Bank for Research in the Field of Tissue Engineering" project (GTE-2002) and the biobank "Osteo" (AN-EK-FRBRG-135/14). Cells from the same donors were used when comparing different culture conditions.

Isolation of SMSC

The cell preparation was described before [7]. Briefly, synovial tissue was gathered during knee operations with arthrotomy and arthroscopies ($n = 4$, male/female 2/2, average age 42.7 ± 15.0 years). The degree of OA was evaluated on X-ray images using Croft's modification of the Kellgren and Lawrence score (KLS). Cells were used only from patients with healthy joints (KLS ≤ 2). The harvested tissue was kept in DMEM F-12 medium (Lonza BioWhittaker, Basel, Switzerland) at 4 °C. Within 2 h the tissue was cut into small pieces, washed, and transferred into DMEM F-12 medium with 10 % FCS (Biochrom, Berlin, Germany), 1 % penicillin/streptomycin (P/S) (Invitrogen, Karlsruhe, Germany), 0.5 % gentamycin (Biochrom), and 3 % collagenase P (Roche, Mannheim, Germany). The suspension was digested during the next 4 h on a shaking incubator (200 rpm) at 37 °C. Subsequently the released cells were centrifuged, washed, and seeded in expansion medium DMEM F-12 (10 % FCS, 1 % P/S, 0.5 % gentamycin). SMSC were seeded on coated T-flasks with a density of 2500–5000 cells/cm^2 for expansion. The cells were frozen after reaching confluence. Thawed cells were grown and used when reaching a log phase of growth (passage 1). These cells were not further enriched and are also known as synovial fibroblasts or type B synoviocytes [11, 12]. Characterization of these cells was done by FACS showing that combined expression of the stem cell markers CD44, CD73, CD90, and CD105 was present in 76 %, but the combined expression of the negative markers CD11b, CD19, CD34, CD45, and HLA-DR reached only 6.9 ± 1.7 %. Osteogenic, adipogenic, and chondrogenic differentiation was possible using standard protocols [13].

Isolation of CHDR

Cell preparation was described before [14]. Briefly, CHDR were gained from femoral heads during hip arthroplasty operations ($n = 6$, male/female 5/1, average age 79.2 ± 8.2 years). The degree of OA was evaluated on X-ray images using Croft's modification of the KLS score. Cells from patients with advanced OA ($\geq$ grade 3) were not used for experiments. Within 8 h after surgery, the cartilage was separated from the bone and cut into small pieces, washed, and transferred into DMEM F-12 10 % FCS, 1 % P/S, 0.5 % gentamycin, and 3 % collagenase CLS type II (Biochrom). Minced cartilaginous tissue was then enzymatically digested during the next 16 h on a shaking incubator at 37 °C with 200 rpm. Subsequently the released CHDR were centrifuged, washed, and seeded in expansion medium DMEM F-12 (Lonza BioWhittaker) (10 % FCS, 1 % P/S, 0.5 % gentamycin). Expansion of CHDR was performed by seeding them on coated T-flasks with a density of 2500–5000 cells/cm^2. The cells were frozen after reaching confluence. Thawed cells were grown and used when reaching a log phase of growth (passage 1).

Coculture conditions
SMSC in the bottom and CHDR on top were separated in a trans-well culture with 0.4 μm inserts. As a basal culture medium, DMEM F-12 medium supplemented with 10 % FCS, 1 % P/S, and 0.5 % gentamycin was used. Cell viability was >95 % before starting the experiment. Half-media changes were performed three times per week. The initial seeding density was 20,000 cells/cm^2. Experiments were repeated at least three times with cells from different donors (one patient for one experimental trial). The cells were not pooled. The total time of coculture was 21 days maximum. There were three different groups: SMSC alone, SMSC with CHDR, and SMSC with CHDR supplemented with transforming growth factor beta (TGFβ)-3. A concentration of 10 ng/ml TGF-β3 (R&D Systems, Minneapolis, MN, USA) was added to the positive controls.

TGFβ enzyme-linked immunosorbent assay
TGFβ levels in supernatants were analyzed by enzyme-linked immunosorbent assay (ELISA) (R&D; and Bio-Source Deutschland GmbH, Solingen, Germany) according to the manufacturers' instructions. Briefly, this assay employs the quantitative sandwich enzyme immunoassay technique. The microplate was precoated with a specific monoclonal antibody. Supernatants were applied to the wells and, after washing, HRP-conjugated specific antibody was added to the wells. Following the next wash, color development was proportional to the protein concentration and calculated by comparison with a standard. A colorimetric method was applied to quantify the total protein amount in the lavage fluids.

Histology
Cover slides were coated with poly-D-lysine (0.1 mg/ml) (Merck Millipore, Billerica, Massachusetts, USA) at 37 °C (5 % CO$_2$) for 60 min, then washed and dried overnight. Afterwards, SMSC were grown on the slides in coculture

in 24-well-plates (Corning Incorporated, Corning, NY, USA). For alcian blue staining, the cells were fixed and stained using the PAS-staining Kit (Merck Millipore, Billerica, Massachusetts, USA). Briefly, staining with alcian blue was followed by incubation with periodic acid, Schiff reagent, and hematoxylin, whereupon each step was followed by washing. After mounting, histological pictures were analyzed using an Olympus BX51 microscope (Olympus Deutschland GmbH, Hamburg, Germany) with the software module Stream Motion adjusting only brightness and contrast.

Immunohistology
Cells were fixed at −20 °C with methanol (Sigma-Aldrich, St. Louis, MO, USA) for 10 min; afterwards, they were washed with Dulbecco's phosphate-buffered saline (DPBS; Gibco Invitrogen, Carlsbad, CA, USA). For blocking of unspecific binding sites, cells were incubated at room temperature with 5 % BSA (AppliChem GmbH, Darmstadt, Germany) in DPBS. Primary antibodies were diluted as follows: collagen type I (mouse monoclonalAb COL-1; abcam, Cambridge, UK) 1:500, collagen type II (rabbit polyclonalAb COL-2; abcam) 1:250, and collagen type X (mouse monoclonalAb COL-10; abcam) 1:750. After washing, the antibody working solutions were applied to the cells and incubated at 4 °C overnight in a humid chamber. After washing three times, the cells were incubated with the working solution of the secondary antibody: Alexa Fluor 488 goat anti-mouse IgG and Alexa Fluor 568 goat anti-rabbit IgG (Life Technologies, Grand Island, NY, USA) in 1 % BSA, dilution 1:250. After washing three times again, color reagent (ProLong® Gold antifade reagent DAPI; Life Technologies) was applied. An Olympus BX51 microscope (Olympus Deutschland GmbH) with special fluorescence filters was used for image acquisition. The program ImageJ (Wayne Rasband, NIH; imagej.nih.gov/ij/download) facilitated overlaying of images.

Real-time PCR
Real-time PCR was carried out for SMSC only. RNA samples from days 7 and 14 were transcribed into cDNA; RNA analysis was carried out for gene expression of aggrecan, TGFβ, and collagen type I, II, and X. Total mRNA was prepared using the Qiagen RNeasy kit according to the manufacturer's instructions (Qiagen, Hilden, Germany). Total RNA (1 μg) was treated with 1 U DNAse I (Invitrogen, Karlsruhe, Germany) to remove genomic DNA. Poly-T primed cDNA synthesis was performed using 1 U reverse transcriptase III (RTIII; Invitrogen) per 1 μg RNA according to the manufacturer's instructions. TaqMan™ PCR assays were performed in 384-well plates in a Roche LightCycler480 (Roche) using the Roche LightCycler Mastermix. For gene expression analyses, Roche's universal ProbeLibrary Probes

and recommended Universal ProbeLibrary Reference Gene Assays were used. The cycling conditions were denaturation (one cycle: 50 °C for 120 sec, 95 °C for 600 sec), followed by 40 amplification cycles (95 °C for 15 sec, 60 °C for 60 sec, 72 °C for 10 sec), followed by melting (one cycle: 95 °C for 10 sec, 50 °C for 30 sec, 70 °C for 1 sec), and cooling (one cycle: 40 °C for 30 sec). Data were quantified via $\Delta\Delta CT$ comparisons. Data were normalized by comparing genes of interest versus reference genes (GAPDH). Reaction efficiency is controlled by a relative standard curve and/or a calibrator per reaction. Real-time PCR was carried out in quadruplicate, each value representing an average of four experiments.

Data analysis and statistics

Concentrations of cytokines determined by the specific ELISA were calculated according to the manufacturers' instructions (R&D; and Thermo Scientific, Rockford, IL, USA), creating a standard curve and reducing data with a four-parameter logistic (4-PL) curve fit using GraphPad Prism 5 software (GraphPad Software, Inc., La Jolla, CA, USA). All values were expressed as mean ± standard error of the mean. Statistical significance was tested nonparametrically using the Mann–Whitney U test. The values of different time points were compared in each group, and the values of one time point were compared between the groups. Statistical significance was defined as $p < 0.05$.

Results

Histology

SMSC were kept in monolayer cultures alone or in coculture with CHDR. As a positive control, these cocultures were supplemented with TGFβ. While SMSC alone stayed separated after 1 week, an aggregation of SMSC was visible in the coculture group. The addition of TGFβ even resulted in sphere formation (Fig. 1a–c). No further time points are shown, because the spheres lost adherence and could no longer be comparably stained. Since spin-downs resulted in cell debris, SMSC were grown on coated cover slides allowing only alcian blue staining and phase-contrast microscopy as shown in Fig. 1d. Again, the phenomenon of cell aggregation, sphere formation, and loss of adherence could be observed after 1 or 2 weeks, first in the TGFβ-treated positive control and then in the coculture. The alcian blue staining, which was more intense in cell aggregates, documents the presence of glycosaminoglycans/mucopolysaccharides. Cell aggregation could be observed with the cells of all different donors. Figure 1a compares the different groups by overlaying immunofluorescence staining and phase-contrast microscopy images. Figure 1b shows the single staining for collagen type I and type II of the different groups. The highest intensity but also the highest

cell density was observed in the TGFβ-treated positive control group. Figure 1c demonstrates representative slides of the single staining for collagen type X and the DAPI staining for the different groups. The percentage-wise estimation of the aggregation extent was 0 % for SMSC alone, ≥40 % for the coculture of SMSC with CHDR, and ≥90 % for the coculture supplemented with TGFβ (overview in phase-contrast microscopy). Additionally, cell spheres per field of view were counted resulting in 0 ± 0 aggregates/field for SMSC alone, 1 ± 0.4 aggregates/field for the coculture of SMSC with CHDR, and 2.25 ± 0.5 aggregates/field for the coculture supplemented with TGFβ ($n = 5$, 20-fold magnification). There was no difference comparing the aggregation after 7 or 14 days.

Role for TGFβ

TGFβ concentrations in the supernatants were measured comparing SMSC in monolayer with the coculture of SMSC and CHDR without or with TGFβ supplementation (Fig. 2). As expected, the highest concentrations were observed in the positive control with TGFβ (777 ± 28 pg/ml). This is lower than the added concentrations indicating receptor immobilization of the cytokine or degradation, because supernatants were collected together with medium change. Concentration levels are followed by the coculture without TGFβ supplementation starting at week 1 with 68 ± 5 pg/ml and steadily increasing up to 183 ± 15 pg/ml at week 3. Although TGFβ was also found in the SMSC monolayer, levels were short over the detection limit. There was a statistically significant difference between the levels of each time point of all groups and each time point within the cocultures ($p < 0.05$). Because the statistical significance reached was marginal, additional comparisons were calculated using a Student's t test, resulting in $p < 0.02$. Furthermore, the values of all groups were merged independent of the time point. The comparison resulted in highly significant differences ($p < 0.001$) using the direct comparison of groups with the Mann–Whitney U test and using the Kruskall–Wallis H test (multiple comparisons). Considering a significance level of the direct group comparisons very close to the defined α and a possibly not complete random sample, the tests may overstate the accuracy of the results. TGFβ mRNA expression was also compared over a 2-week interval (Fig. 3). A statistically significant difference was found between the TGFβ-supplemented group and both other groups ($p = 0.021$), but not between the different time points within each group.

mRNA regulation of aggrecan and collagen type I, II, and X

Aggrecan mRNA expression was compared in SMSC monolayer with the coculture of SMSC and CHDR without or with TGFβ supplementation (Fig. 4). The highest

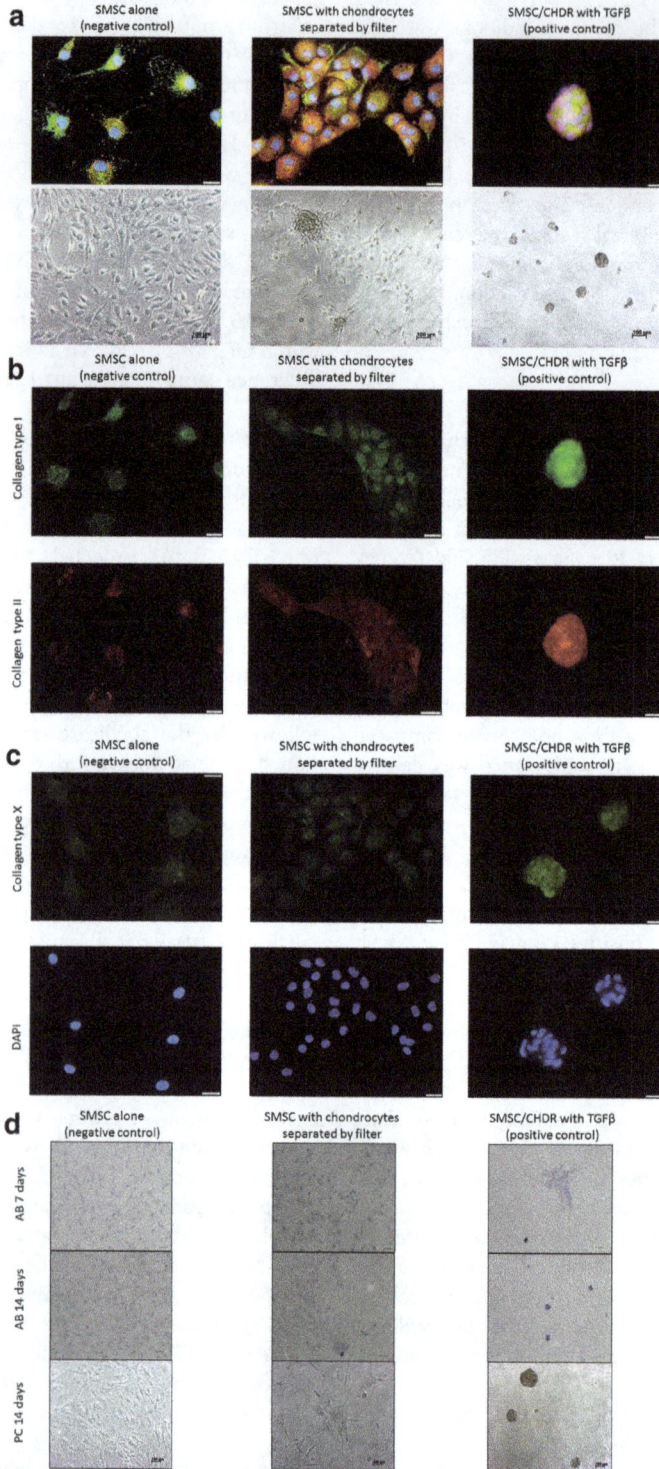

a

SMSC alone (negative control)   SMSC with chondrocytes separated by filter   SMSC/CHDR with TGFβ (positive control)

b

SMSC alone (negative control)   SMSC with chondrocytes separated by filter   SMSC/CHDR with TGFβ (positive control)

Collagen type I

Collagen type II

c

SMSC alone (negative control)   SMSC with chondrocytes separated by filter   SMSC/CHDR with TGFβ (positive control)

Collagen type X

DAPI

d

SMSC alone (negative control)   SMSC with chondrocytes separated by filter   SMSC/CHDR with TGFβ (positive control)

AB 7 days

AB 14 days

PC 14 days

Fig. 1 (See legend on next page.)

(See figure on previous page.)
Fig. 1 a Self-organization of SMSC after 7 days in monolayer culture. SMSC alone (*left*) stay separated, but in coculture with CHDR (*middle*) an aggregation of cells is visible, and addition of TGFβ (*right*) results in sphere formation. *Upper row* shows overlaying immunofluorescence staining (*green*: collagen type I, *red*: collagen type II, *blue*: DAPI; scale 20 μm), and *lower row* the phase-contrast microscopy. **b** Single staining for collagen type I (*upper row*) and type II (*lower row*) of the different groups. **c** Single staining for collagen type X and DAPI of the different groups. **d** Alcian blue (*AB*) staining of the different groups and time points. *Upper row*: AB 7 days (marker 50 μm), *middle row*: AB 14 days (marker 100 μm), *lower row*: phase-contrast (*PC*) microscopy 14 days (marker 100 μm). *CHDR* chondrocytes, *SMSC* synovial mesenchymal stem cells, *TGFβ* transforming growth factor beta (Color figure online)

levels were found in both cocultures at both investigated time points (without TGFβ: 4.7 ± 1.2-fold at week 1 and 6.7 ± 1.1-fold after 2 weeks, with TGFβ: 6.5 ± 0.9-fold at week 1 and 4.9 ± 1.1-fold at week 2). There is a statistically significant difference between both cocultures and the monolayer, but not between the different time points within each group. Collagen type I (col1) mRNA expression was also compared between SMSC monolayer with the coculture of SMSC and CHDR without or with supplemented TGFβ (Fig. 5). The highest values were measured in the TGFβ-supplemented coculture (up to 1625 ± 219-fold). There is a statistically significant difference between the TGFβ-supplemented group and both other groups ($p = 0.021$), but not between the different time points within each group or SMSC alone and the non-supplemented coculture. Collagen type II (col2) mRNA expression was also examined (Fig. 6), showing the highest values in both cocultures (up to 8.9 ± 3.2-fold). There was no statistically significant difference between the TGFβ-supplemented group and day 14 of the coculture, but these values were higher compared with SMSC alone and day 7 of the coculture ($p = 0.021$). There is no difference between the two time points within each group except for the coculture without TGFβ. Here the

col2 mRNA increased between the first week and the second week significantly ($p = 0.043$). Since collagen type II is considered the main marker for differentiated cartilage, the results indicate increasing chondrogenic differentiation induced by the presence of CHDR in coculture and the addition of TGFβ. Furthermore, the mRNA of the chondrogenic hypertrophy marker collagen type X (col10) was analyzed (Fig. 7). Although we found significant values at day 7 in the plain coculture, the highest values were measured when TGFβ was supplemented (up to 3.1 ± 0.7-fold). The value in the TGFβ-supplemented group at day 7 is statistically significantly higher than at all other time points ($p = 0.014$). There is no difference between the different time points within the SMSC (0 caused by rounding) and the coculture groups.

Discussion

The main finding of the study is that human CHDR are able to induce a chondrogenic phenotype in human SMSC in a trans-well coculture. Although SMSC were kept in a monolayer, chondrogenesis leads to loss of adherence and formation of spheres. This was sufficiently regulated by paracrine factors; no direct cell–cell

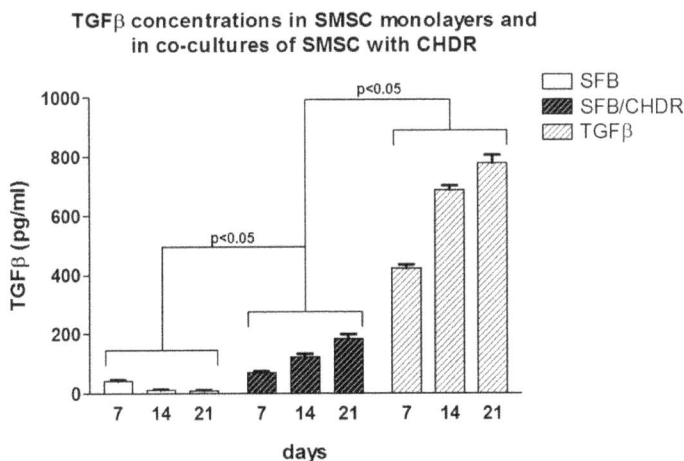

Fig. 2 TGFβ concentrations in supernatants comparing SMSC in monolayer with the coculture of SMSC and CHDR without or with TGFβ supplementation. There is a statistically significant difference between the groups and each time point within the cocultures. *CHDR* chondrocytes, *SMSC* synovial mesenchymal stem cells or SFB synovial fibroblasts, *TGFβ* transforming growth factor beta

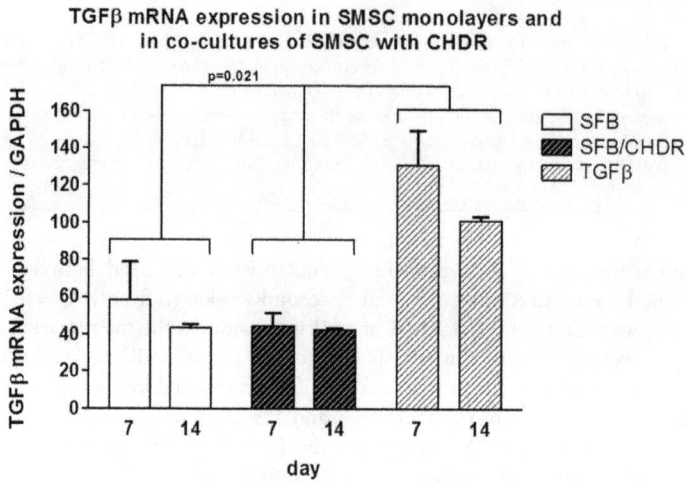

Fig. 3 TGFβ mRNA expression comparing SMSC in monolayer with the coculture of SMSC and CHDR without or with TGFβ supplementation. There is a statistically significant difference between the TGFβ-supplemented group and both other groups, but not between the different time points within each group. *CHDR* chondrocytes, *SMSC* synovial mesenchymal stem cells or *SFB* synovial fibroblasts, *TGFβ* transforming growth factor beta

interactions were necessary. The degree of cell self-assembly and sphere formation and the collagen type II and aggrecan expression were associated with the levels of TGFβ found in the supernatants. Collagen type X, a marker of chondrocyte hypertrophy, was expressed in the coculture of SMSC and CHDR when TGFβ was additionally supplemented.

Cocultures of mesenchymal stem cells and CHDR can result in improved chondrogenic differentiation in pellets [9] and immunological interactions including anti-inflammatory regulations attributed to mesenchymal stem cells [15]. The possible high impact of coculture systems on cartilage tissue engineering is documented by the increasing number of described technical solutions [16] and experimental designs [17]. In contrast to previous studies, this investigation focused on an experimental trans-well design using human synovial stem cells. This set-up was inspired by the idea that—unlike the previously suggested predominant role of bone marrow-derived progenitor cells for cartilage regeneration [1]—synovial stem

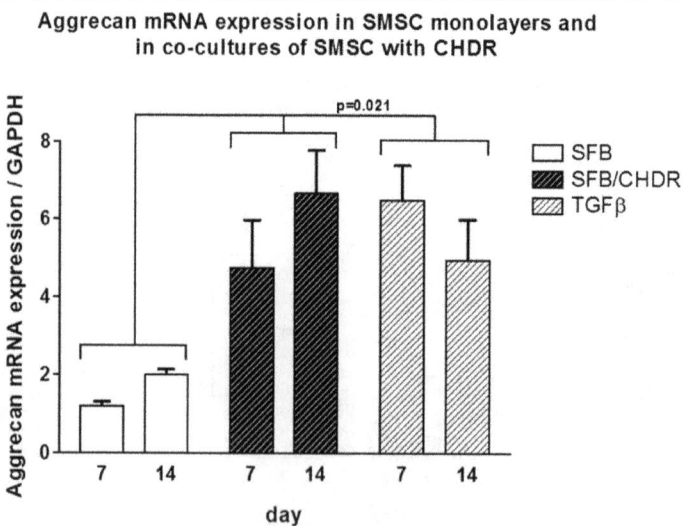

Fig. 4 Aggrecan mRNA expression comparing SMSC in monolayer with the coculture of SMSC and CHDR without or with TGFβ supplementation. There is a statistically significant difference between both cocultures and the monolayer, but not between the different time points within each group. *CHDR* chondrocytes, *SMSC* synovial mesenchymal stem cells or *SFB* synovial fibroblasts, *TGFβ* transforming growth factor beta

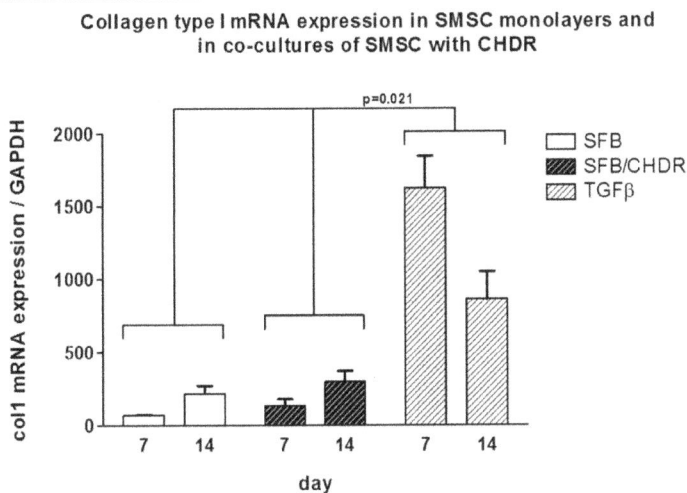

Fig. 5 Collagen type I (col1) mRNA expression comparing SMSC in monolayer with the coculture of SMSC and CHDR without or with TGFβ supplementation. There is a statistically significant difference between the TGFβ-supplemented group and both other groups, but not between the different time points within each group. *CHDR* chondrocytes, *SMSC* synovial mesenchymal stem cells or *SFB* synovial fibroblasts, *TGFβ* transforming growth factor beta

cells might be able to contribute to natural cartilage regeneration. This phenomenon has been described, but the cellular sources for repair are controversially discussed. The potential of CHDR themselves appears to be very limited, as previously published outgrowth experiments have shown [18]. Since synovial fibroblasts are located in the direct vicinity of cartilage and cartilage lesions and also exhibit stem cell characteristics [19] with the best chondrogenic potential of different mesenchymal stem sources

[5], these cells are an interesting candidate for the cellular origin of natural and spontaneous cartilage regeneration. The natural environment of cells, especially stem cells, determines their histological and biochemical phenotype [20]. Therefore, we hypothesized that CHDR are able to secret paracrine signals inducing a chondrogenic differentiation of SMSC. This was confirmed by showing spontaneous formation of first cell aggregation and then cell sphere formation by SMSC in coculture with CHDR. This

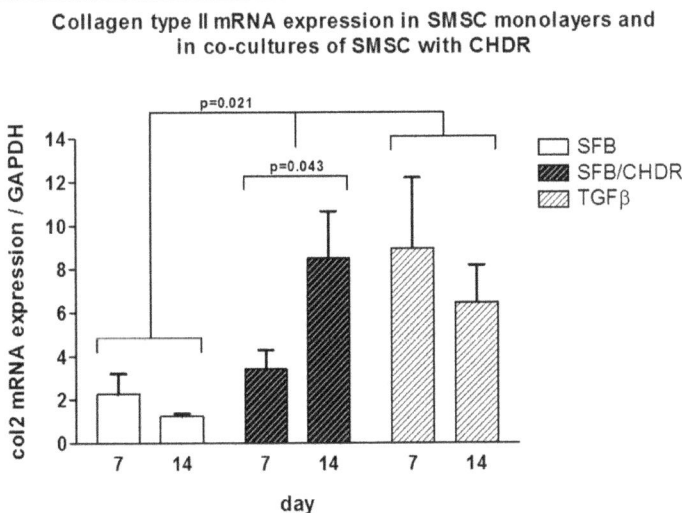

Fig. 6 Collagen type II (col2) mRNA expression comparing SMSC in monolayer with the coculture of SMSC and CHDR without or with TGFβ supplementation. There is no statistically significant difference between the TGFβ-supplemented group and day 14 of the coculture, but these values are higher compared with SMSC alone and day 7 of the coculture. There is no difference between the different time points within each group except for the coculture. *CHDR* chondrocytes, *SMSC* synovial mesenchymal stem cells or *SFB* synovial fibroblasts, *TGFβ* transforming growth factor beta

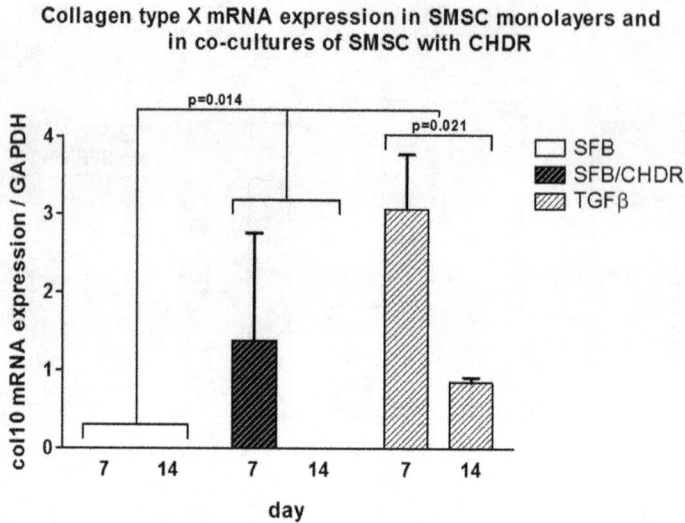

Fig. 7 Collagen type X (col10) mRNA expression comparing SMSC in monolayer with the coculture of SMSC and CHDR without or with TGFβ supplementation. The value of the TGFβ-supplemented group at day 7 is statistically significantly higher than at all other time points. There is no difference between the different time points within the SMSC and the coculture groups. *CHDR* chondrocytes, *SMSC* synovial mesenchymal stem cells or *SFB* synovial fibroblasts, *TGFβ* transforming growth factor beta

was accompanied by expression of cartilage markers as aggrecan and collagen type II on the protein and mRNA levels. The key role for TGFβ in the described and observed process of chondrogenic differentiation was not only demonstrated for BMSC [21], but also for SMSC [22]. TGFβ has been described to upregulate collagen type I mRNA in osteoblasts [23] or CHDR [24]. Furthermore, TGFβ seems to induce the cartilage hypertrophy marker collagen type X. This was confirmed by our results, but high levels of either collagen type I or type X were mainly found in the control group that was supplemented with additional TGFβ. In accordance with these data, the hypertrophic status of chondrogenically differentiated mesenchymal stem cells has previously been described to be associated with overexpression of osteogenic markers as collagen type I and alkaline phosphatase [25], explaining our positive immunostaining and the higher collagen type I mRNA expression in the TGFβ-supplemented group.

The fact that exogenous TGFβ supplementation upregulated collagen type I and type X in the SMSC without providing further enhancement of collagen type II or aggrecan expression is probably a dosage effect, assuming that the CHDR are promoting chondrogenesis in SMSC by releasing TGFβ; or it could suggest that CHDR are releasing other paracrine factors modulating the effect. This has to be considered also in the light of the only marginal statistical significance regarding the increased TGFβ concentrations in the supernatants of the coculture. However, the CHDR alone had a chondrogenic effect (at least in terms of gene expression and

sphere formation), causing minimal fibrochondrogenesis (col1) or hypertrophy (col10).

The observed missing accordance of TGFβ mRNA and protein regulation in the supernatants can have different causes. First, TGFβ protein might also significantly be secreted by CHDR, which in the current experimental set-up cannot be measured separately. Secondly, protein formation in SMSC might undergo further regulatory processing and not only depend on RNA regulation. The higher levels of TGFβ mRNA in the supplemented coculture suggest a positive feedback regulation.

Chondrogenesis of SMSC was also demonstrated in a pellet coculture model using rabbit CHDR that overexpressed TGFβ after adenoviral transfection [26]. Based on the results of our study it may be concluded that gene transfer might not be necessary, because the coculture itself provides sufficient paracrine stimuli for chondrogenic differentiation of SMSC. Similarly, TGFβ induced the chondrogenesis of SMSC with high levels of collagen type II, aggrecan, and Sox 9, and low levels of dedifferentiation or hypertrophy markers in a coculture pellet model using nucleus pulposus cells in serum-free medium [27]. All of the data emphasize the key role for TGFβ in chondrogenic differentiation for both in-vitro cultures of mesenchymal stem cells of different origin [13] and in the natural articular environment. Although the role of TGFβ seems striking, there are a few limitations attributed to the presented experimental set-up. First, the measured TGFβ levels in the supernatants are a summary response of all cells, which makes it impossible to differentiate the true origin. This is the nature of a coculture, and therefore mRNA levels were determined in SMSC. Unfortunately, the data were not completely conclusive, because TGFβ mRNA could also be found in SMSC alone, indicating separate regulation pathways for protein and mRNA. Secondly, the simple presence of TGFβ does not allow drawing functional conclusions. This means we observed an association of TGFβ in the supernatant and chondrogenic differentiation of SMSC, but the biological or mechanistic relation is not necessarily evident. However, the stimulatory success of the coculture with CHDR on the chondrogenic differentiation of SMSC underlines the potential role of these cells in natural and artificially supported cartilage repair.

Conclusions

The vicinity of CHDR in a trans-well coculture induces a chondrogenic phenotype in SMSC. This process is associated with increased TGFβ secretion and offers possible implications for cartilage repair. This effect may play a significant role in natural and surgically induced cartilage repair, especially in cell therapeutic approaches.

Abbreviations
BMSC: bone marrow-derived mesenchymal stem cells; BSA: bovine serum albumin; CHDR: chondrocytes; col1: collagen type I; col10: collagen type X; col2: collagen type II; DMEM: Dulbecco's modified Eagle's medium; DPBS: Dulbecco's phosphate-buffered saline; ELISA: enzyme-linked immunosorbent assay; FCS: fetal calf serum; KLS: Kellgren and Lawrence score; OA: osteoarthritis; P/S: penicillin/streptomycin; PAS-staining: Periodic acid–Schiff stain; SMSC: synovial mesenchymal stem cells; TGFβ: transforming growth factor beta.

Competing interests
All authors declare that they have no financial and personal relationships with other people, or organizations that could inappropriately influence (bias) their work.

Authors' contributions
EJK was involved in the conception and the design of the study, and the analysis and interpretation of the data, contributed to the article draft, and contributed to the final approval of the submitted article. EH was responsible for the collection, assembly, and management of data, performed the experiments, contributed to the article draft, calculated the ELISA values and some statistics, and approved the final version of the article. AB supported the laboratory work, supervised the sample analysis, was responsible for sample storage, revised the article draft critically, and approved the final version of the article. KB supported the laboratory work, revised the article draft critically, and approved the final version of the article. HS was responsible for the conception and the design of the study, obtaining funding, the analysis and interpretation of the data, tutorial of EH and EJK, writing the article, and the final approval of the submitted article. All authors take responsibility for the integrity of the work as a whole, from inception to finished article.

Acknowledgements and funding source
The Research Commission of the Albert-Ludwigs University Medical Center Freiburg (SCHM1014/14) and the Federal Ministry of Education and Research Germany (01EC1001D) funded the study. The article processing charge was funded by the open access publication fund of the Albert Ludwigs University Freiburg.

Author details
[1]Department of Orthopedics and Trauma Surgery, Albert-Ludwigs University Medical Center Freiburg, Freiburg, Germany. [2]Department of Orthopaedics and Traumatology, Odense University Hospital, Sdr. Boulevard 29, 5000 Odense C, Denmark. [3]Department of Clinical Research, University of Southern Denmark, Odense, Denmark.

References

1. Shapiro F, Koide S, Glimcher MJ. Cell origin and differentiation in the repair of full-thickness defects of articular cartilage. J Bone Joint Surg Am. 1993;75:532–53.
2. de Sousa EB, Casado PL, Moura Neto V, Duarte MEL, Aguiar DP. Synovial fluid and synovial membrane mesenchymal stem cells: latest discoveries and therapeutic perspectives. Stem Cell Res Ther. 2014;5:112.
3. Davatchi F, Sadeghi Abdollahi B, Mohyeddin M, Nikbin B. Mesenchymal stem cell therapy for knee osteoarthritis: 5 years follow-up of three patients. Int J Rheum Dis. 2016;19:219.
4. Sekiya I, Muneta T, Horie M, Koga H. Arthroscopic transplantation of synovial stem cells improves clinical outcomes in knees with cartilage defects. Clin Orthop Relat Res. 2015;473:2316.
5. Sakaguchi Y, Sekiya I, Yagishita K, Muneta T. Comparison of human stem cells derived from various mesenchymal tissues: superiority of synovium as a cell source. Arthritis Rheum. 2005;52:2521–9.
6. Baboolal TG, Mastbergen SC, Jones E, Calder SJ, Lafeber FPJG, McGonagle D. Synovial fluid hyaluronan mediates MSC attachment to cartilage, a potential novel mechanism contributing to cartilage repair in osteoarthritis using knee joint distraction. Ann Rheum Dis. 2015;75:908.
7. Pilz IH, Mehlhorn A, Dovi-Akue D, Langenmair ER, Südkamp NP, Schmal H. Development and retranslational validation of an in vitro model to characterize acute infections in large human joints. BioMed Res Int. 2014;2014:848604.
8. Ryu J-S, Jung Y-H, Cho M-Y, Yeo JE, Choi Y-J, Kim YI, Koh Y-G. Co-culture with human synovium-derived mesenchymal stem cells inhibits inflammatory activity and increases cell proliferation of sodium nitroprusside-stimulated chondrocytes. Biochem Biophys Res Commun. 2014;447:715–20.
9. Lettry V, Hosoya K, Takagi S, Okumura M. Coculture of equine mesenchymal stem cells and mature equine articular chondrocytes results in improved chondrogenic differentiation of the stem cells. Jpn J Vet Res. 2010;58:5–15.
10. Wu L, Prins H-J, Helder MN, van Blitterswijk CA, Karperien M. Trophic effects of mesenchymal stem cells in chondrocyte co-cultures are independent of culture conditions and cell sources. Tissue Eng Part A. 2012;18:1542–51.
11. Jones BA, Pei M. Synovium-derived stem cells: a tissue-specific stem cell for cartilage engineering and regeneration. Tissue Eng Part B Rev. 2012;18:301–11.
12. Kurth TB, Dell'accio F, Crouch V, Augello A, Sharpe PT, De Bari C. Functional mesenchymal stem cell niches in adult mouse knee joint synovium in vivo. Arthritis Rheum. 2011;63:1289–300.
13. Pittenger MF, Mackay AM, Beck SC, Jaiswal RK, Douglas R, Mosca JD, Moorman MA, Simonetti DW, Craig S, Marshak DR. Multilineage potential of adult human mesenchymal stem cells. Science. 1999;284:143–7.
14. Schmal H, Mehlhorn AT, Fehrenbach M, Müller CA, Finkenzeller G, Südkamp NP. Regulative mechanisms of chondrocyte adhesion. Tissue Eng. 2006;12:741–50.
15. van Buul GM, Villafuertes E, Bos PK, Waarsing JH, Kops N, Narcisi R, Weinans H, Verhaar J a. N, Bernsen MR, van Osch GJVM. Mesenchymal stem cells secrete factors that inhibit inflammatory processes in short-term osteoarthritic synovium and cartilage explant culture. Osteoarthr Cartil OARS Osteoarthr Res Soc. 2012;20:1186–96.
16. Morita Y, Yamamoto S, Ju Y. Development of a new co-culture system, the "separable-close co-culture system", to enhance stem-cell-to-chondrocyte differentiation. Biotechnol Lett. 2015;37:1911–8.
17. Wu L, Post JN, Karperien M. Engineering cartilage tissue by pellet coculture of chondrocytes and mesenchymal stromal cells. Methods Mol Biol Clifton NJ. 2015;1226:31–41.

18. Zingler C, Carl H-D, Swoboda B, Krinner S, Hennig F, Gelse K. Limited evidence of chondrocyte outgrowth from adult human articular cartilage. Osteoarthr Cartil OARS Osteoarthr Res Soc. 2016;24:124–8.
19. Pei M, He F, Boyce BM, Kish VL. Repair of full-thickness femoral condyle cartilage defects using allogeneic synovial cell-engineered tissue constructs. Osteoarthr Cartil OARS Osteoarthr Res Soc. 2009;17:714–22.
20. Leyh M, Seitz A, Dürselen L, Schaumburger J, Ignatius A, Grifka J, Grässel S. Subchondral bone influences chondrogenic differentiation and collagen production of human bone marrow-derived mesenchymal stem cells and articular chondrocytes. Arthritis Res Ther. 2014;16:453.
21. Mehlhorn AT, Schmal H, Kaiser S, Lepski G, Finkenzeller G, Stark GB, Südkamp NP. Mesenchymal stem cells maintain TGF-beta-mediated chondrogenic phenotype in alginate bead culture. Tissue Eng. 2006;12:1393–403.
22. Kim YI, Ryu J-S, Yeo JE, Choi YJ, Kim YS, Ko K, Koh Y-G. Overexpression of TGF-β1 enhances chondrogenic differentiation and proliferation of human synovium-derived stem cells. Biochem Biophys Res Commun. 2014;450:1593–9.
23. Glueck M, Gardner O, Czekanska E, Alini M, Stoddart MJ, Salzmann GM, Schmal H. Induction of osteogenic differentiation in human mesenchymal stem cells by crosstalk with osteoblasts. BioResearch Open Access. 2015;4:121–30.
24. Perrier-Groult E, Pasdeloup M, Malbouyres M, Galéra P, Mallein-Gerin F. Control of collagen production in mouse chondrocytes by using a combination of bone morphogenetic protein-2 and small interfering RNA targeting Col1a1 for hydrogel-based tissue-engineered cartilage. Tissue Eng Part C Methods. 2013;19:652–64.
25. Mueller MB, Fischer M, Zellner J, Berner A, Dienstknecht T, Prantl L, Kujat R, Nerlich M, Tuan RS, Angele P. Hypertrophy in mesenchymal stem cell chondrogenesis: effect of TGF-beta isoforms and chondrogenic conditioning. Cells Tissues Organs. 2010;192:158–66.
26. Varshney RR, Zhou R, Hao J, Yeo SS, Chooi WH, Fan J, et al. Chondrogenesis of synovium-derived mesenchymal stem cells in gene-transferred co-culture system. Biomaterials. 2010;31:6876–91.
27. Chen S, Emery SE, Pei M. Coculture of synovium-derived stem cells and nucleus pulposus cells in serum-free defined medium with supplementation of transforming growth factor-beta1: a potential application of tissue-specific stem cells in disc regeneration. Spine. 2009;34:1272–80.

16

Retinoic acid receptor signaling preserves tendon stem cell characteristics and prevents spontaneous differentiation in vitro

Stuart Webb[†], Chase Gabrelow[†], James Pierce, Edwin Gibb and Jimmy Elliott[*]

Abstract

Background: Previous studies have reported that adult mesenchymal stem cells (MSCs) tend to gradually lose their stem cell characteristics in vitro when placed outside their niche environment. They subsequently undergo spontaneous differentiation towards mesenchymal lineages after only a few passages. We observed a similar phenomenon with adult tendon stem cells (TSCs) where expression of key tendon genes such as *Scleraxis* (*Scx*), are being repressed with time in culture. We hypothesized that an environment able to restore or maintain *Scleraxis* expression could be of therapeutic interest for in vitro use and tendon cell-based therapies.

Methods: TSCs were isolated from human cadaveric Achilles tendon and expanded for 4 passages. A high content imaging assay that monitored the induction of Scx protein nuclear localization was used to screen ~1000 known drugs.

Results: We identified retinoic acid receptor (RAR) agonists as potent inducers of nuclear Scx in the small molecule screen. The upregulation correlated with improved maintenance of tendon stem cell properties through inhibition of spontaneous differentiation rather than the anticipated induction of tenogenic differentiation. Our results suggest that histone epigenetic modifications by RAR are driving this effect which is not likely only dependent on Scleraxis nuclear binding but also mediated through other key genes involved in stem cell self-renewal and differentiation. Furthermore, we demonstrate that the effect of RAR compounds on TSCs is reversible by revealing their multi-lineage differentiation ability upon withdrawal of the compound.

Conclusion: Based on these findings, RAR agonists could provide a valid approach for maintaining TSC stemness during expansion in vitro, thus improving their regenerative potential for cell-based therapy.

Keywords: Retinoic acid receptor, Scleraxis, Tendon stem cells, Proliferation, Spontaneous differentiation

Background

Tendon stem cells (TSCs) have been characterized in adult tendons and show similar differentiation properties to bone marrow mesenchymal stem cells (BM-MSCs) with the exception of expressing tendon specific markers [1, 2]. Previous studies on mesenchymal stem cells (MSCs) have also shown in vitro aging and spontaneous differentiation during expansion over several passages [3–5]. Similar to MSCs removed from their niche, the absence of tendon extracellular matrix (ECM) in culture interferes with TSC self-renewal and differentiation behavior. This could explain the appearance of bone, adipose, and cartilage markers upon expansion in vitro. TSC expansion for in vitro use, or potentially cell therapy still remain to be optimized in order to preserve their stem cell characteristics while preventing premature differentiation.

Scleraxis (*Scx*) is a basic helix–loop–helix transcription factor highly specific to tendon and has been shown to be both sufficient and necessary to promote the tendon cell fate [6–8]. Although little is known about Scx function in stem cells, TSCs isolated from adult tendons express high levels of Scx but gradually lose this expression after being

* Correspondence: JElliott@gnf.org
[†]Equal contributors
Genomics Institute of the Novartis Research Foundation, 10675 John Jay Hopkins Drive, San Diego, CA 92121, USA

cultured. *Scx* is expressed throughout development and decreases significantly during adulthood. It is known that mechanical stimulation can upregulate *Scx* expression under physiological loading both in vitro and in vivo, suggesting that it might be required for tendon homeostasis [9, 10]. A recent study also showed that overexpression of Scx in TSCs promoted better repair compared with untransduced cells when transplanted in a patellar tendon injury model [11]. Therefore, we performed a high-content imaging screen on TSCs to discover small molecules able to induce Scx nuclear protein levels in later passage TSCs when they have low nuclear signal. We hypothesized that a drug increasing Scx signaling would be beneficial in vitro for cell therapy aimed at promoting tendon regeneration and may also provide a suitable ex vivo culture environment for TSCs.

Methods
Ethical approval
Achilles and patellar tendons from a 55-year-old human cadaver were obtained following informed consent from the subject or the subject's legally authorized representative by Asterand Bioscience (Asterand, Detroit, MI, USA) according to the Department of Health and Human Services regulations for the protection of human subjects (45 CFR §46.116 and §46.117) and Good Clinical Practice (ICH E6). United States Postmortem Sites, except studies being performed by Veterans Affairs (VA) investigators at VA facilities or off-site VA locations, are exempt from institutional review board review because deceased donors are not considered "human subjects" under federal regulations for live donors (CFR 45 part 46). In compliance with the federal regulation stated for postmortem tissues, no approval from an ethics committee was necessary for this study. For rat TSC isolation, Achilles tendons from Sprague Dawley rats 3–4 months old were used following approval by the Institutional Animal Care and Use Committee (IACUC-Protocol P14-357) at the Genomics institute of the Novartis Research Foundation. The experimental animals received care in compliance with the Guide for the Care and Use of Laboratory Animals.

Isolation and expansion of TSCs
Surrounding connective tissue was removed and the tissue was washed in Hank's Balanced Salt Solution (HBSS) and then transferred to 5 % dispase, 3 mg/ml collagenase type 1 (Worthington, Lakewood, NJ, USA) with antibiotics/antimycotics. The tissue was digested at 37 °C for 12 hours. The solution was then passed through a 50 μm cell strainer and the cells were centrifuged at $350 \times g$ for 20 minutes. Pelleted cells were resuspended in growth media and plated at a density of ~750 cells/cm^2 in MSC expansion media (Lonza, Basel, Switzerland). Cells were cultured for 7–10 days at 37 °C, 5 % CO_2. After colony

formation, the cells were trypsinized and expanded. Cell lines from both human and rat origin were able to be maintained in culture for >30 passages.

mRNA extraction and quantitative reverse transcription-PCR analysis
mRNA was extracted using the RNeasy Plus kit (Qiagen, Germantown, MD, USA) as per the manufacturer's protocol. cDNA was synthesized from the isolated mRNA using qScript cDNA SuperMix (Quanta Biosciences, Gaithersburg, MD, USA) as per the manufacturer's protocol. cDNA and primers were transferred using the Echo liquid handler system (Labcyte, Sunnyvale, CA, USA). Quantitative PCR was performed using SYBR Green (Roche, Indianapolis, IN, USA) in a 5 ul reaction and analyzed with the Roche 480 Lightcycler. Raw c(t) values were converted to $2^{\Delta\Delta c(t)}$ for comparison between samples. The average of three different housekeeping genes was used as calibrators for the experiment: GADPH (metabolic), 36B4 (ribosomal), and beta-actin (cytoskeleton). Three biological replicates were used and standard deviation (SD) was calculated for each condition. For primer sequences, see Additional file 1: Table S1.

Immunocytochemistry and high-content imaging
Cells were plated at different densities and grown for the indicated amounts of time. Cells were then fixed with 4 % paraformaldehyde, electron microscopy grade (Electron Microscopy Sciences, Hatfield, PA, USA), for 20–30 minutes at room temperature. Permeabilization and blocking was performed with 3 % bovine serum albumin (Sigma, St-Louis, MO, USA) and 0.2 % Triton X-100 (Sigma) in phosphate-buffered saline (PBS) for 1 hour at room temperature. Primary antibodies were diluted in blocking buffer and stained overnight at 4 °C. Primary antibodies against the following proteins were used: anti-SCX (Abgent, San Diego, CA, USA), anti-Oct4 (Reprocell Inc., Boston, MA, USA), and anti-aggrecan (Millipore, Billerica, MA, USA). After three washes with PBS, cells were incubated with Alexa-Fluor 488 (Life Technologies, Carlsbad, CA, USA) conjugated antibody and Hoechst 33342 dye (Life Technologies) for 1 hour at room temperature. After three washes with PBS, cells were then imaged.

In addition, cells treated with CD1530 (Tocris Bioscience, Bristol, UK), all-trans retinoic acid (Tocris Bioscience, Bristol, UK), CD2665 (Tocris Bioscience), transforming growth factor beta-2 (TGFβ2; R&D Systems, Minneapolis, MN, USA), BIX-01294 (Sigma) and C646 (Sigma) were stained using the same method. High-content imaging was performed in an ImageXpress Ultra Confocal System (Molecular Devices, Sunnyvale, CA, USA) and staining for Scx, aggrecan, and nuclei was analyzed using MetaXpress 5.0 software (Molecular Devices).

Differentiation assays

Cells were plated in MSC media (Lonza) at 80 % confluence into six-well dishes and incubated for 8–12 hours to allow cell attachment. After the cells were attached, the media were changed to the respective differentiation cocktails ± 100 nM tazarotene (Sigma) for 14 days in vitro (DIV). Commercially available differentiation cocktails used were StemPro® Adipogenesis, Osteogenesis, and Chondrogenesis Differentiation Kits (Life Technologies). After 14 days, cells were fixed with 4 % paraformaldehyde for 30 minutes and stained for lineage specific markers. Akaline phosphatase activity in osteoblasts was revealed using Fast Blue RR (Sigma). Adipocytes were stained for lipid accumulation with LipidTOX-Green (Life Technologies). After fixation, cells were incubated with PBS containing LipidTOX for 1 hour and then imaged. Chondrocytes were examined for aggrecan accumulation using the immunofluorescence protocol already described.

Western blots

Cells were plated at confluence grown in the presence or absence of 100 nM tazarotene for 4 days. Cells were harvested by scraping and were pelleted before extraction of total crude protein or nuclear and cytoplasmic protein fractions (NE-PER Reagents, Thermo Scientific, Waltham, MA, USA). Proteins were run on 4–12 % Bis-Tris Gels and transferred as recommended for Polyvinylidene difluoride membrane (Life Technologies). Protein blots were blocked with 3 % ECL Prime blocking buffer for 1 hour. All subsequent antibody incubations were performed in 3 % ECL Prime blocking buffer. Antibodies used were anti-SCXA (Abgent, San Diego, CA, USA) and anti-rabbit horseradish peroxidase (GE Healthcare, Chicaco, IL, USA). The signal was developed using SuperSignal West Pico substrate (Thermo Scientific, Waltham, MA, USA) and visualized using the ChemiDoc MP Imaging System (Bio-Rad, Hercules, CA, USA).

Statistical analysis

Dose–response curves and statistical analysis were done using Graphpad Prism software (San Diego, CA, USA). To determine statistical significance between treatments, an unpaired, two-tailed Student's t test was used. For all graphs, data are presented as mean ± SD. $p < 0.05$ was considered significant.

Results

A small low molecular weight screen to identify Scx modulators

TSCs were isolated from human Achilles cadaveric tendons and plated at low density for colony formation (Fig. 1a). As reported previously, colonies were observed following 7–10 days in culture and displayed heterogeneity in size and density but produced a morphologically homogeneous population of polygonal shaped cells upon expansion. Scx immunofluorescence on early passage cells revealed strong nuclear immunoreactivity while nuclear localization was gradually lost upon expansion (Fig. 1b, c). The Scx antibody specificity was confirmed using a construct overexpressing human SCX protein in 293 cells which are negative for SCX expression, and their TSC identity was confirmed using Oct4 staining (Additional file 2: Figure S1). The decrease in nuclear localization upon culturing was used to design a low molecular weight (LMW) screen (~1000 different known drugs) to identify Scx inducers. Cells were expanded up to passage 4, a time when the vast majority of cells have low nuclear localized Scx protein, and plated in a 384-well format for high-throughput screening. The following day, the cells were stimulated with the small molecules. Cells were then grown for an additional 3 days, and then fixed and stained for Scx. Using high-content imaging, each individual well was visualized and analyzed using an algorithm that quantifies Scx nuclear translocation. TGFβ2, already known to induce Scx expression in TSCs, was used as a positive control in our screen (Fig. 1d, e).

RAR agonists induce strong SCX nuclear localization in human TSCs

Tazarotene, a RARβ and RARγ agonist, was identified as a strong hit in our LMW screen due to intense nuclear staining observed in treated cells (Fig. 2a, b). Further reconfirmation was carried out using two other molecules that also act as RAR agonists (CD1530 and ATRA). Both RAR agonists recapitulated the effect seen with tazarotene and showed expected half-maximal effective concentration which induces a response (EC_{50}) values for such compounds (Fig. 2c–e). Interestingly, RAR antagonist CD2665 was able to reduce SCX to levels lower than dimethyl sulfoxide (DMSO) controls, suggesting that endogenous RAR activity might play a role in SCX maintenance (Fig. 2e). Cellular analysis revealed no decrease in total cell number or obvious change in morphology with any of the compounds (up to 10 μM), confirming that the nuclear localization observed was not caused by nonspecific cytotoxicity (i.e., rounding of the cells) (Fig. 2f, h, i). When added in combination, CD2665 was also able to induce a shift in the tazarotene EC_{50} curve, confirming RAR as the target (Fig. 2g). Although human cells treated with tazarotene exhibited an increase in nuclear Scx protein, the total protein levels remained unchanged, suggesting activation through nuclear translocation rather than general protein increase (Fig. 2j). We also observed a similar phenotype with TSCs isolated from human patellar tendon, suggesting the effect is not unique for Achilles TSCs but is also conserved in TSCs isolated from other tendons (Additional file 3: Figure S2).

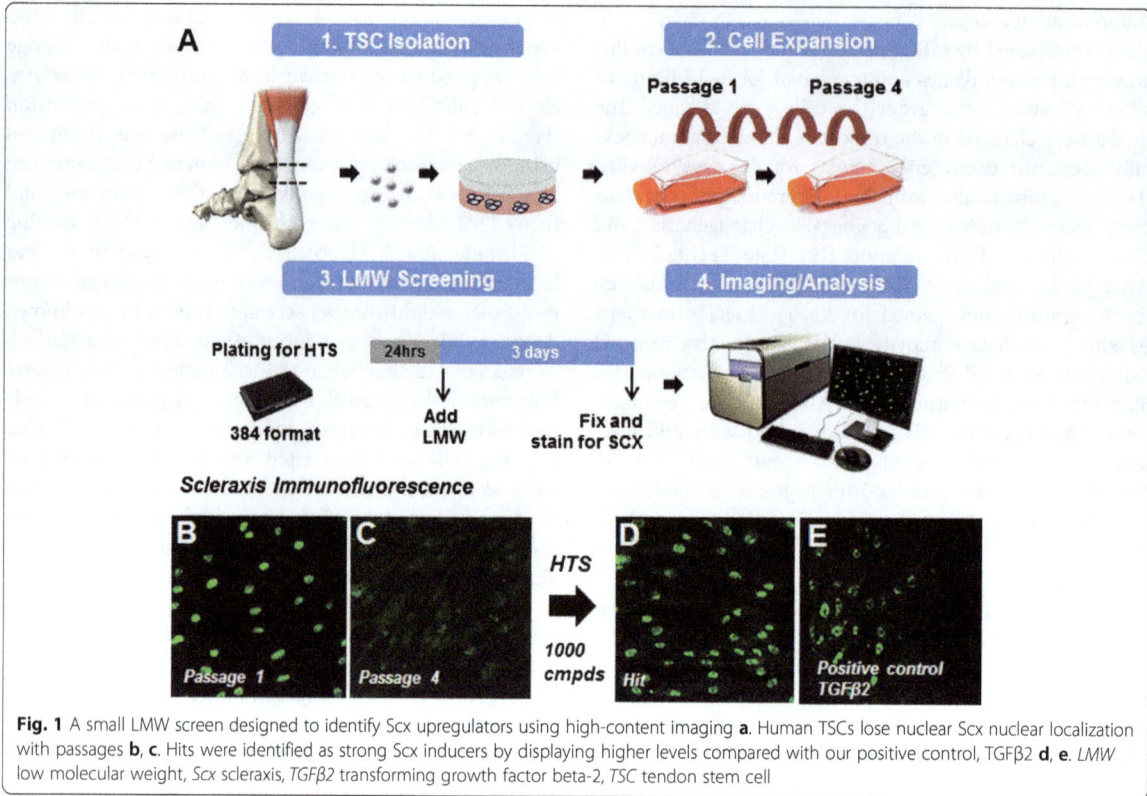

Fig. 1 A small LMW screen designed to identify Scx upregulators using high-content imaging **a**. Human TSCs lose nuclear Scx nuclear localization with passages **b**, **c**. Hits were identified as strong Scx inducers by displaying higher levels compared with our positive control, TGFβ2 **d**, **e**. *LMW* low molecular weight, *Scx* scleraxis, *TGFβ2* transforming growth factor beta-2, *TSC* tendon stem cell

Tazarotene can maintain TSC identity and block differentiation into different mesenchymal lineages

Because of the limited availability of human cells at early passage, the differentiation experiments and further characterization were performed using rat TSCs isolated using an identical protocol. The increase in the nuclear Scx protein following stimulation with tazarotene was confirmed in rat TSCs by western blot using a different Scx commercial antibody that cross-reacts with rat scleraxis protein (Additional file 4: Figure S3). Tazarotene, which was the most potent RAR agonist tested in our assay, was used at 100 nM for subsequent experiments because this concentration gives close to 100 % efficacy at inducing Scx nuclear localization (Fig. 2e).

Even though *Scx* is expressed in TSCs, its increase in expression has also been associated with tenogenic differentiation both in vitro and in vivo [6, 7, 12–14]. To determine whether tazarotene could potentially induce tenogenesis, a differentiation time-course assay was performed in vitro. Following confluency, a time-dependent increase in tendon gene expression (*EphA4*, *Col1a1*, and *Fmod*) was observed, reaching a maximum at 7 days. However, cells treated with tazarotene did not show a greater increase in expression and instead displayed mRNA levels at day 7 similar to those seen at day 1, suggesting that tazarotene is perhaps maintaining the cells in an undifferentiated state (Fig. 3a–c). TSCs have the ability to differentiate into each of the major MSC lineages when stimulated in the appropriate conditions. When treated with osteogenic, adipogenic, or chondrogenic induction cocktails, the cells responded as predicted—shown by specific histological staining and quantitative PCR for each lineage. Osteogenic induction media caused an upregulation of the alkaline phosphatase (*Alp*) messenger RNA as well as by fast blue staining (Fig. 3d, g). Adipogenic induction media caused an upregulation of *Foxo1* and lipid accumulation shown with LipidTOX-Green staining, while chondrogenic media increased expression of both aggrecan and collagen type 2 (Fig. 3e, f, h, i). Each of the lineages also took on the characteristic morphology of their respective lineage. Addition of tazarotene was able to suppress the induction of differentiation for all lineages, as seen by negative staining and low mRNA levels for up to 7 days (Fig. 3a–i). Inversely, treatment with tazarotene could maintain stem cell markers (*Oct4* and *Ssea-1*) for up to 7 days, confirming the inhibition of differentiation and maintenance of a stem cell phenotype (Fig. 4a–d).

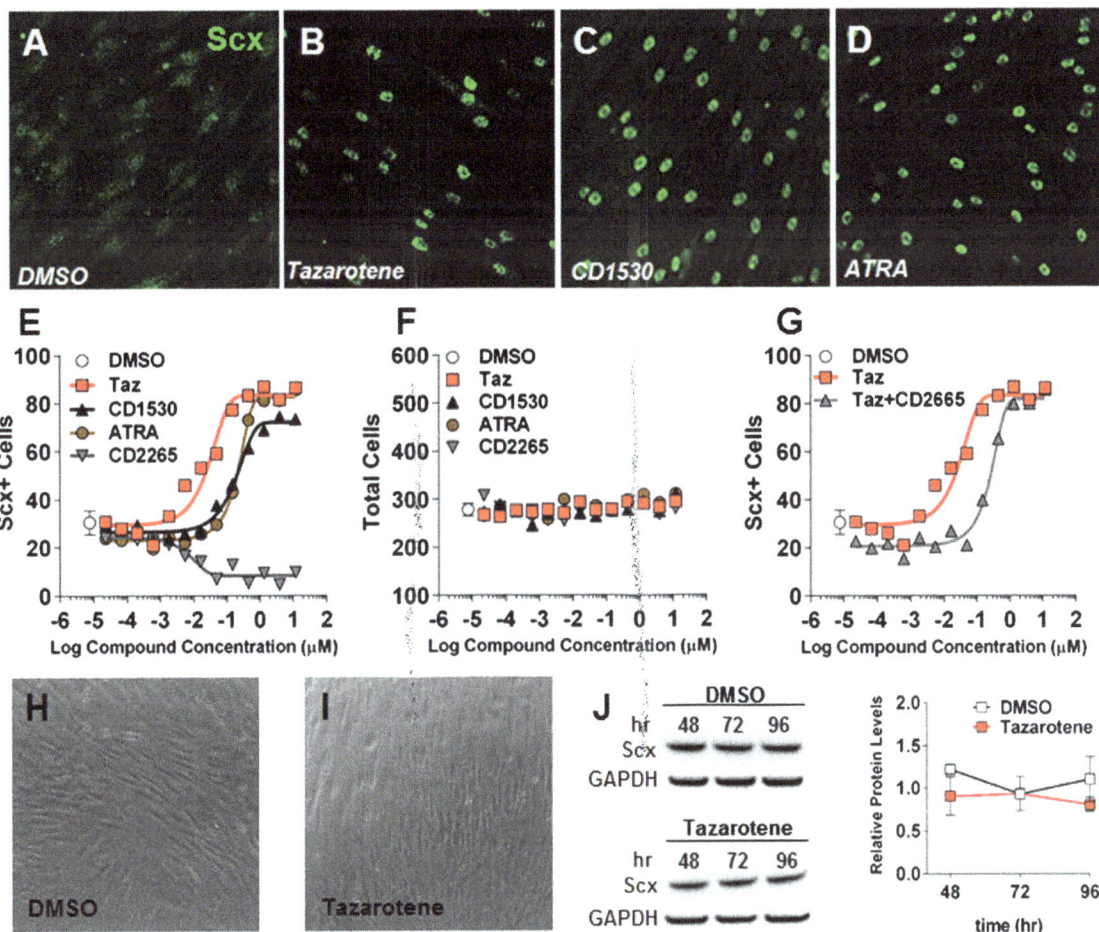

Fig. 2 RAR agonists induce SCX nuclear localization in human TSCs **a–d**. The effect was observed in a dose–response manner and EC$_{50}$ curves could be generated from a SCX nuclear localization algorithm looking at nuclear translocation and corresponding cell counts for different RAR agonists (tazarotene, CD1530, retinoic acid) and an antagonist (CD2665) **e**, **f**. Co-treatment with the RARγ antagonist CD2665 (1 μM) was able to induce a shift in the tazarotene EC$_{50}$ curve **g**. Phase-contrast images showing different cell morphologies between control and tazarotene-treated cells **h**, **i**. Western blot quantification of total SCX protein remaining unchanged following treatment with tazarotene **j** *ATRA* all-trans retinoic acid, *DMSO* dimethyl sulfoxide, *hr* hours, *Scx* scleraxis, *Taz* tazarotene

Tazarotene prevents spontaneous differentiation during expansion

It is known that MSCs from different origins change progressively in culture, with some cells undergoing spontaneous or premature differentiation during successive passages [3–5]. Since TSCs are relatively similar to MSCs, we hypothesized that the expansion of cells in the presence of tazarotene would prevent this process and preserve their stem cell characteristics. To test this hypothesis, TSCs were isolated from naïve rat Achilles tendons and plated at clonal density for 7 days. Colonies were then trypsinized and plated at subconfluency with or without tazarotene. Every 3–4 days, the same of amount of cells from either condition were replated and treated for four consecutive passages. mRNA was harvested at the

first and fourth passages for gene expression analysis (Fig. 5a). As hypothesized, cells grown in the absence of tazarotene showed an increase in several markers associated with differentiation towards multiple mesenchymal lineages. Gene expression level comparison from passage 1 with passage 4 revealed significant upregulation of tenogenic (*EphA4*, *Col1a1*, *Col3a1*, *Bgn*), osteogenic (*Runx2*), adipogenic (*Foxo1*), and chondrogenic (*Col2a1*) genes (Fig. 5b, d). Inversely, stem cell marker expression (*Oct4* and *Ssea-1*) was higher in tazarotene expanded cells (Fig. 5c). The preservation of stem cell marker expression could be observed over several passages (Additional file 5: Figure S4). These findings suggest tazarotene could maintain the TSC phenotype and prevent spontaneous differentiation during cell expansion.

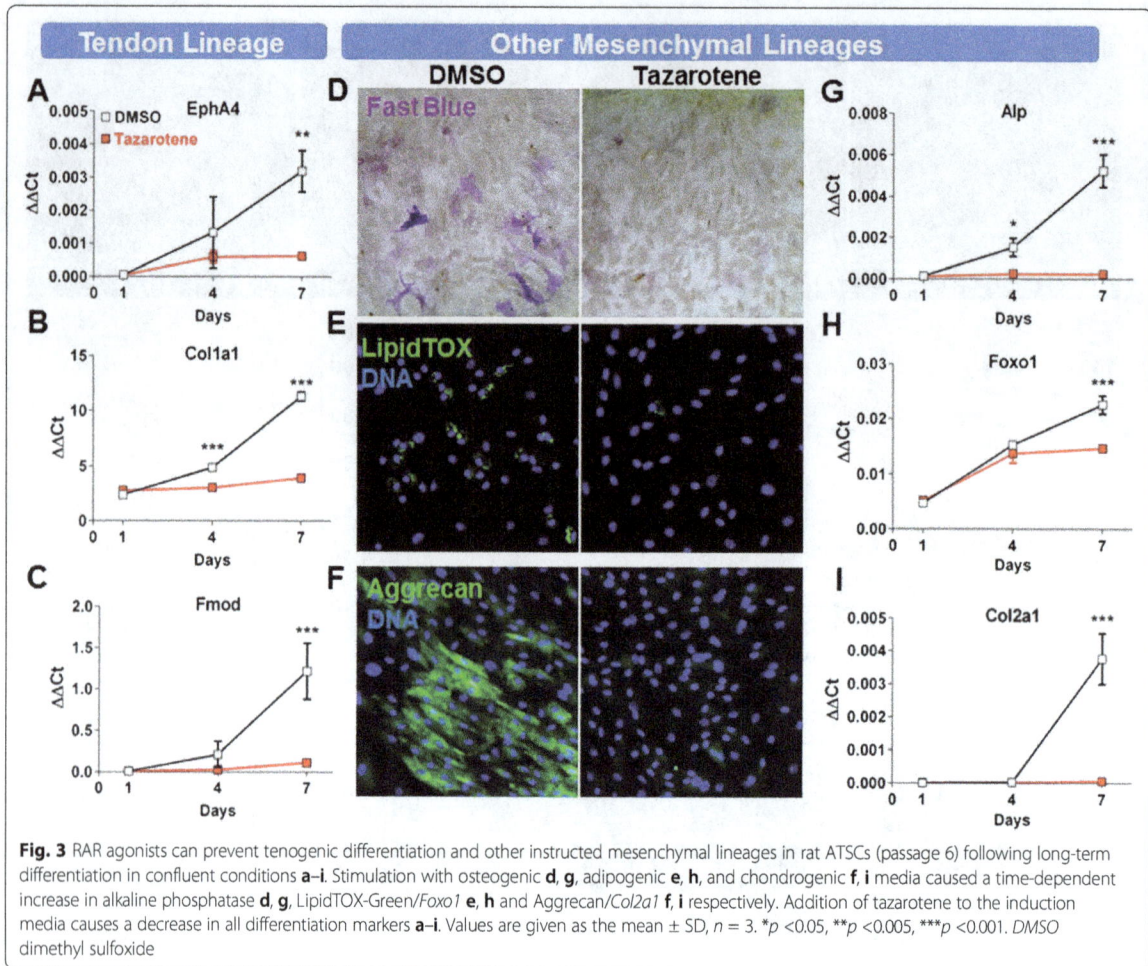

Fig. 3 RAR agonists can prevent tenogenic differentiation and other instructed mesenchymal lineages in rat ATSCs (passage 6) following long-term differentiation in confluent conditions **a–i**. Stimulation with osteogenic **d**, **g**, adipogenic **e**, **h**, and chondrogenic **f**, **i** media caused a time-dependent increase in alkaline phosphatase **d**, **g**, LipidTOX-Green/*Foxo1* **e**, **h** and Aggrecan/*Col2a1* **f**, **i** respectively. Addition of tazarotene to the induction media causes a decrease in all differentiation markers **a–i**. Values are given as the mean ± SD, $n = 3$. *$p < 0.05$, **$p < 0.005$, ***$p < 0.001$. *DMSO* dimethyl sulfoxide

Inhibition of differentiation by tazarotene is reversible upon withdrawal

Because tazarotene appeared to prevent differentiation in subconfluent conditions, we aimed to determine whether this blockade was reversible upon removal of the compound. TSCs grown in the presence of tazarotene for nine passages were further expanded with or without the compound for an additional three passages and induced to differentiate into different lineages (Fig. 6a). Cells from both of these conditions were then plated in the absence of compound and induced to differentiate into osteocytes, adipocytes, or chondrocytes for 14 DIV while using the untreated passage 12 cells as a control. Control cells displayed alkaline phosphatase reactivity in addition to LipidTOX-Green and aggrecan-positive staining when stimulated with osteogenic, adipogenic, and chondrogenic media respectively (Fig. 6b, e, h). Similar to the control cells, withdrawal of the compound for three consecutive passages was sufficient to restore multiple lineage differentiation capabilities to the

cells (Fig. 6c, f, i). Cells expanded in the presence of tazarotene until the same passage number (passage 12) were not able to differentiate despite the absence of the compound in the induction media, suggesting that additional cell divisions following withdrawal might be required to fully restore their competence to differentiate (Fig. 5d, g, j). We also confirmed that tazarotene treatment does not have a major impact on cell proliferation at efficacious doses. The population doubling time at concentrations between 20 and 500 nM did not show significant difference compared with untreated cells and was also unchanged following withdrawal of the compound (Additional file 6: Figure S5).

Histone methylation appears necessary for SCX nuclear induction by tazarotene

To verify any possible epigenetic histone modifications induced by RAR, we took advantage of the Scx nuclear translocation assay and tested different histone methyltransferase and acetyltransferase inhibitors (BIX-01294

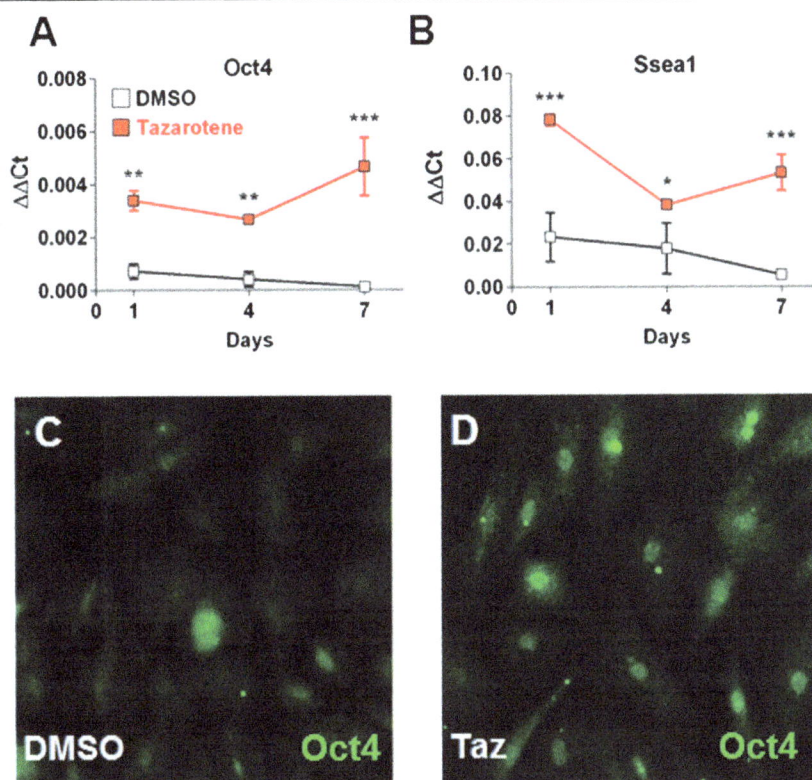

Fig. 4 Stem cell marker maintenance following treatment with tazarotene **a–d**. *Oct4* and *Ssea-1* mRNA levels are higher in tazarotene-treated cells compared with DMSO control **a**, **b**. *Oct4* immunofluorescence staining after 4 days confirms higher nuclear levels in tazarotene-treated cells **c**, **d**. Values are given as the mean ± SD, $n = 3$. *$p < 0.05$, **$p < 0.005$, ***$p < 0.001$. *DMSO* dimethyl sulfoxide, *Taz* tazarotene

and C646, respectively) to assess whether inhibiting such modifications at the chromatin level would be sufficient to block the effect of tazarotene on Scx nuclear translocation. For this experiment, cells were plated in the presence of tazarotene while being co-treated with each inhibitor for 4 DIV. As shown previously, Scx immunofluorescence revealed strong nuclear translocation in tazarotene-treated cells (Fig. 7a). Co-treatment with C646 did not have an effect on Scx induction, whereas BIX-01294 was able to suppress the SCX nuclear localization. This indicates that histone methylation plays a role Scx nuclear binding via RAR signaling (Fig. 7b, c).

Discussion

Using a high-content imaging screen, we were able to identify RAR agonists as potent inducers of Scx nuclear binding in human TSCs. Stimulation of RARs on stem cells from different origins have shown both pro-differentiating and anti-differentiating activities [15–18]. However, the role of RAR in TSCs has not yet been characterized. Although Scx appears to be necessary for tendon development, it remains expressed in freshly isolated TSCs from adult tendons. In our study we have shown that maintaining Scx localization in the nucleus using a RAR agonist does not necessarily induce tenogenic differentiation but is rather associated with the maintenance of the TSC phenotype, consistent with the inhibition of differentiation towards multiple mesenchymal lineages. Even though RAR agonists were identified in our screen, it does not appear that their inhibitory activity on differentiation is entirely mediated through Scx. Knockdown experiments using small interfering mRNA against *Scx* transcript did not revert the inhibition of differentiation induced by tazarotene, suggesting that additional transcription factors might be involved (Additional file 7: Figure S6).

TSCs could be cultured in presence of tazarotene for up to 20 passages with a similar population doubling time compared with untreated cells. The maintenance of stem cell marker expression levels remained similar for up to 10 passages but slowly decreased afterwards, while still remaining higher than untreated cells (Additional file 5: Figure S4). It is possible that TSCs slowly become irresponsive to tazarotene treatment following prolonged exposure. At concentrations higher than 500 nM, we

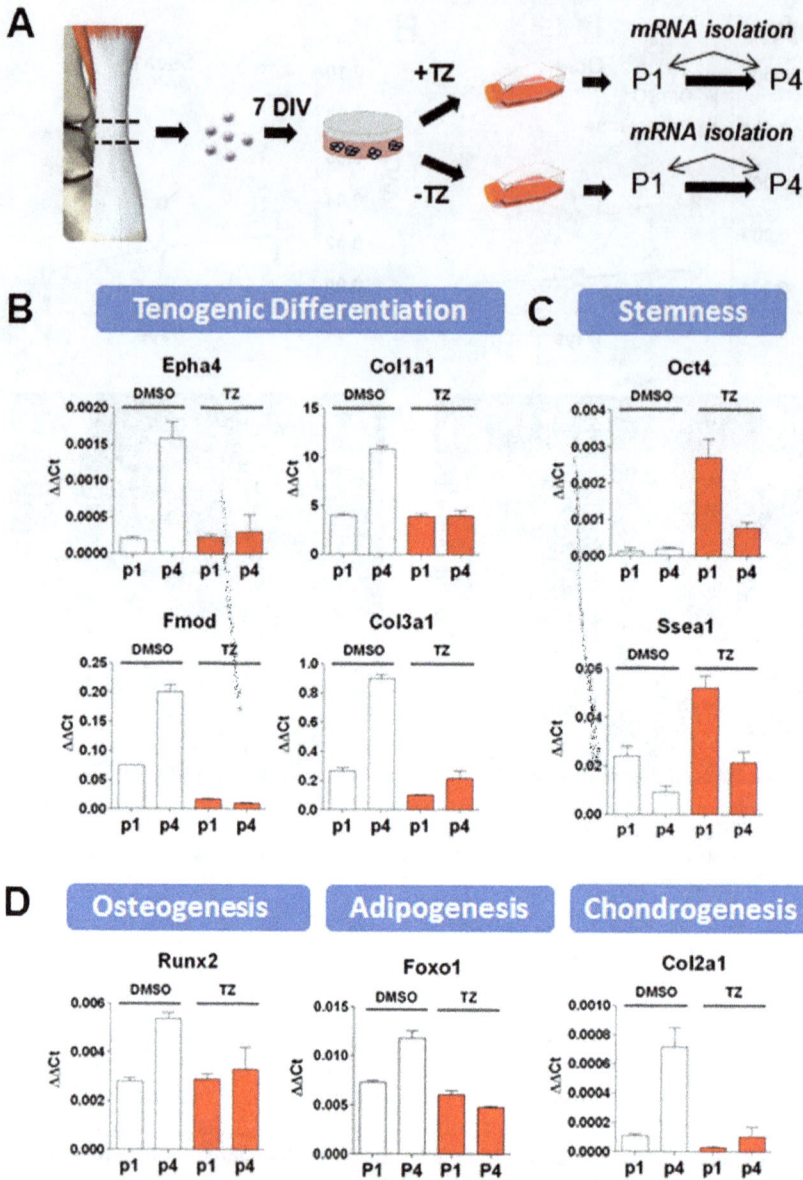

Fig. 5 Tazarotene prevents spontaneous differentiation arising with extended culture time. TSCs were isolated from adult rat Achilles tendons and plated at low density for colony formation before being trypsinized and grown with or without tazarotene (*TZ*). Messenger RNA was collected at early (*P1*) and late passages (*P4*) for gene expression analysis **a**. In control conditions, several genes typical of TSC differentiation towards multiple lineages were found upregulated in late passage cells (P1 vs P4), confirming spontaneous differentiation during in vitro cell expansion **b**, **d**. Addition of tazarotene to the culture media was able to preserve TSC's stemness as seen by cells from passage 4 having similar expression levels to those of passage 1 **c**. Values are given as the mean ± SD, n = 3

could also observe a reduction in proliferation of about 50–75 %. Tazarotene is a selective RARβ and RARγ agonist but it can also bind RARα at higher concentrations [19]. Activation of RARα has been previously shown to inhibit cell proliferation in other cell types, which could explain the slow decrease in cell division we observed at higher concentrations [20]. The anti-differentiation effect of tazarotene appears to be partially conserved among different MSC types depending on their origin. We found that tazarotene has similar activity on MSCs derived from adipose tissue but has the opposite effect on MSCs isolated from bone marrow by being pro-osteogenic (data not shown). This suggests that RAR signaling might have a different role in BM-MSCs.

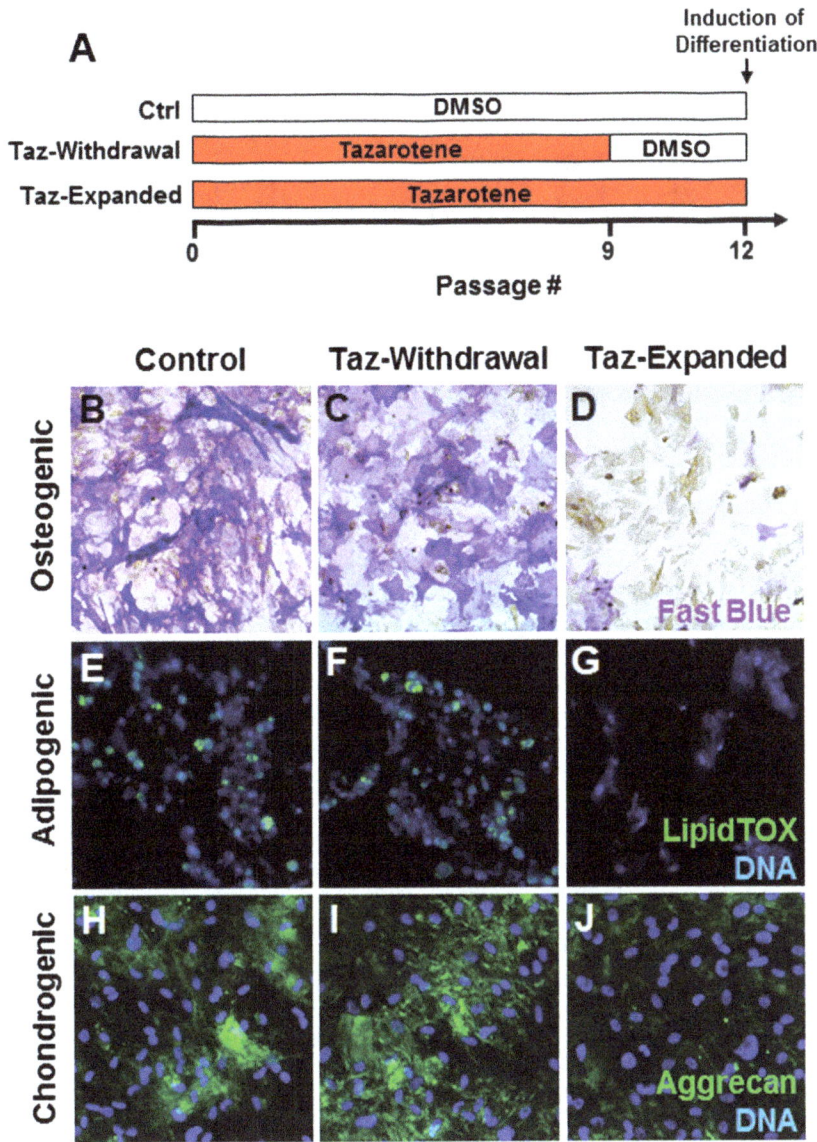

Fig. 6 Differentiation potential of tazarotene-expanded TSCs is conserved upon withdrawal. Tazarotene-expanded cells from passage 9 were cultured with or without the compound for three further passages and induced to differentiate towards different mesenchymal lineages **a**. Both conditions were equally able to differentiate into osteocytes, adipocytes, and chondrocytes as seen by upregulation of alkaline phosphatase, LipidTOX, and aggrecan staining **b, c, e, f, h, i**. However, TSCs at the same passage cultured in a constant presence of tazarotene did not differentiate **d, g, j**. *DMSO* dimethyl sulfoxide, *Taz* tazarotene

RARs have multiple functions at the nuclear level by interacting with retinoid X receptors and other co-repressor or activator proteins leading to epigenetic and gene expression changes [21, 22]. Stem cell homeostasis is usually maintained through these mechanisms that are highly dynamic in regulating chromatin structure as well as specific gene expression programs involved in self-renewal and differentiation [23]. In general, stem cells show a more decondensed chromatin which contributes to an open or accessible state as compared with differentiated cells [24]. The overall increased levels of histone modifications, which are commonly transcriptionally active regions, are enriched in stem cells while silenced regions are strongly reduced compared with more differentiated cells [25]. Our results suggest that RAR agonists could preserve TSC stemness through epigenetic modifications, specifically involving histone methylation. Co-treatment with a histone methyltransferase

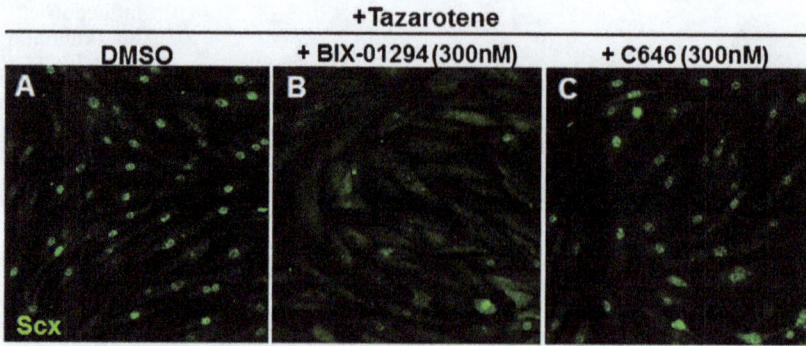

Fig. 7 Tazarotene induces Scx nuclear localization through histone modifications. Human TSCs from passage 4 were co-treated for 4 DIV with histone methyltransferase and acetyltransferase inhibitors in the presence of tazarotene **a–c**. Addition of BIX-01294, a histone methyltransferase inhibitor, was able to suppress Scx nuclear localization induced by tazarotene, while C646, a histone acetyltransferase inhibitor, did not have any effect **b**, **c**. *DMSO* dimethyl sulfoxide, *Scx* scleraxis

inhibitor was sufficient to block tazarotene-induced Scx nuclear localization, while an acetyltransferase inhibitor did not have any effect.

It is possible that the binding elements and/or promoter regions of Scx and other transcription factors critical for TSC identity are gradually repressed with passages due to changes in chromatin structure. Interestingly, hypoxia has been shown to preserve TSC stemness in culture and further reduce differentiation in culture conditions [26]. Another possible explanation could be that mechanical loading in vitro is necessary for maintaining TSC properties similar to what has been shown for the maintenance Scx nuclear localization [9]. We can hypothesize that expansion in normoxia conditions and/or in the absence of mechanical stimulation might induce epigenetic changes responsible for the decrease in Scx nuclear binding and spontaneous differentiation occurring during the expansion process. These changes could be driven by an overall decrease in histone methylation and could be potentially reversed using RAR agonists. DNA methylation has also been shown to prevent spontaneous differentiation of mesenchymal progenitors in culture where removing methyl groups using 5-azacytidine causes differentiation towards the osteogenic and adipogenic cell fate [27]. The direct link between RAR and Scx nuclear localization still remains to be investigated. Further epigenetic analyses on TSCs treated with RAR agonists should help elucidate this process. Additionally, analyzing the methylation status of the SCX gene as well as other key genes involved in self-renewal and differentiation could reveal more details on how RAR agonists affect the epigenetics of TSCs.

Conclusion

Cell-based therapies using adult stem cells harvested from the target tissue represent a potential strategy to address the unmet medical need for tendon regeneration

[28–30]. Recent advances in such an approach show promising results but additional work is essential to understand how best to expand and prepare these cells for treatment and avoid premature differentiation during the expansion process. These non-tendon committed progenitors could potentially engraft following transplantation to a site of tendon injury and interfere with tendon healing and biomechanical properties by producing the wrong ECM in vivo. Tendon calcification and cartilage-like differentiation has been observed in clinical samples of tendinopathy, reinforcing the idea that engrafting such non-tendon progenitors would be detrimental [31]. In this study, we demonstrated that TSCs, like MSCs, undergo spontaneous differentiation in vitro and that addition of a RAR agonist to the culture media was able to prevent this process. In vivo studies comparing TSCs expanded with or without a RAR agonist should reinforce the use of molecules that preserve their stem cell characteristics during the expansion phase.

Additional files

Additional file 1: is Table S1 presenting the quantitative PCR primer list. (TIF 2735 kb)

Additional file 2: is Figure S1 showing Scx antibody specificity and TSC identity. Transduced 293T cells using a lentivirus overexpressing Scx with Zsgreen reporter show clear nuclear staining for Scx protein while control lentiviral vector did not show any staining **A**, **B**. Only TSCs but not tenocytes from the same tendon tissue were positive for Oct4, confirming that our protocol for stem cell isolation was successful **C**, **D**. (TIF 4337 kb)

Additional file 3: is Figure S2 showing that tazarotene treatment increases nuclear Scx translocation in human patellar TSCs. TSCs isolated from human patellar tendon lose nuclear Scx localization with passages **A**, **C**. Treatment with tazarotene at 100 nM is able to induce Scx nuclear translocation similar to Achilles TSCs **B**, **D**, **E**. (TIF 1345 kb)

Additional file 4: is Figure S3 showing that tazarotene treatment increases nuclear Scx translocation in rat Achilles TSCs. Western blot showing enrichment in Scx protein in the nuclear extract **A**, **B**. (TIF 734 kb)

Retinoic acid receptor signaling preserves tendon stem cell characteristics and prevents...

185

Additional file 5: is Figure S4 showing that tazarotene can maintain stem cell marker expression in TSCs for up to 17 passages. mRNA levels of both Oct4 and Ssea1 remain high in the presence of the drug for up to 17 passages **A**. Withdrawal of the compound at passage 8 is followed by a rapid decrease of both Oct4 and Ssea1 which is already visible at passage 10 **B**. (TIF 719 kb)

Additional file 6: is Figure S5 showing TSC expansion in the presence of tazarotene does not affect the population doubling time. A growth curve for up to passage 17 was performed and did not show statistical difference with control DMSO-treated cells for concentrations between 20 and 500 nM **A**. Withdrawal of the compound at passage 8 is not followed by a change in population doubling time **B**. (TIF 811 kb)

Additional file 7: is Figure S6 showing that Scx siRNA knockdown does not suppress the inhibitory effect of tazarotene on osteogenic differentiation. TSCs were transfected with siRNA against Scx or with a siRNA control and were induced to differentiate towards the osteogenic lineage and visualized by alkaline phosphatase staining **A**, **B**. The inhibition of osteogenic differentiation in presence of tazarotene was not blocked following Scx siRNA knockdown **C**, **D**. (TIF 3491 kb)

Abbreviations
ATRA: All-trans retinoic acid; BM-MSC: Bone marrow mesenchymal stem cell; DIV: Days in vitro; DMSO: Dimethyl sulfoxide; EC_{50}: Half-maximal effective concentration which induces a response; ECM: Extracellular matrix; HBSS: Hank's Balanced Salt Solution; LMW: Low molecular weight; MSC: Mesenchymal stem cell; PBS: Phosphate-buffered saline; RAR: Retinoic acid receptor; Scx: Scleraxis; TGFβ2: Transforming growth factor beta-2; TSC: Tendon stem cell.

Competing interest
The authors declare that they have no competing interests relevant to this work. JE, CG, and JP are current employees of Novartis and may own stock or hold stock options in the company. SW and EG may own stock or hold stock options in the company.

Authors' contributions
SW, CG, JP, EG, and JE designed experiments, analyzed the data, and wrote the manuscript. All authors discussed the results and their implications and approved the final manuscript.

Acknowledgments
This work was funded by Novartis Institutes for Biomedical Research. The authors are thankful to Jennifer L Harris, Olivier Leupin, Shea Carter, and Kristen Johnson for reviewing the manuscript.

References
1. Bi Y, Ehirchiou D, Kilts TM, Inkson CA, Embree MC, Sonoyama W, et al. Identification of tendon stem/progenitor cells and the role of the extracellular matrix in their niche. Nat Med. 2007;13(10):1219–27.
2. Zhang J, Wang JH. Characterization of differential properties of rabbit tendon stem cells and tenocytes. BMC Musculoskelet Disord. 2010;11:10.
3. Li Z, Liu C, Xie Z, Song P, Zhao RC, Guo L, et al. Epigenetic dysregulation in mesenchymal stem cell aging and spontaneous differentiation. PLoS One. 2011;6(6):e20526.
4. Gou S, Wang C, Liu T, Wu H, Xiong J, Zhou F, et al. Spontaneous differentiation of murine bone marrow-derived mesenchymal stem cells into adipocytes without malignant transformation after long-term culture. Cells Tissues Organs. 2010;191(3):185–92.
5. Tsai CC, Chen CL, Liu HC, Lee YT, Wang HW, Hou LT, et al. Overexpression of hTERT increases stem-like properties and decreases spontaneous differentiation in human mesenchymal stem cell lines. J Biomed Sci. 2010;17:64.
6. Alberton P, Popov C, Pragert M, Kohler J, Shukunami C, Schieker M, et al. Conversion of human bone marrow-derived mesenchymal stem cells into tendon progenitor cells by ectopic expression of scleraxis. Stem Cells Dev. 2012;21(6):846–58.
7. Murchison ND, Price BA, Conner DA, Keene DR, Olson EN, Tabin CJ, et al. Regulation of tendon differentiation by scleraxis distinguishes force-transmitting tendons from muscle-anchoring tendons. Development. 2007;134(14):2697–708.
8. Chen X, Yin Z, Chen JL, Shen WL, Liu HH, Tang QM, et al. Force and scleraxis synergistically promote the commitment of human ES cells derived MSCs to tenocytes. Sci Rep. 2012;2:977.
9. Maeda T, Sakabe T, Sunaga A, Sakai K, Rivera AL, Keene DR, et al. Conversion of mechanical force into TGF-beta-mediated biochemical signals. Curr Biol. 2011;21(11):933–41.
10. Mendias CL, Gumucio JP, Bakhurin KI, Lynch EB, Brooks SV. Physiological loading of tendons induces scleraxis expression in epitenon fibroblasts. J Orthop Res. 2012;30(4):606–12.
11. Tan C, Lui PP, Lee YW, Wong YM. Scx-transduced tendon-derived stem cells (TDSCs) promoted better tendon repair compared to mock-transduced cells in a rat patellar tendon window injury model. PLoS One. 2014;9(5):e97453.
12. Pryce BA, Brent AE, Murchison ND, Tabin CJ, Schweitzer R. Generation of transgenic tendon reporters, ScxGFP and ScxAP, using regulatory elements of the scleraxis gene. Dev Dyn. 2007;236(6):1677–82.
13. Schweitzer R, Chyung JH, Murtaugh LC, Brent AE, Rosen V, Olson EN, et al. Analysis of the tendon cell fate using Scleraxis, a specific marker for tendons and ligaments. Development. 2001;128(19):3855–66.
14. Scott A, Danielson P, Abraham T, Fong G, Sampaio AV, Underhill TM. Mechanical force modulates scleraxis expression in bioartificial tendons. J Musculoskelet Neuronal Interact. 2011;11(2):124–32.
15. Hisada K, Hata K, Ichida F, Matsubara T, Orimo H, Nakano T, et al. Retinoic acid regulates commitment of undifferentiated mesenchymal stem cells into osteoblasts and adipocytes. J Bone Miner Metab. 2013;31(1):53–63.
16. Solmesky L, Lefler S, Jacob-Hirsch J, Bulvik S, Rechavi G, Weil M. Serum free cultured bone marrow mesenchymal stem cells as a platform to characterize the effects of specific molecules. PLoS One. 2010;5(9):e12689.
17. Oeda S, Hayashi Y, Chan T, Takasato M, Aihara Y, Okabayashi K, et al. Induction of intermediate mesoderm by retinoic acid receptor signaling from differentiating mouse embryonic stem cells. Int J Dev Biol. 2013;57(5):383–9.
18. Shimono K, Tung WE, Macolino C, Chi AH, Didizian JH, Mundy C, et al. Potent inhibition of heterotopic ossification by nuclear retinoic acid receptor-gamma agonists. Nat Med. 2011;17(4):454–60.
19. Nagpal S, Athanikar J, Chandraratna RA. Separation of transactivation and AP1 antagonism functions of retinoic acid receptor alpha. J Biol Chem. 1995;270(2):923–7.
20. Neuville P, Yan Z, Gidlof A, Pepper MS, Hansson GK, Gabbiani G, et al. Retinoic acid regulates arterial smooth muscle cell proliferation and phenotypic features in vivo and in vitro through an RARalpha-dependent signaling pathway. Arterioscler Thromb Vasc Biol. 1999;19(6):1430–6.
21. Kashyap V, Gudas LJ. Epigenetic regulatory mechanisms distinguish retinoic acid-mediated transcriptional responses in stem cells and fibroblasts. J Biol Chem. 2010;285(19):14534–48.
22. Cheong HS, Lee HC, Park BL, Kim H, Jang MJ, Han YM, et al. Epigenetic modification of retinoic acid-treated human embryonic stem cells. BMB Rep. 2010;43(12):830–5.
23. Zhou Y, Kim J, Yuan X, Braun T. Epigenetic modifications of stem cells: a paradigm for the control of cardiac progenitor cells. Circ Res. 2011;109(9):1067–81.
24. Fisher CL, Fisher AG. Chromatin states in pluripotent, differentiated, and reprogrammed cells. Curr Opin Genet Dev. 2011;21(2):140–6.
25. Meshorer E, Misteli T. Chromatin in pluripotent embryonic stem cells and differentiation. Nat Rev Mol Cell Biol. 2006;7(7):540–6.
26. Zhang J, Wang JH. Human tendon stem cells better maintain their stemness in hypoxic culture conditions. PLoS One. 2013;8(4):e61424.
27. Hupkes M, van Someren EP, Middelkamp SH, Piek E, van Zoelen EJ, Dechering KJ. DNA methylation restricts spontaneous multi-lineage differentiation of mesenchymal progenitor cells, but is stable during growth factor-induced terminal differentiation. Biochim Biophys Acta. 2011;1813(5):839–49.
28. Lui PP, Ng SW. Cell therapy for the treatment of tendinopathy—a systematic review on the pre-clinical and clinical evidence. Semin Arthritis Rheum. 2013;42(6):651–66.
29. Wang A, Breidahl W, Mackie KE, Lin Z, Qin A, Chen J, et al. Autologous tenocyte injection for the treatment of severe, chronic resistant lateral epicondylitis: a pilot study. Am J Sports Med. 2013;41(12):2925–32.
30. Chen J, Yu Q, Wu B, Lin Z, Pavlos NJ, Xu J, et al. Autologous tenocyte therapy for experimental Achilles tendinopathy in a rabbit model. Tissue Eng Part A. 2011;17(15-16):2037–48.
31. Rui YF, Lui PP, Rolf CG, Wong YM, Lee YW, Chan KM. Expression of chondro-osteogenic BMPs in clinical samples of patellar tendinopathy. Knee Surg Sports Traumatol Arthrosc. 2012;20(7):1409–17.

17

Angiopoietin-1 receptor Tie2 distinguishes multipotent differentiation capability in bovine coccygeal nucleus pulposus cells

Adel Tekari[1][*], Samantha C. W. Chan[1,2], Daisuke Sakai[3,5], Sibylle Grad[4,5] and Benjamin Gantenbein[1,5]

Abstract

Background: The intervertebral disc (IVD) has limited self-healing potential and disc repair strategies require an appropriate cell source such as progenitor cells that could regenerate the damaged cells and tissues. The objective of this study was to identify nucleus pulposus-derived progenitor cells (NPPC) and examine their potential in regenerative medicine in vitro.

Methods: Nucleus pulposus cells (NPC) were obtained from 1-year-old bovine coccygeal discs by enzymatic digestion and were sorted for the angiopoietin-1 receptor Tie2. The obtained Tie2− and Tie2+ fractions of cells were differentiated into osteogenic, adipogenic, and chondrogenic lineages in vitro. Colony-forming units were prepared from both cell populations and the colonies formed were analyzed and quantified after 8 days of culture. In order to improve the preservation of the Tie2+ phenotype of NPPC in monolayer cultures, we tested a selection of growth factors known to have stimulating effects, cocultured NPPC with IVD tissue, and exposed them to hypoxic conditions (2 % O_2).

Results: After 3 weeks of differentiation culture, only the NPC that were positive for Tie2 were able to differentiate into osteocytes, adipocytes, and chondrocytes as characterized by calcium deposition ($p < 0.0001$), fat droplet formation ($p < 0.0001$), and glycosaminoglycan content ($p = 0.0095$ vs. Tie2− NPC), respectively. Sorted Tie2− and Tie2+ subpopulations of cells both formed colonies; however, the colonies formed from Tie2+ cells were spheroid in shape, whereas those from Tie2− cells were spread and fibroblastic. In addition, Tie2+ cells formed more colonies in 3D culture ($p = 0.011$) than Tie2− cells. During expansion, a fast decline in the fraction of Tie2+ cells was observed ($p < 0.0001$), which was partially reversed by low oxygen concentration ($p = 0.0068$) and supplementation of the culture with fibroblast growth factor 2 (FGF2) ($p < 0.0001$).

Conclusions: Our results showed that the bovine nucleus pulposus contains NPPC that are Tie2+. These cells fulfilled formally progenitor criteria that were maintained in subsequent monolayer culture for up to 7 days by addition of FGF2 or hypoxic conditions. We propose that the nucleus pulposus represents a niche of precursor cells for regeneration of the IVD.

Keywords: Intervertebral disc, Nucleus pulposus, Nucleus pulposus progenitor cells, Tie2, Hypoxia, Fibroblast growth factor 2, Growth factors

* Correspondence: adel.tekari@istb.unibe.ch
[1]Tissue and Organ Mechanobiology, Institute for Surgical Technology & Biomechanics, Medical Faculty, University of Bern, Bern, Switzerland
Full list of author information is available at the end of the article

Background

The intervertebral disc (IVD) has limited regenerative potential and disc degeneration is a major cause of chronic low back pain. This represents a leading cause of disability with significant economic and social burdens [1–3]. The IVD consists of an inner nucleus pulposus (NP) surrounded by the annulus fibrosus (AF) tissue, and hyaline articular cartilage is located at the endplates between the IVD and the vertebral bodies. The gelatinous NP is an avascular tissue containing a highly organized extracellular matrix rich in proteoglycans and collagens with few dispersed cells [4]. In this respect, the NP cells reside within hypoxic conditions, since no vasculature enters the NP [5]. Furthermore, disc cells actively regulate the homeostasis of the extracellular matrix by several cytokines and growth factors acting in an autocrine and paracrine fashion. Members of the transforming growth factor (TGF) superfamily, including TGFβ1, growth and differentiation factor, fibroblast growth factor 2 (FGF2), and vascular endothelial growth factor (VEGF) were identified previously as anabolic regulators within the IVD [6].

IVD degeneration implies a degradation of the extracellular matrix in the NP and the AF resulting in a reduced disc height. The exact mechanism by which IVD degeneration is induced is still unknown. Some risk factors were identified and include aging, genetic predisposition, and stress factors [7]. The degenerative changes of the IVD take place early in life and the cellular turnover rate is much slower compared with other tissues [8–10].

Current treatments aim to repair the degenerated disc by replacement of the injured tissue with a functional biological substitute or prosthesis. Conventional treatments for IVD degeneration are limited, since conservative or surgical therapies do not restore IVD tissue properties. Since the IVD possesses very limited healing capacity, regenerative medicine by injection of cells may represent promising therapy for treatment of disc degeneration [11]. As such, IVD repair strategies require an appropriate cell source that is able to regenerate the damaged NP tissue such as progenitor and stem cells. Cell-based therapies by injection of IVD cells, chondrocytes, or stem cells have gained significant insight and progressed to clinical trials for treatment of spinal disorders [12]. Progenitor cells do have the advantage over terminally differentiated cells that they maintain their multipotent differentiation and self-renewal potential in vivo and in vitro under appropriate conditions. Furthermore, these cells play an important role in the development and homeostasis of the IVD tissue. Recently, progenitor cells that are positive for the angiopoietin-1 receptor (Tie2) were identified in the mouse and human NP [13]. These cells, which express aggrecan and

collagen type II, were shown to have progenitor-like multipotency. Tie2, also known as CD202b, is a cellular membrane receptor tyrosine kinase of the Tie family. This receptor contains immunoglobulin-like loops and an epidermal growth factor (EGF)-similar domain 2 [14]. Expressed mainly in endothelial cells, the angiopoietin groups of ligands, upon binding to their receptor Tie2, are known to regulate angiogenesis [15]. Tie2 signaling appears to be critical for endothelial smooth muscle communication and vascular maturation. Deletion of Tie2 or its ligand in transgenic mice is embryonic lethal and mice die from cardiac failure [16]. The contribution of Tie2 to IVD homeostasis, however, is still poorly understood. Here, we isolated primary nucleus pulposus cells (NPC) from bovine coccygeal discs and sorted these for the Tie2 marker, where the Tie2+ fraction of cells is suggested to represent the nucleus pulposus progenitor cells (NPPC) population. To demonstrate the stemness of the Tie2+ cells, we performed differentiation assays for the Tie2– and Tie2+ cell populations and then addressed their ability to form colonies in methylcellulose-based medium. Presence of these NPPC has never been demonstrated in bovine coccygeal IVD, a leading ex-vivo animal model for studying disc degeneration and regenerative approaches [17]. A second aim was to address the reported difficulties to maintain the phenotype of NPPC in culture [13] and to test different cell culture conditions to maintain and eventually expand these cells in vitro in monolayer culture.

Methods
NPC isolation

NPC were obtained from 1-year-old bovine tail discs within 4 hours post mortem (no ethical permit required) by sequential digestion of NP tissue with 1.9 mg/ml pronase (Roche, Basel, Switzerland) for 1 hour and 80 μg/ml collagenase II (260 U/mg; Worthington, London, UK) on a plate shaker at 37 °C overnight. The remaining undigested tissue debris was removed by filtration through a 100 μm cell strainer (Falcon, Becton Dickinson, Allschwil, Switzerland); subsequently the cell viability was determined by trypan blue exclusion. The isolated NPC were used for further analysis.

Cell sorting and characterization by flow cytometry

To isolate the fraction of Tie2 expressing cells, NPC were labeled as described previously [13]. Briefly, the NPC population obtained after enzymatic digestion of 6-8 IVDs (about 8×10^6 cells for one bovine tail) was resuspended in 100 μl of fluorescence-activated cell sorting (FACS) buffer (phosphate-buffered saline containing 0.5 % bovine serum albumin (Sigma-Aldrich, Buchs, Switzerland) and 1 mM EDTA (Fluka, Buchs, Switzerland)) and was incubated with anti-rat Tie2/CD202b polyclonal rabbit antibody (10 μg/ml, clone bs-1300R; Bioss Antibodies, Woburn, MA, USA) for 30 min at 4 °C. Incubation was performed for a further 30 min at 4 °C with goat anti-rabbit antibody (Molecular Probes, Life Technologies, Zug, Switzerland) labeled with the fluorochrome Alexa 488. Isotype-matched antibody (Invitrogen, Life Technologies) was used as negative control to set the appropriate gate for positive Tie2 cells (Fig. 1). Sorting was performed on FACS Diva III (BD Biosciences, San Diego, USA); only living cells were considered by using the propidium iodide (PI)-negative gate.

To characterize the NPC by Tie2 expression after expansion in monolayer culture, the cells were labeled in a similar way. Briefly, 2×10^5 NPC in 100 µl of FACS buffer were stained with the anti-rat Tie/CD202b antibody for 30 min at 4 °C and further incubated with the goat anti-rabbit secondary antibody for 30 min at 4 °C. Fluorescence was measured on an LSR II flow cytometry system (Becton Dickinson), and the data were analyzed using FlowJo software (version 10.1 for MacOS X; LLC, Ashland, OR, USA).

NPPC proliferation

To identify proliferating cells, NPPC were expanded for 7 days in proliferation medium (alpha minimum essential medium (α-MEM; Gibco, Life Technologies) containing 10 % fetal bovine serum (FBS; Sigma-Aldrich) and penicillin/streptomycin (P/S, 100 units/ml and

100 µg/ml, respectively; Merck, Darmstadt, Germany)), whereby 10 µM bromodeoxyuridine (BrdU) was added at the beginning of the experiment with one medium change. The incorporated BrdU was detected by flow cytometry according to manufacturer's instructions (APC BrdU Flow Kit; Becton Dickinson).

Colony-forming assay

To assess the formation of colonies, single-cell suspensions of 10^3 NPC were seeded in 1 ml of methylcellulose-based medium (MethoCult H4230; Stem Cell Technologies, Vancouver, Canada) in Petri dishes (35 mm in diameter) and cultured for 8 days. The colonies formed (>10 nuclei) were quantified under a light microscope.

Osteogenic differentiation

Differentiation of NPC into osteogenic lineage was performed for cells immediately after digestion of the NP and sorting for Tie2, and was conducted in α-MEM containing 5 % FBS, P/S, 100 nM dexamethasone, 10 mM β-glycerophosphate, and 0.1 mM L-ascorbic acid-2-phosphate (all from Sigma-Aldrich) for 21 days with medium change twice a week. The serum concentration was chosen according to a pilot study (data not shown) showing a better differentiation of NPPC into osteogenic lineage during the given time period. To evaluate the cells' ability for calcium deposition, Alizarin red staining

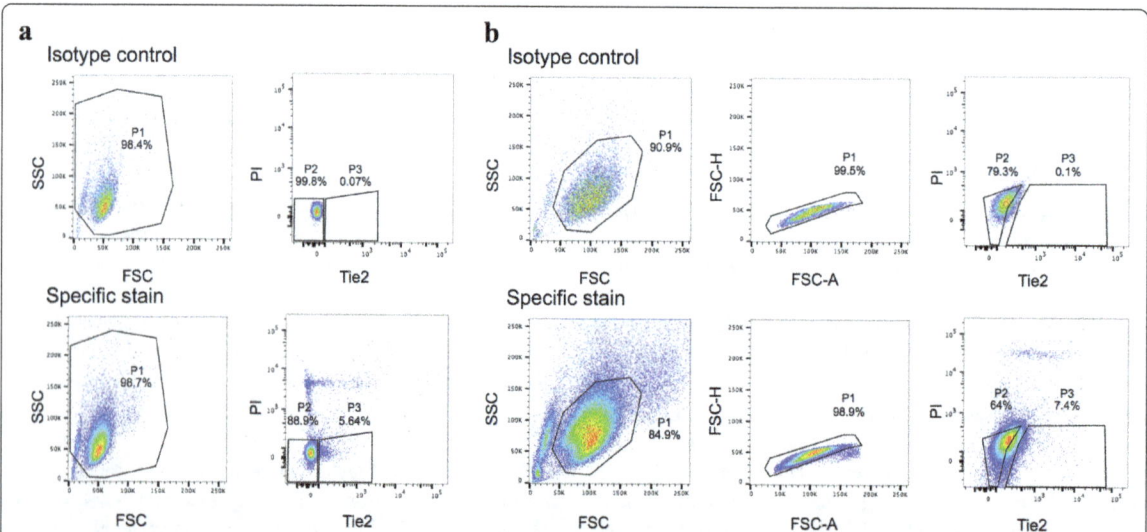

Fig. 1 Sorting and gating strategies for Tie2+ cells from a whole NPC population. The NPC suspension after enzymatic digestion was colabeled with the Tie2 antibody and PI and sorted for the Tie2 marker. **a, b** Two examples show gating of the whole cell population for forward and side scatter (*FSC* and *SSC*, P1; *left panel*). It is important to mention that primary NPC after enzymatic digestion contain tissue fragments, granules of dead cells, and debris, which are removed by a selective gating for FSC and SSC (*left panel*, **b**). In addition doublets are excluded by a FSC-H versus FSC-A gating (*middle panel*, **b**). Proper gating for Tie2 is shown for the two examples and was performed by a negative selection of cells in isotype-matched control with less than 0.1 % (*top right panel*, P3) and setting the gate at the left for the Tie2– cells (P2). The same gating was then applied for the specific Tie2 staining and by excluding PI-positive cells. *P1* whole NPC population, *P2* Tie2– cell population, *P3* Tie2+ cell population, *PI* propidium iodide, *Tie2* angiopoietin-1 receptor

was performed. The cell layers were fixed in 4 % formaldehyde, rinsed with distilled water, and subsequently exposed to 2 % Alizarin red solution for 45 min. The Alizarin red staining was released from the cell layers by addition of 10 % cetylpyridinium chloride solution (Sigma-Aldrich) and incubation for 1 hour with vigorous agitation. The samples were diluted 10-fold, transferred into a 96-well plate, and the optical density was measured at 570 nm using a microplate reader (SpectraMax M5; Bucher Biotec, Basel, Switzerland).

Adipogenic differentiation
Immediately after digestion of the NP and sorting for Tie2, NPC were grown in adipogenic medium consisting of α-MEM with 5 % FBS, P/S, 12.5 µM insulin, 100 nM dexamethasone, 0.5 mM isobutylmethylxanthine, and 60 µM indomethacin (all from Sigma-Aldrich) with medium change twice a week. Adipogenic differentiation was evaluated after 3 weeks of induction by the cellular accumulation of lipid vacuoles that were stained with Oil red O (Merck). The cell layers were fixed in 4 % formaldehyde, rinsed with 50 % ethanol, subsequently stained with Oil red O solution for 20 min, and counterstained with Mayer's Hematoxylin (Fluka) for 3 min. The cellular accumulation of lipids was quantified from the wells by counting the Oil red O-positive cells under a light microscope.

Chondrogenic differentiation
The NPC were expanded in proliferation medium in 6-well plates to compensate for the low number of Tie2+ cells obtained after sorting. Near confluency (1.93 ± 0.32 (mean $\pm$ SD) population doublings), the NPC were resorted and the different NPC populations (Tie2−, Tie2 +, and unsorted NPC) were induced towards chondrogenic differentiation. Briefly, 2.5×10^5 cells in Dulbecco's modified Eagle's medium–high glucose (with 4.5 g/l glucose; Gibco) containing P/S, ITS+, 0.1 mM L- ascorbic acid-2-phosphate, 0.3 mM L-proline, 100 nM dexamethasone (all from Sigma-Aldrich), and 10 ng/ml TGFβ1 (Peprotech, London, UK) were transferred into 15 ml polypropylene tubes and centrifuged at $500 \times g$ for 5 min [18]. After 3 weeks of culture, the pellet cultures were fixed with 4 % formaldehyde solution for 4 hours at room temperature and embedded in paraffin for subsequent preparation of 5 µm-thick sections. Sulfated glycosaminoglycans (GAG) were stained with 0.2 % Safranin-O for 10 min and sections counterstained with 0.04 % Fast Green for 2 min.

To quantify the GAG content, the pellets were recovered by melting the paraffin blocks and subsequently digested with a 3.9 U/ml papain solution containing 5 mM sodium citrate, 150 mM cysteine hydrochloride, and 5 mM EDTA (Sigma-Aldrich) at 60 °C overnight.

The total GAG content was quantified from the lysates using a bovine cartilage chondroitin sulfate standard (Sigma-Aldrich) and normalized to the DNA content (Picogreen ds DNA Assay kit; Molecular Probes, Life Technologies).

Immunohistochemical staining for proteoglycans was performed by incubation of the sections with a monoclonal mouse anti-human proteoglycan antibody (10 µg/ml, clone EFG-4; Millipore, Billerica, MA, USA) at 4 °C overnight after permeabilization with 100 % methanol for 2 min and blocking with 10 % FBS for 1 hour. Incubation was performed for a further 1 hour with a goat anti-mouse secondary antibody (Alexa 488; Molecular Probes, Life Technologies). The tissues were visualized with a confocal laser-scanning microscope (cLSM 710; Carl Zeiss, Jena, Germany).

Expansion of Tie2+ cells and culture conditions
The freshly isolated Tie2+ cells after sorting were treated with various growth factors and oxygen concentrations to test for culture conditions that could amplify and maintain the Tie2+ cells. Growth factors (Peprotech), including growth differentiation factor 5 (GDF5), GDF6, EGF, VEGF, FGF2 (100 ng/ml), and TGFβ1 (10 ng/ml), or coculture with IVD tissue using culture inserts (Becton Dickinson) for 6-well plates were applied to Tie2+ cells after sorting for 7 days in normoxia. The concentrations of the growth factors were selected according to previously published results showing a beneficial effect on NPC and/or maintenance and proliferation of stem cells in vitro [19–25]. Hypoxic conditions at 2 % O_2 have been shown in multiple studies [26, 27], including by our group [19, 28], to have a stimulatory effect on aggrecan expression by NPC. To test for cell proliferation and the conservation of Tie2 markers under hypoxia, Tie2− and Tie2+ cells were cultured in normoxia (atmospheric O_2, ~21 %) or in hypoxia using a C-274-2 shelf chamber inside a standard incubator and 1× Pro-Ox controller (Biospherix, Union Street Parish, New York, USA) adjusted to 2 % O_2 by addition of N_2.

Real-time RT-PCR
Relative gene expression of Tie2 (*TEK*), collagen type II (*COL2*), aggrecan (*ACAN*), hypoxia-inducible factor 1 alpha (*HIF1α*), and ribosomal *18S* RNA as a reference gene were monitored on expanded NPC. In order to determine the baseline expression levels of selected genes, bovine-specific oligonucleotide primers (Table 1) (Microsynth, Balgach, Switzerland) were newly designed with Beacon Designer™ software (Premier Biosoft, Palo Alto, CA, USA) based on nucleotide sequences from GenBank. All primers were tested for efficiency and melting curves of amplicons were performed to determine specific amplification. Relative gene expression was determined by application of a threshold cycle and normalization to the

Table 1 Custom-designed DNA primers used in real-time quantitative PCR study

Gene	Forward sequence	Reverse sequence
18S	ACGGACAGGATTGACAGATTG	CCAGAGTCTCGTTCGTTATCG
TEK	GGACAGGCAATAAGGATACG	ACCGAGTGGATGAAGGAA
COL2	CGGGTGAACGTGGAGAGACA	GTCCAGGGTTGCCATTGGAG
ACAN	GGCATCGTGTTCCATTACAG	ACTCGTCCTTGTCTCCATAG
HIF1α	AGGTGGATATGTCTGGATA	CAAGTCGTGCTGAATAATAC

Amplicons were generated using a two-step amplification cycling (95 °C for 15 s and 57 °C for 30 s for 45 cycles) and SYBR-green mastermix
TEK angiopoietin-1 receptor gene, COL2 collagen type II gene, ACAN aggrecan gene, HIF1α hypoxia-inducible factor 1 alpha gene, 18S ribosomal 18S RNA

reference sample (primary Tie2– NPC on day 0) using the $2^{-\Delta\Delta Ct}$ method according to Livak and Schmitten [29].

Statistical analysis

Differences in the number of colonies ($N = 6$ animals), BrdU-positive cells ($N = 3$), and expression of Tie2 ($N = 3$) were evaluated by Student's t test; histological quantifications ($N = 5$), levels of transcripts ($N = 5$), and Tie2+ cell fractions ($N = 3$) were evaluated by one-way ANOVA with Bonferroni's post-hoc test, using GraphPad Prism (version 6.0 h for Mac OS; GraphPad Software Inc., La Jolla, CA USA). $p < 0.05$ was considered significant.

Results

Sorting of Tie2+ cells from isolated NPC

The fraction of sorted Tie2+ cells after isolation of NPC accounted for 8.66 ± 3.94 % (values presented as mean ± SD) of the entire NPC population ($N = 10$ animals). The amount of Tie2+ cells showed slight variation (variation coefficient = 45.6 %) among the donors.

Differentiation of NPC in vitro

For the differentiation assays of NPC into osteogenic, adipogenic, and chondrogenic lineages, we considered the sorted Tie2– cells, the sorted Tie2+ cells, and a mixed NP population of cells (unsorted) for comparison. After 3 weeks of osteogenic induction, the cell layer formed with Tie2– cells was negative for Alizarin red and no calcium deposition was observed (Fig. 2). By

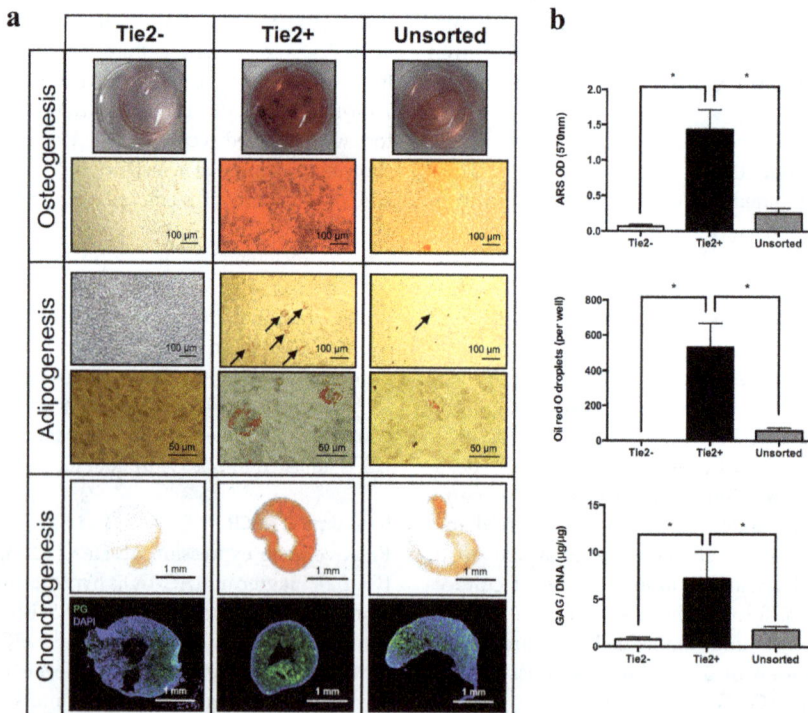

Fig. 2 Osteogenic, adipogenic, and chondrogenic differentiation assays. **a** Differentiation assays were performed in Tie2– cells and Tie2+ cells (i.e., NPPC) after sorting and a mixed cell population (unsorted NPC). *Top panel* Macroscopic and microscopic images of osteogenesis (Alizarin red staining). *Middle panel* Adipogenic differentiation (Oil red O staining), *arrows* highlighting the formation of fat droplets. *Lower panel* Chondrogenic differentiation: Safranin-O staining and proteoglycans (*PG, green*) immunohistochemistry counterstained with 4',6-diamidino-2-phenylindole (*DAPI, blue*). Results of one representative experiment of at least three repeats are shown. Scale bars are indicated on the images. **b** Quantification of Alizarin red staining (*ARS*), Oil red O fat droplet-positive cells, and *GAG*/DNA content. Individual cell populations were cross-compared to determine significance with *$p < 0.05$. Bars represent mean ± SD ($N = 5$). *GAG* glycosaminoglycans, *Tie2* angiopoietin-1 receptor (Color figure online)

contrast, Tie2+ cells deposited an extensive mineralized matrix in osteogenic medium, as demonstrated by strong Alizarin red staining ($p < 0.0001$). Interestingly, some mineralized nodular formation was observed with a mixed cell population; however, the amount of Alizarin red staining did not significantly differ ($p = 0.37$) from Tie2– cells. The adipogenic differentiation of NPC showed that Tie2– cells could not form adipocytes; however, cellular accumulation of lipid vacuoles was detected within the Tie2+ cells as demonstrated by a positive staining with Oil red O. The number of Oil red O-positive cells was significantly higher in Tie2+ cells ($p < 0.0001$) as compared with Tie2– cells. Some fat droplets were detected within the culture of unsorted cells but to a lesser extent compared with Tie2+ cells ($p < 0.001$). However, this did not significantly differ from Tie2– cells ($p = 0.85$). For chondrogenic differentiation, the tissue formed with Tie2– cells stained very weakly for GAG (by Safranin-O) and the cells showed a fibroblastic morphology. However, the cultures with Tie2+ cells stained intensely for GAG with lacunae formation observed, a characteristic of a cartilaginous phenotype, and a higher GAG/DNA content ($p = 0.0095$) compared with Tie2– cells. Similarly, the unsorted cells were able to form a cartilage-like tissue, although staining was less intense compared with the tissue of Tie2+ cells ($p = 0.02$). Similar results were observed for the proteoglycan immunohistochemistry staining, where the highest amount was detected within tissue formed from Tie2+ cells and lower amounts were observed for unsorted and Tie2– cells.

Colony formation

The Tie2– and Tie2+ isolated cell populations were able to form colonies after 8 days of culture in methylcellulose-based medium. However, the colonies formed with Tie2– cells were spread, plastic adherent, and fibroblastic, whereas the Tie2+ colonies formed were spheroid and rounded as observed macroscopically (Fig. 3a). The colonies of Tie2+ cells were quantitatively more abundant ($p = 0.011$) compared with Tie2– colonies (Fig. 3b).

Proliferation of Tie2+ cells in monolayer cultures

After 3 days of culture, 18.49 ± 4.30 % of the NPPC were positive for Tie2 (Fig. 4), while this fraction dropped to 0.61 ± 0.31 % after 7 days. The fraction of BrdU-positive cells increased from 36.56 ± 1.01 % to 93.36 ± 1.56 % when the cells were exposed to BrdU for 3–7 days. The fraction of Tie2+ cells showed a higher proliferative capacity on day 3 compared with Tie2– cells (69.2 ± 8.26 % vs. 29.1 ± 8.26 %, values defined as the ratio of BrdU-positive cells of total Tie2– or Tie2+ cells), while Tie2+ cells were less proliferative on day 7 (64.3 ± 13.4 % vs. 93.5 ± 1.52 %). Cells that incorporated BrdU were found to be either Tie2+ or Tie2–.

Expression of Tie2 during expansion of NPC

The expression of Tie2 was monitored during expansion of primary NPC in monolayer cultures. Therefore, Tie2– and Tie2+ cells after sorting were plated in 6-well plates at a density of 3×10^4 cells/well and kept in the proliferation medium for 7 days in a normoxia or hypoxia environment (2 % O_2). The cells were harvested and processed for flow cytometry analysis by staining for the Tie2 marker. It was found that the fraction of Tie2+ cells was rapidly lost in monolayer cultures in both normoxic and hypoxic conditions (Fig. 5), although culture of the NPPC in hypoxic conditions better maintained the Tie2+ pool of cells (3.34 ± 0.78 %) compared with normoxia (0.83 ± 0.12 %). The proportion of Tie2+ cells of the expanded Tie2– cells was nearly absent after

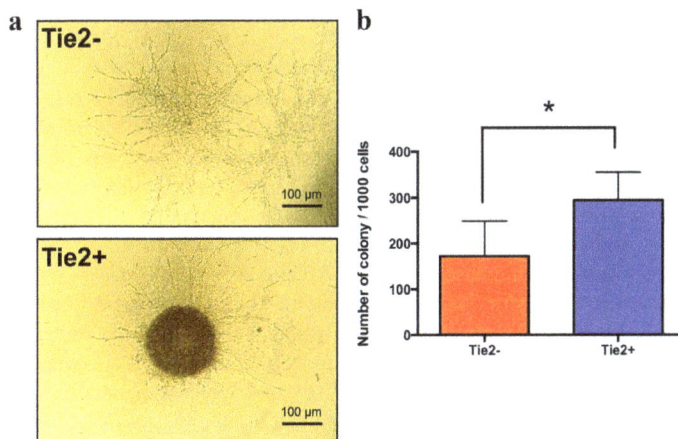

Fig. 3 Colony-forming assay of NPC versus NPPC. **a** Macroscopic images and (**b**) quantification of colonies (>10 cells) formed in Tie2– and Tie2+ cells after 8 days of culture in methylcellulose-based medium ($N = 6$). *$p < 0.05$ compared with Tie2– colonies. *Tie2* angiopoietin-1 receptor

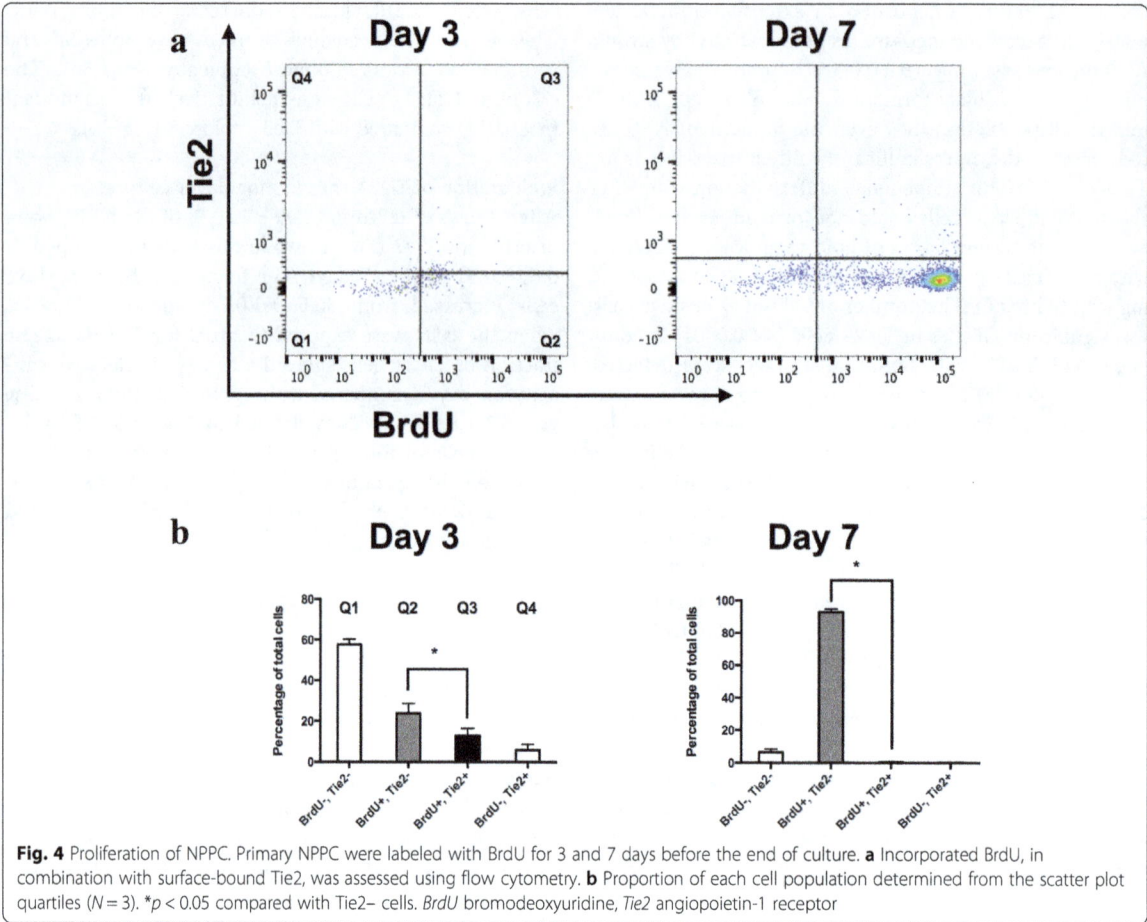

Fig. 4 Proliferation of NPPC. Primary NPPC were labeled with BrdU for 3 and 7 days before the end of culture. **a** Incorporated BrdU, in combination with surface-bound Tie2, was assessed using flow cytometry. **b** Proportion of each cell population determined from the scatter plot quartiles ($N = 3$). *$p < 0.05$ compared with Tie2− cells. *BrdU* bromodeoxyuridine, *Tie2* angiopoietin-1 receptor

7 days of culture, which accounted for 0.31 ± 0.08 % in normoxia and 0.63 ± 0.14 % in hypoxic conditions. More than 95 % of the cells were viable in both culture conditions as detected by negative PI staining.

Gene expression

The isolated Tie2+ NPPC were cultured in the proliferation medium in the presence of various growth factors or cocultured with IVD tissue for 7 days in normoxic conditions. Alternatively, cells were cultured under hypoxic conditions with/without FGF2. Treatment of the cells with FGF2 (100 ng/ml) and/or culture under hypoxic conditions resulted in a significant increase of *TEK* gene expression to levels comparable with Tie2+ after sorting (Fig. 6a). FGF2, EGF, VEGF (100 ng/ml), coculture with IVD tissue, and hypoxia increased collagen type 2 (Fig. 6b) and aggrecan expression (Fig. 6c) compared with Tie2− after sorting or cultures of NPPC for 7 days in normoxia without growth factor or IVD tissue. No such effect was detected when NPPC were treated with GDF5, GDF6 (100 ng/ml), or TGFβ1 (10 ng/ml).

HIF1α was significantly increased in hypoxic conditions (Fig. 6d). A synergistic effect of FGF2 and hypoxia on the transcript (Fig. 6a) and protein levels (Fig. 6e) of Tie2 was observed.

Discussion

Cell-based treatment of disc degeneration represents a promising approach to restore the IVD tissue function and to relieve pain [30–32]. Extensive research in the past decade using different animal models and clinical trials has improved our knowledge on the effects of cell-based therapies. In these studies, different cell types including IVD-derived cells [33–35], chondrocytes [36–38], and stem and progenitor cells [39–43] were used for transplantation into the degenerated IVD either alone or in combination with a biomaterial. The success rate of such treatments was variable and highly dependent on the model used, indicating that the selection of the cell source is a crucial parameter for treatment of disc degeneration. Bone marrow or adipose tissue-derived stem and progenitor cells might have the advantage over committed cells in

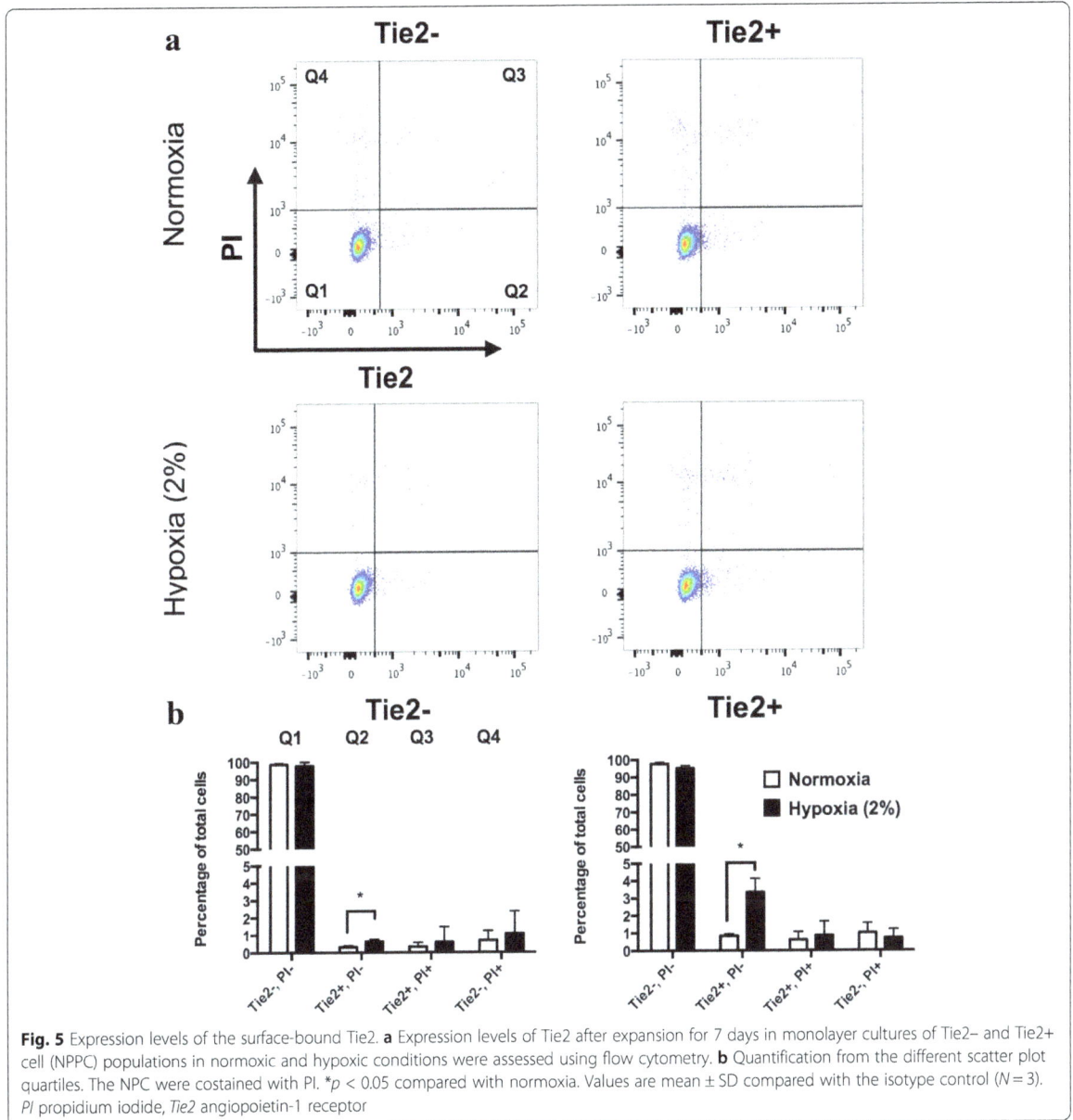

Fig. 5 Expression levels of the surface-bound Tie2. **a** Expression levels of Tie2 after expansion for 7 days in monolayer cultures of Tie2– and Tie2+ cell (NPPC) populations in normoxic and hypoxic conditions were assessed using flow cytometry. **b** Quantification from the different scatter plot quartiles. The NPC were costained with PI. *$p < 0.05$ compared with normoxia. Values are mean ± SD compared with the isotype control ($N = 3$). *PI* propidium iodide, *Tie2* angiopoietin-1 receptor

that they can be isolated in large quantities and without donor site morbidity. Importantly, these cells possess multipotent properties and have a proliferative capacity, which make these cells attractive for delivery into degenerated discs. Preclinical studies showed that cells from the mesenchymal origin can participate in disc regeneration by differentiating into chondrocyte-like cells and producing NP tissue-specific extracellular matrix, namely aggrecan and collagen type 2. Because NPC share some similarities in phenotype and molecular content with cartilage-specific cells, the chondrocytes [44], those cells with the ability to differentiate into chondrocytes are considered a

potential target for the regeneration of the IVD tissue. Resident progenitor cells within the IVD were documented previously [45–47]. Some of these cells were shown to maintain multipotent and self-renewal potential when cultured in vitro; however, little is known about their role in the homeostasis of the IVD.

Within this study, we demonstrated that Tie2+ cells from the bovine coccygeal discs are progenitor-like/multipotent cells, which are able to differentiate into osteogenic, adipogenic, and chondrogenic lineages in vitro. Sakai et al. [13] were the first to identify NP progenitor cells in the Tie2+ and disialoganglioside

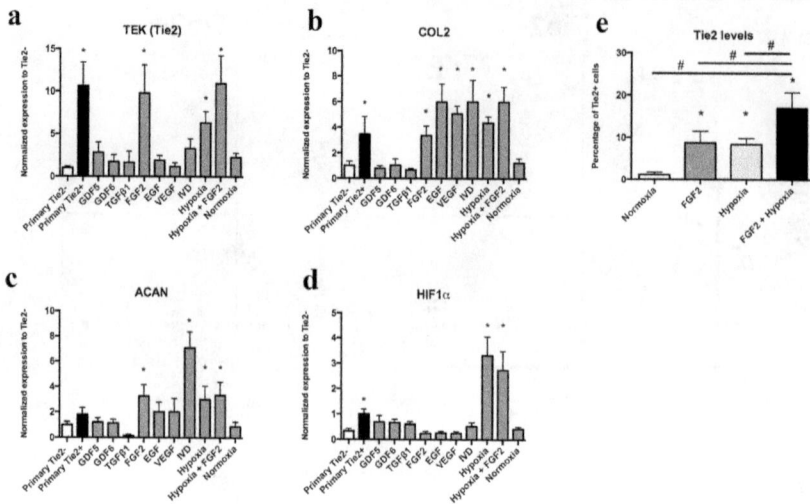

Fig. 6 Maintenance of NPPC (Tie2+) phenotype. Primary NPPC were stimulated with various growth factors (growth and differentiation factor 5 (GDF5), GDF6, transforming growth factor β1 (TGFβ1), fibroblast growth factor 2 (FGF2), epidermal growth factor (EGF), and vascular endothelial growth factor (VEGF)), cocultured with IVD tissue or subjected to hypoxic conditions for 7 days, and the transcript levels of Tie2 were measured: endothelial tyrosine kinase (*TEK*) (**a**), collagen type 2 (*COL2*) (**b**), aggrecan (*ACAN*) (**c**), and hypoxia-inducible factor 1 alpha (*HIF1α*) (**d**). Values are mean ± SD (*N* = 5). *$p < 0.05$ compared with Tie2− of primary NPC. Protein level of Tie2 was monitored by flow cytometry in NPPC that were subjected to hypoxic conditions and/or FGF2 (**e**). Values are mean ± SD (*N* = 3). *$p < 0.05$ compared with normoxic conditions. #$p < 0.05$ compared with FGF2 + hypoxic conditions. *IVD* intervertebral disc, *Tie2* angiopoietin-1 receptor

2 positive (GD2+) cell fraction from human and mouse IVD tissues. These cells were described to derive from the Tie2+ and GD2− precursor cells and are capable of differentiating into multiple mesenchymal and NP lineages. GD2 was described as an additional marker for progeny, whose expression is increased with activation and commitment of the NP progenitor cells. In this study, the expression of GD2 and its contribution to differentiation of the disc cells was not investigated. Our findings show the presence of Tie2+ NPPC in the bovine coccygeal discs and further support previous results on NPPC in human and in mice [13, 48]. Additionally, we showed that in contrast to Tie2−, only Tie2+ cells have a multipotent potential as characterized by their differentiation capacity in vitro and their ability to form spheroid colonies. Tie2− cells within the disc tissue could therefore be considered NP committed cells. The NPPC may represent the key cells for the regenerative capacity of the disc and maintenance of these cells could contribute to the homeostasis of the IVD.

NPPC were first described from human and mouse IVDs [13]. Here, we could successfully isolate them from bovine coccygeal IVDs. These IVDs have been established as a reliable model to assess the biology and biomechanics of the disc [49–51]. The present findings allow further investigations and subsequent translation into human samples, which are clinically more relevant.

Applying flow cytometry to detect the surface-bound Tie2 marker allowed us to investigate the two phenotypes present within a pool of expanded NPC. It should be noted that setting the appropriate gate for Tie2 during sorting of the NPC is highly sensitive and should be made very stringent in order to avoid isolation of Tie2− cells, which may not demonstrate multipotent differentiation potential. In primary NPC, 8.66 ± 3.94 % of the cells stained positive for Tie2. During expansion, the proportion of Tie2+ cells was rapidly lost in subsequent monolayer cultures and less than 1 % could be detected after 2.31 ± 0.28 (mean ± SD) population doublings. In support of our data are studies investigating molecular changes of progenitor cells during in vitro monolayer cultures, where they found that cellular morphology, self-renewal, and differentiation capacity of these cells are altered during expansion [52–54].

When the primary NPC were subjected to monolayer cultures, they adhered and started to proliferate within a few days. Because 8.66 ± 3.94 % of the freshly isolated NPC population expressed Tie2, we wondered whether the proliferating pool of cells comprised Tie2+ cells or whether this pool is restricted to Tie2− cells. The present data confirmed that proliferating cells showed both Tie2+ and Tie2− phenotypes. These experiments showed that cells harvested from the NP tissue are able to maintain, at least for a short period, synthesis of Tie2 while proliferating in monolayer cultures. Furthermore, we found that Tie2+ cells have a higher proliferative capacity after 3 days compared with the Tie2− cell fraction, while the Tie2+ fraction showed less proliferative activity on day 7. The proliferation dynamics of Tie2+ cells could therefore be explained by the massive increase of the Tie2− pool of cells and the loss of the Tie2+ fraction during expansion.

We addressed protocols for enrichment of Tie2+ cells in vitro by application of growth factors known to be beneficial for NPC or for inducing angiogenesis, by varying oxygen concentrations, or by coculture with IVD tissue. Supplementation of the cultures with FGF2 increased the TEK expression to levels similar to primary Tie2+ NPC. FGF2 is known as a potent inducer of angiogenesis [55] and was described as a crucial factor for the successful maintenance of the undifferentiated state and self-renewal of stem cells. Lotz et al. [56] reported that stabilization of FGF2 using controlled poly(lactic-co-glycolic acid) (PLGA) microsphere delivery improves the expression of stem cell markers and cell amplification, and decreases spontaneous differentiation. In addition to FGF2, low oxygen concentrations (2 % O_2) better maintained the Tie2+ pool of cells compared with normoxia; while simultaneous supplementation of the cultures with FGF2 and hypoxic conditions showed a synergistic effect and better maintained the Tie2 expression in NPPC after 7 days of culture. Physiological hypoxic conditions were previously suggested to maintain the undifferentiated state of many precursor cells, including embryonic, hematopoietic, mesenchymal, and neural stem cells [57]. Furthermore, cells of the IVD reside within a hypoxic environment and are preserved throughout their lifespan.

To characterize the NPPC during expansion, and following supplementation with various growth factors and coculture with IVD tissue, we performed a gene expression analysis of two key genes for the NP, namely aggrecan and collagen type 2. It was found that VEGF, EGF, FGF2, or coculture with IVD increased the expression of NP markers, suggesting the contribution of these factors to differentiation of the NPPC towards the NP phenotype. Surprisingly, exposure of NPPC to recombinant GDF5, GDF6, and TGFβ1 could not increase the expression of aggrecan or collagen type 2. An explanation may derive from the fact that these factors might be active in committed NPC rather than in progenitor cells. These growth factors were shown previously to enhance the discogenic phenotype of bone marrow-derived mesenchymal stromal cells in vitro [21] while a stage-dependent TGFβ1-induced chondrogenic differentiation of embryonic stem cells was observed [58].

Conclusions

The data presented herein demonstrate the presence of a progenitor cell population within the NP expressing the cell surface marker Tie2 and being able to differentiate into osteogenic, adipogenic, and chondrogenic lineages in vitro. Strategies to maintain the Tie2+ pool of the NPC merit further evaluation, and sorting for Tie2 may contribute to a more suitable source for cell therapy for regeneration of the IVD.

Abbreviations

AF: annulus fibrosus; α-MEM: alpha minimum essential medium; BrdU: bromodeoxyuridine; EGF: epidermal growth factor; FACS: fluorescence-activated cell sorting; FBS: fetal bovine serum; FGF2: fibroblast growth factor 2; GAG: glycosaminoglycans; GD2: disialoganglioside 2; GDF: growth differentiation factor; HIF1α: hypoxia-inducible factor 1-alpha; IVD: intervertebral disc; NP: nucleus pulposus; NPC: nucleus pulposus cells; NPPC: nucleus pulposus progenitor cells; PI: propidium iodide; P/S: penicillin/streptomycin; PLGA: Poly (lactic-co-glycolic acid); TEK: tyrosine kinase (Tie2); TGFβ1: transforming growth factor beta-1; Tie2: angiopoietin-1 receptor; VEGF: vascular endothelial growth factor.

Acknowledgements

The authors thank Eva Roth and Daniela A Frauchiger for technical assistance. Karin Wuertz-Kozak provided primary bovine endothelial cells to test the specificity and cross-reactivity of the Tie2 antibody to bovine samples. The FACS was conducted at the University of Bern FACSlab core facility. The project was supported by two Swiss National Science Foundation projects: "International Short Research Visit" grant #IZK0Z3_154384 (to SCWC and DS) and project-based funding #310030_153411 (to BG).

Authors' contributions

AT designed the experiments, collected the data, and drafted the manuscript. SCWC established FACS protocols for sorting of bovine cells, provided funding, and edited the manuscript. DS assisted in the experimental design, provided funding, and edited the manuscript. SG analyzed the data and edited the manuscript. BG provided funding, assisted in the experimental design, and edited the manuscript. All authors contributed to final approval of the manuscript.

Competing interests

The authors declare that they have no competing interests.

Author details

[1]Tissue and Organ Mechanobiology, Institute for Surgical Technology & Biomechanics, Medical Faculty, University of Bern, Bern, Switzerland. [2]Biointerfaces, Empa, Swiss Federal Laboratories for Materials Science and Technology, St Gallen, Switzerland. [3]Department for Orthopaedic Surgery, Tokai University School of Medicine, Isehara, Kanagawa, Japan. [4]AO Research Institute Davos, Davos, Switzerland. [5]AO Spine Research Network, AO Spine International, Davos, Switzerland.

References

1. Balagué F, Mannion AF, Pellisé F, Cedraschi C. Non-specific low back pain. Lancet. 2012;379:482–91.
2. Hoy D, March L, Brooks P, Blyth F, Woolf A, Bain C, et al. The global burden of low back pain: estimates from the Global Burden of Disease 2010 study. Ann Rheum Dis. 2014;73:968–74.
3. Fourney DR, Andersson G, Arnold PM, Dettori J, Cahana A, Fehlings MG, et al. Chronic low back pain: a heterogeneous condition with challenges for an evidence-based approach. Spine (Phila Pa 1976). 2011;36:S1–9.
4. Urban JPG, Roberts S, Ralphs JR. The nucleus of the intervertebral disc from development to degeneration. Am Zool. 2000;40:53–61.
5. Agrawal A, Gajghate S, Smith H, Anderson DG, Albert TJ, Shapiro IM, et al. Cited2 modulates hypoxia-inducible factor-dependent expression of vascular endothelial growth factor in nucleus pulposus cells of the rat intervertebral disc. Arthritis Rheum. 2008;58:3798–808.
6. Masuda K, Oegema TR, An HS. Growth factors and treatment of intervertebral disc degeneration. Spine (Phila Pa 1976). 2004;29:2757–69.
7. Hassett G, Hart DJ, Manek NJ, Doyle DV, Spector TD. Risk factors for progression of lumbar spine disc degeneration: the Chingford Study. Arthritis Rheum. 2003;48:3112–7.

8. Johnson WE, Eisenstein SM, Roberts S. Cell cluster formation in degenerate lumbar intervertebral discs is associated with increased disc cell proliferation. Connect Tissue Res. 2001;42:197–207.

9. Roberts S, Evans H, Trivedi J, Menage J. Histology and pathology of the human intervertebral disc. J Bone Joint Surg Am. 2006;88 Suppl 2:10–4.

10. Erwin WM, Islam D, Inman RD, Fehlings MG, Tsui FW. Notochordal cells protect nucleus pulposus cells from degradation and apoptosis: implications for the mechanisms of intervertebral disc degeneration. Arthritis Res Ther. 2011;13:R215.

11. Yim RL, Lee JT, Bow CH, Meij B, Leung V, Cheung KM, et al. A systematic review of the safety and efficacy of mesenchymal stem cells for disc degeneration: insights and future directions for regenerative therapeutics. Stem Cells Dev. 2014;23:2553–67.

12. Sakai D, Andersson GB. Stem cell therapy for intervertebral disc regeneration: obstacles and solutions. Nat Rev Rheumatol. 2015;11:243–56.

13. Sakai D, Nakamura Y, Nakai T, Mishima T, Kato S, Grad S, et al. Exhaustion of nucleus pulposus progenitor cells with ageing and degeneration of the intervertebral disc. Nat Commun. 2012;3:1264.

14. Koblizek TI, Runttng AS, Stacker SA, Wilks AF, Risau W, Deutsch U. Tie2 receptor expression and phosphorylation in cultured cells and mouse tissues. Eur J Biochem. 1997;244:774–9.

15. Loughna S, Sato TN. Angiopoietin and Tie signaling pathways in vascular development. Matrix Biol. 2001;20:319–25.

16. Suri C, Jones PF, Patan S, Bartunkova S, Maisonpierre PC, Davis S, et al. Requisite role of angiopoietin-1, a ligand for the TIE2 receptor, during embryonic angiogenesis. Cell. 1996;87:1171–80.

17. Gantenbein B, Illien-Jünger S, Chan SC, Walser J, Haglund L, Ferguson SJ, et al. Organ culture bioreactors—platforms to study human intervertebral disc degeneration and regenerative therapy. Curr Stem Cell Res Ther. 2015;10:339–52.

18. Barbero A, Grogan SP, Mainil-Varlet P, Martin I. Expansion on specific substrates regulates the phenotype and differentiation capacity of human articular chondrocytes. J Cell Biochem. 2006;98:1140–9.

19. Stoyanov JV, Gantenbein-Ritter B, Bertolo A, Aebli N, Baur M, Alini M, et al. Role of hypoxia and growth and differentiation factor-5 on differentiation of human mesenchymal stem cells towards intervertebral nucleus pulposus-like cells. Eur Cell Mater. 2011;21:533–47.

20. Yang X, Li X. Nucleus pulposus tissue engineering: a brief review. Eur Spine J. 2009;18:1–9.

21. Clarke LE, McConnell JC, Sherratt MJ, Derby B, Richardson SM, Hoyland JA. Growth differentiation factor 6 and transforming growth factor-beta differentially mediate mesenchymal stem cell differentiation, composition and micromechanical properties of nucleus pulposus constructs. Arthritis Res Ther. 2014;16:R67.

22. Taupin P, Ray J, Fischer WH, Suhr ST, Hakansson K, Grubb A, et al. FGF-2-responsive neural stem cell proliferation requires CCg, a novel autocrine/paracrine cofactor. Neuron. 2000;28:385–97.

23. Galas RJ, Liu JC. Vascular endothelial growth factor does not accelerate endothelial differentiation of human mesenchymal stem cells. J Cell Physiol. 2014;229:90–6.

24. Levenstein ME, Ludwig TE, Xu RH, Llanas RA, VanDenHeuvel-Kramer K, Manning D, et al. Basic fibroblast growth factor support of human embryonic stem cell self-renewal. Stem Cells. 2006;24:568–74.

25. Tsai TL, Wang B, Squire MW, Guo LW, Li WJ. Endothelial cells direct human mesenchymal stem cells for osteo- and chondro-lineage differentiation through endothelin-1 and AKT signaling. Stem Cell Res Ther. 2015;6:88.

26. Feng G, Li L, Liu H, Song Y, Huang F, Tu C, et al. Hypoxia differentially regulates human nucleus pulposus and annulus fibrosus cell extracellular matrix production in 3D scaffolds. Osteoarthritis Cartilage. 2013;21:582–8.

27. Mwale F, Ciobanu I, Giannitsios D, Roughley P, Steffen T, Antoniou J. Effect of oxygen levels on proteoglycan synthesis by intervertebral disc cells. Spine (Phila Pa 1976). 2011;36:E131–8.

28. Gantenbein B, Calandriello E, Wuertz-Kozak K, Benneker LM, Keel MJ, Chan SC. Activation of intervertebral disc cells by co-culture with notochordal cells, conditioned medium and hypoxia. BMC Musculoskelet Disord. 2014;15:422.

29. Livak KJ, Schmittgen TD. Analysis of relative gene expression data using real-time quantitative PCR and the 2(-Delta Delta C(T)) method. Methods. 2001;25:402–8.

30. Oehme D, Goldschlager T, Ghosh P, Rosenfeld JV, Jenkin G. Cell-based therapies used to treat lumbar degenerative disc disease: a systematic review of animal studies and human clinical trials. Stem Cells Int. 2015;2015:946031.

31. Benneker LM, Andersson G, Iatridis JC, Sakai D, Härtl R, Ito K, et al. Cell therapy for intervertebral disc repair: advancing cell therapy from bench to clinics. Eur Cell Mater. 2014;27:5–11.

32. Arkesteijn IT, Smolders LA, Spillekom S, Riemers FM, Potier E, Meij BP, et al. Effect of coculturing canine notochordal, nucleus pulposus and mesenchymal stromal cells for intervertebral disc regeneration. Arthritis Res Ther. 2015;17:60.

33. Okuma M, Mochida J, Nishimura K, Sakabe K, Seiki K. Reinsertion of stimulated nucleus pulposus cells retards intervertebral disc degeneration: an in vitro and in vivo experimental study. J Orthop Res. 2000;18:988–97.

34. Sato M, Asazuma T, Ishihara M, Ishihara M, Kikuchi T, Kikuchi M, et al. An experimental study of the regeneration of the intervertebral disc with an allograft of cultured annulus fibrosus cells using a tissue-engineering method. Spine (Phila Pa 1976). 2003;28:548–53.

35. Iwashina T, Mochida J, Sakai D, Yamamoto Y, Miyazaki T, Ando K, et al. Feasibility of using a human nucleus pulposus cell line as a cell source in cell transplantation therapy for intervertebral disc degeneration. Spine (Phila Pa 1976). 2006;31:1177–86.

36. Gorensek M, Jaksimović C, Kregar-Velikonja N, Gorensek M, Knezevic M, Jeras M, et al. Nucleus pulposus repair with cultured autologous elastic cartilage derived chondrocytes. Cell Mol Biol Lett. 2004;9:363–73.

37. Henriksson H, Hagman H, Horn H, Lindahl L, Brisby B. Investigation of different cell types and gel carriers for cell-based intervertebral disc therapy, in vitro and in vivo studies. J Tissue Eng Regen Med. 2011;6:738–47.

38. Coric D, Pettine K, Sumich A, Boltes MO. Prospective study of disc repair with allogeneic chondrocytes presented at the 2012 Joint Spine Section Meeting. J Neurosurg Spine. 2013;18:85–95.

39. Sakai D, Mochida J, Yamamoto Y, Nomura T, Okuma M, Nishimura K, et al. Transplantation of mesenchymal stem cells embedded in Atelocollagen gel to the intervertebral disc: a potential therapeutic model for disc degeneration. Biomaterials. 2003;24:3531–41.

40. Jeong JH, Lee JH, Jin ES, Min JK, Jeon SR, Choi KH. Regeneration of intervertebral discs in a rat disc degeneration model by implanted adipose-tissue-derived stromal cells. Acta Neurochir (Wien). 2010;152:1771–7.

41. Tam V, Rogers I, Chan D, Leung VY, Cheung KM. A comparison of intravenous and intradiscal delivery of multipotential stem cells on the healing of injured intervertebral disk. J Orthop Res. 2014;32:819–25.

42. Sheikh H, Zakharian K, De La Torre RP, Facek C, Vasquez A, Chaudhry GR, et al. In vivo intervertebral disc regeneration using stem cell-derived chondroprogenitors. J Neurosurg Spine. 2009;10:265–72.

43. Wang H, Zhou Y, Huang B, Liu LT, Liu MH, Wang J, et al. Utilization of stem cells in alginate for nucleus pulposus tissue engineering. Tissue Eng Part A. 2014;20:908–20.

44. Lee CR, Sakai D, Nakai T, Toyama K, Mochida J, Alini M, et al. A phenotypic comparison of intervertebral disc and articular cartilage cells in the rat. Eur Spine J. 2007;16:2174–85.

45. Shi R, Wang F, Hong X, Wang YT, Bao JP, Cai F, et al. The presence of stem cells in potential stem cell niches of the intervertebral disc region: an in vitro study on rats. Eur Spine J. 2015;24(11):2411-24. doi: 10.1007/s00586-015-4168-7. [Epub ahead of print].

46. Blanco JF, Graciani IF, Sanchez-Guijo FM, Muntión S, Hernandez-Campo P, Santamaria C, et al. Isolation and characterization of mesenchymal stromal cells from human degenerated nucleus pulposus: comparison with bone marrow mesenchymal stromal cells from the same subjects. Spine (Phila Pa 1976). 2010;35:2259–65.

47. Feng G, Yang X, Shang H, Marks IW, Shen FH, Katz A, et al. Multipotential differentiation of human anulus fibrosus cells: an in vitro study. J Bone Joint Surg Am. 2010;92:675–85.

48. Sakai D, Grad S. Advancing the cellular and molecular therapy for intervertebral disc disease. Adv Drug Deliv Rev. 2014;84:159–71.

49. Maroudas A, Stockwell RA, Nachemson A, Urban J. Factors involved in the nutrition of the human lumbar intervertebral disc: cellularity and diffusion of glucose in vitro. J Anat. 1975;120:113–30.

50. Miyazaki T, Kobayashi S, Takeno K, Meir A, Urban J, Baba H. A Phenotypic comparison of proteoglycan production of intervertebral disc cells isolated from rats, rabbits, and bovine tails; which animal model is most suitable to study tissue engineering and biological repair of human disc disorders? Tissue Eng Part A. 2009;15:3835–46.

51. Showalter BL, Beckstein JC, Martin JT, Beattie EE, Espinoza Orías AA, Schaer TP, et al. Comparison of animal discs used in disc research to human lumbar disc: torsion mechanics and collagen content. Spine (Phila Pa 1976). 2012;37:E900–7.

52. Kim YH, Yoon DS, Kim HO, Lee JW. Characterization of different subpopulations from bone marrow-derived mesenchymal stromal cells by alkaline phosphatase expression. Stem Cells Dev. 2012;21:2958–68.

53. Sun HJ, Bahk YY, Choi YR, Shim JH, Han SH, Lee JW. A proteomic analysis during serial subculture and osteogenic differentiation of human mesenchymal stem cell. J Orthop Res. 2006;24:2059–71.

54. Prockop DJ, Sekiya I, Colter DC. Isolation and characterization of rapidly self-renewing stem cells from cultures of human marrow stromal cells. Cytotherapy. 2001;3:393–6.

55. Seghezzi G, Patel S, Ren CJ, Gualandris A, Pintucci G, Robbins ES, et al. Fibroblast growth factor-2 (FGF-2) induces vascular endothelial growth factor (VEGF) expression in the endothelial cells of forming capillaries: an autocrine mechanism contributing to angiogenesis. J Cell Biol. 1998;141:1659–73.

56. Lotz S, Goderie S, Tokas N, Hirsch SE, Ahmad F, Corneo B, et al. Sustained levels of FGF2 maintain undifferentiated stem cell cultures with biweekly feeding. PLoS One. 2013;8:e56289.

57. Mohyeldin A, Garzón-Muvdi T, Quiñones-Hinojosa A. Oxygen in stem cell biology: a critical component of the stem cell niche. Cell Stem Cell. 2010;7:150–61.

58. Yang Z, Sui L, Toh WS, Lee EH, Cao T. Stage-dependent effect of TGF-beta1 on chondrogenic differentiation of human embryonic stem cells. Stem Cells Dev. 2009;18:929–40.

18

Properties of internalization factors contributing to the uptake of extracellular DNA into tumor-initiating stem cells of mouse Krebs-2 cell line

Evgeniya V. Dolgova[1*], Ekaterina A. Potter[1], Anastasiya S. Proskurina[1], Alexandra M. Minkevich[1], Elena R. Chernych[2], Alexandr A. Ostanin[2], Yaroslav R. Efremov[1,3], Sergey I. Bayborodin[1], Valeriy P. Nikolin[1], Nelly A. Popova[1,3], Nikolay A. Kolchanov[1] and Sergey S. Bogachev[1]

Abstract

Background: Previously, we demonstrated that poorly differentiated cells of various origins, including tumor-initiating stem cells present in the ascites form of mouse cancer cell line Krebs-2, are capable of naturally internalizing both linear double-stranded DNA and circular plasmid DNA.

Methods: The method of co-incubating Krebs-2 cells with extracellular plasmid DNA (pUC19) or TAMRA-5′-dUTP-labeled polymerase chain reaction (PCR) product was used. It was found that internalized plasmid DNA isolated from Krebs-2 can be transformed into competent *Escherichia coli* cells. Thus, the internalization processes taking place in the Krebs-2 cell subpopulation have been analyzed and compared, as assayed by *E. coli* colony formation assay (plasmid DNA) and cytofluorescence (TAMRA-DNA).

Results: We showed that extracellular DNA both in the form of plasmid DNA and a PCR product is internalized by the same subpopulation of Krebs-2 cells. We found that the saturation threshold for Krebs-2 ascites cells is 0.5 μg DNA/10^6 cells. Supercoiled plasmid DNA, human high-molecular weight DNA, and 500 bp PCR fragments are internalized into the Krebs-2 tumor-initiating stem cells via distinct, non-competing internalization pathways. Under our experimental conditions, each cell may harbor 340–2600 copies of intact plasmid material, or up to $3.097 \pm 0.044 \times 10^6$ plasmid copies (intact or not), as detected by quantitative PCR.

Conclusion: The internalization dynamics of extracellular DNA, copy number of the plasmids taken up by the cells, and competition between different types of double-stranded DNA upon internalization into tumor-initiating stem cells of mouse ascites Krebs-2 have been comprehensively analyzed. Investigation of the extracellular DNA internalization into tumor-initiating stem cells is an important part of understanding their properties and possible destruction mechanisms. For example, a TAMRA-labeled DNA probe may serve as an instrument to develop a target for the therapy of cancer, aiming at elimination of tumor stem cells, as well as developing a straightforward test system for the quantification of poorly differentiated cells, including tumor-initiating stem cells, in the bulk tumor sample (biopsy or surgery specimen).

Keywords: Ascites Krebs-2, Extracellular DNA, DNA internalization factors, Tumor-initiating stem cells

* Correspondence: dolgova.ev@mail.ru
[1]Institute of Cytology and Genetics, Siberian Branch of the Russian Academy of Sciences, 10 Lavrentieva Ave., Novosibirsk 630090, Russia
Full list of author information is available at the end of the article

Background

Studies of DNA internalization by eukaryotic cells has become increasingly popular. It is well known that extracellular DNA (eDNA) can be directed into cells using transfection agents, such as lipofectamine and polyplexes [20, 37, 39]. However, eDNA has also been reported to become directly delivered to cells without the use of additional transfection tools [5, 11, 12, 14, 19, 34, 40]. Furthermore, upon internalization, eDNA sequences may become expressed [6, 14, 19].

Neither the exact mechanisms of how eDNA becomes internalized, nor the factors aiding in this process, have been comprehensively characterized. One mechanism established to mediate internalization of eDNA is endocytosis [2, 22]. In this scenario, eDNA molecules interact either with cell membrane components (adsorption endocytosis) or with appropriate cell receptors (receptor-mediated endocytosis) [28]. Experimental data are available showing that DNA molecules are transported into lymphocytes, monocytes, neutrophils, and skeletal myocytes via interaction with cell surface proteins [4, 15, 18, 35, 40]. However, most of these DNA-binding cell surface proteins have received surprisingly little attention and so their exact role in the internalization process still remains obscure. It has also been demonstrated that short nucleic acid fragments can penetrate the cell interior using special channels [7, 17, 24, 30, 36].

Our earlier studies ([12, 33] and Krebs-2 transcriptome analysis (data not shown)) indicate that Krebs-2 cells capable of internalizing eDNA (TAMRA+ cells) display properties of tumor-initiating cancer stem cells. Notably, this property is also shared by clonogenic glioma [12], multiple myeloma, and lymphoma cells (data not shown).

Both linear double-stranded DNA (dsDNA) and supercoiled plasmid DNA were shown to be taken up by Krebs-2 cells [12]. Whether these DNA species may compete with each other during internalization, and which mechanisms and factors are involved, is the focus of our research.

In the present paper, we describe internalization of pUC19 plasmid DNA by ascites cells of mouse cell line Krebs-2. We explore the efficiency of plasmid DNA internalization depending on its concentration in the medium and the incubation time. Next, we estimate whether various DNA species may competitively influence the internalization of each other (human dsDNA (0.3–6 kb) versus polymerase chain reaction (PCR) fragment (0.5 kb) versus supercoiled plasmid DNA) upon co-incubation with Krebs-2 ascites cells.

Methods

Laboratory animals and tumor model

We used 2- to 3-month-old CBA/Lac mice bred in the animal facility at the Institute of Cytology and Genetics, Siberian Branch of the Russian Academy of Sciences.

Animals were grown in groups of 5–10 mice per cage with free access to food and water. The Ascites form of the mouse carcinoma Krebs-2 (derived from the solid form) was used as a model [21]. This cancer cell line was obtained from the cell depository of the Institute of Cytology and Genetics (Novosibirsk, Russia) and is maintained in mice as a transplanted tumor. To obtain ascites, Krebs-2 cells were diluted 1:10 in 200 µl normal saline and inoculated intraperitoneally (2×10^6 cells).

TAMRA labeling of human Alu repeat DNA and incubation of extracellular DNA with ascites Krebs-2 cells

DNA was labeled with TAMRA-5'-dUTP (N-90100, Biosan, Novosibirsk) using PCR. The PCR template was human *Alu* repeat material cloned in pBlueScript SK(+) (Alu-pBS), this repeat encompassing the tandemly repeated AluJ and AluY sequences (NCBI: AC002400.1, 53494–53767). Standard M13 primers were used for amplification. PCR purification was done by standard phenol-chloroform extraction followed by ethanol precipitation using ammonium acetate as a salt. The quantity of eDNA being added to the cells (*Alu*-TAMRA DNA, pUC19 (#440060, Medigen, Novosibirsk), pEGFP-N1 (#6085-1, Clontech), sonicated pEGFP-N1) was 1 µg plasmid DNA/10^6 cells and 0.2 µg *Alu*-TAMRA DNA/10^6 cells. The cells that incorporated the fluorescently labeled DNA probe were analyzed by either FACS (BD FACSAria, Becton Dickinson) or by fluorescence microscopy (laser scanning microscope LSM 510 META (Zeiss), ZEN software or AxioImager ZI microscope (Zeiss), ISIS software).

Saturation of Krebs-2 cells with pUC19 plasmid DNA and estimates of the copy number of internalized pUC19

One million Krebs-2 cells were incubated for 1 h with pUC19 plasmid DNA (0.01; 0.1; 1; 5; 10; 20 µg). Next, to eliminate non-internalized DNA, the cells were treated with DNaseI (#18525, Serva) (10 µg/ml, 37 °C, 1 h), washed once with RPMI-1640 medium (#1.3.4, Biolot, St. Petersburg, Russia), and resuspended in 50 mM EDTA. SDS was added to a final concentration of 1 % and proteinase K (BIO-405010, Bioron GmbH, Germany) treatment (100 µg/ml) was performed at 58 °C for 1 h. Cell lysate was subjected to phenol-chloroform extraction, and the DNA was re-precipitated with 0.6 volumes of isopropanol, washed with 70 % ethanol and dissolved in water (15–40 µl). The DNA thus obtained was transformed into chemically competent XL1Blue MRF' *E. coli* cells. The cells were spread on agar-Amp plates. Colonies were counted, and this information was used to estimate plasmid copy number per cell. To verify that the transformed cells indeed carried the intended pUC19 plasmid, several individual colonies were grown in LB-Amp overnight. Plasmid DNA was purified and its identity was confirmed by gel electrophoresis.

Plasmid copy number estimate

The following input data were available to us: 1) *E. coli* transformation efficiency (transformation of 10 pg pUC19 plasmid DNA produced 200 colonies upon transformation); 2) 10 pg of pUC19 plasmid (2.9 kb) translates into 4.6×10^6 plasmid copies; 3) the number of colonies formed upon transformation of DNA isolated from Krebs-2 cells incubated with pUC19; 4) the percentage of DNA-internalizing cells among all Krebs-2 cells is 3 % on average. Thereby, we can estimate how many cells in fact internalize DNA—3 % of 1 million cells equals 3×10^4 cells,

Based on the proportion between 200 colonies and 4.6×10^6 plasmid molecules, and the known number of colonies obtained in the experimental point (N), one can estimate how many plasmid molecules were present (X). Therefore, each cell internalized on average $X/3 \times 10^4$ plasmid molecules.

Analysis of co-internalization of pUC19 and Alu-TAMRA DNA by Krebs-2 ascites cells

The cells were incubated with a mixture of 1 μg pUC19 and 0.2 μg *Alu*-TAMRA DNA (per one million Krebs-2 cells). Following a single wash, TAMRA+ cells were sorted using BD FACSAria flow cytometer (Becton Dickinson). Flow-sorted cells were subjected to all the treatments (starting from the DNaseI step) described in the above section of the Methods.

Analysis of competition between different types of eDNA

Alu-TAMRA DNA, pUC19 (see "TAMRA labeling of human Alu repeat DNA and incubation of extracellular DNA with ascites Krebs-2 cells" section above), and human dsDNA (300–6000 bp) were used. Human DNA was isolated from placentas of healthy women using a phenol-free method, and sonicated as described in [1].

Krebs-2 cells were incubated for 1 h with the first type of eDNA, according to the above protocol. Next the cells were washed once with RPMI-1640 and resuspended in this medium. The second type of DNA was either added immediately, or the cells were left at 37 °C in the medium supplemented with 10 % fetal bovine serum (FBS; SH30071.03, HyClone, USA) until the second type of DNA was added. After the incubation with the second type of eDNA, Krebs-2 cells were washed with medium and subjected to all the procedures described in the "TAMRA labeling of human Alu repeat DNA and incubation of extracellular DNA with ascites Krebs-2 cells" section above.

Cell cycle analysis of TAMRA+ and unsorted Krebs-2 cells

Sorted TAMRA+ or unsorted Krebs-2 cells were centrifuged at 400 g, 4 °C for 5 min, and fixed in 60 % methanol for 1 h at 4 °C. The cell suspension was pelleted, washed with phosphate-buffered saline (PBS), and treated with 200 μg/ml RNase (LLC Samson-Med, St. Petersburg, Russia) for 30 min at 37 °C. Next, propidium iodide (P4170, Sigma-Aldrich) was added to the cell suspension for 10 min at room temperature. Cell cycle profiling was performed using BD FACSAria flow cytometer (Becton Dickinson).

Quantitative PCR quantification of eDNA copy number in Krebs-2 cells

Isolation of DNA

Following incubation of Krebs-2 cells with eDNA and DNaseI treatment, the cell membrane was lysed with 0.5 % Triton-X100 (#37238, Serva) (15 min on ice). Nuclei were pelleted by centrifugation at 200 g for 5 min at 4 °C. Supernatants were collected and DNA was precipitated by adding 0.6 volumes of isopropanol. The pellets were re-dissolved in a small volume of water. Nuclear pellets were resuspended in 50 mM EDTA, SDS was added to 1 %, and samples were treated with proteinase K. DNA was purified by phenol-chloroform extraction and re-precipitated as described above.

Quantitative PCR and calibration curve

DNA molecules were quantified by real-time PCR using SYBR Green PCR Master Mix (#4309155, Applied Biosystems, UK). To generate the quantitative PCR (qPCR) calibration curve, standard M13 primers or primers from the Amp gene (forward: 5'-ATGAGTATTCAACATTTCCG-3'; reverse: 5'-GATCTTACCGCTGTTGAGAT-3') were used, and 0, 0.5, 5, 50, 500, 5000 and 50,000 pg of each pUC19 and *Alu*-repeat DNA were added to the reactions. Each concentration was run in triplicate. The linear fit of Ct versus eDNA content was plotted using StepOne Software v2.3. pUC19 and *Alu* DNA present in the nuclear or cytoplasmic fractions of Krebs-2 cells was quantified using StepOne Software v2.3. Template DNA (100 ng) was added to each qPCR reaction. DNA isolated from intact Krebs-2 cells was used as a negative control (and no product whatsoever was observed). All real-time PCR experiments were performed in triplicate and repeated twice on a Step One Real-Time PCR System (Applied Biosystems).

Conversion of qPCR data into eDNA copy numbers

Calibration curve-based qPCR data were converted into absolute plasmid or *Alu*-repeat molecule numbers as follows: 100 ng of Krebs-2 DNA added to each qPCR equals ~8333 cells (12 pg/cell). Given than TAMRA+ cells were shown to be the same cells as those internalizing plasmid DNA, we could estimate the exact percentage of cells that internalized both types of eDNA. From fluorescence microscopy analysis, we know that 2 % of cells were eDNA-internalizing (i.e., ~167 cells). Hence, by dividing the eDNA copy number measured by 167, one obtains the number of eDNA molecules per cell.

Statistical analysis

Statistical analysis was performed using Statistica 10 software. In the figures, bars show standard deviation ($n = 3–4$ at different experimental points). The level of significance was estimated using Students t tests.

Results

Internalization of Alu-TAMRA dsDNA and supercoiled plasmid pUC19 DNA by Krebs-2 cells

Previously, passaging the ascites in a grafted form was demonstrated not to affect the ability of a subpopulation of ascites cells (tumor-initiating stem cells (TISCs)) to internalize extracellular dsDNA in the absence of additional transfection factors [12] (Fig. 1). The percentage of Krebs-2 cells that internalized *Alu*-TAMRA DNA has been analyzed by confocal microscopy and cytometry and was within the previously reported experimental range (1–7 %), namely 3 %. Performing the incubation of Krebs-2 cells with *Alu*-TAMRA DNA at 4 °C, 25 °C, or 37 °C did not influence the efficiency of internalization.

First and foremost, we wanted to understand whether Krebs-2 cells internalizing linear *Alu*-TAMRA DNA are the same cells that internalize supercoiled plasmid DNA. To do this, ascites Krebs-2 cells were co-incubated with TAMRA-labeled *Alu* DNA and supercoiled pUC19 plasmid DNA. The cells were flow-sorted into TAMRA-positive and -negative subpopulations. Their DNA was isolated and transformed into competent *E. coli* cells. Upon transformation, only TAMRA+ cells produced *E. coli* colonies. Plasmid DNA isolated from these colonies was identical to the original pUC19 plasmid, which was used for co-incubation experiments (Fig. 2).

Copy number analysis of dsDNA molecules internalized by Krebs-2 TISCs

Selective targeting of TISCs by delivery of cell-killing genes carried on the plasmids is an attractive approach with a clear therapeutic application. For this approach to work, it is important to know whether non-degraded DNA molecules are delivered to the cells and how many native plasmid DNA molecules can be found in the cell at a time. To address these questions, two series of experiments were carried out. In the first series, we estimated the number of plasmid DNA molecules internalized by Krebs-2 cells by transformation of DNA from such cells into competent *E. coli*. Also, using an *E. coli* transformation assay, we determined the saturation threshold of Krebs-2 cells, which is 1 µg/10^6 cells (Fig. 3a). In the second series, the internalized plasmid DNA was quantified directly using qPCR. Importantly, unlike transformation, qPCR quantifies the entire pool of DNA molecules, regardless of whether they are nicked, partially degraded, or integrated into the genome or not. The results expectedly differ by three orders of magnitude: transformation experiments give an estimate from 340 (Fig. 3a) to 2600 (data not shown) plasmid copies/cell, with qPCR producing an estimate of $1.070 \pm 0.054 \times 10^6$ copies/cell and $3.097 \pm 0.044 \times 10^6$ copies/cell (Amp and M13 primer sets, respectively) (see Methods) (Fig. 3b-1). We also analyzed the copy number of *Alu* fragments internalized by the cells, which was $0.425 \pm 0.011 \times 10^6$ (Fig. 3b-2). Additionally, we monitored the dynamics of *Alu*-TAMRA internalization by Krebs-2 TISCs. The cells become saturated with eDNA fragments by the end of the first hour of incubation. According to the confocal imaging analysis (data not shown), internalized labeled DNA material appears to double every 10 min.

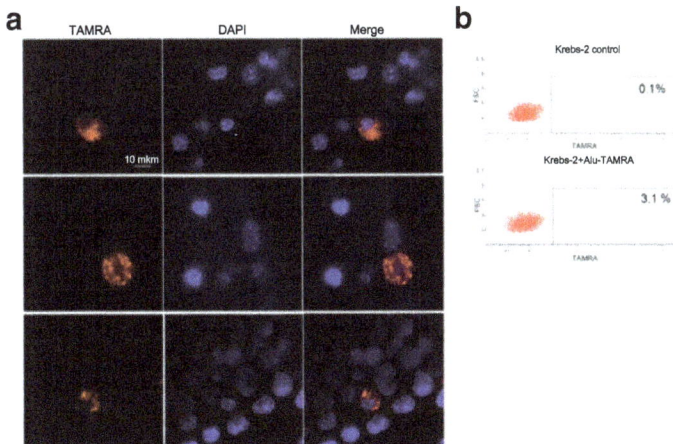

Fig. 1 Cytofluorescence (**a**) and flow cytometry (**b**) analyses of *Alu*-TAMRA DNA internalization by ascites Krebs-2 cells. DNA is stained with DAPI (*blue*); TAMRA (*red*) corresponds to *Alu*-TAMRA DNA

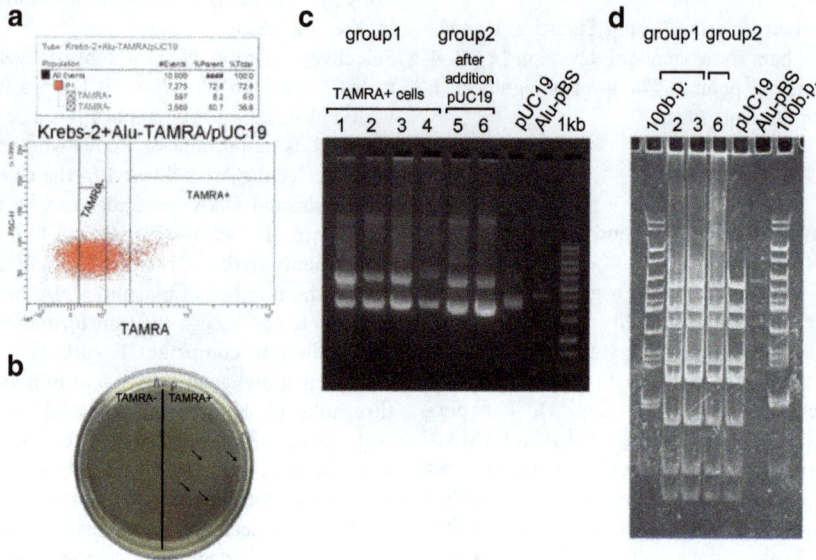

Fig. 2 Analysis of plasmids isolated from the colonies obtained by transformation of competent *E. coli* cells with DNA from Krebs-2 ascites pre-incubated with different types of eDNA (pUC19 only or pUC19 + *Alu*-TAMRA dsDNA). **a** Krebs-2 ascites cells were incubated with a mixture of plasmid pUC19 DNA and linear *Alu*-TAMRA dsDNA. TAMRA+ and TAMRA– subpopulations were gated as shown. **b** Image of colony growth on an LB-amp plate seeded with *E. coli* cells transformed with DNA from TAMRA+ or TAMRA– Krebs-2 subpopulations. No colonies are formed in the latter group. **c** Agarose gel electrophoresis analysis of the plasmids recovered. 1–4, Plasmids obtained from TAMRA+ material; 5, 6, plasmids obtained from the control transformation (Krebs-2 cells incubated with pUC19 DNA only); pUC19, Alu-pBS, original plasmids; 1 kb, DNA molecular weight 1 kb ladder. **d** Restriction analysis of plasmid DNA with a 4-cutter *Hae*III. Plasmids derived from TAMRA+ cells (2, 3) correspond to pUC19, much as the plasmid (6) isolated from control Krebs-2 cells incubated with just pUC19

Competitive interactions between human dsDNA, yeast RNA, bovine serum albumin, and heparin during internalization of pUC19 by Krebs-2 cells

Next, we proceeded to characterize possible competition between supercoiled plasmid DNA (pUC19) and linear DNA (fragmented human genomic DNA), total yeast RNA, bovine serum albumin (BSA), and heparin. Neither linear dsDNA (300–6000 bp), nor RNA, nor BSA compete with plasmid DNA for internalization factors (Fig. 4a, b and c). Amazingly, the addition of 10 U heparin (7.7 µg) abolished the internalization by Krebs-2 cells (Fig. 4d). In parallel, heparin was also demonstrated to abolish the uptake of *Alu*-TAMRA dsDNA by human mesenchymal stem cells (Fig. 4e).

Competition between various eDNA types during internalization by Krebs-2 cells

Our earlier studies showed that internalization of extracellular DNA by human MCF-7 cells proceeds in two waves, and peaks at 1 and 3 h after the beginning of incubation [34]. In the present work, a series of experiments was performed to characterize the dynamics of eDNA internalization when Krebs-2 cells have been preincubated with various types of dsDNA. Specifically, one million Krebs-2 cells were incubated for 1 h with

different dsDNA preparations. Next, additional DNA samples were added to the cells (once every hour, 3–4 h in total). To eliminate non-internalized DNA (unbound or membrane-associated DNA), the cells were finally treated with DNaseI. All experiments involved either pUC19 plasmid DNA or TAMRA-labeled human *Alu* repeat. These served as references for comparisons (*E. coli* transformation assay for pUC19, and flow cytometry for *Alu*-TAMRA). Several types of competition experiments were performed; each was run several times to achieve statistical significance. The following combinations were tested: pUC19 + pUC19 (Fig. 5), *Alu*-TAMRA + *Alu*-TAMRA (Fig. 6a), pUC19 + *Alu*-TAMRA (Fig. 6b), total human DNA + *Alu*-TAMRA (Fig. 7a), and sonicated pEGFP-N1 + *Alu*-TAMRA (Fig. 7b). The reasons for choosing these specific DNA combinations are explained below.

1. *pUC19 + pUC19*

   First, we wanted to confirm or challenge the existence of a second wave of internalization 3–4 h after the beginning of incubation, as was demonstrated for MCF-7 cells (see Methods). Indeed, we observed that adding more plasmid DNA 3–4 h following the incubation results in significantly more plasmids capable of forming colonies upon

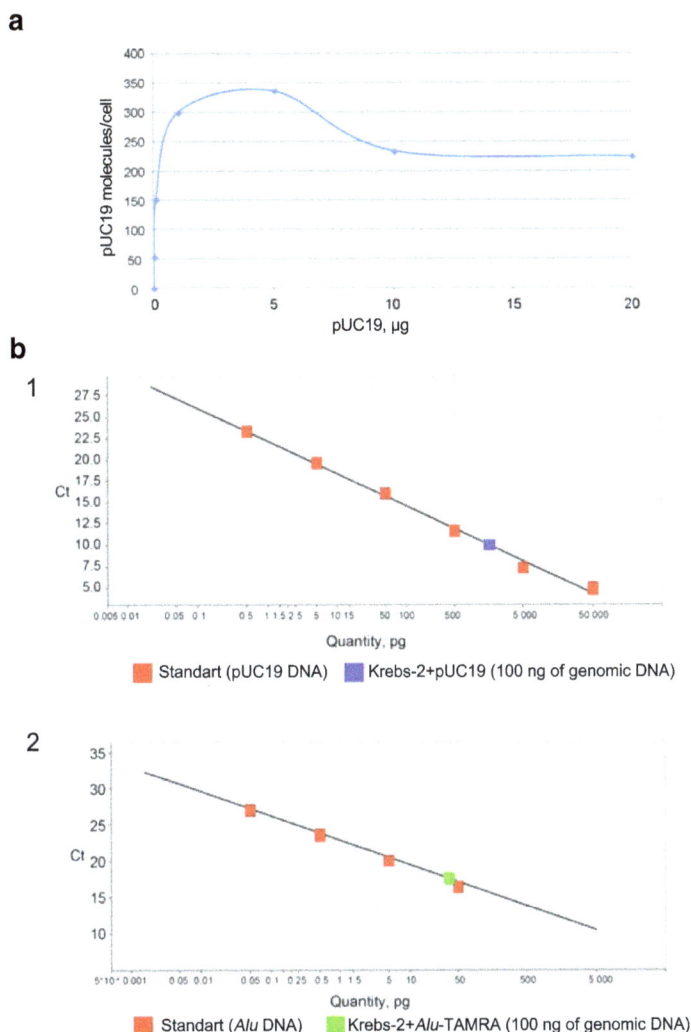

Fig. 3 Copy number analysis of eDNA internalized by Krebs-2 cells. **a** saturation of Krebs-2 cells with increasing amounts of pUC19 DNA in the incubation medium. **b** qPCR quantification of pUC19 (1) and *Alu* DNA (2) present in Krebs-2 cells after co-incubation with eDNA. Linear calibration plot was constructed using StepOne v2.3 software; each datapoint was run in triplicate

transformation into competent *E. coli* (Fig. 5a and b). This is formally compatible with two basic scenarios: either more internalization-competent cells become available with time, or more plasmid copies accumulate in the same cells. To understand which scenario is correct, fluorescently labeled DNA was included in the analysis (*Alu*-TAMRA). Using flow cytometry, we show that there is a 20 % gain in the TAMRA+ cell subpopulation (from 1.4 % to 1.7 %) when *Alu*-TAMRA DNA is added 3 h after beginning the first incubation (Fig. 5c-1). This observation is consistent with our earlier report [12] where the

percentage of TAMRA+ cells was found to increase upon a longer incubation time (Fig. 5c-2). We performed an additional experiment wherein the cells were first incubated with *Alu*-TAMRA DNA, followed by *Alu*-FITC DNA 3 h later. The cells were analyzed under a fluorescence microscope. This analysis showed that the second portion of eDNA was largely absent from TAMRA+ cells, and is restricted to TAMRA cells (Fig. 5d-1). Cells that were double-positive for *Alu*-TAMRA and *Alu*-FITC (Fig. 5d-2) were present, but were rather rare. Taken together, these data suggest that 3–4 h

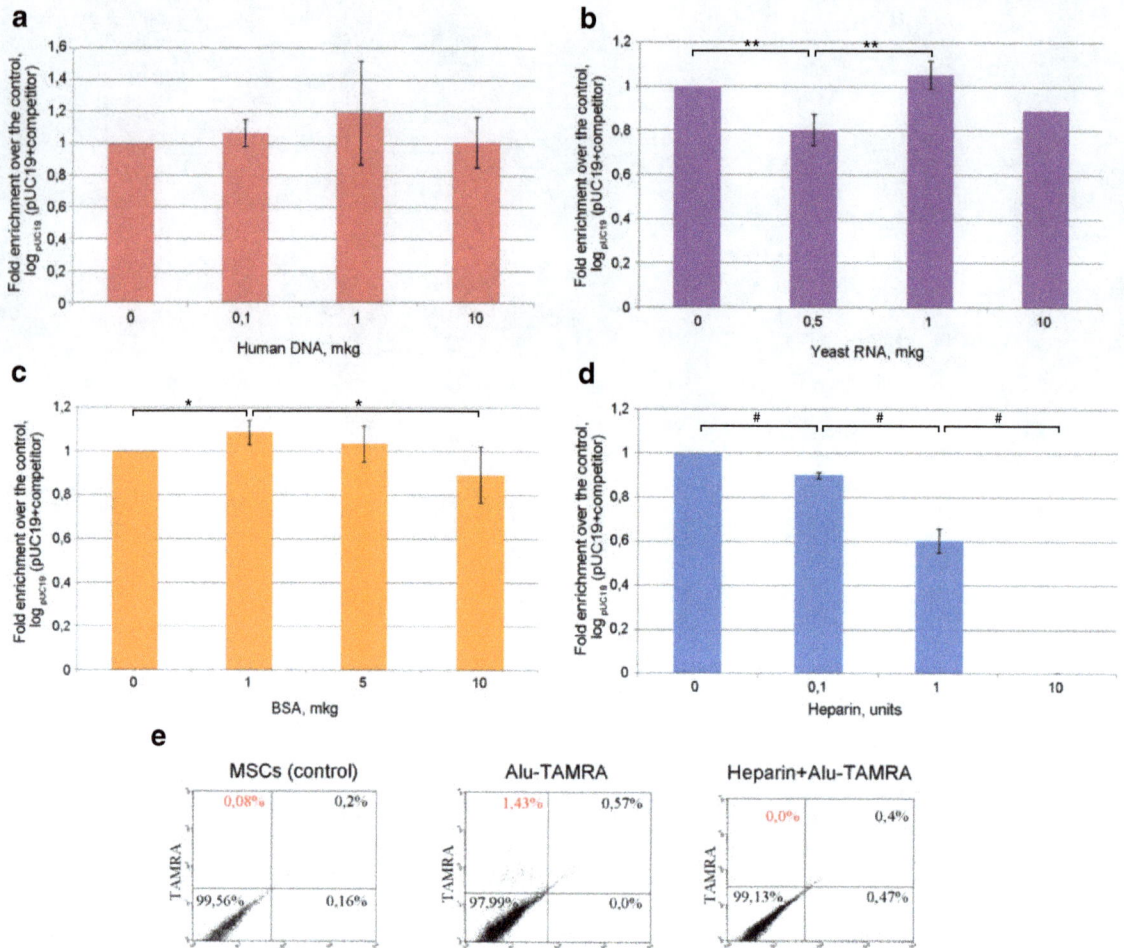

Fig. 4 Internalization of supercoiled pUC19 plasmid DNA by Krebs-2 cells in the presence of various competitors. The relative internalization efficiency of pUC19 plasmid DNA in the presence of different amounts of human dsDNA (**a**), yeast RNA (**b**), bovine serum albumin (*BSA*) (**c**), and heparin (**d**). The values are shown on a log $_{pUC19}$ (pUC19 + competitor) scale. **e** FACS analysis of *Alu*-TAMRA dsDNA internalization by human mesenchymal stem cells (*MSCs*) in the presence or absence of heparin. The data obtained suggest that heparin is a dose-dependent inhibitor of plasmid DNA internalization and suppresses linear DNA internalization by the cells. *$p < 0.05$; **$p < 0.01$; #$p < 0.001$

following the beginning of the incubation there is a 13–50 % increase in the number of cells internalizing exogenous DNA.

Next, we compared cell cycle distributions of unsorted cells and eDNA-internalizing cells (Fig. 5e). The two profiles were indistinguishable, with about 20 % of cells found in the G2-M phases. One can speculate that the cells undergoing mitosis are unlikely candidates for a cell-internalizing subpopulation. However, upon exiting the M phase such cells may become internalization-competent. The timing of the second wave of internalization events matches the length of the M phase, which is known to be 3–4 h. Data on the percentage of dividing cells and the increase in TAMRA+

subpopulation at a 4-h timepoint indicate that the increase in the plasmid copy number is attributable to the availability of an additional pool of internalization-competent cells that were undergoing mitosis and were incapable of internalizing DNA.

Next, we aimed to address the question whether *Alu*-TAMRA PCR fragments and supercoiled DNA are internalized using the same factors or not. Two combinations of DNA were tested: *Alu*-TAMRA + *Alu*-TAMRA (Fig. 6a) and pUC19 + *Alu*-TAMRA (Fig. 6b). The first combination was necessary to control for the effect of the second wave of internalization. The second combination allowed estimating the contribution of both types of DNA into internalization, so as to understand whether

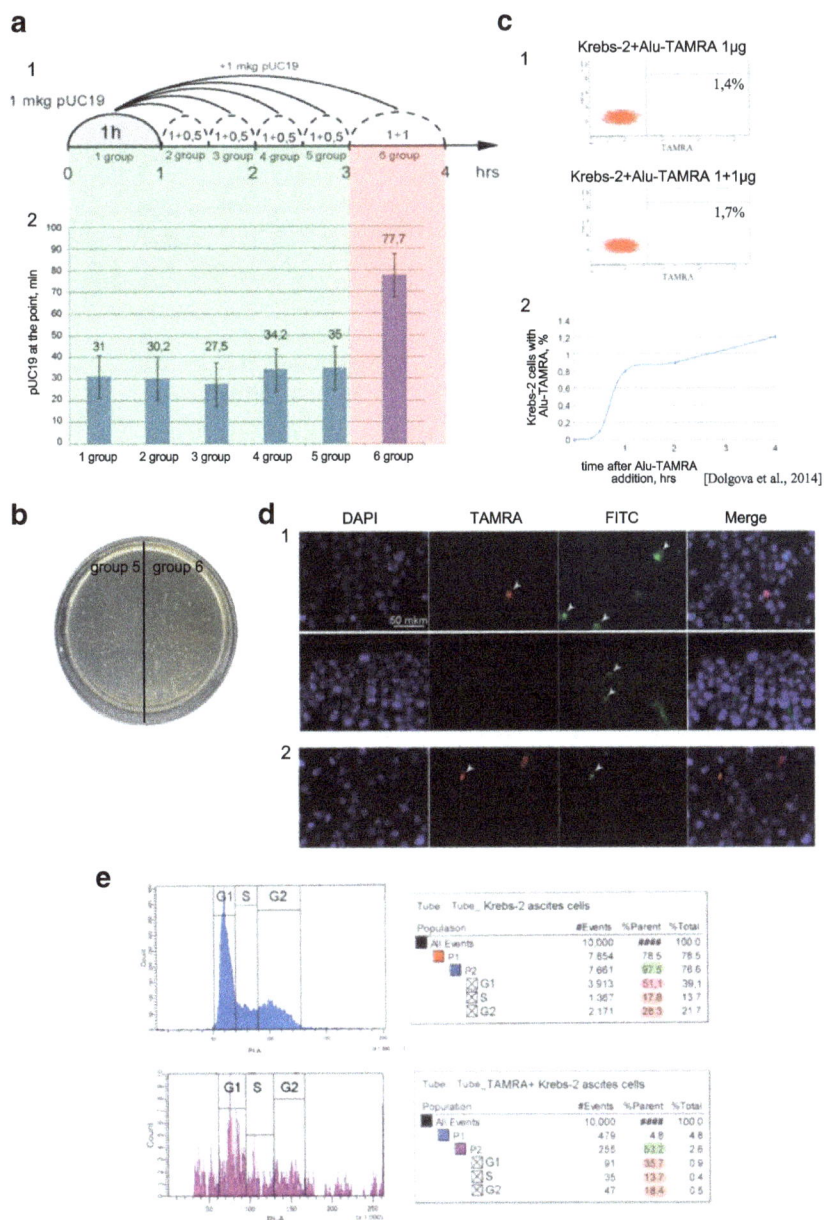

Fig. 5 Internalization of eDNA by Krebs-2 cells upon additional rounds of incubation. **a** Internalization of pUC19 plasmid was analyzed when this plasmid was added every 30 min (groups 2–6) following 1-h pre-incubation with 1 µg pUC19 (group 1). Shown is the total number of colonies obtained by transforming competent *E. coli* with DNA from 10^6 Krebs-2 cells incubated with pUC19 DNA; **b** representative image of *E. coli* colonies that formed as a result of transformation of DNA isolated from Krebs-2 cells that received additional pUC19 DNA 2.5 (group 5) and 3 h (group 6) after the pre-incubation step. Many more colonies are visible on the group 6 half of the plate; **c** FACS analysis of *Alu*-TAMRA dsDNA internalization by Krebs-2 cells 3 h following the pre-incubation with the same DNA; **d** fluorescence microscopy analysis of Krebs-2 cells pre-incubated with *Alu*-TAMRA DNA, followed by addition of *Alu*-FITC DNA. Cells internalizing fluorescent probes are marked with *arrows*; **e** cell cycle profiling of the total (unsorted) cell population (*top*), and TAMRA+ Krebs-2 cells (*bottom*). Percentages of cells found in G1, S, and G2/M phases are shaded in *red*

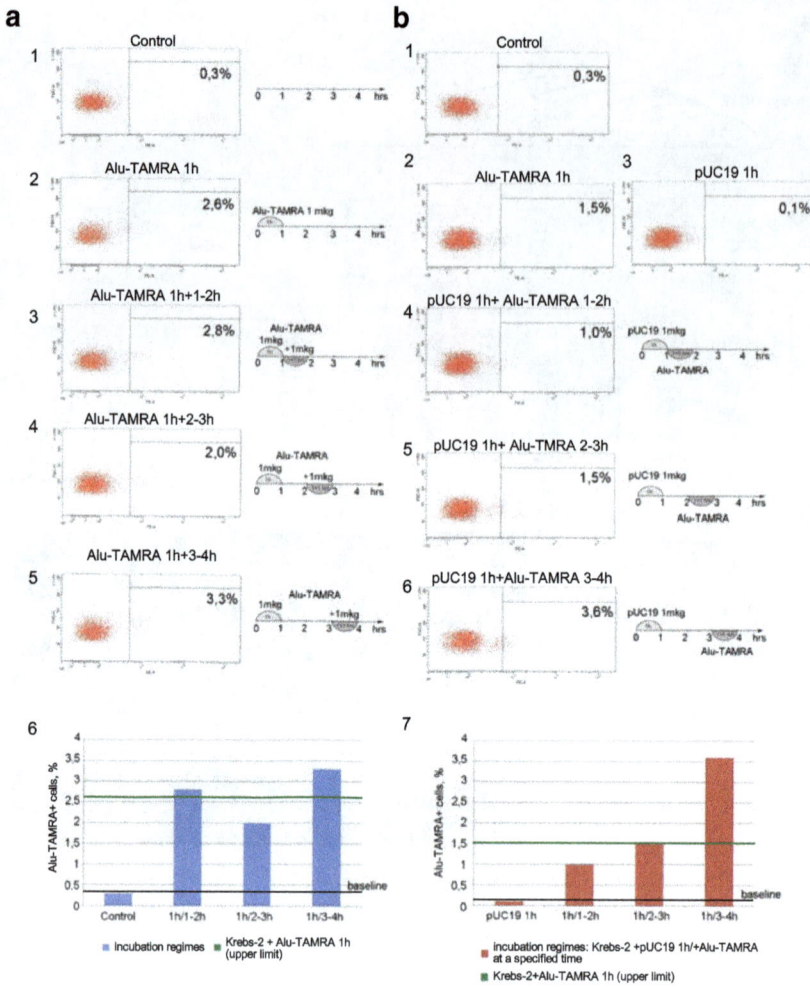

Fig. 6 Competition between *Alu*-TAMRA dsDNA and supercoiled pUC19 plasmid DNA for internalization into Krebs-2 cells. **a** *Alu*-TAMRA DNA is added 1 h after pre-incubation with the same DNA. **b** *Alu*-TAMRA DNA is added after pre-incubation with pUC19. 1a-5a, 1b-6b, Flow cytometry plots and schematics of each experimental point analyzed; 6a, 7b, percentage of TAMRA+ cells analyzed at each point. Baseline equals background level of fluorescence, i.e., either when no DNA was added (**a**), or when cells were incubated for 1 h with non-fluorescent plasmid DNA (**b**); upper limit represents percentage of TAMRA+ cells following incubation with *Alu*-TAMRA DNA

internalization factors used for *Alu*-TAMRA and plasmid DNA are the same.

2. *Alu-TAMRA + Alu-TAMRA*
   When Krebs-2 cells are pre-incubated with *Alu*-TAMRA DNA and then additional portions of the same DNA are introduced, the pre-incubation step should saturate all the G1-S cells available and no increase in TAMRA-positive cells should be observed during the first 3–4 h. The increase in internalizing cells (about 20 %) should, however, be observed after the 3–4 h timepoint, which corresponds to the percentage of cells found in G2-M phases in the beginning of the experiment. Data obtained for

Alu-TAMRA DNA appear similar to those obtained for the plasmid DNA, i.e., there is no internalization during the first 3 h and a second wave of internalization events follows immediately after (Fig. 6a 1–6). The plot shown in Fig. 6a indicates that the percentage of TAMRA+ cells remains unchanged at 1-h, 2-h, and 3-h timepoints (upper limit). However, it does increase some 13 % later on (from 2.6 % to 3.3 %) (Fig. 6a 2,5,6).

3. *pUC19 + Alu-TAMRA*
   Should the factor controlling internalization of *Alu*-TAMRA and plasmid DNA be the same, pre-incubation of Krebs-2 cells with plasmid DNA

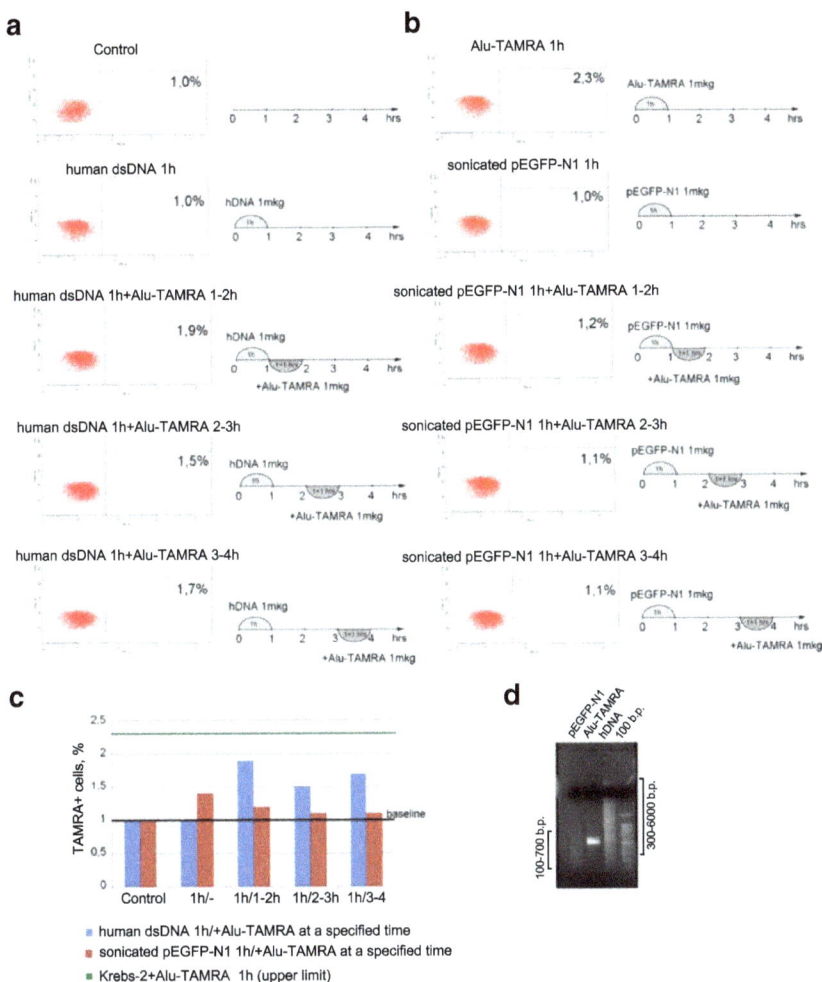

Fig. 7 Competition between *Alu*-TAMRA DNA and high-molecular weight human DNA or sonicated pEGFP-N1 plasmid for internalization into Krebs-2 cells. **a** Addition of *Alu*-TAMRA DNA to Krebs-2 cells pre-incubated with human dsDNA for 1 h; **b** addition of *Alu*-TAMRA DNA to Krebs-2 cells pre-incubated sonicated pEGFP-N1 plasmid DNA. FACS plots and schematics of the experimental points analyzed are shown; **c** bar plot summarizing the percentages of TAMRA+ cells. Baseline equals background fluorescence of untreated cells; control represents cells incubated for 1 h with 1 µg of either human dsDNA (*blue*) or sonicated pEGFP-N1 plasmid (*red*); upper limit represents percentage of TAMRA+ cells after 1 h incubation with 1 µg *Alu*-TAMRA DNA; **d** results of an agarose gel electrophoresis showing fragment size distributions of the three eDNAs tested. Brackets on the sides of the gel indicate average sizes of sonicated pEGFP-N1 (*left*) and human dsDNA (*right*) fragments. *dsDNA* double-stranded DNA

should prevent TAMRA+ cells from appearing following the incubation steps, and thus internalization above the "upper limit" should only become detectable 3–4 h after the beginning of the experiment. Alternatively, if internalization factors are different, incubation with *Alu*-TAMRA DNA should immediately result in a TAMRA+ cell subpopulation, with the percentage matching that of the control (*Alu*-TAMRA, 1 h).

What we observed when testing this combination was that, following pre-incubation with pUC19, the percentage of *Alu*-TAMRA cells equaled the control level (as in *Alu*-TAMRA 1 h). After 4 h, the second wave of internalization occurs. No decrease in the percentage of *Alu*-TAMRA-internalizing cells was observed, when they were saturated by the pre-incubation with plasmid DNA (Fig. 6b 2,4,5). These results indicate that Krebs-2 cells use at least two distinct internalization pathways for the uptake of supercoiled plasmid DNA and linear TAMRA-labeled *Alu* DNA (500 bp) (Fig. 6b). The conclusions of these experiments are in line with co-incubation experiments using plasmid DNA and human dsDNA (300–6000 bp) showing lack of competition between the two species for internalization factors (Fig. 4a).

The final experimental series was performed to estimate the competition between various types of linear DNA. Two combinations were selected: human DNA + *Alu*-TAMRA (Fig. 7a), and sonicated pEGFP-N1 + *Alu*-TAMRA (Fig. 7b), and analyzed as above.

4. *human DNA+ Alu-TAMRA*

Testing this combination helped address possible competition with human genomic DNA sonicated to the size of 300–6000 bp (as this DNA species was previously demonstrated to internalize into Krebs-2 TISCs [12]). The results obtained show that internalization factors for 300–6000 bp human DNA (hereafter referred to as high-molecular weight DNA) and human *Alu* repeat provided as a PCR fragment are distinct (Fig. 7a and c). We observed that there was partial competition between *Alu* DNA and sonicated human DNA for internalization factor(s), as demonstrated by the slight decrease in *Alu*-TAMRA-internalizing cells. This is likely attributable to the presence of trace amounts of DNA below 500 bp in the sonicated human DNA preparation, which serves as a competitor.

5. *Sonicated pEGFP-N1 + Alu-TAMRA*

Use of this combination focused the analysis on the competition between DNA species similar in structure and size (Fig. 7d). Internalization of 500 bp *Alu*-TAMRA PCR product by Krebs-2 cells pre-incubated with sonicated pEGFP-N1 plasmid (<700 bp) was significantly suppressed, which suggests competition for the internalization factor (Fig. 7b and c).

Taken together, our data suggest that supercoiled plasmid DNA reaches internal compartments of Krebs-2 TISCs essentially intact, given that functional plasmids can be recovered upon transformation into competent *E. coli* cells. Hence, it was interesting to test whether GFP could be expressed from the pEGFP-N1 plasmid internalized by Krebs-2 TISCs. Cells and the plasmid DNA were co-incubated for 1 h. Then incubation medium was replaced with a 10 % FBS-supplied medium, and cells were kept in the CO_2 incubator for another 24 h. We ensured pEGFP-N1 plasmid was fully functional by incubating it with mouse fibroblasts. In the presence of lipofectamine, about 70 % of cells were scored GFP-positive. Two incubation protocols were tested for Krebs-2 TISCs, one including protamine, and one with protamine omitted. When plasmid DNA was pre-incubated with protamine, this resulted in a 17-fold increase in internalization

efficiency. Notably, in neither protocol was any GFP fluorescence observed in Krebs-2 TISCs (Fig. 8a and b). Control experiments involving *Alu*-TAMRA DNA or *Alu*-TAMRA "pre-complexed" with protamine indicate that the percentage of DNA-internalizing cells remains unchanged (Fig. 8c). This argues for the existence of an additional internalization factor for DNA-protamine complexes.

Discussion

In the present study, we performed comprehensive analysis of plasmid DNA internalization by Krebs-2 cells, and established several interesting aspects of competition between distinct DNA species upon internalization. To do so, we developed a special approach wherein Krebs-2 cells were first incubated or co-incubated with plasmid DNA; total DNA was isolated from such cells and used for transformation of competent *E. coli* cells. The rationale behind this approach is that it is important to analyze whether plasmid DNA can be used as a versatile vector to deliver cell-killing genes to TISCs. Efficiency of internalization was also assayed independently using qPCR.

About 3 % of mouse ascites Krebs-2 cells were shown to internalize eDNA, such as TAMRA-5'-dUTP human *Alu* DNA fragment (Fig. 1). Saturation of the cells by *Alu* DNA fragments was achieved during the first hour of incubation, and the quantity of internalized DNA material doubled every 10 min. The efficiency of internalization remained unaffected across a wide range of temperatures tested (4 °C, 25 °C, and 37 °C).

We demonstrate that internalization of dsDNA (500 bp) and supercoiled pUC19 plasmid DNA occurs in exactly the same tumor cells (Fig. 2). The cell saturation threshold with plasmid DNA was defined. When 1 µg plasmid DNA is co-incubated with 10^6 Krebs-2 cells, ascites cells cannot internalize more eDNA. Such cells encompass 340–2600 copies of the plasmid, depending on the transformation experiment, or up to $3.097 \pm 0.044 \times 10^6$ molecules, as assayed by qPCR (Fig. 3). The approximate 1000-fold difference between the results obtained using these two quantification methods is explained by the distinct sensitivities of these approaches. The number of colonies obtained after transformation is affected by many parameters, such as the competence level of bacterial cells, sterical interactions of circular plasmid DNA with chromatin, association of cellular proteins with the internalized plasmids, and structural integrity of the plasmid. In contrast, these factors are of little importance when PCR quantification is used, as it only measures the number of target molecules delimited by the primer binding sites.

When eDNA is condensed by forming complexes with a positively charged protamine [9], this results in a 17-fold

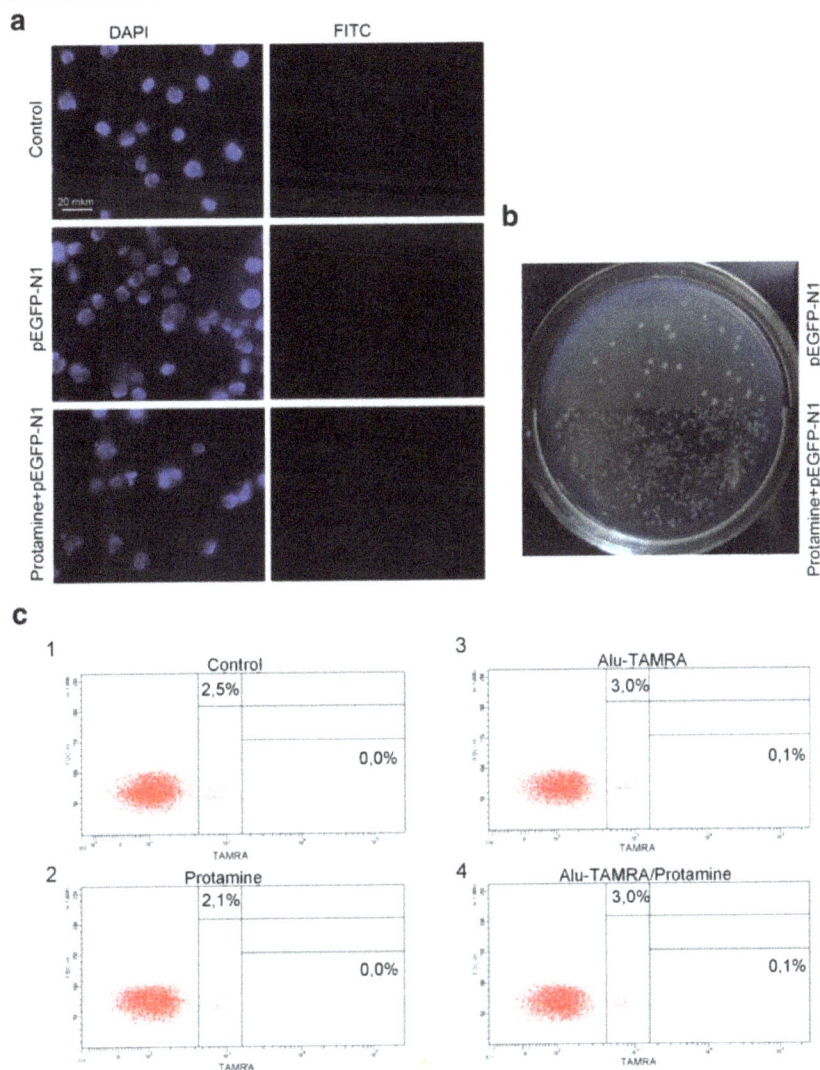

Fig. 8 Internalization of protamine/extracellular DNA complexes by Krebs-2 cells. **a** Analysis of GFP fluorescence in Krebs-2 cells following incubation with pEGFP-N1 plasmid. DAPI represents chromatin *(blue channel)*; FITC represents GFP *(green channel,* no signal). **b** Transformation of *E. coli* competent cells with DNA isolated from Krebs-2 cells incubated with pEGFP-N1 or pEGFP-N1/protamine complexes. **c** FACS analysis of *Alu*-TAMRA DNA/protamine internalization by Krebs-2 cells: 1: Negative control, intact Krebs-2 cells; 2: Krebs-2 + protamine; 3: Krebs-2 + *Alu*-TAMRA DNA; 4: Krebs-2 + *Alu*-TAMRA DNA/protamine complexes. Percentage values correspond to autofluorescence levels observed in the control and to the fraction of TAMRA+ cells at experimental points

increase in internalization, with the percentage of DNA-internalizing cells remaining unchanged.

Experiments were performed to analyze possible competition between plasmid pUC19 and various forms of nucleic acids (human dsDNA (300–6000 bp), PCR fragment (500 bp), yeast RNA), as well as other agents (BSA, heparin) for internalization by Krebs-2 cells. Of these, only heparin worked as a potent inhibitor of internalization (Fig. 4). These observations are consistent with the idea that heparin interacts with the same internalization

factor(s) as plasmid DNA. Heparin was reported to interact with a number of cell surface molecules, such as heparin binding protein and CAP37/azurocidin [16]. Bennett and colleagues showed that heparin and DNA molecules compete for binding to a 30-kDa protein in the context of lymphocyte membranes, whereas RNA, poly-dA/dT, or dNTP mix do not [3]. Also, heparin is known to interact with a family of fibroblast growth factor receptors [13]. Exosomes have been reported to contain dsDNA, and so their cargo can be easily internalized—in a process that is inhibited by heparin [8]. This could have pointed to the possible involvement of exosomes in eDNA internalization. We checked whether the TAMRA+ probe could be effluxed by Krebs-2 TISCs and observed that, upon internalization, the DNA did not leave the cells in any form, including exosomal (data not shown). This indicates that internalized DNA does not become part of exosomes, at least in the given context. The formal possibility that an exosome would internalize extracellular dsDNA appears rather unlikely, nor has such a property been reported in the literature. Given the above-described range of cell receptors bound by heparin and its property to inhibit eDNA internalization, one can expect that in the case of Krebs-2 cells these cell surface proteins are bona fide eDNA internalization factors.

Depending on the experiment, we observed increased internalization (13–50 %) of plasmid DNA 3 h after the beginning of incubation (Fig. 5). Formally, this observation is compatible with two scenarios. The first implies that internalization is mediated by specific cell receptors that shuttle back and forth between the cell surface and the cell interior [40]. The other relies on the greater availability of internalization-competent cells. We showed that TAMRA+ cells are the same cells as those internalizing plasmid DNA, and that their percentage ranges 1–3 % in our experiments. Under the first scenario, the percentage of Alu-TAMRA internalizing cells should remain unchanged at the 3- to 4-h timepoint; however, this was not the case, and more TAMRA+ cells were observed 3 h following the beginning of incubation (Fig. 5c). Thus, the second scenario appears the likeliest. Internalization of pUC19 translates into more colonies because more internalization-competent Krebs-2 cells become available, rather than because more plasmid copies are internalized.

Krebs-2 TISCs are not expected to internalize eDNA during mitosis, yet they become internalization-competent soon thereafter. Importantly, all other cells of this particular cell subpopulation already contain eDNA molecules. Thus, there should be a gain in DNA-internalizing cells that should be equal to the proportion of cells found in the G2-M phases in the beginning of the experiment, which in turn should translate into more E. coli colonies formed upon transformation. The cells that internalized DNA in G1 persist until mitosis, which explains why cell cycle profiling experiments failed to uncover significant accumulation of TAMRA+ cells in any of the cell cycle phases (Fig. 5a, c and e; [12]).

Finally, we tested competition between supercoiled plasmid DNA (pUC19), short PCR product (human Alu PCR fragment, 500 bp), sonicated pEGFP-N1 plasmid (100–700 bp), and larger-sized sonicated human dsDNA (300–6000 bp). If internalization factors for distinct types of DNA are the same, then saturating pre-treatment with one DNA species should block

internalization of the other. Alternatively, should internalization factors be distinct, the second DNA type should be internalized regardless of the pre-treatment type of DNA. It turned out that pre-treatment with Alu PCR fragment abolishes subsequent internalization of Alu DNA and sonicated pEGFP-N1 by Krebs-2 cells (Figs. 6a and 7b). We attribute this to the fact that sonicated plasmid DNA is similar in size (100–700 bp) to Alu fragment (500 bp), thereby efficiently competing for binding to the receptor that mediates internalization of small (up to 500 bp) DNA fragments. The same was observed for plasmid DNA that blocked its own internalization. Efficiency of protamine-assisted eDNA internalization was also analyzed.

Pre-treatment of Krebs-2 cells with supercoiled plasmid DNA does not affect subsequent internalization of Alu PCR fragment (Fig. 6b). This is unlike pre-treatment with sonicated human dsDNA, which shows partial inhibition of Alu-TAMRA DNA internalization by Krebs-2 cells (Fig. 7a). We believe this partial inhibition is due to the presence of small amounts of low-molecular weight DNA fragments (<500 bp) in the preparation of sonicated human DNA (300–6000 bp). Most of the fragments in this preparation, however, are longer than 1 kb, and so they do not compete for binding to the internalization factor that mediates the uptake of fragments below 500 bp.

Biological meaning and clinical significance of the phenomenon observed

The phenomenon of internalization of extracellular dsDNA by stem cells (SCs) and TISCs is broadly related to the functioning of an organism as a single entity. We speculate that continuous flow of a highly dynamic set of extracellular DNA fragments through various SCs may function as a mechanism that helps such cells sense the genetic "image" of the body. Quantitative as well as qualitative changes among these eDNA molecules are interpreted by the multipotent cells as the clues setting the direction of differentiation.

Upon internalization and reaching the internal compartments of SCs and TISCs, eDNA may participate in many molecular processes. For instance, it is known that: 1) Any free dsDNA ends, including the ends of eDNA molecules, potently induce repair cascades [23, 29, 41–43]. Our studies as well as other reports indicate that, upon internalization, linear plasmid DNA is partially digested at its termini and is ligated to form a circle [11, 25, 32, 34]. 2) When the cell is undergoing the repair of interstrand DNA crosslinks induced by an earlier treatment with a cytostatic drug, eDNA fragments delivered at this point into the cell interfere with nucleotide excision repair and homologous recombination repair phases. In the context of murine CD34+ hematopoietic stem cells

(HSCs), this results in the failure to give rise to lymphoid cell lineage [10], whereas Krebs-2 TISCs either die or become non-tumorigenic [33]. 3) When the cells repairing radiation-induced dsDNA breaks receive eDNA fragments, the latter interfere with the non-homologous end-joining step, yet HSCs are rescued from the aberrant recovery of chromosomal integrity and apoptosis. These surviving HSCs help repopulate the murine hematopoietic system, thereby protecting the animals from developing radiation sickness and death, and increasing the survival of lethally irradiated mice to 60–90 % [26]. Thus, "planting" eDNA to TISCs in a timely manner, when these cells are trying to repair the DNA, may potently interfere with the repair process, and so poorly differentiated cell types including TISCs either die or profoundly alter their properties. Clearly, this approach relies on fundamentally different principles of targeting SCs and TISCs. One of the most intriguing questions then is whether extracellular fragmented dsDNA may also target solitary cancer cells that give rise to metastases and which are known to be largely quiescent (and represent a form of TISCs). The very existence of these cells prompts for the careful analysis of many interesting questions related to cancer biology and therapy [31, 38]. In the absence of experimental data, we propose several ideas on the possible interplay between solitary cancer cells and fragmented dsDNA. First, these cells may be imagined not to internalize eDNA, and it is important to understand why poorly differentiated cells otherwise classifiable as TISCs fail to do so [12]. Alternatively, dormant cancer cells do internalize dsDNA, but whether their quiescence may prevent the fragmented DNA from efficiently interfering with the aforementioned molecular processes remains unclear. A number of so-called stress signals have been reported to induce proliferation of dormant cancer cells, which leads to the development of a metastasis [38]. dsDNA fragments internalized by solitary cancer cells may be viewed as a classical stress signal. Our work, as well as the data from other research groups, is consistent with the following scenario. Free ends of internalized dsDNA molecules launch a kinase cascade thereby leading to the cell cycle arrest and block of cell division [23, 25, 29, 32, 41–43]. Our unpublished ex vivo studies indicate that dsDNA fragments are delivered into CD34+ HSCs, most of which are non-dividing, and paradoxically induce their proliferation. We speculate that stalling of the cell cycle prevents the HSC from reverting to the quiescent state, thereby forcing the dsDNA-activated cell to proliferate [27]. This scenario may point to the possibility that solitary cancer cells may be induced to proliferate by extracellular DNA fragments, which should translate into a clinically negative effect. Nevertheless, if dormant cancer cells do internalize eDNA this may well be used as a tool

to selectively kill such cells, essentially in the same way as recently proposed [33]. Thus, our approach primarily targets poorly differentiated cell types including TISCs. Expression of a toxic protein or other protein of interest delivered to such cells in the form of DNA may be used to selectively alter their phenotype or eliminate them altogether.

Conclusion

Upon reaching the cell interior, eDNA molecules do not behave as an inert cargo. They are actively integrated in multiple on-going cellular processes and frequently compromise their correct progression. Additionally, extracellular DNA may also induce a number of cellular processes. Finally, the very property of dsDNA internalization may serve to develop a versatile approach to detect and target any type of poorly differentiated cells, such as TISCs.

Abbreviations
Alu-TAMRA, human Alu repeat DNA labeled by the fluorescently modified nucleotide TAMRA-5'-dUTP; BSA, bovine serum albumin; dsDNA, double-stranded DNA; eDNA, extracellular DNA; FBS, fetal bovine serum; HSC, hematopoietic stem cell; PCR, polymerase chain reaction; qPCR, quantitative polymerase chain reaction; SC, stem cell; TISC, tumor-initiating stem cell.

Acknowledgements
The authors are thankful for the human dsDNA preparation to Natalya V. Yudina. The authors are grateful to Artem V. Kozel for assistance in pEGFP-N1/protamine experiments. The authors express their gratitude to Dr. Andrey Gorchakov for translating the paper, as well as to the Flow Cytometry and Microscopy Centers for Collective Use at the Institute of Cytology and Genetics, Siberian Branch of the Russian Academy of Sciences.

Funding
The work was funded by the LLC "BA-Pharma", LLC "Panagen", RFBR grant N 15-04-03386/16 and the State scientific project N 0324-2015-0003.

Authors' contributions
EVD performed the analysis, interpreted the data, and drafted the manuscript. EAP, ASP, and AMM carried out the molecular studies. ERC participated in the design of the study. AAO performed the analysis and interpreted the data. YRE supervised work related with FACS. SIB directed the work on confocal microscopy. VPN and NAP carried out the mouse experiments, performed the analysis, and interpreted the data. NAK participated in the study coordination and provide the technical conditions for the performed works. SSB conceived the study, participated in its design, and coordinated and drafted the manuscript. All authors reviewed, critically revised, and approved the final manuscript.

Competing interests
The authors declare that they have no competing interests.

Consent for publication
Not applicable.

Ethical approval and consent to participate
All animal experiments were performed in accordance with protocols approved by the Animal Care and Use Committee of the Institute of Cytology and Genetics, Siberian Branch of the Russian Academy of Sciences (protocol #8, 19/03/2012).

Author details
[1]Institute of Cytology and Genetics, Siberian Branch of the Russian Academy of Sciences, 10 Lavrentieva Ave., Novosibirsk 630090, Russia. [2]Institute of Clinical Immunology, Siberian Branch of the Russian Academy of Medical Sciences, 14 Yadrintsevskaya Street, Novosibirsk 630099, Russia. [3]Novosibirsk State University, 2 Pirogova Street, Novosibirsk 630090, Russia.

References

1. Alyamkina EA, Dolgova EV, Likhacheva AS, Rogachev VA, Sebeleva TE, Nikolin VP, et al. Exogenous allogenic fragmented double-stranded DNA is internalized into human dendritic cells and enhances their allostimulatory activity. Cell Immunol. 2010;262:120–6.

2. Amyere M, Mettlen M, Van Der Smissen P, Platek A, Payrastre B, Veithen A, et al. Origin, originality, functions, subversions and molecular signalling of macropinocytosis. Int J Med Microbiol. 2002;291:487–94.

3. Bennett RM, Gabor GT, Merritt MM. DNA binding to human leukocytes. Evidence for a receptor-mediated association, internalization, and degradation of DNA. J Clin Invest. 1985;76:2182–90.

4. Bennett CB, Lewis AL, Baldwin KK, Resnick MA. Lethality induced by a single site-specific double-strand break in a dispensable yeast plasmid. Proc Natl Acad Sci U S A. 1993;90:5613–17.

5. Bergsmedh A, Szeles A, Henriksson M, Bratt A, Folkman MJ, Spetz AL, et al. Horizontal transfer of oncogenes by uptake of apoptotic bodies. Proc Natl Acad Sci U S A. 2001;98:6407–11.

6. Bergsmedh A, Szeles A, Spetz AL, Holmgren L. Loss of the p21(Cip1/Waf1) cyclin kinase inhibitor results in propagation of horizontally transferred DNA. Cancer Res. 2002;62(2):575–9.

7. Budker V, Budker T, Zhang G, Subbotin V, Loomis A, Wolff JA. Hypothesis: naked plasmid DNA is taken up by cells in vivo by a receptor-mediated process. J Gene Med. 2000;2:76–88.

8. Christianson HS, Svensson KJ, van Kuppevelt TH, Li J, Belting M. Cancer cell exosomes depend on cell-surface heparan sulfate proteoglycans for their internalization and functional activity. Proc Natl Acad Sci U S A. 2013; 110(43):17380–5.

9. Dolgova EV, Rogachev VA, Nikolin VP, Popova NA, Likhacheva AS, Aliamkina EA, et al. Leukocyte stimulation by DNA fragments shored up protamine in cyclophosphamide-induced leukopoiesis in mice. Vopr Onkol. 2009;55(6): 761–64 [Article in Russian].

10. Dolgova EV, Proskurina AS, Nikolin VP, Popova NA, Alyamkina EA, Orishchenko KE, et al. "Delayed death" phenomenon: a synergistic action of cyclophosphamide and exogenous DNA. Gene. 2012;495(2):134–45.

11. Dolgova EV, Efremov YR, Orishchenko KE, Andrushkevich OM, Alyamkina EA, Proskurina AS, et al. Delivery and processing of exogenous double-stranded DNA in mouse CD34+ hematopoietic progenitor cells and their cell cycle changes upon combined treatment with cyclophosphamide and double-stranded DNA. Gene. 2013;528:74–83.

12. Dolgova EV, Alyamkina EA, Efremov YR, Nikolin VP, Popova NA, Tyrinova TV, et al. Identification of cancer stem cells and a strategy for their elimination. Cancer Biol Ther. 2014;15:1378–94.

13. Fannon M, Forsten KE, Nugent MA. Potentiation and inhibition of bFGF binding by heparin: a model for regulation of cellular response. Biochemistry. 2000;39:1434–45.

14. Filaci G, Gerloni M, Rizzi M, Castiglioni P, Chang HD, Wheeler MC, et al. Spontaneous transgenesis of human B lymphocytes. Gene Ther. 2004;11:42–51.

15. Gabor G, Bennett RM. Biotin-labelled DNA: a novel approach for the recognition of a DNA binding site on cell membranes. Biochem Biophys Res Commun. 1984;122:1034–39.

16. Gautam N, Olofsson AM, Herwald H, Iversen LF, Lundgren-Akerlund E, Hedqvist P, et al. Heparin-binding protein (HBP/CAP37): a missing link in neutrophil-evoked alteration of vascular permeability. Nat Med. 2001;7:1123–27.

17. Hanss B, Leal-Pinto E, Bruggeman LA, Copeland TD, Klotman PE. Identification and characterization of a cell membrane nucleic acid channel. Proc Natl Acad Sci U S A. 1998;95:1921–26.

18. Hefeneider SH, Bennett RM, Pham TQ, Cornell K, McCoy SL, Heinrich MC. Identification of a cell-surface DNA receptor and its association with systemic lupus erythematosus. J Invest Dermatol. 1990;94:79S–84S.

19. Holmgren L, Szeles A, Rajnavölgyi E, Folkman J, Klein G, Ernberg I, et al. Horizontal transfer of DNA by the uptake of apoptotic bodies. Blood. 1999; 93(11):3956–63.

20. Kamiya H, Fujimura Y, Matsuoka I, Harashima H. Visualization of intracellular trafficking of exogenous DNA delivered by cationic liposomes. Biochem Biophys Res Commun. 2002;298:591–7.

21. Klein G, Klein E. The transformation of a solid transplantable mouse carcinoma into an "ascites tumor". Cancer Res. 1951;11(6):466–9.

22. Khalil IA, Kogure K, Akita H, Harashima H. Uptake pathways and subsequent intracellular trafficking in nonviral gene delivery. Pharmacol Rev. 2006;58:32–45.

23. Kumagai A, Dunphy WG. Claspin, a novel protein required for the activation of Chk1 during a DNA replication checkpoint response in Xenopus egg extracts. Mol Cell. 2000;6(4):839–49.

24. Leal-Pinto E, Teixeira A, Tran B, Hanss B, Klotman PE. Presence of the nucleic acid channel in renal brush-border membranes: allosteric modulation by extracellular calcium. Am J Physiol Renal Physiol. 2005; 289(1):F97–106.

25. Lin J, Krishnaraj R, Kemp RG. Exogenous ATP enhances calcium influx in intact thymocytes. J Immunol. 1985;135(5):3403–10.

26. Likhacheva AS, Nikolin VP, Popova NA, Rogachev VA, Prokhorovich MA, Sebeleva TE, et al. Exogenous DNA can be captured by stem cells and be involved in their rescue from death after lethal-dose γ-radiation. Gene Ther Mol Biol. 2007;11:305–14.

27. Likhacheva AS, Rogachev VA, Nicolin VP, Popova NA, Sebeleva TE, Strunkin DN, et al. Involvement of exogenous DNA in the molecular processes in somatic cell. Vestnik VOGiS. 2008;12(3):426–73 [Article in Russian].

28. Loke SL, Stein CA, Zhang XH, Mori K, Nakanishi M, Subasinghe C, et al. Characterization of oligonucleotide transport into living cells. Proc Natl Acad Sci U S A. 1989;86:3474–8.

29. MacDougall CA, Byun TS, Van C, Yee MC, Cimprich KA. The structural determinants of checkpoint activation. Genes Dev. 2007;21(8):898–903.

30. Munkonge FM, Dean DA, Hillery E, Griesenbach U, Alton EW. Emerging significance of plasmid DNA nuclear import in gene therapy. Adv Drug Deliv Rev. 2003;55:749–60.

31. Páez D, Labonte MJ, Bohanes P, Zhang W, Benhanim L, Ning Y, et al. Cancer dormancy: a model of early dissemination and late cancer recurrence. Clin Cancer Res. 2012;18(3):645–53.

32. Perucho M, Hanahan D, Wigler M. Genetic and physical linkage of exogenous sequences in transformed cells. Cell. 1980;22(1 Pt 1):309–17.

33. Potter EA, Dolgova EV, Proskurina AS, Minkevich AM, Efremov YR, Taranov OS, Omigov VV, Nikolin VP, Popova NA, Bayborodin SI, Ostanin AA, Chernykh ER, Kolchanov NA, Shurdov MA, Bogachev SS. A strategy to eradicate well-developed Krebs-2 ascites in mice. Oncotarget. 2016. doi:10. 18632/oncotarget.7311.

34. Rogachev VA, Likhacheva A, Vratskikh O, Mechetina LV, Sebeleva TE, Bogachev SS, et al. Qualitative and quantitative characteristics of the extracellular DNA delivered to the nucleus of a living cell. Cancer Cell Int. 2006;6:23.

35. Schubbert R, Renz D, Schmitz B, Doerfler W. Foreign (M13) DNA ingested by mice reaches peripheral leukocytes, spleen, and liver via the intestinal wall mucosa and can be covalently linked to mouse DNA. Proc Natl Acad Sci U S A. 1997;94:961–6.

36. Shi F, Gounko NV, Wang X, Ronken E, Hoekstra D. In situ entry of oligonucleotides into brain cells can occur through a nucleic acid channel. Oligonucleotides. 2007;17:122–33.

37. Vaughan EE, DeGiulio JV, Dean DA. Intracellular trafficking of plasmids for gene therapy: mechanisms of cytoplasmic movement and nuclear import. Curr Gene Ther. 2006;6:671–81.

38. Wan L, Pantel K, Kang Y. Tumor metastasis: moving new biological insights into the clinic. Nat Med. 2013;19(11):1450–64.

39. Wilschut KJ, van der Aa MA, Oosting RS, Hennink WE, Koning GA, Crommelin DJ, et al. Fluorescence in situ hybridization to monitor the intracellular location and accessibility of plasmid DNA delivered by cationic polymer-based gene carriers. Eur J Pharm Biopharm. 2009;72(2):391–6.

40. Yakubov LA, Deeva EA, Zarytova VF, Ivanova EM, Ryte AS, Yurchenko LV, et al. Mechanism of oligonucleotide uptake by cells: involvement of specific receptors? Proc Natl Acad Sci U S A. 1989;87:6454–8.

41. Yoo HY, Shevchenko A, Shevchenko A, Dunphy WG. Mcm2 is a direct substrate of ATM and ATR during DNA damage and DNA replication checkpoint responses. J Biol Chem. 2004;279(51):53353–64.
42. Yoo HY, Jeong SY, Dunphy WG. Site-specific phosphorylation of a checkpoint mediator protein controls its responses to different DNAstructures. Genes Dev. 2006;20(7):772–83.
43. Zou L. Single- and double-stranded DNA: building a trigger of ATR-mediated DNA damage response. Genes Dev. 2007;21(8):879–85.

19

Prospective purification of perivascular presumptive mesenchymal stem cells from human adipose tissue: process optimization and cell population metrics across a large cohort of diverse demographics

C. C. West[1,2†], W. R. Hardy[3†], I. R. Murray[1†], A. W. James[3], M. Corselli[3,4], S. Pang[3], C. Black[3,5], S. E. Lobo[3,6], K. Sukhija[3,7], P. Liang[3,8], V. Lagishetty[3,9], D. C. Hay[1], K. L. March[10], K. Ting[3,11], C. Soo[3,12,13] and B. Péault[1,3*]

Abstract

Background: Adipose tissue is an attractive source of mesenchymal stem cells (MSC) as it is largely dispensable and readily accessible through minimally invasive procedures such as liposuction. Until recently MSC could only be isolated in a process involving *ex-vivo* culture and their *in-vivo* identity, location and frequency remained elusive. We have documented that pericytes (CD45-, CD146+, and CD34-) and adventitial cells (CD45-, CD146-, CD34+) (collectively termed perivascular stem cells or PSC) represent native ancestors of the MSC, and can be prospectively purified using fluorescence activated cell sorting (FACS). In this study we describe an optimized protocol that aims to deliver pure, viable and consistent yields of PSC from adipose tissue. We analysed the frequency of PSC within adipose tissue, and the effect of patient and procedure based variables on this yield.

Methods: Within this twin centre study we analysed the adipose tissue of $n = 131$ donors using flow cytometry to determine the frequency of PSC and correlate this with demographic and processing data such as age, sex, BMI and cold storage time of the tissue.

Results: The mean number of stromal vascular fraction (SVF) cells from 100 ml of lipoaspirate was 34.4 million. Within the SVF, mean cell viability was 83 %, with 31.6 % of cells being haematopoietic (CD45+). Adventitial cells and pericytes represented 33.0 % and 8 % of SVF cells respectively. Therefore, a 200 ml lipoaspirate would theoretically yield 23.2 million viable prospectively purified PSC - sufficient for many reconstructive and regenerative applications. Minimal changes were observed in respect to age, sex and BMI suggesting universal potential application.

(Continued on next page)

* Correspondence: bpeault@mednet.ucla.edu
†Equal contributors
[1]British Heart Foundation Centre for Vascular Regeneration & Medical Research Council Centre for Regenerative Medicine, University of Edinburgh, Edinburgh, UK
[3]Orthopaedic Hospital Department of Orthopaedic Surgery and the Orthopaedic Hospital Research Center, University of California, Los Angeles, CA, USA
Full list of author information is available at the end of the article

(Continued from previous page)

Conclusions: Adipose tissue contains two anatomically and phenotypically discreet populations of MSC precursors – adventitial cells and pericytes – together referred to as perivascular stem cells (PSC). More than 9 million PSC per 100 ml of lipoaspirate can be rapidly purified to homogeneity using flow cytometry in clinically relevant numbers potentially circumventing the need for purification and expansion by culture prior to clinical use. The number and viability of PSC are minimally affected by patient age, sex, BMI or the storage time of the tissue, but the quality and consistency of yield can be significantly influenced by procedure based variables.

Keywords: Mesenchymal stem cells, Adipose tissue, Adipose-derived stem cell, Cell sorting, Flow cytometry, Pericyte, Tunica adventitia

Background

In a series of pioneering experiments in the 1960s, Alexander Friedenstein et al. [1] identified a population of cells from rodent bone marrow that adhered to culture vessels, formed colonies, could differentiate into osteoblasts in culture, and generated bone when implanted ectopically in vivo [2, 3]. Friedenstein termed these cells colony forming unit fibroblast (CFU-F), until Arnold Caplan [4], in the early 1990s, coined the term mesenchymal stem cells (MSC). Since their initial description, these cells have been the focus of much attention for their ability to differentiate into multiple mesodermal cell lineages, to modulate the immune system, and to stimulate regeneration through trophic support and the secretion of cytokines [5].

Despite the ultimate desire to translate MSC research into novel therapies, our understanding of these cells had been based principally on observations made in vitro on cells of undocumented purity and homogeneity, in ignorance of their anatomical location and physiological role in natural and pathological processes. MSC have been enlisted from bone marrow and multiple other tissues, according to their ability to adhere and grow on plastic [6]. This suggested that these cells possess a common identity and a widespread anatomical distribution but provided little insight into location, cellular phenotype, frequency, and specific properties of these cells. Some of these questions were answered when Crisan et al. [7] demonstrated that microvascular pericytes in multiple human fetal and adult tissues express MSC markers, and that when purified to homogeneity by fluorescence-activated cell sorting (FACS) and cultured they are identical to conventional MSCs in terms of morphology, phenotype, and function.

This led to the conclusion that pericytes, defined as $CD31^-CD45^-CD34^-CD146^+$, represent an origin of the MSC grown in culture, a finding that has been validated by other groups [8, 9]. Subsequently, a second population of anatomically and phenotypically distinct $CD31^-CD45^- CD34^+CD146^-$ cells with identical function to conventional MSCs has been identified that reside in the adventitial layer of larger blood vessels [10, 11]. Henceforth, we will refer to these two populations collectively as perivascular stem cells (PSC). Since their description, PSC have been confirmed to behave in vitro and in vivo like MSC, and to be equal in function, if not superior in some instances, to other stem and progenitor populations (reviewed in [12]).

Despite promising findings, there are challenges to address prior to the successful translation and wider clinical use of MSC. Because these cells are procured in low yield from tissues such as bone marrow, they are typically expanded and "purified" based on adherence to plastic under Good Manufacturing Practice compliant conditions—a costly, labor-intensive, and extended process requiring ex vivo culture prior to transplantation. The potential risks entailed by in vitro expansion include infection and immunogenicity due to the exposure of cultured cells to animal-based supplements [13]. Extended periods of in vitro expansion adversely impact the function of these cells, resulting in reductions in their chondrogenic, adipogenic, and osteogenic potentials [14–16]. Higher passage cells show modified and diminished expression of chemokine receptors and adhesion molecules resulting in lower response to chemokines and increased senescence [17]. Concerns have also been raised about the development of genetic instability and the potential for malignant transformation in cultured cells. In addition there are significantly more stringent regulatory hurdles that must be addressed with the use of cultured cells compared with those that have been minimally manipulated [18].

In an attempt to eliminate many of these issues, some groups have investigated the use of an adipose-derived stromal vascular fraction (SVF) as a source of MSC/progenitor cells that bypasses the need for in vitro culture and may be delivered at the point of use, requiring only basic preparation such as enzymatic digestion, washing, and centrifugation. Whilst the SVF may eliminate the requirement of ex vivo expansion, it is a very heterogeneous cell population containing endothelial cells, hematopoietic and inflammatory cells, and fibroblasts as well as cellular debris. This cellular heterogeneity may limit the regenerative potential of the SVF when compared with a more homogeneous MSC population, as has been demonstrated in models of osteogenesis [19].

It is therefore clear that there is a dichotomy in the current approaches to delivering MSC for clinical use. Methods relying on in vitro culture as a means of selection provide a relatively enriched but still undefined cell population and in addition incur the practical, financial, ethical, and regulatory problems this method manifests. Those methods which use the SVF do so at the expense of product identity, purity, and function. We have therefore sought to develop methods for the prospective purification of homogeneous populations of MSC based on our understanding of the exact anatomic location and identity of their native ancestors. Using multicolor FACS, we have purified cells from a range of human adult and fetal tissues [7, 12]. This early work was able to deliver distinct populations of cells that were subsequently expanded in vitro for further characterization, analysis, and experimental work. As our interest in these cells has developed and their potential for immediate clinical use was explored, we established that large numbers of the cells recovered immediately from FACS were of poor quality, and were in the process of dying. This had been previously overlooked in our in-vitro populations because only the healthy cells would adhere and expand. For PSC to be used immediately after FACS, we would need to demonstrate that it is possible to recover pure, viable, and consistent yields of cells. We therefore sought to refine our protocols to maximize not only total cell yield, but maximum viable cell yield, purity, and consistency leading to the development of an optimized protocol that we describe in this work. Using this optimized protocol, we document the cellular composition of human adipose tissue using flow cytometry across a wide demography of donors, and to establish whether PSC can be prospectively purified in clinically relevant numbers, circumventing the need for ex vivo expansion. Furthermore we examined the patient and procedure based variables that may influence this yield. Results were collected from 131 individual lipoaspirates processed in two distinct research centers by different investigators.

Methods

This was a twin-center study based at the University of California at Los Angeles (UCLA), USA and the University of Edinburgh, UK. Adipose tissue was collected with prior written consent from patients undergoing cosmetic lipectomy procedures. Permission for the collection of tissue and subsequent research was granted in Edinburgh by the South East Scotland Research Ethics Committee (Reference: 10/S1103/45), and was unnecessary in Los Angeles because fat was collected under an institutional review board exemption, being considered medical waste.

Tissue was processed and flow cytometry performed using our previously published protocols [12] and the

modifications detailed in this manuscript (Additional file 1). Briefly, lipoaspirate was washed in phosphate buffered saline and centrifuged to separate the fat from the oil and liquid phases. Fat was combined vol/vol with 125 CDU/ml type II collagenase (collagenase from *Clostridium histolyticum*, C6885; Sigma Aldrich, St Louis, Mo, USA) in Dulbecco's modified Eagle's medium + 3.5 % bovine serum albumin (BSA) (Cohn Fraction V A7906; Sigma) and digested for 45 minutes at 37 °C in a shaking water bath (200 rpm). Samples were centrifuged to isolate the SVF which was subsequently filtered through a 100 μm filter and then a 70 μm filter before red cells were lysed. Finally, the SVF was filtered through a 40 μm filter prior to manually counting live cells using trypan blue staining and a hemocytometer. The SVF was then stained with the following antibodies: CD31, CD34, and CD45 (all from BD Biosciences, San Jose, CA, USA) and CD146 (AbD Serotec, Raleigh, NC, USA). Flow cytometry was performed using a BD FacsAria II or III sorter fitted with a 100 μm nozzle following adequate compensation controls using either single stained cells or compensation beads. An initial forward scatter (FSC) vs side scatter (SSC) gate was used to identify cells, followed by gates to select for single cells. Cells not stained by 4′,6-diamidino-2-phenylindole (DAPI) were gated as viable, and hematopoietic and endothelial cells were eliminated based on CD45 and CD31 expression respectively. PSC were subsequently purified from the $CD31^-/CD45^-$ cell fraction according to their differential expression of CD34 and CD146 (pericytes: $CD146^+$, $CD34^-$, $CD31^-$, $CD45^-$; adventitial cells: $CD146^-$, CD34+, $CD31^-$, $CD45^-$) [7, 10].

Flow cytometry data were analyzed using FlowJo (v. 10.0, FlowJo, Ashland, OR, USA) or Diva (v. 6.0, BD Biosciences, San Jose, CA, USA) software, and reviewed by CCW, IRM, WRH, and BP to ensure that accurate and standardized gating strategies were employed across sites and between users. Statistical analysis was performed using JMP10 software (SAS Institute Inc., Cary, NC, USA). Normally distributed nominal data were analyzed using a paired t test, while bivariate fit using regression analysis was used for continuous data. Multivariate analysis was performed using multiple linear least-squares regression. Data were considered significant when $p < 0.05$.

Results

Demographic and PSC parameter analysis

Demographic information as well as cell yield, viability, and subpopulation sort statistics are summarized in Table 1 for 131 unique donor samples.

Analysis of SVF

The SVF was isolated from total fat by collagenase digestion. The mean yield of nucleated cells was 34.4×10^6

Table 1 Demographic data of the 131 donors

Demographic data ($n = 131$)	
Sex	Female = 112
	Male = 19
Age (years)	Mean = 41 (range 22–64)
BMI (kg/m^2)	Mean = 26.5 (range 19–43)
SVF (cells × 10^6)	Mean = 34.4 (range 4.7–120)
Viability (%)	Mean = 83 % (range 36–99)
Pericytes (%)	Mean = 8 % (range 0–55)
Adventitial cells (%)	Mean = 33.0 % (range 3–72)
PSC total (%)	Mean = 41 % (range 6–78)
PSC yield (cells × 10^6)	Mean = 11.6 (range 1.1–47.2)

BMI body mass index, *PSC* perivascular stem cell, *SVF* stromal vascular fraction

per 100 ml of lipoaspirate (median: 30.0×10^6; standard deviation (SD): $\pm 21.0 \times 10^6$; range: 4.7×10^6–120×10^6; $n = 131$). Upon FACS analysis, an initial FSC vs SSC gate was set to delimit PSC that occupy a characteristic region of the cytogram in terms of size and internal complexity amongst the diverse mixture of cells that comprise the SVF. Significantly, this demonstrated that

the majority of the SVF in many samples is comprised of dead and dying cells and cellular debris. Gating to select single cells is necessary to prevent the collection of non-PSC due to incomplete collagenase digestion and/or cell aggregation post isolation and during the processing and staining of the cells prior to sorting. Since collagenase dissociated cells approximate a sphere, the height-to-width ratio should maintain a constant value—or said differently, the height and width of the cells should scale proportionally to the area. Single cell selection by FACS is possible using the gating parameters shown in Fig. 1. Viable cells possess an intact plasma membrane and exclude DAPI, allowing cell viability to be assessed (mean cell viability: 83 %; median: 84 %; SD: ±12 %; range: 36–99 %; $n = 131$). CD45$^+$ haematopoietic cells and CD31$^+$ endothelial cells averaged 34 % (median: 33 %; SD: ±16 %; range: 1–73 %; $n = 113$) and 4 % (median: 3 %; SD: ±4 %; range: 0.1–16 %; $n = 17$), respectively, of total live cells in the SVF. The mean proportion of PSCs was 41 % (median 42 %; SD: ±16 %; range: 6–78 %; $n = 124$) with a mean proportion of pericytes of 8 % (median 5 %; SD: 8 %; range: 0–55 %; $n = 131$), and adventitial cells of 33 % (median 34 %; SD: ±16 %;

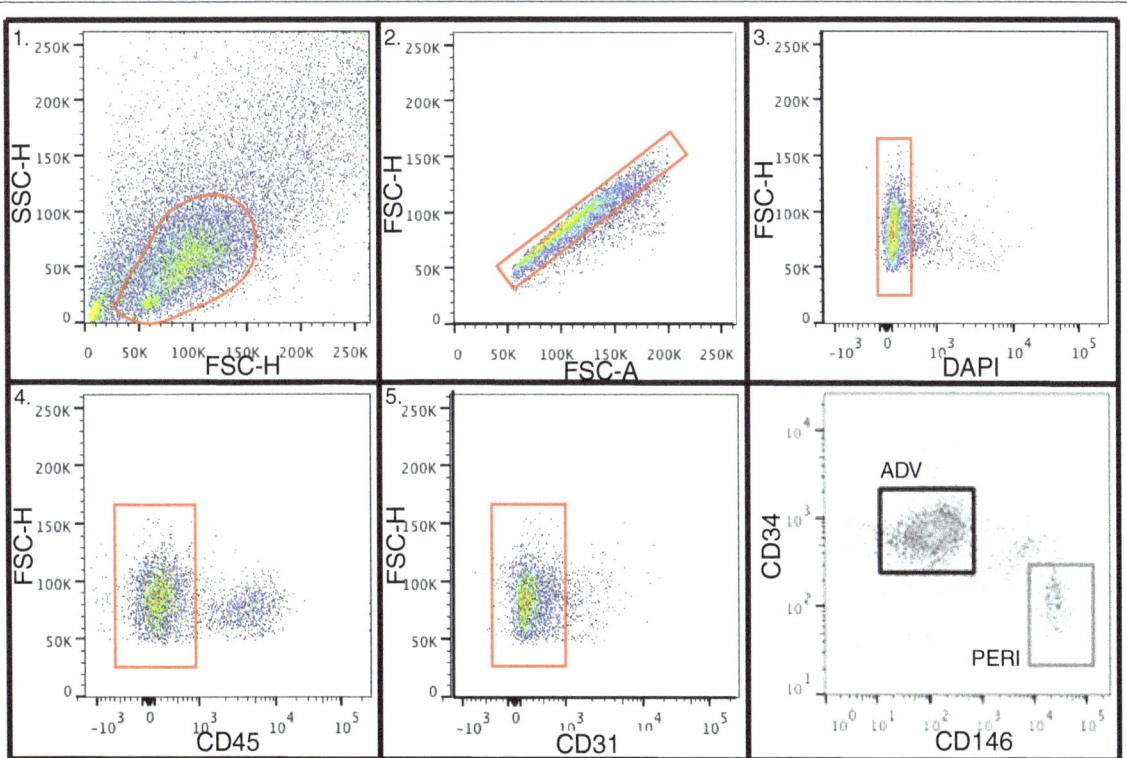

Fig. 1 Gating strategies for the isolation of PSC from the SVF. Gate 1 = cells, Gate 2 = single cells, Gate 3 = live cells, Gate 4 = CD45$^-$ (nonhaematopoietic cells), Gate 5 = CD31$^-$ (nonendothelial cells), Gate ADV = adventitial cells (CD31$^-$, CD45$^-$, CD34$^+$, CD146$^-$), and Gate PERI = pericytes (CD31$^-$, CD45$^-$, CD34$^+$, CD146$^-$). *DAPI* 4′,6-diamidino-2-phenylindole, *FSC* forward scatter, *SSC* side scatter

range: 3–72 %; $n = 131$). This represents a mean yield of 11.6×10^6 PSCs (median 10.0×10^6; SD: 8.6×10^6; range: 1.1×10^6–47.2×10^6; $n = 124$) per 100 ml of lipoaspirate.

Effects of demographics on cell yield

The mean age of donors was 41 years (range: 22–64, $n = 124$ in Table 1). There were no differences observed in either the total number of viable SVF cells or the proportion of PSC as a percentage of live cells with age, with linear correlation coefficients (R) of 0.07 and 0.09, respectively (Fig. 2a, b). No statistical difference was observed in the yield of SVF cells ($p = 0.34$) or viable PSC ($p = 0.79$) between genders (Fig. 2c, 2d, respectively). Although the

proportion of PSC (as a percentage of live cells) was significantly higher in male vs. female donors, with mean PSC proportions of 47 % vs 40 %, respectively ($p = 0.05$, one-tailed t test), the average SVF yield (male: 30×10^6 cells vs female: 35×10^6 cells) and cell viabilities (male: 82 % vs female: 83 %) were correspondingly lower, although not significantly, in males, resulting in a zero sum scenario (data not shown). Body mass index (BMI) had no significant effect on either the total yield of SVF cells or the proportion of PSC as a percentage of live cells ($R = 0.05$ and 0.01, respectively; Fig. 2e, f). Multivariate analysis was performed using multiple linear least-squares regression for all demographic data and cold storage times, resulting in F ratio = 2.99, $p = 0.035$. The only factor

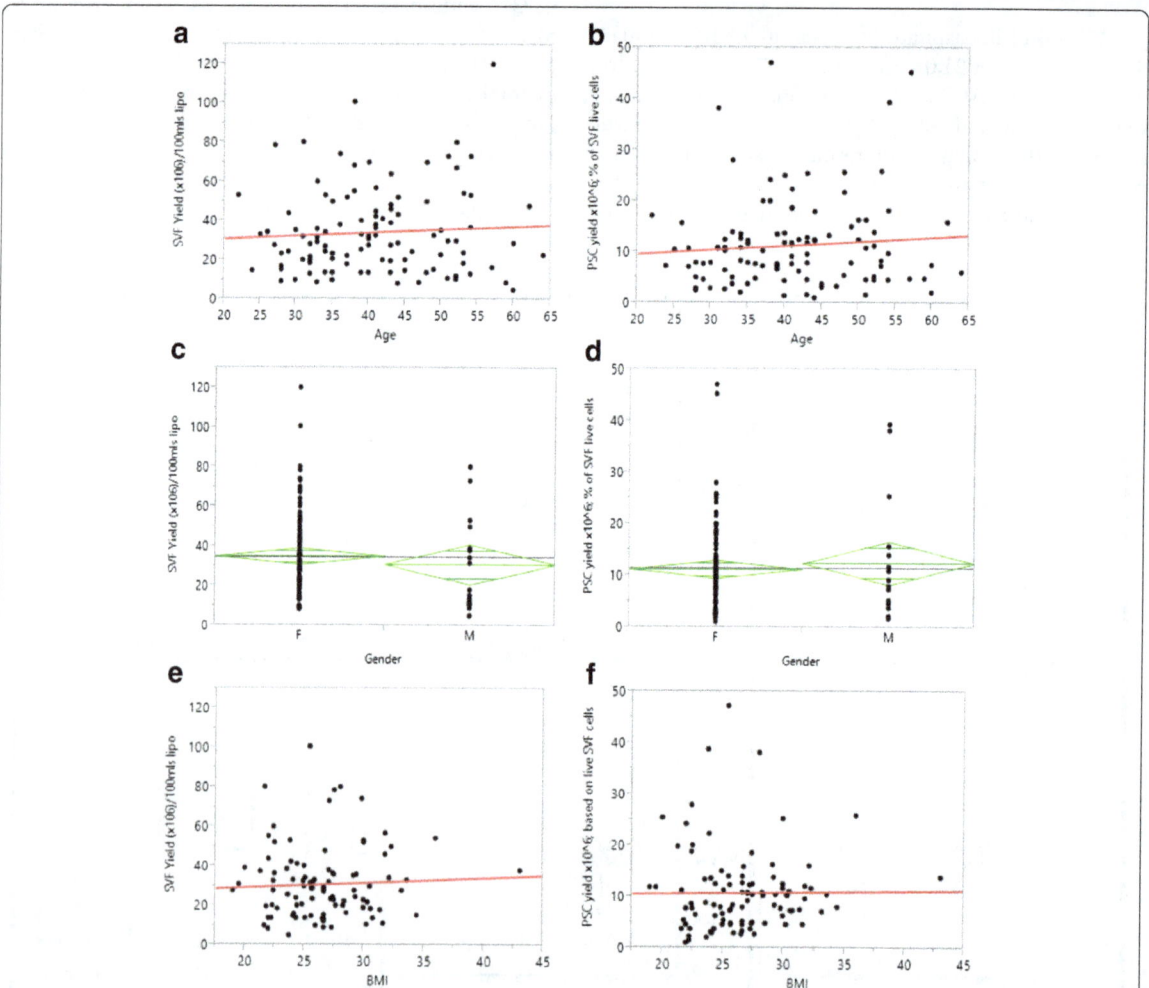

Fig. 2 a, b. Linear fit of SVF yield and PSC yield ($\times 10^6$) per 100 ml of lipoaspirate with respect to donor age ($n = 124$, $R = 0.07$ and 0.09 respectively). **c, d.** One-way analysis of variance of SVF and PSC yields ($\times 10^6$) vs donor gender ($n = 131$; male = 19, female = 112): *green diamonds* reflect the mean yield (center line) and 95 % confidence interval (vertical span) for each gender, and the grand mean in *gray*. Linear fit of SVF yield **e** and proportion of PSCs comprising the SVF **f** with respect to donor BMI ($n = 97$, $R = 0.05$ and 0.01 respectively). *BMI* body mass index, *F* female, *lipo* lipoaspirate, *M* male, *PSC* perivascular stem cell, *SVF* stromal vascular fraction (Color figure online)

within these parameters that demonstrated any significant correlation to cell yield was the cold storage time (t ratio = 2.99, $p = 0.004$), which is discussed further below.

Effect of cold storage time on cell yield

After surgical removal, adipose tissue was stored at 4 °C until analyzed. The majority of samples were processed within 24 hours following surgery; however, some samples were stored for up to 7 days. When split into discreet time points and analyzed using the Tukey–Kramer (honest significant difference) test, we observed a general increase in the proportion of PSC recovered from the SVF with increasing time (not significant) (Fig. 3a); however, the absolute numbers remained consistent and the relative rise was in fact due to a reduction in the proportion of CD45$^+$ haematopoietic cells (Fig. 3b).

Statistical process control over the prospective isolation of PSCs

Since 2013, both groups at UCLA and Edinburgh have been using the same protocol based on the improvements and developments made during the process of optimization. To assess the extent to which our process was consistent, reproducible, and under statistical control we used Levey-Jennings charts that show control limits 3 SDs above (upper control limit) and below (lower control limit) the process mean. When PSC yields (per 100 ml of lipoaspirate) obtained using the optimized protocol were compared with the non optimized earlier protocols, it was apparent that our optimized protocol had resulted in improvements in the reproducibility and purity of PSC isolation (Figs 4 and 5). Levey-Jennings charts depicting individual data points (Fig. 4) showed that mean PSC yields

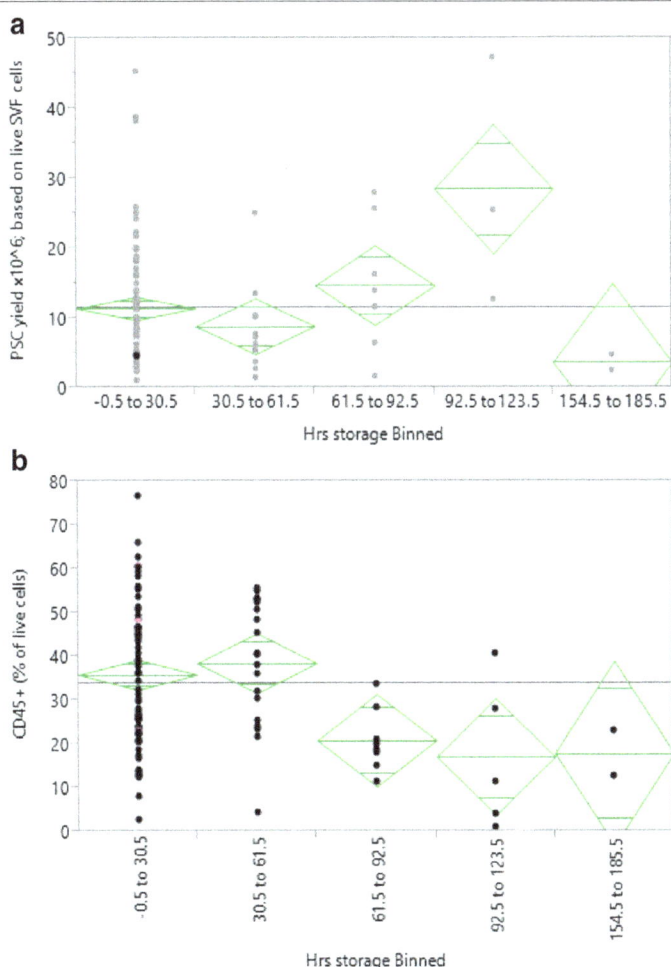

Fig. 3 One-way analysis of variance analysis of the number and proportion of cells recovered with respect to storage time at 4 °C. **a** PSC yield (× 10^6) per 100 ml of lipoaspirate showing a gradual increase with longer storage times; however, this is a relative effect due to the deceasing number of CD45$^+$ hematopoietic cells seen in **b**. *Green diamonds* indicate the mean and 95 % confidence interval for each storage time interval, while the *gray line* represents the grand or overall mean. *Hrs* hours, *PSC* perivascular stem cell, *SVF* stromal vascular fraction (Color figure online)

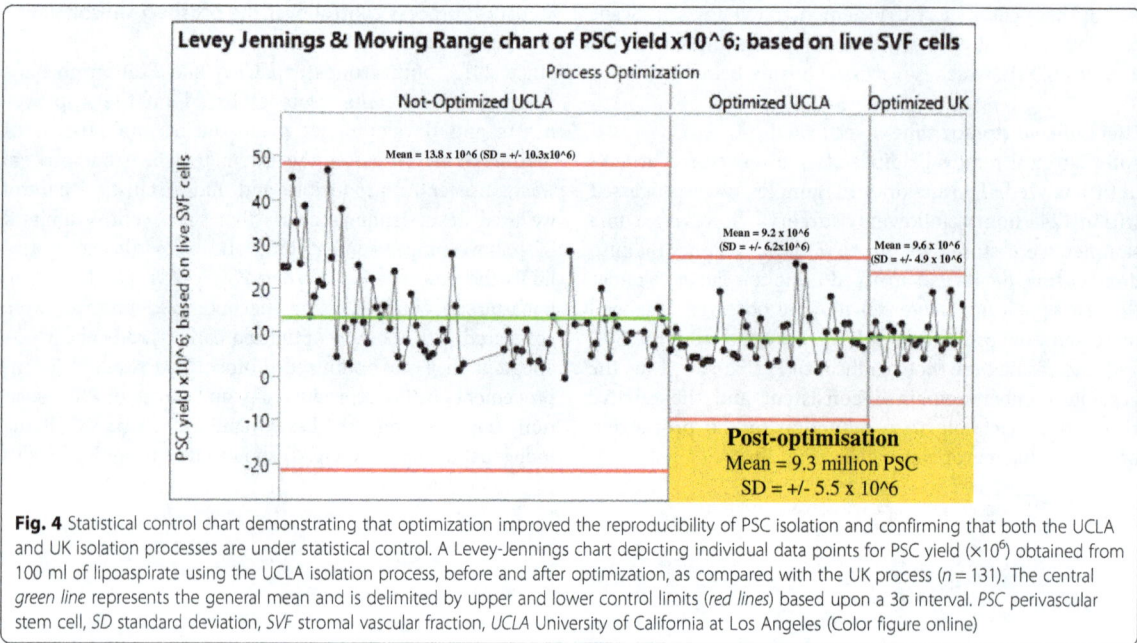

Fig. 4 Statistical control chart demonstrating that optimization improved the reproducibility of PSC isolation and confirming that both the UCLA and UK isolation processes are under statistical control. A Levey-Jennings chart depicting individual data points for PSC yield ($\times 10^6$) obtained from 100 ml of lipoaspirate using the UCLA isolation process, before and after optimization, as compared with the UK process ($n = 131$). The central *green line* represents the general mean and is delimited by upper and lower control limits (*red lines*) based upon a 3σ interval. *PSC* perivascular stem cell, *SD* standard deviation, *SVF* stromal vascular fraction, *UCLA* University of California at Los Angeles (Color figure online)

were reduced following process optimization (from 13.8 to 9.3 million PSC per 100 ml of lipoaspirate) but resulted in quite similar PSC yields between the UCLA and Edinburgh groups (9.2 and 9.6 million PSC)—it should be noted that the viability of cells post optimization was higher. Furthermore, upper and lower control limits were greatly improved following process optimization, resulting in a 53 % reduction in sample SD: 5.5×10^6 vs 10.3×10^6 PSC for the optimized and nonoptimized process, respectively. The statistical control charts thus demonstrate that our current protocol has improved the reproducibility of PSC isolation and confirm that both the UCLA and Edinburgh isolation processes are under statistical control. At the same time, contamination of sorted PSC by CD45+ haematopoietic cells and CD31+ endothelial cells has decreased, yielding greater than 99.5 % depletion of these cells based upon the detection of protein tyrosine phosphatase, receptor type C (CD45) and platelet endothelial cell adhesion molecule PECAM1 (CD31) transcripts by real time quantitative PCR (Fig. 5).

Discussion

Regenerative medicine is defined as the process of replacing, engineering, or regenerating cells, tissues, or organs to restore or establish normal function. Paramount to the clinical translation of basic research into adult stem cell therapies is the identification of cell sources and methods that minimize any potential risks to the patient. These risks are mitigated when pure, well-defined populations of cells are used that have not undergone the extensive ex vivo manipulation that culturing entails with its

associated risks due to infection, immunogenicity, and genetic instability. Additionally, the future clinical adoption of MSC for regenerative therapies requires a precise delineation of cellular identity, purity, and potency as specified by the Food and Drug Administration in the April 2008 Content and Review of Chemistry, Manufacturing, and Control Information for Human Somatic Cell Therapy Investigational New Drug Applications [20], and as reinforced in the November 2013 Guidance for Industry: Preclinical Assessment of Investigational Cellular and Gene Therapy Products [21]. Clearly, progress toward the clinical use of MSC awaits a standardized manufacturing process that will consistently produce sufficient numbers of MSC of defined identity, purity, and potency safely and from easily dispensable human tissues uncompromised by conditions that might alter their sterility, stability, immunogenicity, and genomic integrity.

Adipose tissue is an abundant source of perivascular associated regenerative cells that can be harvested in large quantities even from individuals of normal BMI with minimal morbidity. These PSC consist of a bipartite population of adventitial cells and pericytes which together embody the entire regenerative potential of adipose tissue and are well defined in their anatomical origin, immunophenotype, and contribution to fat tissue composition [7, 10–12, 14, 19, 22, 23]. Notwithstanding their tissue origin, purified PSC are indistinguishable from conventional MSCs in their multidifferentiation potential [7, 10–12, 14, 19, 24–29], and in vitro and in small animal studies have been shown to function in hematopoietic stem cell support [30], fibrosis [31–33], and bone [14, 19], muscle [34], pulmonary [35],

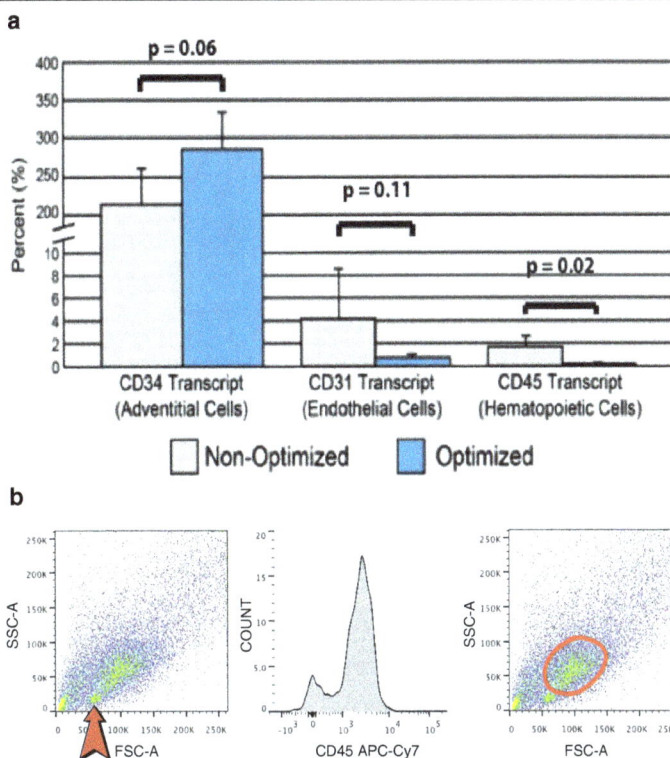

Fig. 5 Improvements in PSC purity. **a** Optimization of the PSC isolation process has led to increased purity of cells, as indicated by enrichment for the adventitial cell antigen CD34 (*left*), and the dramatic reduction in the endothelial cell antigen CD31 (*middle*) and haematopoietic marker CD45 (*right*) by real-time quantitative PCR (*n* = 7; pre optimization = 4, post optimization = 3). **b** (*Left*) FSC vs SSC demonstrating the population of lymphocytes (*arrow*). (*Centre*) Confirmation of lymphocytes by demonstration of CD45$^+$ phenotype of the subpopulation. (*Right*) Selection of the CD45$^-$ depleted cellular fraction for subsequent analysis. *FSC* Forward scatter, *PSC* perivascular stem cell, *SSC* side scatter

peripheral nerve [36], white adipocyte [24], follicular dendritic cell [37], and cardiovascular [38–42] regeneration. In some of these models, the function of PSCs was equivalent or superior to that of conventional MSCs or other progenitor populations (reviewed in [12]). Furthermore, unlike conventional MSC, PSC do not require culture for isolation and or purification, and have been convincingly shown to induce greater bone formation than their unpurified counterpart cell populations in the SVF [19, 43]. In this present study, we have optimized a process to rapidly and consistently purify healthy and viable PSC from moderate amounts of lipoaspirate to high homogeneity by FACS [10, 12, 19, 44] in quantities theoretically sufficient to address many clinical needs.

Currently, PSC isolation relies on sophisticated FACS flow cytometers that are uncommon in clinical settings and costly to acquire, operate, and maintain. Although clinical-grade cell sorters are available [45], the widespread adoption of PSC and other sorted cell therapies will require the codevelopment of stem cell treatment centers like the recently funded California Alpha-stem Cell Clinics, which are technically staffed and equipped with such instruments [46].

This analysis of 131 samples represents the largest and most comprehensive analysis of adipose tissue stem cells to date. We demonstrate a theoretical yield of 11.6 million PSC per 100 ml of lipoaspirate; however, an average yield of 9.3 million PSC per 100 ml is a more realistic and consistent estimate of PSC based on process optimization and data compiled from two different laboratories at UCLA (9.2 million PSC from *n* = 109 samples) and Edinburgh (9.6 million PSC from *n* = 22 samples). Given that cosmetic liposuction procedures often exceed volumes of 1 l, more than sufficient cells could be prospectively isolated to satisfy a range of regenerative applications (Table 2) based on estimates obtained from clinical trials data (www.clinicaltrials.gov) and our own group. Nevertheless, *ex vivo* culture would still be necessary for systemic conditions and where multiple doses are required, such as graft versus host disease.

The protocol presented in this work is based on the one originally developed to purify PSC from a variety of fetal and adult tissues [7], which was subsequently modified specifically for adipose tissue [19]. Through examining the health, quality, and purity of these PSC

Table 2 Potential clinical applications and the amount of adipose tissue required to provide sufficient numbers of prospectively purified perivascular stem cells to eliminate the need for ex-vivo expansion

Clinical use	Estimated number of cells required	Amount of fat needed
Tissue engineered tendon	1.5 million/cm	12 ml/cm
Tissue engineered mandible	10 million/cm	80 ml/cm
Cartilage for nasal reconstruction	20 million	160 ml
Scaphoid nonunion	25 million	200 ml
Tibial nonunion[a]	40 million	320 ml
Total ear reconstruction	50 million	400 ml
Critical limb ischemia[a]	2 million cells/kg	16 ml/kg
Graft versus host disease[a]	0.5–13 million cells/kg	4.8–125 ml/kg
		(repeat doses required)

[a]Data from clinicaltrials.gov

populations immediately after FACS, we noted that large numbers of PSCs were of poor quality and in the process of dying. We therefore sought to refine and optimize our protocol with emphasis on consistently delivering maximum numbers of pure and viable PSC.

There are a number of modifications that have been made to this current protocol when compared with that previously published by James et al. [19]. The most significant change made was to reduce the enzymatic digestion time down from 70 minutes to 45 minutes On close examination of the cells following digestion, we noted that although the length of digestion did not seem to have a major effect on total SVF yield (45 minutes = 34.4×10^6 vs 70 minutes = 39.4×10^6 nucleated cells/100 ml of lipoaspirate), the health of those cells recovered was adversely affected by longer digestion times—such that significant proportions of cells recovered immediately following FACS were dead or apoptotic. Two further filtration steps (100 and 40 μm) were introduced immediately following enzymatic digestion to remove debris that were more apparent with the shorter digestion time. When comparing the number and type of cells recovered between the current protocol and that published previously [19] it appears that there has been a large reduction in the percentage of pericytes retrieved (8 % vs 19.5 %), and an increase in the adventitial cells (33 % vs 23.8 %). It is our experience that this change is more a reflection on the number of healthy cells that can be recovered, and although the total proportion of PSC remains similar (41 % vs 43.2 %), the current protocol delivers healthier cells that are therefore more suitable for immediate use. We have noted that the pericyte population is especially prone to damage due to the longer digestion. Further changes were made through the addition of an endothelial cell specific antibody (CD31) to eliminate these cells as part of the FACS purification process, resulting in greater purity of the samples (Fig. 5a). This protocol relies on the selection of pericytes being CD146+ and adventitial cells being CD34+ but

both of these groups were probably contaminated with endothelial cells because subsets express both CD34 and CD146. These cells were not present in subsequent analysis of in-vitro populations due to the unfavorable culture conditions for endothelial cells. We have also noted reductions in other contaminating populations such as the CD45+ hematopoietic cells (Fig. 5). Both protocols utilize a CD45 antibody to deplete these cells; however, we believe that as our experience of this process has improved we have also been able to set tighter FACS gates to eliminate these populations with more accuracy (Fig. 5b). In addition to the importance of being able to deliver a pure and defined clinical product, it has been shown that contamination of MSC by endothelial cells also inhibits their function [47, 48].

The ultimate goal is to translate our work to novel therapies that address a wide range of clinical needs. The protocol described here utilizes research-grade reagents and therefore consideration will need to be given to translating the work for clinical use. Wherever possible we have sought to utilize reagents that can also be bought at clinical grade; however, there are some reagents where alternatives may need to be sought, such as substituting human serum albumin for BSA. Further optimization will probably be required for this process, but the current work will act as a benchmark for subsequent studies.

One part of the process over which we had no control was the liposuction methods used by the surgeons recovering the adipose tissue. Whilst we have not specifically investigated the effect of different liposuction procedures on PSC yield, it is likely that variations in the type of liposuction cannula and hence the size of fat within the lipoaspirate will influence the digestion process and recovery of PSC. In all of the samples we processed, the adipose tissue was the by-product of a cosmetic procedure and so the type of liposuction performed was at the discretion of the surgeon. In future, if liposuction is to be performed with the primary intention of recovering cells for clinical use we would suggest that the liposuction procedure should also be examined, optimized, and standardized.

There are a number of limitations of this current study that need to be addressed. Whilst this study represents the single largest analysis of SVF and stem/progenitor cell content, the demographics of this cohort reflect the unique type of patient undergoing cosmetic plastic surgery, and might not reflect the full demographics of people requiring stem cell therapies. Generally patients undergoing cosmetic surgery are young to middle-age women and free from any significant comorbidities. Whilst the results suggest that age is not a factor in PSC yield, the oldest patient in our study was aged 64. Further studies are therefore required on patients at later ages. Furthermore, our study was limited to looking only at the number and viability of cells and did not examine their function or potency for particular regenerative purposes; however, the developmental potential of fat-derived PSC has been addressed in many other publications [14, 30, 49] and shown not to be significantly affected by gender, age, and BMI [19]. Although we have qualitative evidence to demonstrate that pericytes and adventitial cells have a similar function and potency [7, 10], detailed assays examining this have not been performed. This is an important issue that needs to be addressed because there is evidence from our group and others that specific subsets of cells within the MSC/PSC family may have specific and unique functions [30, 50].

By selecting a purified population of stem/progenitor cells, we are more likely to increase the efficacy of these cells by eliminating contaminating cells; however, the effects of patient lifestyle, genetic background, and other variables on the function of the resulting populations should be examined because both age and disease have been implicated in reduced function of conventional MSC [51]. This is particularly relevant if an allogeneic source of MSCs is to be proposed and defended as a viable alternative to autologous cells.

Conclusions

Adipose tissue contains two anatomically and phenotypically discrete populations of MSC precursors—adventitial cells and pericytes—together referred to as PSC. More than 9 million PSCs per 100 ml of lipoaspirate can be rapidly purified to homogeneity using flow cytometry in clinically relevant numbers, potentially circumventing the need for purification and expansion by culture prior to clinical use. Modern high-speed flow cytometers can theoretically process cells at rates well in excess of 20,000 cells/second, thus meaning tissue could be processed and purified PSC returned and administered for therapeutic use within a single operative procedure. The number and viability of these cells are minimally affected by factors such as age, sex, BMI, and storage time in this cohort; however, further studies are required to examine the effects of age and pathology on the number and efficacy of these cells.

Abbreviations

BMI: Body mass index; BSA: Bovine serum albumin; CD: Cluster of differentiation; DAPI: 4′,6-Diamidino-2-phenylindole; FACS: Fluorescence-activated cell sorting; FSC: Forward scatter; MSC: Mesenchymal stem cell; PSC: Perivascular stem cell; SD: Standard deviation; SSC: Side scatter; SVF: Stromal vascular fraction; UCLA: University of California at Los Angeles.

Competing interests

KT, BP, and CS are inventors of PSC-related patents filed from UCLA. KT and CS are founders of Scarless Laboratories Inc (Beverley Hills, CA, USA). which sublicenses PSC-related patents from University of California Regents, and who also hold equity in the company. CS is also an officer of Scarless Laboratories, Inc.

Authors' contributions

CCW, WRH, IRM, AWJ, MC, SP, CB, SEL, KS, PL, and VL performed the experimental work and collected the data. KLM, KT, CS, DCH, and BP designed the study and supervised the project. CCW, WRH, IRM, and BP analyzed the data and wrote the manuscript. All authors reviewed, revised, and approved the manuscript prior to submission.

Acknowledgements

The authors would like to acknowledge the support they have received from the British Heart Foundation (Research Grant # R42775), the Medical Research Council, and the California Institute for Regenerative Medicine. CCW was supported by a Chief Scientist Office Clinical Academic Training Fellowship, a research grant from the William Rooney Plastic Surgery and Reconstructive Surgery Trust, and a Research Fellowship from the Royal College of Surgeons of England. KLM and WRH received support from a VA Merit Review grant, VA Center for Regenerative Medicine, NIH CCTRN grant UM1 5UM1HL113457-04, and NIH T32 grant 5T32HL079995-10. IRM was supported by a Wellcome Trust funded Edinburgh Clinical Academic Track (ECAT) Lectureship (ref. 097483).

Author details

[1]British Heart Foundation Centre for Vascular Regeneration & Medical Research Council Centre for Regenerative Medicine, University of Edinburgh, Edinburgh, UK. [2]Department of Plastic and Reconstructive Surgery, St Johns Hospital, Howden Road West, Livingston, UK. [3]Orthopaedic Hospital Department of Orthopaedic Surgery and the Orthopaedic Hospital Research Center, University of California, Los Angeles, CA, USA. [4]BD Biosciences, San Diego, CA, USA. [5]Bone and Joint Research Group, Institute of Developmental Sciences, University of Southampton, Southampton, UK. [6]Department of Surgery, School of Veterinary Medicine and Animal Science, University of São Paulo, São Paulo, Brazil. [7]Department of Emergency Medicine, Kaweah Delta Health Care District, Visalia, CA, USA. [8]Department of Urology, David Geffen School of Medicine, University of California at Los Angeles, Los Angeles, CA, USA. [9]Department of Pathology and Laboratory Medicine, David Geffen School of Medicine, University of California at Los Angeles, Los Angeles, CA, USA. [10]Indiana Center for Vascular Biology and Medicine, Krannert Institute of Cardiology, and Vascular and Cardiac Center for Adult Stem Cell Research, Indiana University, Bloomington, IN, USA. [11]Division of Growth and Development and Section of Orthodontics, School of Dentistry, University of California, Los Angeles, CA 90095, USA. [12]Division of Plastic and Reconstructive Surgery, Department of Surgery and the Orthopaedic Hospital Research Center, University of California, Los Angeles, CA 90095, USA. [13]Department of Orthopaedic Surgery and the Orthopaedic Hospital Research Center, University of California, Los Angeles, CA 90095, USA.

References

1. Friedenstein AJ, Chailakhjan RK, Lalykina KS. The development of fibroblast colonies in monolayer cultures of guinea-pig bone marrow and spleen cells. Cell Tissue Kinet. 1970;3(4):393–403.

2. Friedenstein AJ, Piatetzky-Shapiro II, Petrakova KV. Osteogenesis in transplants of bone marrow cells. J Embryol Exp Morphol. 1966;16(3):381–90.

3. Friedenstein AJ, Chailakhyan RK, Latsinik NV, Panasyuk AF, Keiliss-Borok IV. Stromal cells responsible for transferring the microenvironment of the hemopoietic tissues. Cloning in vitro and retransplantation in vivo. Transplantation. 1974;17(4):331–40.

4. Caplan AI. Mesenchymal stem cells. J Orthop Res Wiley Online Library. 1991; 9(5):641–50.

5. Caplan AI, Dennis JE. Mesenchymal stem cells as trophic mediators. J Cell Biochem. 2006;98(5):1076–84.

6. da Silva Meirelles L. Mesenchymal stem cells reside in virtually all post-natal organs and tissues. J Cell Sci. 2006;119(11):2204–13.

7. Crisan M, Yap S, Casteilla L, Chen C-W, Corselli M, Park TS, et al. A perivascular origin for mesenchymal stem cells in multiple human organs. Cell Stem Cell. 2008;3(3):301–13.

8. Covas DT, Panepucci RA, Fontes AM, Silva Jr WA, Orellana MD, Freitas MCC, et al. Multipotent mesenchymal stromal cells obtained from diverse human tissues share functional properties and gene-expression profile with CD146+ perivascular cells and fibroblasts. Exp Hematol. 2008;36(5):642–54.

9. Zannettino ACW, Paton S, Arthur A, Khor F, Itescu S, Gimble JM, et al. Multipotential human adipose-derived stromal stem cells exhibit a perivascular phenotype in vitro and in vivo. J Cell Physiol. 2007;214(2):413–21.

10. Corselli M, Chen C-W, Sun B, Yap S, Rubin JP, Péault B. The tunica adventitia of human arteries and veins as a source of mesenchymal stem cells. Stem Cells Dev. 2012;21(8):1299–308.

11. Traktuev DO, Merfeld-Clauss S, Li J, Kolonin M, Arap W, Pasqualini R, et al. A population of multipotent CD34-positive adipose stromal cells share pericyte and mesenchymal surface markers, reside in a periendothelial location, and stabilize endothelial networks. Circ Res. 2008;102(1):77–85.

12. Corselli M, Crisan M, Murray IR, West CC, Scholes J, Codrea F, et al. Identification of perivascular mesenchymal stromal/stem cells by flow cytometry. Cytometry A. 2013;83(8):714–20.

13. Gad SC. Pharmaceutical manufacturing handbook. Hoboken, NJ, USA: John Wiley & Sons; 2008. p. 1

14. James AW, Zara JN, Zhang X, Askarinam A, Goyal R, Chiang M, et al. Perivascular stem cells: a prospectively purified mesenchymal stem cell population for bone tissue engineering. Stem Cells Transl Med. 2012;1(6): 510–9.

15. Zhao Y, Waldman SD, Flynn LE. The effect of serial passaging on the proliferation and differentiation of bovine adipose-derived stem cells. Cells Tissues Organs. 2012;195(5):414–27.

16. Wall ME, Bernacki SH, Loboa EG. Effects of serial passaging on the adipogenic and osteogenic differentiation potential of adipose-derived human mesenchymal stem cells. Tissue Eng. 2007;13(6):1291–8.

17. Ma T. Mesenchymal stem cells: from bench to bedside. WJSC. 2010;2(2):13.

18. Mahalatchimy AA, Rial-Sebbag EE, Tournay VV, Faulkner AA. The legal landscape for advanced therapies: material and institutional implementation of European Union rules in France and the United Kingdom. J Law Soc. 2012;39(1):131–49.

19. James AW, Zara JN, Corselli M, Askarinam A, Zhou AM, Hourfar A, et al. An abundant perivascular source of stem cells for bone tissue engineering. Stem Cells Transl Med. 2012;1(9):673–84.

20. Center for Biologics Evaluation and Research. Xenotransplantation Guidances—guidance for FDA reviewers and sponsors: content and review of chemistry, manufacturing, and control (CMC) information for human somatic cell therapy investigational new drug applications (INDs). fdagov. Center for Biologics Evaluation and Research; 2008.

21. Center for Biologics Evaluation and Research. Cellular & Gene Therapy Guidances—guidance for industry: preclinical assessment of investigational cellular and gene therapy products. fdagov. Center for Biologics Evaluation and Research; 2013.

22. Zimmerlin L, Donnenberg VS, Pfeifer ME, Meyer EM, Péault B, Rubin JP, et al. Stromal vascular progenitors in adult human adipose tissue. Cytometry A. 2010;77(1):22–30.

23. Zimmerlin L, Donnenberg VS, Rubin JP, Donnenberg AD. Mesenchymal markers on human adipose stem/progenitor cells. Donnenberg VS, Ulrich H, Tárnok A, editors. Cytometry. 2012;83A(1):134–40.

24. Tang W, Zeve D, Suh JM, Bosnakovski D, Kyba M, Hammer RE, et al. White fat progenitor cells reside in the adipose vasculature. Science. 2008; 322(5901):583–6.

25. Feng J, Mantesso A, De Bari C, Nishiyama A, Sharpe PT. Dual origin of mesenchymal stem cells contributing to organ growth and repair. Proc Natl Acad Sci U S A. 2011;108(16):6503–8.

26. Dellavalle A, Maroli G, Covarello D, Azzoni E, Innocenzi A, Perani L, et al. Pericytes resident in postnatal skeletal muscle differentiate into muscle fibres and generate satellite cells. Nat Commun. 2011;2:499.

27. Bouacida A, Rosset P, Trichet V, Guilloton F, Espagnolle N, Cordonier T, et al. Pericyte-like progenitors show high immaturity and engraftment potential as compared with mesenchymal stem cells. Camussi G, editor. PLoS One. 2012;7(11):e48648.

28. Chen C-W, Montelatici E, Crisan M, Corselli M, Huard J, Lazzari L, et al. Perivascular multi-lineage progenitor cells in human organs: regenerative units, cytokine sources or both? Cytokine Growth Factor Rev. 2009; 20(5–6):429–34.

29. Crisan M, Chen C-W, Corselli M, Andriolo G, Lazzari L, Péault B. Perivascular multipotent progenitor cells in human organs. Ann N Y Acad Sci. 2009;1176: 118–23.

30. Corselli M, Chin CJ, Parekh C, Sahaghian A, Wang W, Ge S, et al. Perivascular support of human hematopoietic stem/progenitor cells. Blood. 2013;121(15): 2891–901.

31. Dulauroy S, Di Carlo SE, Langa F, Eberl G, Peduto L. Lineage tracing and genetic ablation of ADAM12+ perivascular cells identify a major source of profibrotic cells during acute tissue injury. Nat Med. 2012;18(8):1262–70.

32. Henderson NC, Arnold TD, Katamura Y, Giacomini MM, Rodriguez JD, McCarty JH, et al. Targeting αv integrin identifies a core molecular pathway that regulates fibrosis in several organs. Nat Med. 2013;19(12): 1617–24.

33. Goritz C, Dias DO, Tomilin N, Barbacid M, Shupliakov O, Frisen J. A pericyte origin of spinal cord scar tissue. Science. 2011;333(6039):238–42.

34. Park TS, Gavina M, Chen C-W, Sun B, Teng P-N, Huard J, et al. Placental perivascular cells for human muscle regeneration. Stem Cells Dev. 2011; 20(3):451–63.

35. Pierro M, Ionescu L, Montemurro T, Vadivel A, Weissmann G, Oudit G, et al. Short-term, long-term and paracrine effect of human umbilical cord-derived stem cells in lung injury prevention and repair in experimental bronchopulmonary dysplasia. Thorax. 2013;68(5):475–84.

36. Lavasani M, Thompson SD, Pollett JB, Usas A, Lu A, Stolz DB, et al. Human muscle-derived stem/progenitor cells promote functional murine peripheral nerve regeneration. J Clin Invest. 2014;124(4):1745–56.

37. Krautler NJ, Kana V, Kranich J, Tian Y, Perera D, Lemm D, et al. Follicular dendritic cells emerge from ubiquitous perivascular precursors. Cell. 2012; 150(1):194–206.

38. Campagnolo P, Cesselli D, Haj Zen Al A, Beltrami AP, Krankel N, Katare R, et al. Human adult vena saphena contains perivascular progenitor cells endowed with clonogenic and proangiogenic potential. Circulation. 2010;121(15): 1735–45.

39. Chen C-W, Okada M, Proto JD, Gao X, Sekiya N, Beckman SA, et al. Human pericytes for ischemic heart repair. Stem Cells. 2013;31(2):305–16.

40. Dar A, Domev H, Ben-Yosef O, Tzukerman M, Zeevi-Levin N, Novak A, et al. Multipotent vasculogenic pericytes from human pluripotent stem cells promote recovery of murine ischemic limb. Circulation. 2012;125(1):87–99.

41. He W, Nieponice A, Soletti L, Hong Y, Gharaibeh B, Crisan M, et al. Pericyte-based human tissue engineered vascular grafts. Biomaterials. 2010;31(32):8235–44.

42. Katare RG, Madeddu P. Pericytes from human veins for treatment of myocardial ischemia. Trends Cardiovasc Med. 2013;23:66–70.

43. James AW, Zara JN, Corselli M, Chiang M, Yuan W, Nguyen V, et al. Use of human perivascular stem cells for bone regeneration. J Vis Exp. 2012;63:e2952.

44. Askarinam A, James AW, Zara JN, Goyal R, Corselli M, Pan A, et al. Human perivascular stem cells show enhanced osteogenesis and vasculogenesis with Nel-like molecule i protein. Tissue Eng A. 2013;19(11–12):1386–97.

45. Fluorescence-activated cell sorting for CGMP processing of therapeutic cells. BioProcess International. 11 November 2014. http://www.bioprocessintl. com/manufacturing/cell-therapies/fluorescence-activated-cell-sorting-for-cgmp-processing-of-therapeutic-cells-297340/. Accessed 11 Mar 2016.

46. Stem Cell Agency unveils CIRM 2.0—its aggressive plan to significantly accelerate promising therapies. Invests $24 million to create Alpha Clinic Network. California's Stem Cell Agency. 11 November 2014. http://www.cirm.ca.gov/about-cirm/newsroom/press-releases/10232014. Accessed 11 Mar 2016.

47. Meury T, Verrier S, Alini M. Human endothelial cells inhibit BMSC differentiation into mature osteoblasts in vitro by interfering with osterix expression. J Cell Biochem. 2006;98(4):992–1006.

48. Rajashekhar G, Traktuev DO, Roell WC, Johnstone BH, Merfeld-Clauss S, Van Natta B, et al. IFATS collection: adipose stromal cell differentiation is reduced by endothelial cell contact and paracrine communication: role of canonical Wnt signaling. Stem Cells. 2008;26(10):2674–81.

49. Chung CG, James AW, Asatrian G, Chang L, Nguyen A, Le K, et al. Human perivascular stem cell-based bone graft substitute induces rat spinal fusion. Stem Cells Transl Med. 2014;3(10):1231–41.

50. Levi B, Wan DC, Glotzbach JP, Hyun J, Januszyk M, Montoro D, et al. CD105 Protein depletion enhances human adipose-derived stromal cell osteogenesis through reduction of transforming growth factor β1 (TGF-β1) signaling. J Biol Chem ASBMB. 2011;286(45):39497–509.

51. El-ftesi S, Chang EI, Longaker MT, Gurtner GC. Aging and diabetes impair the neovascular potential of adipose-derived stromal cells. Plast Reconstr Surg. 2009;123(2):475–85.

Permissions

All chapters in this book were first published in SCRT, by BioMed Central; hereby published with permission under the Creative Commons Attribution License or equivalent. Every chapter published in this book has been scrutinized by our experts. Their significance has been extensively debated. The topics covered herein carry significant findings which will fuel the growth of the discipline. They may even be implemented as practical applications or may be referred to as a beginning point for another development.

The contributors of this book come from diverse backgrounds, making this book a truly international effort. This book will bring forth new frontiers with its revolutionizing research information and detailed analysis of the nascent developments around the world.

We would like to thank all the contributing authors for lending their expertise to make the book truly unique. They have played a crucial role in the development of this book. Without their invaluable contributions this book wouldn't have been possible. They have made vital efforts to compile up to date information on the varied aspects of this subject to make this book a valuable addition to the collection of many professionals and students.

This book was conceptualized with the vision of imparting up-to-date information and advanced data in this field. To ensure the same, a matchless editorial board was set up. Every individual on the board went through rigorous rounds of assessment to prove their worth. After which they invested a large part of their time researching and compiling the most relevant data for our readers.

The editorial board has been involved in producing this book since its inception. They have spent rigorous hours researching and exploring the diverse topics which have resulted in the successful publishing of this book. They have passed on their knowledge of decades through this book. To expedite this challenging task, the publisher supported the team at every step. A small team of assistant editors was also appointed to further simplify the editing procedure and attain best results for the readers.

Apart from the editorial board, the designing team has also invested a significant amount of their time in understanding the subject and creating the most relevant covers. They scrutinized every image to scout for the most suitable representation of the subject and create an appropriate cover for the book.

The publishing team has been an ardent support to the editorial, designing and production team. Their endless efforts to recruit the best for this project, has resulted in the accomplishment of this book. They are a veteran in the field of academics and their pool of knowledge is as vast as their experience in printing. Their expertise and guidance has proved useful at every step. Their uncompromising quality standards have made this book an exceptional effort. Their encouragement from time to time has been an inspiration for everyone.

The publisher and the editorial board hope that this book will prove to be a valuable piece of knowledge for researchers, students, practitioners and scholars across the globe.

List of Contributors

Fa-Ming Chen, Guang-Ying Dong, Hong Lu, Bei-Min Tian, Xi-Yu Zhang, Qing Chu, Jie Xu, Rui-Xin Wu and Yuan Yin
State Key Laboratory of Military Stomatology, Department of Periodontology, School of Stomatology, Fourth Military Medical University, Xi'an, Shannxi, P. R. China

Li-Na Gao and Yang Yu
State Key Laboratory of Military Stomatology, Department of Periodontology, School of Stomatology, Fourth Military Medical University, Xi'an, Shannxi, P. R. China
State Key Laboratory of Military Stomatology, Research and Development Center for Tissue Engineering, School of Stomatology, Fourth Military Medical University, Xi'an, Shannxi, P. R. China

Yong-Jie Zhang and Yan Jin
State Key Laboratory of Military Stomatology, Research and Development Center for Tissue Engineering, School of Stomatology, Fourth Military Medical University, Xi'an, Shannxi, P. R. China

Songtao Shi
Department of Anatomy and Cell Biology, School of Dental Medicine, University of Pennsylvania, 240 South 40th Street, Philadelphia, PA 19104, USA

Xiangchun Liu, Haiying Liu, Lina Sun, Zhixin Chen, Huibin Nie, Aili Sun, Gang Liu and Guangju Guan
Department of Nephrology, The Second Hospital of Shandong University, Shandong University, Jinan, PR. China

Zhanhai Yin
Department of Orthopedics, First Affiliated Hospital, College of Medicine, Xi'an Jiaotong University, Xi'an 710061, P. R. China.

Qi Wang and Ye Li
Department of Periodontology, Stomatological Hospital, College of Medicine, Xi'an Jiaotong University, Xi'an 710004, P. R. China

Hong Wei and Jianfeng Shi
Research Center for Stomatology, Stomatological Hospital, College of Medicine, Xi'an Jiaotong University, Xi'an 710004, P. R. China

Ang Li
Department of Periodontology, Stomatological Hospital, College of Medicine, Xi'an Jiaotong University, Xi'an 710004, P. R. China
Research Center for Stomatology, Stomatological Hospital, College of Medicine, Xi'an Jiaotong University, Xi'an 710004, P. R. China

Martin Leahy
Tissue Optics & Microcirculation Imaging Group, School of Physics, National University of Ireland (NUI), Galway, Ireland
Chair of Applied Physics, National University of Ireland (NUI), Galway, Ireland

Kerry Thompson and Peter Dockery
Centre for Microscopy and Imaging, Anatomy, School of Medicine, National University of Ireland (NUI), Galway, Ireland

Haroon Zafar and Sergey Alexandrov
Tissue Optics & Microcirculation Imaging Group, School of Physics, National University of Ireland (NUI), Galway, Ireland

Mark Foley
Medical Physics Research Cluster, School of Physics, National University of Ireland (NUI), Galway, Ireland

Cathal O'Flatharta
Regenerative Medicine Institute (REMEDI), National University of Ireland (NUI), Galway, Ireland

You Zhang and Jie Hui
Department of Cardiology of the First Affiliated Hospital, Soochow University, Suzhou, China

Wei Lei, Weiya Yan, Zhenao Zhao, Zhenya Shen and Junjie Yang
Institute for Cardiovascular Science & Department of Cardiovascular Surgery of The First Affiliated Hospital, Soochow University, Suzhou, China

Xizhe Li
Department of Cardiovascular Surgery, Affiliated Shanghai 1st People's Hospital, Shanghai Jiaotong University, Shanghai, China

Xiaolin Wang
Department of Thoracic and Cardiovascular Surgery, Northern Jiangsu People's Hospital, Yangzhou, China

Shuai Gao
Institute of Molecular Medicine, Health Science Center, Shenzhen University, Shenzhen 518060, China National Institute of Biological Sciences, NIBS, Beijing 102206, China
Translational Medical Center for Stem Cell Therapy, Shanghai East Hospital, School of Medicine, Tongji University, Shanghai 200120, China

Li Tao and Jianhui Tian
Ministry of Agriculture Key Laboratory of Animal Genetics, Breeding and Reproduction; National Engineering Laboratory for Animal Breeding; College of Animal Sciences and Technology, China Agricultural University, Beijing 100193, China

Xinfeng Hou, Zijian Xu and Tao Cai
National Institute of Biological Sciences, NIBS, Beijing 102206, China

Wenqiang Liu, Shaorong Gao and Hong Wang
National Institute of Biological Sciences, NIBS, Beijing 102206, China
School of Life Sciences and Technology, Tongji University, Shanghai 200092, China

Kun Zhao and Mingyue Guo
School of Life Sciences and Technology, Tongji University, Shanghai 200092, China

Gang Chang
Institute of Molecular Medicine, Health Science Center, Shenzhen University, Shenzhen 518060, China National Institute of Biological Sciences, NIBS, Beijing 102206, China

Zhuoyue Chen, Jihong Cui, Hongmin Li and Fulin Chen
Laboratory of Tissue Engineering, Faculty of Life Science, Northwest University, 229 TaiBai North Road, Xi'an, Shaanxi Province 710069, P.R. China
Provincial Key Laboratory of Biotechnology of Shaanxi, Northwest University, 229 TaiBai North Road, Xi'an, Shaanxi Province 710069, P.R. China

Jing Wei, Jun Zhu and Wei Liu
Laboratory of Tissue Engineering, Faculty of Life Science, Northwest University, 229 TaiBai North Road, Xi'an, Shaanxi Province 710069, P.R. China

Soraia C. Abreu and Patricia R. M. Rocco
Laboratory of Pulmonary Investigation, Carlos Chagas Filho Institute of Biophysics, Federal University of Rio de Janeiro, Av. Carlos Chagas Filho, 373, Ilha do Fundão, Rio de Janeiro, RJ 21941-902, Brazil

Daniel J. Weiss
Department of Medicine, Vermont Lung Center, College of Medicine, University of Vermont, 89 Beaumont Ave Given, Burlington, VT 05405, USA

Eric Farrell
Department of Oral and Maxillofacial Surgery, Special Dental Care and Orthodontics, Erasmus MC, University Medical Centre, Room Ee1614, Erasmus MC, Wytemaweg 80, Rotterdam 3015CN, The Netherlands

Niamh Fahy
Regenerative Medicine Institute, National University of Ireland Galway, Galway, Ireland
Musculoskeletal Regeneration, AO Research Institute Davos (ARI), Davos, Switzerland

Aideen E Ryan
Regenerative Medicine Institute, National University of Ireland Galway, Galway, Ireland
College of Medicine, Nursing and Health Sciences, National University of Ireland Galway, Galway, Ireland
Discipline of Pharmacology and Therapeutics, National University of Ireland Galway, Galway, Ireland

Cathal O Flatharta
Regenerative Medicine Institute, National University of Ireland Galway, Galway, Ireland

Lisa O'Flynn
Regenerative Medicine Institute, National University of Ireland Galway, Galway, Ireland
College of Medicine, Nursing and Health Sciences, National University of Ireland Galway, Galway, Ireland
Orbsen Therapeutics Ltd, Galway, Ireland

Thomas Ritter
Regenerative Medicine Institute, National University of Ireland Galway, Galway, Ireland
College of Medicine, Nursing and Health Sciences, National University of Ireland Galway, Galway, Ireland

J Mary Murphy
Regenerative Medicine Institute, National University of Ireland Galway, Galway, Ireland

Aleksandra Maziarz, Beata Kocan and Agnieszka Banas
Laboratory of Stem Cells' Biology, Department of Immunology, Chair of Molecular Medicine, Faculty of Medicine, University of Rzeszow, ul. Kopisto 2a, 35-310 Rzeszow, Poland
Centre for Innovative Research in Medical and Natural Sciences, Faculty of Medicine, University of Rzeszow, ul. Warzywna 1a, 35-310 Rzeszow, Poland

Mariusz Bester, Sylwia Budzik and Marian Cholewa
Department of Biophysics, Faculty of Mathematics and Natural Sciences, University of Rzeszow, ul. Pigonia 1, 35-310 Rzeszow, Poland

Takahiro Ochiya
Division of Molecular and Cellular Medicine, National Cancer Center Research Institute, 5-1-1 Tsukiji, Chuo-ku 104-0045Tokyo, Japan

Chang Youn Lee
Department of Integrated Omics for Biomedical Sciences, Graduate School, Yonsei University, Seoul 03722, Republic of Korea

Jin Young Kang
Department of Rehabilitation Medicine, National Traffic Injury Rehabilitation Hospital, College of Medicine, The Catholic University of Korea, Yangpyeong-gun 12564, Republic of Korea.

Soyeon Lim
Institute for Bio-Medical Convergence, College of Medicine, Catholic Kwandong University, Gangneung-si 25601Gangwon-do, Republic of Korea

Onju Ham
Catholic Kwandong University International St. Mary's Hospital, Incheon Metropolitan City 22711, Republic of Korea

Woochul Chang
Department of Biology Education, College of Education, Pusan National University, Busan 46241, Republic of Korea

Dae-Hyun Jang
Department of Rehabilitation Medicine, Incheon St. Mary's Hospital, College of Medicine, The Catholic University of Korea, Dongsu-ro 56, Bupyeong-gu, Incheon 21431, Republic of Korea

Xiao Ke and Xiao-Rong Shu
Department of Cardiology, Sun Yat-sen Memorial Hospital of Sun Yat-sen University, No. 107, Yanjiangxi Road, Guangzhou, China
Guangdong Province Key Laboratory of Arrhythmia and Electrophysiology, Guangzhou 510120, China

Fang Wu
Department of Geriatric, The First Affiliated Hospital of Sun Yat-sen University, Guangzhou 510080, China

Qing-Song Hu, Bing-Qing Deng, Jing-Feng Wang and Ru-Qiong Nie
Department of Cardiology, Sun Yat-sen Memorial Hospital of Sun Yat-sen University, No. 107, Yanjiangxi Road, Guangzhou, China
Guangdong Province Key Laboratory of Arrhythmia and Electrophysiology, Guangzhou 510120, China

Yoolhee Yang, Hyunju Choi and Mira Seon
Department of Plastic Surgery, Samsung Medical Center, Sungkyunkwan University School of Medicine, Seoul, Korea

Daeho Cho
Department of Life Science, Sookmyung Women's University, Seoul, Korea

Sa Ik Bang
Department of Plastic Surgery, Samsung Medical Center, Sungkyunkwan University School of Medicine, Seoul, Korea
Bio-Med Translational Research Center, Samsung Medical Center, Seoul, Korea

Irina Arutyunyan, Evgeniya Kananykhina and Andrey Elchaninov
Research Center for Obstetrics, Gynecology and Perinatology of Ministry of Healthcare of the Russian Federation, 4 Oparina Street, Moscow 117997, Russia

Scientific Research Institute of Human Morphology, 3 Tsurupa Street, Moscow 117418, Russia

Timur Fatkhudinov
Research Center for Obstetrics, Gynecology and Perinatology of Ministry of Healthcare of the Russian Federation, 4 Oparina Street, Moscow 117997, Russia
Pirogov Russian National Research Medical University, Ministry of Healthcare of the Russian Federation, 1 Ostrovitianov Street, Moscow 117997, Russia
Laboratory of Regenerative Medicine, Research Center for Obstetrics, Gynecology and Perinatology, 4 Oparin Street, Moscow 117997, Russia

Natalia Usman and Andrey Makarov
Research Center for Obstetrics, Gynecology and Perinatology of Ministry of Healthcare of the Russian Federation, 4 Oparina Street, Moscow 117997, Russia
Pirogov Russian National Research Medical University, Ministry of Healthcare of the Russian Federation, 1 Ostrovitianov Street, Moscow 117997, Russia

Galina Bolshakova and Gennady Sukhikh
Research Center for Obstetrics, Gynecology and Perinatology of Ministry of Healthcare of the Russian Federation, 4 Oparina Street, Moscow 117997, Russia

Dmitry Goldshtein
Research Center of Medical Genetics, 1 Moskvorechie Street, Moscow 115478, Russia

Eva Johanna Kubosch, Emanuel Heidt, Anke Bernstein and Katharina Böttiger
Department of Orthopedics and Trauma Surgery, Albert-Ludwigs University Medical Center Freiburg, Freiburg, Germany

Hagen Schmal
Department of Orthopaedics and Traumatology, Odense University Hospital, Sdr. Boulevard 29, 5000
Odense C, Denmark
Department of Clinical Research, University of Southern Denmark, Odense, Denmark

Stuart Webb, Chase Gabrelow, James Pierce, Edwin Gibb and Jimmy Elliott
Genomics Institute of the Novartis Research Foundation, 10675 John Jay Hopkins Drive, San Diego, CA 92121, USA

Adel Tekari
Tissue and Organ Mechanobiology, Institute for Surgical Technology & Biomechanics, Medical Faculty, University of Bern, Bern, Switzerland
Biointerfaces, Empa, Swiss Federal Laboratories for Materials Science and Technology, St Gallen, Switzerland

Samantha C. W. Chan
Tissue and Organ Mechanobiology, Institute for Surgical Technology & Biomechanics, Medical Faculty, University of Bern, Bern, Switzerland
Biointerfaces, Empa, Swiss Federal Laboratories for Materials Science and Technology, St Gallen, Switzerland

Daisuke Sakai
Department for Orthopaedic Surgery, Tokai University School of Medicine, Isehara, Kanagawa, Japan
AO Spine Research Network, AO Spine International, Davos, Switzerland

Sibylle Grad
AO Research Institute Davos, Davos, Switzerland
AO Spine Research Network, AO Spine International, Davos, Switzerland

Benjamin Gantenbein
Tissue and Organ Mechanobiology, Institute for Surgical Technology & Biomechanics, Medical Faculty, University of Bern, Bern, Switzerland
AO Spine Research Network, AO Spine International, Davos, Switzerland

Evgeniya V. Dolgova, Ekaterina A. Potter, Anastasiya S. Proskurina and Alexandra M. Minkevich
Institute of Cytology and Genetics, Siberian Branch of the Russian Academy of Sciences, 10 Lavrentieva Ave., Novosibirsk 630090, Russia

Elena R. Chernych and Alexandr A. Ostanin
Institute of Clinical Immunology, Siberian Branch of the Russian Academy of Medical Sciences, 14 Yadrintsevskaya Street, Novosibirsk 630099, Russia

Sergey I. Bayborodin, Valeriy P. Nikolin, Nikolay A. Kolchanov and Sergey S. Bogachev
Institute of Cytology and Genetics, Siberian Branch of the Russian Academy of Sciences, 10 Lavrentieva Ave., Novosibirsk 630090, Russia

Nelly A. Popova and Yaroslav R. Efremov
Institute of Cytology and Genetics, Siberian Branch of the Russian Academy of Sciences, 10 Lavrentieva Ave., Novosibirsk 630090, Russia
Novosibirsk State University, 2 Pirogova Street, Novosibirsk 630090, Russia

C. C. West
British Heart Foundation Centre for Vascular Regeneration & Medical Research Council Centre for Regenerative Medicine, University of Edinburgh, Edinburgh, UK
Department of Plastic and Reconstructive Surgery, St Johns Hospital, Howden Road West, Livingston, UK

I. R. Murray and D. C. Hay
British Heart Foundation Centre for Vascular Regeneration & Medical Research Council Centre for Regenerative Medicine, University of Edinburgh, Edinburgh, UK

A. W. James, S. Pang and W. R. Hardy
Orthopaedic Hospital Department of Orthopaedic Surgery and the Orthopaedic Hospital Research
Center, University of California, Los Angeles, CA, USA

M. Corselli
Orthopaedic Hospital Department of Orthopaedic Surgery and the Orthopaedic Hospital Research
Center, University of California, Los Angeles, CA, USA
BD Biosciences, San Diego, CA, USA

C. Black
Orthopaedic Hospital Department of Orthopaedic Surgery and the Orthopaedic Hospital Research
Center, University of California, Los Angeles, CA, USA
Bone and Joint Research Group, Institute of Developmental Sciences, University of Southampton, Southampton, UK

S. E. Lobo
Orthopaedic Hospital Department of Orthopaedic Surgery and the Orthopaedic Hospital Research
Center, University of California, Los Angeles, CA, USA
Department of Surgery, School of Veterinary Medicine and Animal Science, University of São Paulo, São Paulo, Brazil

K. Sukhija
Orthopaedic Hospital Department of Orthopaedic Surgery and the Orthopaedic Hospital Research
Center, University of California, Los Angeles, CA, USA
Department of Emergency Medicine, Kaweah Delta Health Care District, Visalia, CA, USA

P. Liang
Orthopaedic Hospital Department of Orthopaedic Surgery and the Orthopaedic Hospital Research
Center, University of California, Los Angeles, CA, USA
Department of Urology, David Geffen School of Medicine, University of California at Los Angeles, Los Angeles, CA, USA

V. Lagishetty
Orthopaedic Hospital Department of Orthopaedic Surgery and the Orthopaedic Hospital Research
Center, University of California, Los Angeles, CA, USA

Department of Pathology and Laboratory Medicine, David Geffen School of Medicine, University of California at Los Angeles, Los Angeles, CA, USA

K. L. March
Indiana Center for Vascular Biology and Medicine, Krannert Institute of Cardiology, and Vascular and Cardiac Center for Adult Stem Cell Research, Indiana University, Bloomington, IN, USA.

K. Ting
Orthopaedic Hospital Department of Orthopaedic Surgery and the Orthopaedic Hospital Research
Center, University of California, Los Angeles, CA, USA
Division of Growth and Development and Section of Orthodontics, School of Dentistry, University of
California, Los Angeles, CA 90095, USA

C. Soo
Orthopaedic Hospital Department of Orthopaedic Surgery and the Orthopaedic Hospital Research
Center, University of California, Los Angeles, CA, USA
Division of Plastic and Reconstructive Surgery, Department of Surgery and the Orthopaedic Hospital Research Center, University of California, Los Angeles, CA 90095, USA
Department of Orthopaedic Surgery and the Orthopaedic Hospital Research Center, University of California, Los Angeles, CA 90095, USA

B. Péault
British Heart Foundation Centre for Vascular Regeneration & Medical Research Council Centre for Regenerative Medicine, University of Edinburgh, Edinburgh, UK
Orthopaedic Hospital Department of Orthopaedic Surgery and the Orthopaedic Hospital Research
Center, University of California, Los Angeles, CA, USA

Index